MANAGING MUSKIES

A Treatise on the Biology and Propagation of Muskellunge in North America

This volume, is dedicated to the memory of Gil Hamm, (1904 to 1984) founder of Muskies, Inc. and principal instigator of this symposium.

A Symposium on

Muskellunge in North America

Co-Sponsors

American Fisheries Society
Muskies, Inc.

Robert Schmidt, General Chairman

Steering Committee

Robert Strand, Chairman

John Casselman	Shyrl Hood
Steve Clarkin	Charles E. Johnson
Michael Dombeck	John Klingbiel
Delano R. Graff	Bruce Menzel
David Hanson	Kevin Richards
Paul Haug	Richard Ryder
	Arden Trandahl

Special Advisors

E.J. Crossman	Arthur Oehmcke
Jim Mayhew	Gil Radonski

Sponsors

Coolwater Symposium	Reed's Tackle Co.
North Central Division -	Berkley and Co.
American Fisheries Society	Tuffy Boat Co.
Lowrance Electronics Co.	Sherm Pehrson (artist)
E.I. Dupont (Stren)	Zebco - Fish America
Plano Molding Co.	Foundation

MANAGING MUSKIES

A Treatise on the Biology and Propagation of Muskellunge in North America

Gordon E. Hall
Editor

Proceedings of an International Symposium
held in LaCrosse, Wisconsin
April 4-6, 1984

Special Publication No. 15

American Fisheries Society
Bethesda, MD
1986

Contributors and Donors

MUSKIES, INC. CHAPTERS

Central Minnesota	Hayward Lakes	Quad Lakes
Chautauqua Lake	Headwaters	Red River
Chicagoland	Lakeland	St. Cloud
Cleveland	Lazy Lunkers	S.E. Wisconsin Muskie Hunters
Cloquet	Michiana	Susquehanna
Fargo-Moorhead	Milwaukee Huskie Muskies	Twin Cities
First Wisconsin	Mississippi Valley	Upper Great Plains
Flatlanders	Paul Bunyan Heartland	Vikingland
God's Country	Pomme de Terre	

MUSKIES, INC. MEMBERS

Rick Anderson	Douglas Hookum	Larry Ramsell
Hugh Becker	John Hotle	Rod Ramsell
Max Case, Jr.	James Kruschek	Al Skaar
Lou Cook	Jerry Meachum	Bob Schmidt
Cathy Costa	Tom Meyers	J. J. Schmidt
Bill Crane	Bill Neely	Frank Schneider, Jr.
Dr. Roland Cull	Helen Ness	Mike Stoner Memorial
Byron Danielson	Keith Ogden	Terry Tassi
Bill Davis	Kurt Ogden	Dean Teal
Gil Hamm	Ed Peterka	Denielle Winskowski
Paul Haug		

Other Donations Received

Bemidji Minnesota Stocking Fund
Dairy Land Power Cooperative
Illinois Muskie Alliance
Kentucky Silver Muskie Club
Lund America
Mike Lyga Taxidermy
Muskies Canada, Inc.
World Championship Muskie Classic
Minnesota Northern Pike Association

Copyright 1986 by the
American Fisheries Society

Library of Congress Catalog Card Number: 86-083239
International Standard Book Number: ISBN 0-913235-33-4
International Standard Serial Number: ISSN 0097-0638

All orders should be addressed to

American Fisheries Society
5410 Grosvenor Lane
Bethesda, Maryland 21044 U.S.A.

Printed in the United States of America
by Echo Printing Co., Alexandria, Minnesota

Contents

Robert Schmidt Robert L. Kendall Gordon E. Hall	ix	Preface
	xii	List of Fish Species

Introduction

E.J. Crossman	1	The Noble Muskellunge: A Review

Biology and Life History

J.M. Casselman E.J. Crossman P.E. Ihssen J.D. Reist H.E. Booke	14	Identification of Muskellunge, Northern Pike and their Hybrids
R.A. Sonstegard T.T. Chen	47	A Review of Lymphosarcoma of Muskellunge and Northern Pike
Marlyn L. Miller Bruce W. Menzel	51	Movement, Activity, and Habitat Use Patterns of Muskellunge in West Okoboji Lake, Iowa
Robert F. Strand	62	Identification of Principal Spawning Areas and Seasonal Distribution and Movements of Muskellunge in Leech Lake, Minnesota
Bruce Gilbertson	74	Winter Rescue of Juvenile Muskellunge from Shallow Winterkill Lakes
Robin E. Craig Ronald M. Black	79	Nursery Habitat of Muskellunge in Southern Georgian Bay, Lake Huron, Canada
J.R. Gammon	87	Sperm Production and Spawning Success for Muskellunge and Northern Pike
J.M. Casselman E.J. Crossman	93	Size, Age, and Growth of Trophy Muskellunge and Muskellunge-Northern Pike Hybrids — The Cleithrum Project 1979-1983
Jeffrey B. Koppelman David P. Philipp	111	Genetic Applications in Muskellunge Management

Community Interactions

Michael P. Dombeck Bruce W. Menzel Paul N. Hinz	122	Natural Muskellunge Reproduction in Midwestern Lakes
Peter D. Inskip	135	Negative Associations Between Abundances of Muskellunge and Northern Pike: Evidence and Possible Explanations
Robert F. Carline Roy A. Stein Lawrence M. Riley	151	Effects of Size at Stocking, Season, Largemouth Bass Predation and Forage Abundance on Survival of Tiger Muskellunge
Stephen R. Mooradian John L. Forney Michael D. Staggs	168	Response of Muskellunge to Establishment of Walleye in Chautauqua Lake, New York
Peter D. Inskip John J. Magnuson	176	Fluctuations in Growth Rate and Condition of Muskellunge and Northern Pike in Escanaba Lake, Wisconsin

Robert Engstrom-Heg Richard T. Colesante Gerald A. Stillings	189	Prey Selection by Three Esocid Species and Hybrid Esocid

Management Considerations

Delano R. Graff	195	Musky Management - A Changing Perspective from Past to Present
Paul J. Wingate	199	Philosophy of Muskellunge Management
Dale A. Brege	203	A Comparison of Muskellunge and Hybrid Muskellunge in a Southern Wisconsin Lake
Michael P. Dombeck	208	Muskellunge Habitat with Guidelines for Habitat Management
David A. Hanson Michael D. Staggs Steven L. Serns Leon D. Johnson Lloyd M. Andrews	216	Survival of Stocked Muskellunge Eggs, Fry, and Fingerlings in Wisconsin Lakes
Steven L. Serns Lloyd M. Andrews	229	Comparative Survival and Growth of Three Sizes of Muskellunge Fingerlings Stocked in Four Wisconsin Lakes
David A. Hanson	238	Population Characteristics and Angler Use in Eight Northern Wisconsin Lakes
Michael H. Hoff Steven L. Serns	249	The Muskellunge Fishery of Escanaba Lake, Wisconsin under Liberalized Angling Regulations, 1946-1981
Dell H. Siler George B. Beyerle	257	Introduction and Management of Northern Muskellunge in Iron Lake, Michigan
James R. Axon Lewis E. Kornman	263	Characteristics of Native Muskellunge Streams in Eastern Kentucky

Culture

John Klingbiel	273	Culture of Purebred Muskellunge
Thomas R. Bender, Jr. Delano R. Graff	279	Pennsylvania's Practices for Intensive Culture of Hybrid Muskellunge
Wallace D. Jorgensen	285	Iowa Culture of Muskellunge on Artificial Diet
James E. Harvey	288	Disease Incidence and Management Implications of Cultured Esocids
James W. Meade Carol A. Lemm	292	Effects of Temperature, Diet Composition, Feeding Rate, and Cumulative Loading Level on Production of Tiger Muskellunge

Special Panels on Muskellunge Management

Catch and Release

Paul Gasbarino	300	Catch and Release of Muskellunge - Philosophy and Methods
Kevin Richards Rod Ramsell	309	Quantifying the Success of Muskellunge Catch and Release Programs: A Summary of Cooperative Angler-Tagging Studies
Ronald J. Dent	316	A Case History of Voluntary Catch and Release of Muskellunge with Management Implications for Fishery Programs

Private Sector Participation

Arthur A. Oehmcke — 323 — Role of Anglers and Private Organizations in Muskellunge Management

Future Research Needs

David A. Hanson — 335 — Improving Musky Management: A Review of Management and Research Needs

Poster Contributions

Abstracts Only

B. Lebeau, G. Pageau, E.J. Crossman — 342 — The Muskellunge as a Multiple Spawner: An Adaptive Strategy for Production of a Large Number of Eggs

Robert R. Jackson — 343 — Joint Participation of Government Agencies, an Angling Organization and Riparians in Management of a Lake Fishery

Bruce C. Adair — 343 — Tiger Muskellunge as a Trophy Species in Iowa Lakes

Zenas Bean — 344 — Muskellunge Transportation Units

J.M. Casselman — 344 — The Esocid Cleithrum - Its Removal, Preparation and Interpretation for the Acquisition of Age-Growth and Other Biological Data

E.J. Crossman, Stephen Campbell, Leslie M. Munto — 345 — The Muskellunge - What's in a Name

Poster Papers

Dennis L. Newman, Ted W. Storck — 346 — Angler Catch, Growth, and Hooking Mortality in Small Centrarchid-Dominated Impoundments

William F. Krise, Joseph T. Fuss — 352 — Managing Water Heating Costs for Intensive Culture of Tiger Muskellunge

John H. Schachte, Jr., Glenn L. Hoffman — 357 — Mortality of Muskellunge Fingerlings Attributed to a Multiple Tapeworm Infection

Summary

Robert F. Strand — 360 — Review of Technical Contributions

Appendix

James A. Ragan, Terry Steinwand, Gene Van Eeckhout — 370 — A Synopsis of Results from a Questionaire on Muskellunge - Resource Agency Survey

Preface

The International Musky Symposium held in LaCrosse, Wisconsin April 4-6, 1984, and its output "Managing Muskies" represent a significant cooperative effort between users, managers, scientists, and businessmen. It all began on August 26, 1981 at Superior, Wisconsin. As International President of Muskies, Inc., I was appearing at a meeting of the Wisconsin Natural Resources Board to present them with a check from Muskies, Inc. for funding of an electrophoretic analysis of genetic variability in selected populations of muskellunge. I remember how impressed the Board was that an angler group would actually "put its money where its mouth was". After the meeting a small group, including Bob Strand, Art Oehmcke and myself, met for coffee. We listened as Muskies, Inc. founder Gil Hamm said "These are the things we need to be associated with if we have any hope of good sport fishing in the future", and the idea of the Symposium was born. Unfortunately Gil passed away on November 8, 1984, but his contributions to musky fishing conservation will live forever.

Muskies, Inc. had supported selected research projects in the past, and we knew that a variety of research and management was being done, but no "master plan" was evident. We recognized a need to synthesize known information on the subject, to give direction to future research and to promote the exchange of information and ideas between anglers and professionals. To accomplish this the American Fisheries Society was enlisted to provide the necessary expertise and assistance to sponsor a Symposium. With Muskies, Inc. providing initial financial support we began to develop the committee structure. We set goals of high quality and diversity of geography, agency and scientific representation, and the gem of the idea of 1981 became what we believe to have been the largest gathering of Musky "experts" ever assembled.

The success of the Musky Symposium is due to the considerable effort of many dedicated people who, collectively, brought the event to fruition. I wish to express my sincere appreciation for the talented efforts of my Steering Committee and Special Advisors. Robert Strand was the Technical Steering Committee Chairman and Michael Dombeck served as Program Chairman for the committee composed of John Casselman, Delano Graff, David Hanson, Shyrl Hood, Charles Johnson, John Klingbiel, Bruce Menzel, Kevin Richards, Richard Ryder and Arden Trandahl. Ed Crossman, Jim Mayhew, Art Oehmcke, and Gil Radonski served as Special Advisors and their quality input was reflected in the conference at LaCrosse. Steve Clarkin was the Steering Committee member in charge of local arrangements and he handled the complex task efficiently. My expression of gratitude would not be complete without recognizing Paul Haug, Steering Committee member, who not only served as Treasurer for the Symposium, but provided moral support to me during the entire project.

We must recognize the financial support of hundreds of anglers, musky clubs, businesses and especially Muskies, Inc. Chapters, who put up the front money to get the project underway. Pete Haug's special fund raisers provided timely moneys. The quality of the Symposium was set superbly by Dr. E. J. Crossman's keynote address "This Noble Beast". The poolside demonstrations coordinated by Dick Pearson, business displays set up by Steve Voss and S. L. "Swede" Stelzer, and the mounted esocids display by regions built by Mike Brandt and Randy Lawry gave attendees much enjoyment. I especially wish to thank Jan Eggers "The Pike Ferret" from Bovenkarspel, Holland for his contributions. Jan provided a real international flavor and continually compared findings with northern pike records. His angling seminars were truly a delight to attend. The hard work and dedications of these people will always be remembered.

So what has been accomplished? I feel the following important conclusions should be noted:
* User, management agency and business can cooperate to perform large scale projects.
* Management can be improved by implementing the findings of existing research.
* Duplication on some aspects of musky research exists which may not be completely necessary, or some of the funds better used for different, needed projects.
* Anglers can positively impact fisheries programs.
* Increased dissemination of reliable information is needed by all responsible parties.

I was particularly pleased to find scientific backing for every program Muskies, Inc. has been supporting. I feel this book represents a beginning of this kind of work. Instead of duplication, projects can be expanded to more comprehensive work with the cooperation of groups and relationships begun at the

Symposium.

I wish to note one such project has already begun with Muskies, Inc. helping to fund "The Wabigoon Muskellunge and Pike Research" with Dr. E. J. Crossman. Articles on research needs are appearing in regular publications. It is my sincere hope that the spirit of this Symposium will continue into the next century.

Bob Schmidt
General Chairman

The International Symposium on Muskellunge built bridges between anglers and fishery professionals. It was conceived by an angler organization -- Muskies, Inc. -- endorsed by a professional organization -- the American Fisheries Society -- and brought to fruition by a thorough mix of aquatic sportsmen, managers, culturists, scientists, and educators. In all respects, from the style of the presentations to the practical workshops, lively trade show, and informal social events, the symposium encouraged direct communication among groups of people who often misunderstand each other. The conference exposed much new information about muskellunge, one of the world's great trophy fishes, as these proceedings attest. More fundamentally, I think, it showed what can be accomplished when users and managers of a public resource work amiably and constructively together.

It is curious, really, that this symposium was such an unusual example of this cooperation. Good relations between fishermen and fishery managers ought to be taken for granted. In the absence of management and control, fisheries are just like any other common-property resources: owned by no one, exploited by anyone, and inevitably degraded. A century before Garrett Hardin popularized the "tragedy of the commons," far-sighted citizens recognized this and called for action to rebuild declining North American fisheries. Subsequent legislative moves to place fisheries under public ownership and management have been broadly popular. Despite set-backs and mistakes, fishery management has been very successful over the years; people catch immensely more fish today than they would if the fisheries had remained truly common property. The fishing community, particularly its recreational sector, has directly supported much of this management, and of the underlying research, through licenses, fees, and equipment taxes. And, most fishery professionals are themselves ardent sportsmen and women. All these are ingredients for harmony.

Too often, however, the sense of shared purpose has been lost, replaced by mistrust, disrespect, and sometimes hostility between user and manager. If this were limited to commercial fisheries, it might be attributed simply to antagonisms that arise whenever constraints are placed on one's livelihood. But is has been a feature of recreational fisheries, too. Normal tensions between police and public might be part of the story, but I suspect it has more to do with the emerging complexity of fisheries science and management. This complexity has turned professionals into a technological elite, keepers of arcane knowledge and jargonesque language, and increasingly paternalistic (if not condescending) toward layfolk. I suspect, too, that the fishing public came to see professionals as unaccountable dispensers of too many restraints, regulations that seemed arbitrary because they were unexplained. Although few may have questioned the professional's dedication to the resource, many have asked if the professionals remember who they are protecting the resource for, and who is paying the bill. People typically go to government only with demands or complaints anyway; when communications deteriorate to un-informed complaints from one side and unintelligible responses from the other, we have all the ingredients for misunderstanding.

I have overdrawn this gloomy picture for effect, but the picture is changing in any case. The muskellunge symposium is an important and conspicuous example of growing rapport between fishermen and managers. From maritime grounds to mountain lakes, from meeting rooms to halls of government, users and trustees are coming together to talk, to listen, to solve problems, to advance programs, and to share responsibilities. It is a young movement with a long way to go, but I think it is a solid one. If the old antagonisms were rooted in disparities of knowledge, the new resolutions owe much to a common fund of knowledge. Muskellunge anglers, for example, have acquired a sophisticated understanding not only of their prized prey but of its ecological and managerial contexts as well. The heightened ecological awareness of North Americans in general underlie this education, no doubt, and it is abetted by a growing number of fine popular writers who link science to sport. Fishery professionals, for their part, have raised their social awarenesses along with everyone else in our time. They are learning more about the human element in fisheries and about its importance. With the help of ever-improving media specialists, they are finding ways to convey complex ideas in understandable terms.

The muskellunge symposium was a showcase for this convergence of attitudes and misunderstanding. On behalf of the American Fisheris Society, I thank Muskies, Inc. -- particularly Bob Schmidt -- and the many selfless people who made the symposium an outstanding success. I commend these proceedings to you not only for the abundant information about muskellunge they contain but also for the new spirit in fisheries management they represent. This spirit is no less than a democratization of resource management, and its cumulative effect will be revolutionary.

Robert L. Kendall
American Fisheries Society

Prior to acceptance for publication in this book, each technical paper, including those contributed from panel and poster sessions, underwent a minimum of three separate peer reviews and a final editorial scrutiny by me. The editorial guidelines and the organizational format of the North American Journal of Fisheries Management were used as a model for technical acceptance and publication. The assistance of two members of the symposium technical committee -- Mike Dombeck and Bob Strand -- who provided me with much needed guidance and support in my editorial efforts is especially acknowledged. Approximately 50 other fishery scientists in various government agencies, universities and private organizations throughout the country reviewed one or more technical papers and my grateful appreciation is extended to them for their volunteer efforts.

The final publication includes the results from 37 scientific investigations, two panel sessions describing research needs and the role of anglers, abstracts of unsubmitted poster presentations and summaries of each technical session. Collectively, these are considered to cover most every known aspect of muskellunge biology, behavior, culture and angling, and current management efforts. This volume should provide any individual or agency interested in muskies with a better understanding of the current state of muskellunge management and needed research and, hopefully, will stimulate the latter toward development of more effective management techniques for optimum utilization of this great fishery resource.

Gordon E. Hall
Editor

LIST OF FISH SPECIES

The colloquial names of most fish in the United States and Canada have been standardized in *A List of Common and Scientific Names of Fishes from United States and Canada,* Fourth Edition, 1980, American Fisheries Society Special Publication No. 12. Throughout this book, species covered by the *List,* and a few foreign species, are cited by their common names only; their respective scientific names follow. For plants, and animals other than fishes, the scientific names are given in the text.

Alewife	*Alosa pseudoharengus*
Amur pike	*Esox reicherti*
Bigmouth buffalo	*Ictiobus cyprinellus*
Bigeye shiner	*Notropis boops*
Black bullhead	*Ictalurus melas*
Black crappie	*Pomoxis nigromaculatus*
Bluegill	*Lepomis macrochirus*
Bluntnose minnow	*Pimephales notatus*
Bowfin	*Amia calva*
Brook silverside	*Labidesthes sicculus*
Brook trout	*Salvelinus fontinalis*
Brown bullhead	*Ictalurus nebulosus*
Burbot	*Lota lota*
Carp	*Cyprinus carpio*
Central mudminnow	*Umbra limi*
Chain pickerel	*Esox niger*
Channel catfish	*Ictalurus punctatus*
Chinook salmon	*Onchorhyncus tshawytscha*
Cisco	*Coregonus artedii*
Estuary grouper	*Epinephelus salmoides*
Eurasian perch	*Perca fluviatilis*
Fathead minnow	*Pimephales promelas*
Freshwater drum	*Aplodinotus grunniens*
Gizzard shad	*Dorosoma cepedianum*
Golden redhorse	*Moxostoma erythrurum*
Golden shiner	*Notemigonus crysoleucas*
Goldfish	*Carassius auratus*
Grass pickerel	*Esox americanus vermiculatus*
Iowa darter	*Etheostoma exile*
Johnny darter	*Etheostoma nigrum*
Lake chubsucker	*Erimyzon sucetta*
Lake trout	*Salvelinus namaycush*
Lake whitefish	*Coregonus clupeaformis*
Largemouth bass	*Micropterus salmoides*
Logperch	*Percina caprodes*
Longear sunfish	*Lepomis megalotis*
Longnose gar	*Lepisosteus osseus*
Mottled sculpin	*Cottus bairdi*
Muskellunge	*Esox masquinongy*
Northern pike	*Esox lucius*
Plaice	*Pleuronectes platessa*
Pumpkinseed	*Lepomis gibbosus*
Puffer	*Fugu rubripes*
Rainbow trout	*Salmo gairdneri*
Redbelly tilapia	*Tilapia zillii*
Redfin pickerel	*Esox americanus americanus*

Rock bass	*Ambloplites rupestris*
Smallmouth bass	*Micropterus dolomieui*
Spotfin shiner	*Notropis spilopterus*
Spotted bass	*Micropterus punctulatus*
Spotted sucker	*Minytrema melanops*
Striped shiner	*Notropis chrysocephalus*
Tadpole madtom	*Noturus gyrinus*
Tiger muskellunge	*(Esox lucius x E. masquinongy)* hybrid
Trout-perch	*Percopsis omiscomaycus*
Walleye	*Stizostedion vitreum*
White crappie	*Pomoxis annularis*
White sucker	*Catostomus commersoni*
Yellow perch	*Perca flavescens*

INTRODUCTION

The Noble Muskellunge: A Review

E. J. CROSSMAN

Department of Ichthyology and Herpetology
Royal Ontario Museum
Toronto, Ontario M5S 2C6

ABSTRACT

Our knowledge and exploitation of the muskellunge have had interesting histories. The species was recognized late, regularly confused with *Esox lucius*, and the changes in its nomenclature and in its taxonomy are almost impossible to follow. The meagre fossil evidence suggests muskellunge occurred in North America at least back to the Miocene, and had a much wider distribution in the past. It may have been a riverine fish originally, only secondarily adapting to standing waters at the end of the Wisconsin glaciers. An extensive, poorly-documented commercial fishery, which ended in 1936, probably contributed to an early decline in availability and eventual apprehension for the survival of the species. Documentation of information began about 1838, increased very slowly to the end of the 1940's, and for some areas of study, doubled during the decade from 1950-1960. The history of the development of information of the muskellunge is summarized in this review paper, and suggestions are made on the types of studies which still need to be carried out.

The trappings we usually associate with nobility include grandeur, titles, folklore, controversy, place names, statuary and an aloof avoidance of publicity. The muskellunge obviously qualifies. It has the stature and stamina. No other fish has so many names and aliases. The muskellunge has not been known as long as the pike (or northern pike), *Esox lucius*, so its folklore is not, as yet, as extensive as that of the pike. However, the stories of the rapaciousness of the muskellunge grow, and it is referred to by such terms as "waterwolf", "tiger of the waters", and "a long, slim, strong, and swift fish, in every way formed for the life it leads, that of a fierce and dauntless marauder" (Hallock 1877). The muskellunge has always been controversial and standing arguments include -whether or not there is a mystique associated with it, how many kinds there are, what its name should be, and whether or not it controls prey fish populations.

As with other nobles, there are places named after the muskellunge. There is a county and a municipality in Quebec named Maskinongé, and as a result of the great numbers of muskellunge in the river on which the town is located, the name Muscallonge was once suggested for what we know today as Racine, Wisconsin (Anon. 1897). Not many noblemen have statues erected in their memory with heroic proportions to equal those of the muskellunge statue at Hayward, Wisconsin. The walls of many private and public places are adorned with mummified remains that would have delighted any ancient Egyptian. We have published references to it probably going back to the Jesuit diaries of the 17th century, yet in the intervening 300 years we have not learned enough about it to properly assure its survival in the face of man's increasing interest in it, and man's increasing pressure on it and its environment.

The purpose of this paper is to introduce the muskellunge, to point out not only some of its history, but the history of our attempt to learn about it, and possibly to suggest areas of study which we should consider. The hybrid or tiger muskellunge will enter into this treatment but not as a major emphasis.

NOMENCLATURE

A noble animal should have a name known to all, with an interesting story associated with its derivation and spelling. Every aspect of naming this fish is storied and controversial. Even today arguments rage about the "proper" spelling of the most often used abbreviation of the approved common name.

The Scientific Name

Esox masquinongy Mitchill 1824 - the presently accepted scientific name - has been established by long practice rather than precise documentation. It is attributed to Samuel Latham Mitchill, a New York physician. The original use of this name, supposedly to describe the fish we now know by this name, is thought to have appeared on page 297 of an 1824 issue of a periodical called "The Mirror," or a periodical with the word Mirror in its title. There is no known copy of this article, and our use of the name is based on reference to it in a publication in 1838 (Kirtland 1838). Publications in the twenty years following the date attributed to Mitchill contain so many errors and contradictions that there is considerable doubt that Mitchill's 1824 article ever existed. By 1842 (DeKay 1842), references to the original description were being based only on Kirtland's use of it in 1838.

By the late 1800's, doubts about the existence of Mitchill's original publication were being expressed as follows: "A search through the files of the Mirror for Mitchill's description has proved unsuccessful, it is not on the page cited by DeKay" (Jordan 1877); "Mirror, 1824, but it is not there; I cannot find the description anywhere" (Jordan 1882); "...but the description is said not to be there" (Meek and Newland 1886); "This reference [Mitchill's] is given on the authority of DeKay. We have carefully searched the files of the Mirror and do not find this description. The name *masquinongy* will, in any case, however, hold from the account given by Kirtland." (Jordan and Evermann 1896-1900). Although Kirtland (1838) cited the 1824 description of Mitchill for the name *E. masquinongy,* Kirtland himself, in his 1838 article, used the name *Esox estor* for that species. These references do not support Mitchill's name as suggested by Jordan and Evermann.

Mitchill did publish extensively on New York fishes between 1814 and 1818. Those publications include several species descriptions under the name *Esox* but many of these were not, in fact, esocids. The biography of Mitchill (Hall 1934) includes a bibliography of his written work, but the 1824 "Mirror" article is not even mentioned.

There was, in the early to middle 1800's, a considerable amount of confusion concerning the distinction between the northern pike, *Esox lucius,* and the muskellunge. This confusion led, starting in 1818 (LeSeur 1818), to the use of names which cannot now be attributed to one or the other of the two species, or which seem to apply to a composite of both species. These names include *Esox atromaculata* (or *atromaculatus*) *E. estor, E. fasciolaris, E. lucioides, E. major, E. nobilior, E. nobilis,* and *E. salmoneus.* Use of these names for the muskellunge must be considered suspect.

The problem of scientific name was compounded by the gradual restriction of the name *Esox masquinongy* to the Great Lakes muskellunge and the description of two additional species, *Esox ohioensis* (Kirtland 1854) and *Esox immaculatus* for fish in the Ohio River and in Wisconsin and Minnesota.

The latter name was apparently derived from an unpublished work of Garrard and possibly first referred to in the scientific literature by Mosher (1892). A suggestion that the northern pike of North America and the muskellunge should be considered distinct from the pike of Europe led to the inclusion of the North American forms in the genus *Lucius*. So, many of the above names re-appeared in the literature towards the end of the 1800's in the genus *Lucius* or the genus *Mascalongus*. Jordan and Evermann (1896-1900) then decided that only one species was involved, *Lucius masquinongy*. They erected three forms or subspecies, the typical form *L. m. masquinongy* (Mitchill) in the Great Lakes, *L. m. ohiensis* [sic] (Kirtland) for those in the Ohio River and tributaries, and *L. m. immaculatus* (Garrard) for those in Wisconsin, Minnesota, and the headwaters of the Mississippi River. Sadly enough, they assigned to *L. m. immaculatus* the common name Great Northern Pike, further confusing the issue. The scientific name gradually stabilized on *Esox masquinongy* but even in recent years the use of a trinomial, especially for muskellunge in the Ohio River system, suggests that some workers still consider various populations of muskellunge to be subspecifically, rather than simply racially, distinct.

The Common Name

Probably no other fish has, in a single language, as many forms or spellings of its common name. Weed (1927) claimed there were "at least 40 legitimate spellings," seeming to imply there were additional aberrations of those. In this regard, he was speaking only of the various forms of the word we now spell as muskellunge. Weed suggested the number increased to 94, if one counted the various other names applied to this species such as pike, tiger, jack, brochet, picaneu and picareau, and added to them the various modifiers for size, color and region. A rather fanciful set of scientific and common names *Esox fasciolaris* Rafinesque (1818) or *E. salmoneus* Rafinesque (1820) - The Salmon Pike, was assigned by the often over-zealous and uncritical naturalist, Constantine Rafinesque (-Schmaltz), who described new species on the basis of the verbal descriptions of others of animals they had "seen".

The genesis of the 40 "legitimate" names was the long-lasting controversy concerning the derivation. It has been derived by various authors from 1) the Latin *masca longus* (Long mask); 2) *masque allonge* or *masque-longue* of the Québécois, and 3) from *mashk* (deformed), *mis, mask* or *mas* (ugly, large or great), *kinonje, kenosha* or *kenoza* (pike or fish), and *maskanonja* (long snout). These are said to be from the Chippewa, and/or Ojibway and Cree languages via French phonetic spelling (Chambers 1923). Nationalism also crept into this argument. The Dominion Commissioner of Fisheries, Prince (1901), said "Thus the name maskinonge commonly but erroneously spelt muskellunge or mascalonge in the United States is really an Indian name, the Chippewa name for pike being "Kenosha" and the prefix 'Mis' or 'Mas' means large or great, so that Maskenosha or Maskinongé corrupted into Maskinonge is really a large deformed pike." Chambers (1923) claimed that Prince and David Star Jordan had agreed at one time that maskinongé would become the official name in the United States as well. The name maskinonge is still a statute name in Ontario, and maskinongé the accepted common name in Quebec. Chambers (1923) gave a unique derivation of lunge, the other often used abbreviation of the presently accepted common name. He claimed that "U.S. anglers in Canada mistook big lake trout [*Salvelinus namaycush*] (called longe or lunge) for maskinongé".

A more recent interpretation (Legendre pers. comm.) is available in a lexicon of the Menomini Algonquin language (Bloomfield 1975). Bloomfield derived the name muscalonge from Algonquin words rendered as *mias* - spotted, and *kenu. si. w* - pickerel (= pike). The typical form of *E. masquinongy,* in the Great Lakes, has a pattern of black or brown spots.

FOSSILS AND PREHISTORY

Our earliest published information on the muskellunge probably dates from the settlement of Canada by the French, and the penetration to the "Western Frontier" (Ohio). However, fossil and archaeological evidence pushes our knowledge of the history of the fish considerably farther back in time.

Fossils attributable to, or (?) only possibly attributable to, *Esox masquinongy* have been reported from the Miocene (25 million years before present) of Oregon (Cavender et al. 1970); (?) the lower Pliocene (approximately 7 MYBP) of Oklahoma (Smith 1962); and several records from the Pleistocene - of Oklahoma (0.3 MYBP, Smith 1954), of Kansas (0.3 MYBP, Smith 1963), and of Michigan (Wilson 1967). There were also indefinite prehistory records for Ontario (700 YBP, McAllister, 1962), and St. Lawrence River (600 YBP, McAllister 1959). Available evidence would suggest that the muskellunge evolved in North America from a pike-like ancestor, possibly *E. tiemani* (Wilson 1980). These fossil records have been used (Cavender et al. 1970; Crossman 1978) to suggest that the muskellunge was very broadly distributed in the past, possibly extending southwest to Oklahoma and northwest to Oregon but not necessarily at the same time.

The present distribution pattern probably reflects 1) events during, and following, the Wisconsin glacial period only; 2) reduction of native range by changes in habitat resulting from settlement and agriculture; and 3) man's later extension of the native range by culture and release. It would appear that *Esox masquinongy* "overwintered" the Wisconsin glacier in only the Mississippi refugium (Underhill 1984). As the glacier "receded", the muskellunge moved north in the Mississippi valley and invaded the present northern parts of its range via the Ohio River, one or more of the early outfalls of precursors to the Great Lakes (lakes Michigan to Erie), and waters in Wisconsin and Minnesota adjacent to the tributaries of the Upper Mississippi River. It would appear that the muskellunge did not use the Red River-Lake Agassiz route used by so many other species which moved north in the same area. The muskellunge may have reached the upper Mississippi only after the connection between the Mississippi River and Lake Agassiz had been broken. If so, it would suggest that the muskellunge arrived in the Upper Mississippi no earlier than 9,000-10,000 YBP.

It is interesting to speculate on the nature of the color pattern and habitat of the muskellunge, and on its distribution, during the height of the Wisconsin glacial period. A number of points in the biology of the muskellunge suggest a riverine species living in situations somewhat similar to those inhabited today by muskellunge in Tennessee. If this is true, then the muskellunge might be only secondarily a lake-dweller. It may have invaded lakes only after the post-Wisconsin move north to those lakes which became available during the "retreat" of that glacier. The muskellunge most often approaches its potential maximum size in large, running waters (e.g., St. Lawrence River).

Muskellunge in the typical, or Great Lakes, race can, with caution, be said to have a pattern of small, dark brown or black spots on a silvery to green, or golden brown, ground color. This pattern does not provide good cryptic camouflage in vegetated, still waters. The patterns of the northern pike and the smaller pickerels, *E. niger* and *E. americanus*, are much better in this regard. The only other living esocid with a pattern of black spots is the Asian Amur pike, *Esox reicherti*. That species does live largely in rivers, and among rocks, not vegetation.

It is often in large river situations that the muskellunge does well in sympatry with the northern pike (Harrison and Hadley 1978). It may be only in the still waters, which the muskellunge has invaded relatively recently, that interaction between these two species usually favors the northern pike. Where the muskellunge occurs today in large rivers (e.g., St. Lawrence River), spawning usually takes place at greater and safer depths than those used for spawning by muskellunge inhabiting lakes in Central Ontario. The dangerous habit of spawning in shallow waters, which are subject to rapid and severe fluctuations in level and temperature, may also have been acquired after a secondary invasion of standing waters.

COMMERCIAL HARVEST

The history of the commercial exploitation of muskellunge has never been adequately examined. There are several obvious questions. What was the real extent of the harvest? What was the effect of the commercial harvest on growth rate, population size, and angling success during, and immediately after, the years in which the harvest took place? What was the contribution of that harvest to the problem of declining availability identified at least by 1850, and to population levels today? Buss (1960) said that the muskellunge was one of the first species to become commercially important in Ohio and that it was one of the first to show a decrease in abundance, becoming scarce by 1850. Many Ontario waters which produce "satisfactory" numbers of moderately large muskellunge today were heavy contributors to the commercial catch. The greatest commercial landings in Ontario were from Lake of the Woods, Eagle Lake, French River, Georgian Bay, lakes St. Clair, Scugog and Simcoe, and the St. Lawrence River. These are still favored angling locations.

The extent of the commercial fishery for muskellunge is not well known today; the records of landings are difficult to assemble and often incomplete. Normally, only weight of fish landed was provided and was often expressed as barrels, a unit equal to 200 pounds. Historical records are expressed in pounds and I will use pounds here except for tentative calculations of past production in certain lakes in Ontario. No figures on numbers of muskellunge landed were recorded. To generate a crude estimate of the number of individuals represented by the commercial catch, I have used a mean weight of 15 pounds per muskellunge. If the factor errs, it overestimates the number taken. The error is, however, probably low.

Statistics on commercial landings in U.S. federal records are apparently not available for the years before 1894. Those statistics include fragmentary records of catches from the U.S. portions of the Great Lakes up to 1917 (Lake Huron, 145 lbs.) and none after that date. The 1899 report of the New York Commissioners of Fisheries, Game and Forests listed landings from the Pennsylvania portion of Lake Erie of 190 lbs. with a value of $13.00, and from Michigan 9 lbs. with a value of $1.00.

Information on the history of commercial fishing for muskellunge was requested from ten U.S. states most likely to have been involved. Seven—Illinois, Kentucky, Minnesota, New York, Pennsylvania, Tennessee and West Virginia—returned positive or qualified answers that no commercial fishing had occurred. New York State is a good example of the extent to which knowledge of the commercial fishery for the muskellunge has become obscured. The New York Department of Environmental Conservation could find evidence of "a depression 'market' hook and line fishery in Chautauqua Lake in the 1930's" and said there was "no note of a legitimate historic commercial fishery in Chautauqua Lake". However, the 1895 Annual Report of the New York Commissioners of Fisheries, Game and Forests listed the commercial catch of muskellunge from "interior waters of the state" as follows: 1894 - 89,000 lbs. valued at $13,350 and 1895 - 106,130 lbs. valued at $15,920. The quantity of muskellunge landed annually, as listed in subsequent New York reports, dwindled rapidly and no muskellunge were included in the report for 1900 or thereafter.

Cobb (1905) described in detail the commercial spearing of muskellunge through the ice at Chautauqua Lake. He stated that, "From a commercial standpoint this lake is one of the most important in the state, and principally on account of one fish, the muskellunge." He reported that until 1902 fishermen were allowed to spear muskellunge for two days a week for five weeks in February and March, but that this was reduced in 1903 to one day per week in February only. This commercial fishery involved ice huts called "fish coops," short-handled spears, and weighted wooden models of bait fishes.

The catch in New York in 1894 and 1895 represented approximately 13,000 muskellunge. The Ohio Department of Natural Resources reported that muskellunge were taken in the commercial fishery of Sandusky Bay, Lake Erie, in the 1830's. Michigan DNR suggested that a commercial fishery existed at one time, collapsed around the turn of the [19th] century, and that the species was not protected until several decades after that time. Wisconsin DNR reported that a fishery had existed in that state, that by 1858 it was limited to waters over 12 mi^2, and that sale of muskellunge was prohibited by 1917. The Pennsylvania Fish Commission stated that early 20th century documents place the muskellunge in the "miscellaneous commercial catch" category, and that a legislative act in 1907 and the New Fish Code of 1909 designated the muskellunge a game species, excluding it from any further commercial harvest.

The records for the commercial harvest of muskellunge in Canada are probably the most complete and easily assembled. They are included in the annual reports of the Commissioner of Fisheries and consist mainly of figures for Ontario and Quebec. However, these records also state, in the 1917 report, that of the fish landed in the Yukon Territory in that year, there were 1,100 pounds of "maskinonge" with a value of $330. These were said to have been taken from Dawson, Forty Mile, Laberge, Teslin, and other Yukon lakes. Pike were also reported, so it should not have been a case of confusion of names. There is no evidence of the muskellunge ever having been otherwise reported from the territory. This leads one to wonder about the dependability of the other records.

Records of muskellunge landed commence in the year 1868; continue to 1901 in Ontario and to 1936 in Quebec. Over that period, approximately 13.2 million pounds (with a landed value of approximately $590,000) were harvested in Ontario, and approximately 2.8 million pounds (with a landed value of approximately $266,000) in Quebec (Table 1). The Canadian total for the 70 years is over 16 million pounds. The reported annual catch in Ontario fluctuated wildly from a low of 600 pounds in 1869 to a high of 774,320 pounds in 1898. In Quebec, the highest catch was 209,415 pounds in 1886, with a somewhat regular decline to 9,300 pounds in 1928. Landings increased again from 1929 to a high of 93,000 pounds in 1935. It is, however, not possible to definitely attribute the decline in harvest to decrease in availability since there are no figures for effort.

No numbers of individual muskellunge, nor mean weight, were provided in the annual Canadian reports. In order to derive a crude estimate of the number of muskellunge represented by the commercial catch, the mean weight of 15 pounds per fish was used. This yielded an estimate of total catch in Canada in the years 1868-1936 of 1,069,958 muskellunge. The landings were recorded by individual lakes so past production for parts of the Trent-Severn system of Ontario, a major muskellunge area today, can be estimated. Muskellunge harvested from Lake Scugog in the 19 years between 1883 and 1901 totalled 3,840,550 pounds yielding an estimate of 256,036 muskellunge of 15 pounds. These figures represent an actual mean annual harvest of 202,134 pounds (91,686 kg) and an estimated 13,475 muskellunge per year. Lake Scugog has 7,030 surface hectares so production was 13 kg/hec/yr. For the Trent River and Rice Lake, in the 21 years between 1881 and 1901, the actual harvest was 1,703,717 pounds (772,793 kg) which is an estimated 113,581 muskellunge. The mean annual figures for those waters are 81,129 pounds (36,799

Table 1. Commercial Landings of Muskellunge in Ontario and Quebec 1867-1936.

Decade	Ontario			Quebec		
	Pounds	$ Value	Est'd No. Fish[a]	Pounds	$ Value	Est'd No. Fish[a]
1867	No muskellunge reported captured					
1868-1877	518,600	Not given	345,733	41,010[b]	19,468[c]	2,734[d]
1878-1887	4,822,220	235,278	321,481	1,063,745	80,287	70,916
1888-1897	5,812,187	231,231	387,479	901,490	54,086	60,099
1898-1907[e]	2,049,341	122,958	136,622	331,790	22,700	22,119
1908-1917				123,210	13,088	8,214
1918-1927				34,500	6,910	2,300
1929-1936[f]				392,300	69,571	26,153
Totals	13,202.348	589,467	1,191,315	2,888,045	266,110	192,535

[a] Based on assigned mean of 15 lbs/fish
[b] Estimated, from 15 lbs/fish
[c] 1874-1877 only
[d] Reported on estimated; no weight given
[e] For Ontario 4 years only, 1898-1901
[f] No muskellunge recorded in 1937 or thereafter

kg) and 5,408 fish. Lakes Simcoe and Couchiching yielded 445,680 pounds (202,157 kg) and an estimated 29,712 muskellunge in the 21 years from 1881 to 1901. This represents 21,222 pounds (9,626 kg) and an estimated 1,414 muskellunge per year. Those two lakes have a total surface area of 77,368 hectares, so mean annual production was 0.12 kg/hec.

There are conflicting reports of past and present attitudes toward the muskellunge as food. A 1900 New York report stated that "epicures consider it one of the best fish in the west" (Ohio, Kentucky). Nevin (1901), speaking of muskellunge in Wisconsin, said "...it has a table value not excelled by any other large fish of our fresh waters ... ranked in quality next to the salt water salmon." It would appear to have attracted a moderately high price per pound for the time. The New York figures for 1894-1895 represent 15 cents per pound. The lowest value was in Ontario, where the mean price per pound per decade (Table 1) fluctuated from 4 cents to 6 cents. In Quebec, mean value per decade rose from 8 to 20 cents. On an annual basis, there is a sudden halving of price per pound in Quebec, possibly associated with the depression, or simply with the increase in landings which occurred during that period. In 1930—1932, prices in Quebec ranged from 22 to 27 cents. In 1933 it fell to 10 cents. From then to the end of the Fishery in 1936, the price was 11 or 12 cents per pound.

The only modern figures on production which can be compared with those of the past commercial fishery must be derived from the sport fishery. The annual angler harvest on Lake Scugog in 1980 was estimated at 720 muskellunge with a total weight of 3,190 kg (pers. comm. C. Lewis). Although muskellunge are still angled regularly in lakes Simcoe and Couchiching, the numbers taken annually are too small to enter the summary of sport fish recorded by creel census. While figures on angler harvest do not truly represent total production, they can certainly be said to indicate that the annual production of muskellunge in these Ontario waters in the mid to late 1900's is drastically reduced from the levels of the late 1800's and early 1900's.

GROWTH OF INTEREST IN THE MUSKELLUNGE AND CONCERN FOR THE SPORT FISHERY

Even though there was confusion in the names, and in the identification of the two large esocids, the merits of the muskellunge as a sport fish were quickly recognized. As early as 1854 (anon. 1854) there was an unsuccessful attempt to introduce the muskellunge on Cape Cod as a new sportfish. Accounts of its importance as a game fish began to appear in the U.S.A. and in Canada (Herbert 1849; Small 1865), and muskellunge were soon the subject of the well-known angling writers of the times (Henshall 1892) in major angling compendia. Anecdotal accounts of angling for muskellunge were abundant, and sizes were recorded and compared as avidly as they are now.

In the beginning, angling pressure was low and largely local, since distant anglers had to travel by train to areas where muskellunge occurred. As early photos attest, regulations on number and on size of muskellunge retained were, at first, non-existent or very liberal. Reduction in numbers of muskellunge was apparently recognized as early as 1850 (Buss 1960), and again in the 1930's (Harkness 1934). Nevin (1901) said "...since the wilderness of Northern Wisconsin was opened by railways ... the pursuit of the muskellunge has been constant and relentless. Its utter extermination has been well nigh accomplished in many of our lakes ..., and nearly all of our waters have been cleaned of this fish to such an extent that its future has become a matter of much concern... ." MacKay and Werner (1934) stated that the commercial exploitation of muskellunge in Ontario "reduced the original stock to such an extent that abundant and widespread distribution were supplanted by reduction and restriction to widespread but more localized areas." They said that the Ontario government then prohibited its capture by commercial means, prohibited the sale of muskellunge, and took steps to protect it during the spawning season.

Before the turn of the century, artificial culture had already become an important tool in fisheries, and beginning in New York in 1890 (Evermann and Goldsborough 1901), artificial culture of muskellunge was seen as a means of supporting populations depleted by commercial fishing and spearing (Cotchefer 1904). Culture of muskellunge began in Wisconsin in 1899 (Nevin 1901) and shortly after in Minnesota. By the 1930's the western states and New York were rearing and releasing advanced fingerlings (7—22 cm) and by 1940 using pituitary extracts to induce ripening of gametes (Hasler et al. 1940). Muskellunge culture began in 1927 in Ontario (MacKay and Werner 1934) and in 1950 in Quebec (Dufour et Paulhus 1979).

The continually growing interest in the muskellunge as a sportfish, which led to the closure of the commercial fishery, was also expressed in increasing apprehension concerning the state of muskellunge populations, and the need to do something if the opportunity to angle for large muskellunge was to be saved. It is true that in the beginning more rigorous regulations, attempts to stop spearing on the spawning grounds, and increased stocking were looked upon as the solution. There was, however, a dawning recognition that we must learn something about the animal, if it was to be saved. Activity in Ontario at that time can be cited as an example of the sequence of events.

In 1934, Harkness said "...lack of knowledge is abysmal. No one has ever made a continuous and intensive study of this fish extending over a number of years", and he requested the appointment of a government biologist to study the muskellunge problem. Harkness (1936) called for the establishment of sanctuaries to save muskellunge. Those pleas, combined with the involvement of J.R. Dymond and the Ontario Federation of Anglers, led to the following: 1) a questionnaire sent in 1937 to anglers and fishery officers across the Canadian range of the muskellunge asking for reports of precise harvest, and comments on present success compared to past success, expressed as number and weight of muskellunge angled. The covering letter with the questionnaire stated, "...we are convinced that the end of the maskinonge fishing is in sight, if present conditions are allowed to continue... One of the most urgent problems facing the organized anglers of Ontario is saving the maskinonge from extermination."; 2) a study of the potential of a natural area used as a nursery for further rearing of hatchery muskellunge (Elson 1941); 3) the late 1940's work, begun by Cameron (1948) and finished by Hourston (1952,1955), on the natural hybrid, the "races", and the food of the muskellunge; and 4) the start in 1951 of the long-standing Maskinonge (later Muskellunge) Research Project at Nogies Creek.

Early work on the muskellunge in the U.S. and in Quebec was largely carried out by government agencies. In contrast, studies in Ontario other than culture (including Elson's nursery work) were largely financed by angler or conservation organizations and much of it carried out by the Royal Ontario Museum rather than by the provincial government.

The serious interest in the muskellunge as a sportfish and the concern for its survival have been manifest both in the literature and in the activity of the anglers. The literature is somewhat summarized in a bibliography (Zack 1982) and the angling history in a thorough and intersting compendium by Ramsell (1982). Angler activity has in part been channelled into organizations, such as Muskies, Incorporated and Muskies Canada, which are committed to education, research, and wise use of the resource. These organizations regularly make suggestions to state and provincial government agencies, and cooperate in various study projects.

HISTORICAL DEVELOPMENT OF KNOWLEDGE

Fortunately, the biological literature on the muskellunge has been well summarized (Buss 1960; MacGregor et al. 1960; Hasse 1976; Porter 1977; Crossman and Goodchild 1978; Hess and Heartwell 1978), and these summaries are useful in an analysis of the development of our knowledge of the muskellunge.

Awareness of the need for good biological information developed slowly. Early scientific works dealt only with taxonomy, distribution, and culture. True biological information began with notes on natural reproduction (Burnett 1854), on culture and biology in New York (Cheney 1896), life history (Kendall 1919) and feeding (Weed 1925). Kendall and Weed were the two most important authors of this period. In the 1930's there were important studies of the growth of muskellunge in Wisconsin (Schloemer 1936) and New York (Greeley 1938). There were early attempts to properly document the annual harvest by anglers in New York (Moore 1938), and the early life history (Fish 1932). The surprising fact that muskellunge of all pattern types (at that time thought to represent sub-species) could occur in a single location was published (Seaborn 1937). MacKay (1931), Dymond (1932) and Harkness (1934) were the first to study muskellunge in Canada.

The war years from 1939 to 1945 seriously slowed the initiative that had begun, and, as a result, the work of Eddy in Minnesota, that of Heacox (1946) at Chautauqua Lake, N.Y., of Krumholz (1949) on Lake St. Clair, and Van Oosten (1946) stand out almost alone. The 1940's were a period of emphasis on introduction and stocking based on artificial culture. This period did, however, see the beginning of the concept of managing muskellunge, but the paucity of solid biological data seriously hampered the opportunity to implement this idea. The 1940's were also the beginning of the awareness of the natural hybrid (Eddy 1940; Cameron 1948) that led ultimately to the recent development of this cross as a sportfish.

The 1950's were the "Wisconsin years" when the studies of Johnson, Oehmcke and Threinen dominated. People seeking information on the biology of the muskellunge still quote a reprinted version of an earlier work by those authors (Oehmcke et al. 1958). The extensive contributions of Williams in Michigan began in that period. Results of work in places like Tennessee, outside the traditional muskellunge angling areas, began to appear at this time (Parsons 1959). An emphasis on studies of the relationships between age, growth, length and weight seems to characterize this period.

In contrast, studies in the 1960's concentrated on population size and structure, as well as a continuation of the documentation of length-weight relationship, creel census, and age and growth. In fact, the highly beneficial increase in rate of data acquisition during the two decades from 1950 to 1970 makes it impossible to cite here all the important contributions of the many workers involved. Most of the publications, however, were listed and indexed to subject by Crossman and Goodchild (1978). The work of Wisconsin biologists Gammon, Hacker, Kempinger, Klingbiel and Snow was added to that of their colleagues mentioned above. Clark (1964) was active in Ohio at that time. The rate at which information on the muskellunge became available in the 1960's increased, in part, as a result of an increase in the number of areas and agencies undertaking studies of this species. Results of work in Kentucky, South Dakota (Applegate), and Tennessee began to appear at this time. The number of studies in Quebec, especially those by Courtemanche, Couture, Mongeau, Pageau, Simard and co-authors, increased considerably. The contributions of the Nogies Creek Project in Ontario began to appear in the publications of Muir (see Muir 1964) and coauthors.

The data base which developed in the 1950's and 1960's made it possible to design different kinds of studies in the 1970's and early 1980's. These studies moved into new fields, and used new tools to attempt to answer quite different questions. A few examples only of these involve the following: habitat requirements (Dombeck et al. 1983), patterns of movement and habitat utilization (Minor and Crossman 1978; Dombeck 1979), stress exerted by capture methods (Miles et al. 1974; Beggs et al. 1980), bioenergetics and utilization of food (Bevelheimer 1983; Gorrie 1984), use of spawning grounds and reproductive success (Strand 1983; Lebeau 1983), interaction with competitors (Harrison and Hadley 1978; Caplin 1982) and thermoregulatory rhythm (Reynolds and Casterlin 1979).

In contrast to these newer studies, basic work was still required in certain areas. Studies involving the status or evaluation of fisheries and native populations were carried out during this period in Tennessee (Riddle 1975), and Kentucky (Axon 1970, Brewer 1980).

As in previous decades, the expansion of our knowledge in the 1970's arose from several sources - 1) sustained productivity by biologists in Wisconsin; 2) renewed productivity in some older centers; and 3) contributions of newer groups in newer areas. In the category of renewed activity we can include New York (Colesante, Hadley, Mooradian, Osterberg, Panek and coworkers); Michigan (Beyerle, Haas, Pecor, Schrouder and coworkers); Pennsylvania (Buss, Graff, Meade and coworkers); Ontario (Casselman, Crossman and coworkers). The newer individuals and centers include Miles in West Virginia; Carline, Stein and coworkers in Ohio; Belusz in Missouri; and Menzel and coworkers in Iowa. The expanding interest, in the 1960's and 1970's, in the tiger muskellunge, and its establishment in areas peripheral to the present natural range of the muskellunge, has resulted in work in places like Indiana (Braun and Pearson 1980), outside the usual concept of the distribution of the species.

One of the surprises in a survey of readily available information on the muskellunge is the short list of contributors from Minnesota. Between the time of the early work by Eddy and the late 1970's there is very little in the published literature. The recent work of Strand in that state is a welcome addition, and possibly information existing in DNR files may appear in future.

A general overview of work on muskellunge up to the late 1970's is available in the subject index provided by Crossman and Goodchild (1978). That summary does not include all the literature available but is fairly complete. Of the 28 subject categories used, I will limit this discussion to 14 (Table 2) which have more direct application to understanding and managing living populations of muskellunge. I have chosen to leave out the long list of useful Dingell-Johnson Project Reports since those, like internal reports in the files of various state and provincial agencies, are not in general circulation and are often represented by a later publication. The number of studies (or publications) in each subject category indicates a more even balance of information than may have been suspected. The dominance of culture and stocking (188 articles) is not unusual considering the dependence that has long been placed on hatchery liberations. The studies represented by the next highest number (72) deal with

Table 2. A survey of the type of work completed up to 1978 derived from the citations listed by Crossman and Goodchild (1978). Any one published work can contribute to the number for one or more of the subject categories.

	No. of Citations
Age and growth (including length-weight)	65
Behavior	5
Culture and stocking	188
Food and feeding	25
Habitat	31
Hybrids	33
Life History	27
Management	72
Marking and tagging	17
Pathology (including parasites)	30
Physiology	11
Population Studies	43
Predator-prey relationship	20
Spawning (natural)	20

aspects best categorized simply as management. These include data on production, yield, catch per unit of effort, and they overlap somewhat with another category called population studies.

Information on age and growth, including length-weight relationships, is the next most abundant (65 articles); this number might seem excessive but the more we learn of muskellunge, the more we realize that the opportunity to apply newer concepts like slotted-size regulations will depend on adequate knowledge of age versus growth rate of local populations. The use of the cleithrum (Casselman 1979), as a more reliable source of information on age of large muskellunge, was a very important adjunct to these studies. There is a central repository of such information at the Royal Ontario Museum, referred to as the Cleithrum Aging Project, to which biologists, anglers and taxidermists contribute cleithra and information. This data repository is now beginning to yield very useful information.

Information on various aspects of population dynamics (43 articles) is equally important, especially for comparisons in any future attempts to manipulate population numbers and to follow the results. There is a second, and lower, level of information (25-30 articles) for such subjects as hybrids, habitat, pathology, life history, food and feeding, predator-prey relationships and natural spawning. A third, and still lower, level (5-20 articles) of information is available on marking and tagging, physiology, and behavior.

Some artificiality in this hierarchy of available information is created by the way the categories were chosen. Articles on tagging and marking as a technique were separated from those on the problems the technique was used to study (e.g. populations, etc.). The figures do, however, indicate strengths and weakness in the data accumulated up to 1978. They indicate as well the upsurge in activity in the 1950's. For many areas of study there is doubling in the amount of published information in the decade from 1950-1960. There appears to be no obvious indication from these summaries that any region is the main source of certain kinds of information, nor that any region is neglecting critical areas of study. The summary does, however, indicate there is much yet to be done.

In contrast to past literature, recent publications indicate an increasing tendency toward both a review and analysis of what has been done (Miller 1983), and toward setting priorities for the future. The Ministry of Natural Resources of Ontario has established an Esocid Advisory Committee. This committee consists of MNR and university personnel who represent both research and managment needs. The function of the committee is to advise government on what kinds of work should be done, and to suggest the priority to be given to the various tasks in regard to support. The Wisconsin DNR is doing much the same thing by developing a Research Prospectus for each important fish species. Although much interactive benefit results from contacts between individual workers, there might be more benefit from interactive meetings of workers from all of the major areas. New York DEC has regularly invited Canadian participation in planning and review sessions concerning the muskellunge in the St. Lawrence River. There is still too much to do, and too little funding, to condone duplication of effort which might result from inadequate communication. Even special formal sessions on the muskellunge at international meetings fail to reveal the breadth of work underway in various regions. An open exchange of questions or problems identified by one agency that might more readily be studied by another would be very beneficial. Newsletters, like those of Systematic Ichthyology and of Underwater Telemetry, are the kinds of approaches to communication which help prevent duplication of effort.

WHAT DOES THE FUTURE HOLD?

The future of the muskellunge will be effected by several factors - 1) increasing pressure on large muskellunge; 2) the threat of increased damage to prime muskellunge habitats by pollution and development; 3) the strong programme of release encouraged by managers and enthusiastically supported by organized anglers; and 4) interest in making this species (or a facsimile in the form of the hybrid tiger muskellunge) available beyond the present limits of distribution. Each of these may dictate how or when certain kinds of study should be carried out. The highly beneficial release program will impose certain restrictions on the opportunity to acquire information. Fewer muskellunge will be landed or be available for post-mortem examination, and this imposes a greater need for anglers to help make certain that as much as possible is learned from each one killed. New techniques will have to be devised which will enable us to derive a maximum of information from muskellunge which can be returned alive to public waters (e.g., Crossman and Hamilton 1978). Money to support research may have to be derived from increased user fees, such as the muskellunge stamp fee in Ontario.

Culture has contributed very significantly to the survival of the muskellunge. We must, however, learn more about the real contribution made by hatchery releases into habitats with different levels of natural reproduction. The development of the hybrid or tiger muskellunge has proved an economic benefit to culture of a large esocid form. I feel, however, a need to express the hope that the zeal to reduce hatchery costs does not lead us into the widespread introduction of the hybrid into situations which will still support populations of the muskellunge. Even if fish culture support is imperative, I feel we should continue with hatchery stock of muskellunge derived from a local indigenous population in any situation for which we can predict probable survival of the natural form.

This review is merely an indication of the history of our interest in the species and the wealth of information available in the literature. In spite of the extent of the information available, much that is available must be done again elsewhere. We have learned that the muskellunge is a variable animal which responds in unusual ways to its environment. As a result, we cannot assume that data derived from the Great Lakes race of the species can be used directly as the basis for management strategies for populations of the Ohio and western races. Even within races, certain populations are distinctive in regard to potential maximum size, size at first attainment of sexual maturity, etc., and require special management for best utilization. More concern must be given to the genetic uniqueness of the races, and to populations with special adaptations to environmental conditions.

Subject areas which might be neglected in a more practical analysis of research needs include the consequences of homing of adults to a traditional spawning habitat which has been destroyed or made inaccessible by man's activities; time-trend analysis to test for amplitude and frequency of natural fluctuations; the biology of very old muskellunge; angling stress; and problems of food for young muskellunge which result from the dissynchrony in hatching and growth of muskellunge and of suitable prey species.

In general terms, recreational angling seems to have shifted toward family groups happy with a mixed bag less than legal limit in weight or number. In contrast, muskellunge angling has shifted toward specialist anglers, selective retention of only trophy fish, and release of even individuals exceeding minimum size regulations. It seems obvious to me that a major emphasis, therefore, must be placed on the opportunity for muskellunge to survive to greater ages. Within areas where natural reproduction is still effective, there is the additional benefit of increasing reproductive potential by the same technique. It seems foolhardy to continue to remove from the population, in increasing numbers per year, animals at one quarter of both their potential size and reproductive capacity.

We may have passed through the ages of dependence on regulations, and then on culture. Possibly we must now enter the age of slot-sizes, minimum sizes specific to populations or groups of populations, and effective live release.

The "musky mystique" is changing. The 100-hour/legal fish barrier, like the four-minute mile, is falling to increased skill and knowledge of the specialist. In an era burgeoned with economic terms, one might say that fishery managers now seem more sensitive to the wishes of their resource-user clients in regard to commodity, unit size, packaging, inventory turn-over time, and cost mark-up. Managers may, in turn, start to press researchers for information on shelf-life, product viability, and for estimates of GLP (gross local product).

ACKNOWLEDGEMENTS

In a review article it is the work of past authors which should be acknowledged most, and I do so in order to express my appreciation to those who took the time to record data, ideas, and dates in our rather short, but intensive, pursuit of information on this "noble" species. I quickly found that a review article meant I would be emphasizing some people's work more than that of others. I hope all such shortcomings will be attributed to my faulty awareness of the literature, rather than to any conscious attempt to disregard the work of anyone who has contributed. I extend my thanks to those who more directly assisted in the preparation of this manuscript - J. Axon, M. Burridge-Smith, C.C. Cooper, B.F. Dowler, P.J. Festa, N.E. Folde, C. Horkey, J. Klingbiel, B. Lebeau, C. Lewis, J. Mick, C.W. Pollock, R. Pugsley, J. Ross and P.J. Wingate.

REFERENCES

Anon. 1854. No title. Proceedings of the Boston Society of Natural History (1851-1854) 4:287.

Anon. 1897. La ville de Racine. Bulletin Recherches Historiques 3:41-42.

Axon, J.R. 1981. Development of a muskellunge fishery at Cave Run Lake, Kentucky, 1974-1979. North American Journal of Fisheries Management 1:134-143.

Beggs, G.L., G.F. Holeton, and E.J. Crossman. 1980. Some physiological consequences of angling stress in muskellunge, *Esox masquinongy* Mitchill. Journal of Fish Biology 17:649-659.

Bevelheimer, M.S. 1983. Assessing significance of physiological differences among three esocids with a bioenergetics model. Master of Science dissertation. The Ohio State University, Columbus Ohio, USA.

Bloomfield, L. 1975. (Charles F. Hockett, editor). Menomini Lexicon. Milwaukee Public Museum Publications in Anthropology and History, No. 3:289pp.

Braun, E.R., and J. Pearson. 1980. An evaluation of musky stockings in northeastern Indiana. Indiana Department of Natural Resources, Division of Fish and Wildlife, Fisheries Section. Indianapolis, Indiana, USA.

Brewer, D.L. 1980. A study of native muskellunge populations in eastern Kentucky streams. Kentucky Department of Fish and Wildlife Resources, Fisheries Bulletin 64. Frankfort, Kentucky, USA.

Burnett, W.I. 1854. (Note on breeding of muskellunge). Proceedings of the Boston Society of Natural History 4:360.

Buss, K. 1960. The muskellunge. Pennsylvania Fish Commission Special Purpose Report, Bellefonte, Pennsylvania, USA.

Cameron, G.S. 1948. An unusual maskinonge from Little Vermillion Lake, Ontario. Canadian Journal of Research 26:223-229.

Caplin, D.L. 1982. An experimental study of the interactions between young of the year pike *Esox lucius* and muskellunge *Esox masquinongy*. Master of Science dissertation. University of Wisconsin, Madison, Wisconsin, USA.

Casselman, J.M. 1979. The esocid cleithrum as an indicator calcified structure. Pages 249-272 *in* J. Dubé and Y. Gravel, editors. Proceedings of the 10th (1977) Warm Water Workshop. Ministère du Loisir, de la Chasse et de la Pêche du Québec, Direction de la recherche faunique. Montréal, Québec, Canada.

Cavender, T.M., J.G. Lundberg, and R.L. Wilson. 1970. Two new fossil records of the genus *Esox* (Teleostei, Salmoniformes) in North America. Northwest Science 44:176-183.

Chambers, E.T.D. 1923. The maskinonge: a question of priority in nomenclature. Transactions of the American Fisheries Society 52(1922):171-177.

Cheney, A.N. 1896. Mascalonge, pike, pickerel, and pike perch. New York Commissioners of Fisheries, Game and Forests. 1st Annual Report (1895). Albany, New York, USA.

Clark, C.F. 1964. Muskellunge in Ohio and its management. Ohio Department of Natural Resources, Division of Wildlife Publication W-329. Columbus, Ohio, USA.

Cobb, J.N. 1905. Commercial fisheries of the interior lakes and rivers of New York and Vermont. Pages 225-232 *in* Report of the United States Commissioner of Fisheries (1903), Part 29, Washington, USA.

Cotchefer, R. 1904. The maskalonge supply. Report of the General Foreman of Hatcheries for 1902. New York Forest, Fish and Game Commission, 8th and 9th Annual Reports (1902-1903).

Crossman, E.J. 1978. Taxonomy and distribution of North American esocids. American Fisheries Society, Special Publication 11:13-26.

Crossman, E.J., and C.D. Goodchild. 1978. An annotated bibliography of the muskellunge, *Esox masquinongy* (Osteichthyes: Salmoniformes). Royal Ontario Museum, Life Sciences Miscellaneous Publication: 131 pp.

Crossman, E.J., and J.G. Hamilton. 1978. An apparatus for sampling gut contents of large, living fishes. Environmental Biology of Fishes 3:297-300.

DeKay, J.E. 1842. Fishes. Part 4, Pages 1-415 *in* Zoology of New York, or the New York fauna...etc. New York Geological Survey, Albany, New York, USA.

Dombeck, M.P. 1979. Movement and behavior of the muskellunge determined by radio telemetry. Wisconsin Department of Natural Resources, Technical Bulletin 113.

Dombeck, M.P., B.W. Menzel and P.N. Hinz. 1983. Muskellunge spawning habitat and reproductive success. Iowa Agriculture and Home Economics Experiment Station Project 2236, Ames, Iowa, USA.

Dufour, M., et P.-J. Paulhus. 1979. L' élevage et l' ensemencement du maskinongé. Pages 117-127 *in* J. Dubé and Y. Gravel, editors. Proceedings of the 10th (1977) Warm Water Workshop. Ministère du Loisir, de la Chasse et de la Pêche du Québec, Direction de la recherche faunique. Montréal, Québec, Canada.

Dymond, J.R. 1932. About the maskinonge. Rod and Gun 34.18.

Eddy, S. 1940. Do muskellunge and pickerel interbreed? Progressive Fish Culturist 48:25-28.

Elson, P.F. 1941. Rearing maskinonge in a protected area. Transactions of the American Fisheries Society 10(1940):421-429.

Evermann, B.W., and E.L. Goldsborough. 1901. Notes on the fishes and mollusks of Lake Chautauqua, New York. New York Forest, Fish and Game Commission, 6th Annual Report (1900), Albany, New York, USA.

Fish, M.P. 1932. Contributions to the early life histories of sixty-two species of fishes from Lake Erie and its tributary streams. United States Department of Commerce, Bureau of Fisheries Bulletin 47(10). Washington, USA.

Gorrie, J.F. 1984. Bioenergetics of the muskellunge, *Esox masquinongy* Mitchill, and a comparison with that of *Esox lucius* Linnaeus. Master of Science dissertation. University of Toronto, Toronto, Ontario, Canada.

Greeley, J.R. 1938. Fishes of the area with annotated list. Section 2,, Pages 48-60 *in* A biological survey of the Allegheny and Chemung watersheds. New York Conservation Department, Supplement to the 27th Annual Report (1937) Biological Survey 12. Albany, New York, USA.

Hall, C.R. 1934. Scientists in the early republic; Samuel Latham Mitchill, 1764-1831. Columbia University Press, New York, New York, USA.

Hallock, C. 1877. The sportsmen's gazateer and general guide ...etc. Forest and Stream Publication. New York, New York, USA.

Harkness, W.J.K. 1934. The maskinonge in Ontario. Rod and Gun in Canada 35:17-19, 29, 30. Toronto, Ontario, Canada.

Harkness, W.J.K. 1936. Sanctuaries are necessary to preserve Nipissing musky. Article 12. Biological study of Lake Nipissing. North Bay Nuggett, April 9, North Bay, Ontario, Canada.

Harrison, E.J., and W.F. Hadley. 1978. Ecological separation of sympatric muskellunge and northern pike. American fisheries Society, Special Publication 11:129-134.

Hasler, A.D., R.K. Meyer, and H.M. Field. 1940. The use of hormones for the conservation of muskellunge, *Esox masquinongy immaculatus* Garrard. Copeia 1940:43-46.

Hasse, J.J. 1976. Bibliography of the muskellunge (*Esox masquinongy*). New York Department of Environmental Conservation, Albany, New York, USA.

Heacox, C.E. 1946. The Chautauqua Lake muskellunge: research and management applied to sport fishery. Transactions of the 11th North American Wildlife Conference: 419-425.

Henshall, J.A. 1892. The mascalonge. Pages 191-199 *in* W.A. Perry, *et al.* American game fishes, their habits, habitat, and peculiarities; how, when, and where to angle for them. Rand, McNally, Chicago, Illinois, USA.

Herbert, H.W. 1849. Frank Forester's fish and fishing of the United States and British provinces of North America. W.A. Townsend, New York, New York, USA.

Hess, L., and C. Heartwell. 1978. Literature review of large esocids (muskellunge, northern pike, hybrid tiger muskellunge). Pages 139-175 *in* J. Dube and Y. Gravel, editors. Proceedings of the 10th Warm Water Workshop. Northeastern Division, American Fisheries Society. Montreal, Quebec, Canada.

Hourston, A.S. 1952. The food and growth of the maskinonge (*Esox masquinongy* Mitchill) in Canadian waters. Journal of the Fisheries Research Board of Canada 8:347-368.

Hourston, A.S. 1955. A study of the variation in the maskinonge from three regions in Canada. Royal Ontario Museum of Zoology and Palaeontology, Contribution 40:13 pp.

Jordan, D.S. 1877. Contributions to North American ichthyology: based primarily on the collection of the United States National Museum. Volume 2. United States National Museum, Bulletin 10:116 pp.

Jordan, D.S. 1882. Report on the fishes of Ohio. Section 4, Pages 785-1002 *in* Zoology and Botany, Part 1, Zoology. Ohio Geological Survey 4, Columbus, Ohio, USA.

Jordan, D.S., and B.W. Evermann. 1896-1900. The fishes of North and Middle America. United States National Museum Bulletin 47:3313 pp.

Kendall, W.C. 1919. The pikes: their geographic distribution, habits, culture, and commercial importance. United States Department of Commerce. Report of the Commissioner of Fisheries (1917). Appendix 5. Bureau of Fisheries Document 853. Washington, USA.

Kirtland, J.P. 1838. Report on the zoology of Ohio. Ohio Geological Survey, 2nd Annual Report, Columbus, Ohio, USA.

Kirtland, J.P. 1854. Revision of the species belonging to the genus *Esox*, inhabiting Lake Erie and the River Ohio. Annals of Science (Transactions of the Cleveland Academy of Natural Science) 2:78-79.

Krumholz, L.A. 1949. Length-weight relationship of the muskellunge, *Esox m. masquinongy*, Lake St. Clair. Transactions of the American Fisheries Society 77(1947):42-48.

LeBeau, B. 1984. La reproduction du maskinongé, *Esox masquinongy* Mitchill, au lac Barrière et au lac Saint-Louis, Mémoire de la maîtrise es sciences. Université de Montréal, Montréal, Québec, Canada.

LeSeur, C.A. 1818. Description of several new species of the genus *Esox*, of North America. Journal of the Academy of Natural Sciences of Philadelphia 1:413-417.

MacGregor, J.M., J.A. Scott, and B.C. Bean. 1960. A review of the life history and the proposed management of the northern. Michigan Department of Conservation, Fisheries Division, Lake and Stream Improvement Report 100. Lansing, Michigan.

MacKay, H.H 1931. The maskinonge and its conservation. Ontario Department of Game and Fisheries, Biology and Fish Culture Branch, Bulletin 1. Toronto, Ontario, Canada.

MacKay, H.H., and W.H.R. Werner. 1934. Some observations on the culture of the maskinonge. Transactions of the American Fisheries Society 64:313-317.

McAllister, D.E. 1959. Fish remains from a 600-year-old St. Lawrence River Iroquois site. National Museum of Canada, Contributions in Zoology, Bulletin 172:34-38.

McAllister, D.E. 1962. Fish remains from Ontario Indian sites 700 to 2500 years old. National Museum of Canada, Natural History Papers 17:6 pp.

Meek, S.E., and R. Newland. 1886. A review of the species of the genus *Esox*. Proceedings of the Academy of Natural Sciences of Philadelphia 3(1885):367-375.

Miles, H.M., S.M. Loehner, D.T. Michaud, and S.L. Salivar. 1974. Physiological responses of hatchery reared muskellunge (*Esox masquinongy*) to handling. Transactions of the American Fisheries Society 103:336-342.

Miller, M.L. 1983. The status of muskellunge management in North America. Ohio Department of Natural Resources, Division of Wildlife, Columbus, Ohio, USA.

Minor, J.D., and E.J. Crossman. 1978. Home range and seasonal movements of muskellunge as determined by radiotelemetry. American Fisheries Society, Special Publication 11:146-158.

Moore, E. 1938. A creel census of Chautauqua Lake, New

York. Transactions of the American Fisheries Society 67(1937):130-138.

Mosher, A.A. 1892. The mascalonge in Wisconsin waters. Pages 199-207 *in* Perry, W.A. et al American game fishes, their habits, habitat, and peculiarities; how, when, and where to angle for them. Rand, McNally, Chicago, Illinois, USA.

Muir, B.S. 1964. Vital statistics of *Esox masquinongy* in Nogies Creek, Ontario. II. Population size, natural mortality, and the effect of fishing. Journal of the Fisheries Research Board of Canada 21:727-746.

Nevin, J. 1901. The propagation of muskellunge in Wisconsin. Transactions of the American Fisheries Society 30:90-93.

Oehmcke, A.A., L. Johnson, J. Klingbiel and W. Wistrom. 1958. The Wisconsin muskellunge. Its life history, ecology, and management. Wisconsin Conservation Department, Publication 225, Madison, Wisconsin, USA.

Parsons, J.W. 1959. Muskellunge in Tennessee streams. Transactions of the American Fisheries Society 88:136-140.

Porter, L.R. 1977. Review of selected literature on muskellunge life history, ecology and management. Minnesota Department of Natrual Resources, Division of Fish and Wildlife, Section of Fisheries, Special Publication 119. Minnesota, USA.

Prince, E.E. 1901. The vernacular names of fishes. Report 2. Pages 12-23 *in* Special appended reports for 1900, Government Printing Bureau, Ottawa, Ontario, Canada.

Rafinesque, C.S. 1818. Museum of natural history. Further discoveries in natural history made during a journey through the western region of the United States. American Monthly Magazine and Critical Review 3:445-447.

Rafinesque, C.S. 1820. Ichthyologia Ohiensis, or Natural history of the fishes inhabiting the river Ohio and its tributary streams. Lexington, Kentucky, USA.

Ramsell, L.A. 1982. A compendium of muskie angling history. Echo Printing, Alexandria, Minnesota, USA.

Reynolds, W.W., and M.E. Casterlin. 1979. Thermoregulatory rhythm in juvenile muskellunge (*Esox masquinongy*). Evidence of a diel shift in the lower setpoint. Comparative Biochemistry, Physiology and Comparative Physiology 63:523-526.

Riddle, Jr. 1975. Status of the native muskellunge, *Esox masquinongy ohioensis*, of the Cumberland Plateau, Tennessee. Master of Science dissertation, Tennessee Technical University, Cookeville, Tennessee, USA.

Schloemer, C.L. 1936. The growth of the muskellunge, *Esox masquinongy immaculatus* (Garrard) in various lakes and drainage areas of northern Wisconsin. Copeia 1936:185-193.

Seaborn, E. 1937. Variations in the maskinonge in the Sauble River, Ontario. Copeia 1937: 237-238.

Small, H.B. 1865. The animals of North America. Series 2. Fresh-water fish. M. Longmoore, Montreal, Quebec, Canada.

Smith, C.L. 1954. Pleistocene fishes of the Berends fauna of Beaver County, Oklahoma. Copeia 1954:176-180.

Smith, C.L. 1962. Some Pliocene fishes from Kansas, Oklahoma, and Nebraska. Copeia 1962:505-520.

Smith, G.R. 1963. A late Illinoian fish fauna from southwestern Kansas and its climatic significance. Copeia 1963:278-285.

Strand, R.F. 1983. Identification of principal muskellunge spawning areas in Leech Lake by radiotelemetry. Minnesota Department of Natural Resources, Division of Fish and Wildlife, Fisheries Section, Investigation Report 337.

Underhill, J.C. 1986. The fish fauna of the Laurentian Great Lakes, the St. Lawrence Lowlands, Newfoundland, and Labrador. pp. 105-136 *in* C.H. Hocutt and E.O. Wiley, editors. The zoogeography of North American freshwater fishes. John Wiley and Sons, New York, New York, USA.

Van Oosten, J. 1946. The pikes. United States Fish and Wildlife Service, Fisheries Leaflet 166. Washington, USA.

Weed, A.C. 1925. Feeding the muskellunge. Copeia 146:46-70.

Weed, A.C. 1927. Pike, pickerel and muskalonge. Chicago Field Museum of Natural History, Zoology Leaflet 9:158-204.

Wilson, M.V.H. 1980. Oldest known *Esox* (Pisces: Esocidae), part of a new Paleocene teleost fauna from western Canada. Canadian Journal of Earth Sciences 17:307-312.

Wilson, R.L. 1967. The Pleistocene vertebrates of Michigan. Michigan Academy of Science, Arts and Letters. Paper 52, Part 1:197-234.

Zack, S.S., 1982. The muskellunge - a bibliography. Published by the author. Denver, Colorado, USA.

BIOLOGY AND LIFE HISTORY

Identification of Muskellunge Northern Pike, and their Hybrids[1]

J.M. CASSELMAN

Ontario Ministry of Natural Resources
Fisheries Branch, Research Section
Box 50, Maple, Ontario L0J 1E0

E.J. CROSSMAN

Department of Ichthyology and Herpetology
Royal Ontario Museum
100 Queen's Park, Toronto, Ontario M5S 2C6

P.E. IHSSEN

Ontario Ministry of Natural Resources
Fisheries Branch, Research Section
Box 50, Maple, Ontario L0J 1E0

J.D. Reist

Department of Fisheries and Oceans
501 University Crescent
Winnipeg, Manitoba R3T 2N6

H.E. BOOKE

Massachusetts Cooperative Fishery Research Unit
204 Holdsworth Hall, University of Massachusetts
Amherst, Massachusetts 01003

ABSTRACT

Discrimination among live and even dead esocids can sometimes be very difficult. In areas where muskellunge and northern pike cohabit, there is inadvertent (sometimes illegal) harvest of the rare and more valuable muskellunge because many anglers can't readily distinguish between them. Muskellunge × northern pike hybrids are recognized only by the most avid muskellunge angler and fisheries specialist. This study reviews criteria for identifying and distinguishing among esocids, and researches new techniques that can be applied specifically to muskellunge, northern pike, and their hybrid, the tiger muskellunge.

Body color pattern and external appearance of young, juveniles, and adults provide discriminating characteristics, as do the submandibular pores, and cheek and gill-cover scalation. External morphology such as body proportions can be used to distinguish between northern pike and muskellunge; however, their hybrids generally show intermediate meristic and morphometric characteristics that somewhat overlap muskellunge. Differences in the shape of dentary teeth and shape and dentition of the vomer and palatine tooth patches provide distinct diagnostic features. Scale pattern, and shape and zonation of both scales and cleithra, can be used to distinguish among these species and their hybrids. Scales and cleithra in large hybrids are larger, relative to body size, than in the parent species. The internal anatomy also has distinguishing characteristics (e.g., location of the dorsal aorta, vertebral column, and centra).

The fish were compared electrophoretically for 18 enzyme systems encoded by 35 loci. Muskellunge and northern pike are electrophoretically distinct for 54% (19) of their loci. The hybrid has intermediate isozyme patterns for all but two of these loci.

The results describe not only better and more easily recognizable characteristics for the angler, but also more sophisticated, annalytical laboratory techniques, which can be used as forensic tools for discrimination among muskellunge, northern pike, and their hybrids.

---

[1] Contribution No. 85-10 of the Ontario Ministry of Natural Resources, Fisheries Branch, Research Section, Box 50, Maple, Ontario L0J 1E0.

Occasionally, even for the professional fisheries worker and law enforcement officer, discrimination among live or even dead esocids can be difficult. Survey responses suggest that most management agencies consider the identification of muskellunge to be a problem. In waters where muskellunge and northern pike cohabit and muskellunge are caught infrequently, most casual anglers are unable either to recognize muskellunge or to distinguish them from northern pike. As a result, there is an unwitting, and sometimes illegal, harvest of the rare and more prized muskellunge. Muskellunge × northern pike hybrids are readily recognized only by the most experienced muskellunge angler and fisheries specialist.

Increasing interest in culture and stocking of the "tiger muskellunge", the hybrid between muskellunge and northern pike, creates a specific need to review means of identifying hybrids and distinguishing them from the parental forms. Since hybrids occur naturally (in some populations up to 17% of the large, muskellunge-like esocids; J. Casselman, unpublished data), it is important that we understand the true extent of their natural occurrence. To facilitate this, there must be convenient, widely-known ways for recognizing the hybrid. The culture and release of these hybrids into public waters will require that anglers, law enforcement officers, and biologists be able to recognize them readily, especially in political jurisdictions in which all three forms occur but may be subject to different management practices and regulations.

This examination provides basic characteristics for recognizing all fishes of the family Esocidae, a key to discriminate among the various species, and more specific details to distinguish among muskellunge, northern pike, and their hybrids. The distinguishing features considered are body coloration and pattern, external body morphometrics and meristics, degree of scalation on cheeks and opercula, scale shape, configurations of circuli in the focus, regenerated scale patterns, type and degree of dentition on the roof of the mouth, shape of the dentary teeth, morphological characteristics of the cleithra, and features of the vertebral column and individual centra. Electrophoretic analyses are conducted, and chromosome morphology is reviewed.

GENERAL DESCRIPTION OF FAMILY CHARACTERISTICS

Features used to distinguish pikes and pickerels from fishes in other families are: body elongate, somewhat laterally compressed; caudal fin deeply forked; dorsal and anal fins far back on body, situated opposite one another, and equal or sub-equal in size; head large, snout elongate, flat, and somewhat like a duck's bill; lower jaw articulated behind posterior edge of orbit; teeth on jaws large and prominent; prominent patches of teeth on vomer, palatines, and tongue; gill rakers reduced to patches of sharp denticles; pectoral fins low; pelvic fins abdominal; all fins soft-rayed; mesocoracoid bone absent from pectoral girdle (the so-called "simple shoulder" used in past to classify the pikes and pickerels in a separate order, Haplomi); and swim bladder physostomous (but there is no known ability to breathe atmospheric oxygen).

The family is composed of one genus and five species: one species, northern pike, is circumpolar; one, Amur pike, is endemic to Siberia and China; and three species, muskellunge, chain pickerel, and grass and redfin pickerels, are endemic to North America. North American distribution maps can be found in Crossman (1978).

The members of the family are moderate to large fishes that are found in small streams, ponds, rivers, lakes, and occasionally in brackish waters. They are known from the Middle Eocene to the present in Eurasia, and Paleocene to the present in North America. They range from subtropical to arctic North America, Europe, and Asia.

KEY TO SPECIES OF THE ESOCIDAE

Salient external characteristics used in the key are illustrated in Figure 1. It may seem inappropriate to include the Asiatic species, the Amur pike, in a North American key to the species of the family. However, in recent years this species was released into the public waters of the state of Pennsylvania, so it is necessary to include it in the key.

1. (a) Mandibular pores usually 5 or more on each side, rarely as few as 4 on one side only (Fig. 1A); infraorbital canal continuous, usually with 8-10 pores; extrascapular canal present, forked, usually with 3 pores (Fig. 1A); number of lateral line scales as high as 176, notched (cardioid) lateral line scales 32-77; vertebrae usually 60-65; pelvic rays usually 10-13; one or both of cheeks and opercula not fully scaled (Fig. 1C). Dark spots or vermiculations on dorsal and anal fins; no prominent vertical pigmented bar below the eye ------------------(the pikes, subgenus *Esox*), 2

   (b) Mandibular pores usually 4 on each side, rarely 3 or 5 on one side only (Fig. 1A); infraorbital canal usually divided into three parts, each with 2-4 pores; extrascapular canal absent (Fig. 1A); lateral line scales rarely over 135, notched lateral line scales 25-51; vertebrae rarely more than 55; pelvic

A HEAD PORES AND CANALS

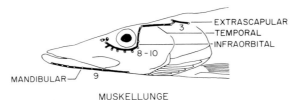

MUSKELLUNGE

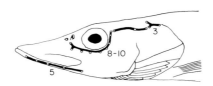

NORTHERN PIKE

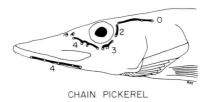

CHAIN PICKEREL

B NUMBER AND LOCATION OF BRANCHIOSTEGAL RAYS

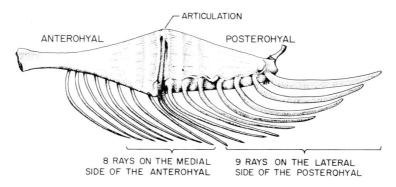

Figure 1. External characteristics used to separate species of the family Esocidae: (A) Head pores and canals--lateral view of the heads of muskellunge, northern pike (Amur pike similar), and chain pickerel (all pickerels similar) showing the diagnostic head pores and canals (after Nelson 1972); (B) Number and location of left branchiostegal rays of muskellunge--the bases of the rays on the lateral side of the posterohyal are depicted as enveloped in soft connective tissue; (C) Scalation on cheek and operculum--the scaled areas are shaded; (D) Relation of snout length to postorbital head length--lateral view of heads of chain pickerel and grass or redfin pickerel (adapted from Legendre 1952); (E) Relation of snout length to nape-dorsal fin length--dorsal and lateral views of redfin and grass pickerels (adapted from Legendre 1952); (F) Number of cardioid, or notched, scales (unshaded) between the pelvic fins--ventral views of grass and redfin pickerels.

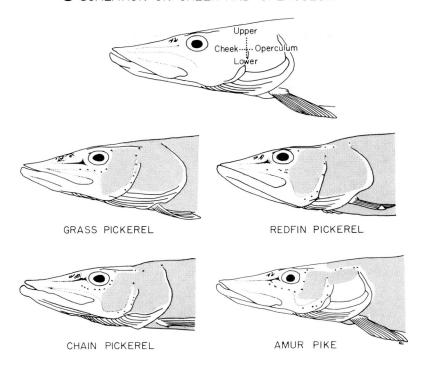
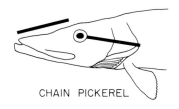

Figure 1 *(continued)*

E SNOUT LENGTH TO NAPE-DORSAL FIN LENGTH

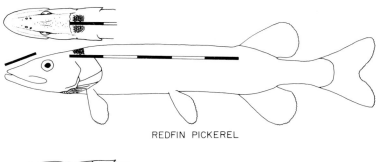

F NUMBER OF CARDIOID SCALES BETWEEN THE PELVIC FINS

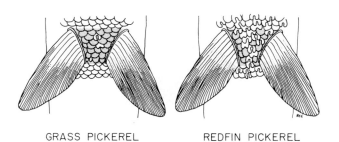

Figure 1 *(continued)*

rays usually 8-10; both cheeks and opercula fully scaled (Fig. 1C). No prominent markings on dorsal or anal fin; a prominent, vertical to oblique, darkly-pigmented bar below the eye ---
---------(the pickerels, subgenus *Kenoza*), 4

2. (a) Total mandibular pores (2 sides) 12-20, 6-10 on each side; branchiostegal rays on one side 16-20, divided count (anterohyal/posterohyal, or ceratohyal/epihyal) most often 8/10 or 8/9 (Fig. 1B); lateral line scales usually more than 145 (140-176), notched lateral line scales 55-77; neither cheeks nor opercula fully scaled. Body color pattern, if present, dark brown to black spots, blotches or vertical bars on paler undercolor (Fig. 2) ---------
---------------------------------------muskellunge

   (b) Total mandibular pores 9-11, usually 5 on each side, rarely 4 or 6 on one side only; branchiostegal rays on one side 11-16; lateral line scales usually fewer than 145 (105-148), notched lateral line scales 42-56; cheeks fully scaled, opercula approximately half-scaled (Fig. 1C). Color pattern in adults either pale spots (white to yellow), or dark brown to black spots approximately the size of pupil of eye ---3

3. (a) Branchiostegal rays 13-16 on one side, divided count (anterohyal/posterohyal) usually 7/8 or 6/8; lateral line scales usually fewer than 145 (105-148), notched lateral line scales 42-55; top of head and snout without scales; in the first circumferential row of scales behind the naked top of the head (the nape), the mid-dorsal scales are one or two, which are larger than others in the row, and each is larger than the pupil of the eye. Pattern on body of adults[1] horizontal rows of white to yellow, oval to bean-shaped spots on a green to almost black ground color (Fig. 2), unpaired fins with dark vermiculations or lines, sides of head with golden vermiculations----
---------------------------------------northern pike

   (b) Branchiostegal rays 11-15 on one side, divided count usually 5/8 or 6/7; lateral line scales usually more than 145 (130-165), notched lateral line scales usually 48-64; top of head and snout said to be scaled in large individuals[2], in the first circumferential row of scales behind the head, the mid-dorsal scales are smaller, not larger than others in the row, and each is smaller than the pupil of the eye. Pattern in adults consists of black to brown randomly placed spots on body, head, and dorsal, anal, and caudal fins. Ground color of back and unpaired fins is sulphur-yellow; the sides are iridescent silver to golden ----Amur pike

4. (a) Branchiostegal rays 14-17 on one side, divided count most often 6/9 or 7/9; usually more than 120 scales in lateral line (114-131); posterior end of maxilla not reaching beyond mid-point of eye (usually not reaching midpoint); snout longer than post-orbital length of head (Fig. 1D). Color pattern on lower part of sides of adults with dark markings forming chain-like reticulations with golden centers---
---------------------------------------chain pickerel

   (b) Branchiostegal rays 11-13 on one side; usually fewer than 110 scales in lateral line (104-114); posterior end of maxilla reaching beyond mid-point of eye; snout shorter than post-orbital length of head (Fig. 1D). Color pattern in adults vertical or oblique dark bars on dark brown undercolor---------------------5

5. (a) Divided count of branchiostegal rays on one side usually 6/8 or 5/8; snout shorter, length more than 4.5 times into distance from nape to origin of dorsal fin (Fig. 1E); dorsal profile of snout flat to convex (Fig. 1C, 1E); more than 5 cardioid scales in the triangle between the pelvic fins (Fig. 1F); usually more than 3 cardioid scales in the oblique row of scales from origin of dorsal fin to origin of anal fin----------------------------------redfin pickerel

   (b) Divided count of branchiostegal rays usually 5/7 or 4/7; snout longer, length fewer than 4.5 times into distance from nape to origin of dorsal fin (Fig. 1E); dorsal profile of snout usually concave (Fig. 1C, 1E); fewer than 5 cardioid scales in the triangle between the pelvic fins (Fig. 1F); usually only 1 or 2 cardioid scales in the oblique row from origin of dorsal fin to origin of anal fin----grass pickerel

DISTINGUISHING CHARACTERISTICS OF MUSKELLUNGE, NORTHERN PIKE, AND THEIR HYBRIDS

Body Coloration and Color Pattern

The information on body coloration and color

---

[2]Excluding the color mutant of northern pike now called silver pike, which is silvery blue to silvery green overall and lacks the characteristic bean-shaped spots.

[3](Berg 1948), but specimens 40-45 cm long currently in the collection at the Royal Ontario Museum had no detectable scales on top of head or snout.

pattern for muskellunge and northern pike, although modified for the young, is largely derived from Scott and Crossman (1973).

Muskellunge

The overall ground color of muskellunge can be extremely variable, but is generally dark markings on a light background (Fig. 2). There are three types of color patterns--clear, or immaculate (Fig. 2Ai), spotted (Fig. 2Aii), and barred (Fig. 2Aiii). The ground color and the nature of the dark markings vary, however, between individuals and within and between areas.

In adults the back, head, and upper sides of the body are iridescent greenish-gold to light brown, and the flanks range from green-gold to brownish to grey or silvery. In very silvery specimens, no dark markings or only faint dark markings are visible (Fig. 2Ai). In others the brown to dark markings on the flanks take the form of spots (Fig. 2Aii), prominent vertical bars (Fig. 2Aiii), blotches or vermiculations, or combinations of these.

All dark patterns are represented with different frequencies across the species' range. The only distinctions that can be made are that muskellunge in the St. Lawrence River-Great Lakes exhibit the spotted pattern more frequently than do muskellunge in other areas. There is also a slight tendency for western fish either to be devoid of markings (clear), or have prominent oblique bars. The largest specimens usually have a silvery overlay that hides the pattern. This pattern is more obvious after death or preservation. Hourston (1955) analysed the relative occurrence of pattern types in three Canadian areas, and found it difficult to attribute any specific type of pattern to any area.

In muskellunge the ventral surface of the body is cream-colored to milky white with small brown to grey spots or blotches. The head, dark above and paler on the sides, is marked with spots or several dark bars radiating back from the eyes. The fins are greenish or buff to red-brown with dark blotches. Blotches on the unpaired fins are more prominent than those on the paired fins. The unpaired fins are often blood red after capture by nets or angling. They apparently hemorrhage very easily during capture and handling.

Young-of-the-year muskellunge, less than approximately 20 cm total length (TL), have very distinct color patterns. At a length of approximately 10 cm, dark pigmentation on the back extends downward with a wavy margin to a point near the lateral line. In the middle of the flank there are irregular dark blotches and dark pigmentation below this on the flanks. At a length of approximately 20 cm, the dark pigmentation on the back is arranged in vertical bars or saddles, which are equal in width to the intervening light areas. There are a few dark spots below the bars and on the unpaired fins. The gold-green mid-dorsal stripe is straight and is more obvious in this species than in the others. Illustrations of young-of-the-year are found in Crossman and Buss (1965).

Northern pike

The basic color pattern of northern pike is light spots on a dark background (Fig. 2C). The dorsal surface, upper sides, and top and upper part of the head are dark, brilliant green through olive green, to almost black. The ground color of the flanks is lighter, and in northern pike over approximately 38 cm long, conspicuously marked with 7-9 irregular, longitudinal rows of yellowish to whitish, bean-shaped spots, some as long as the diameter of the eye. The body appears to be flecked with gold, which results from a tiny gold spot on the tip of the exposed edge of most body scales. The ventral surface of the body and head is cream to milky white, with streaks of white projecting up into the ground coat of the flanks. These streaks are all that remain of the characteristic juvenile pattern of this species, which consists of long, wavy, white to yellow, vertical to oblique bars extending almost to the lateral line. These vertical bars break into rows of spots in the adult pattern. Because of their distinctly different color pattern, juvenile northern pike are often thought to be a different species, called "grass pike."

The sides of the head are vermiculated with bright golden marks, and the eye bars are inconspicuous. The eyes are usually bright yellow. The dorsal, caudal, and anal fins are green to yellow, or sometimes orange or pale red, blotched with large, irregular, dark markings. The paired fins are often unmarked and buff to dusky, but sometimes have a color pattern similar to the other fins.

Juveniles have similar ground color, but the white of the ventral surface extends dorsally into 8-12 irregular, vertical bars. The bars are much narrower (1/3 to 1/5) than the intervening dark area. The mid-dorsal gold to green stripe and subocular bar are present as in other esocids, but are less obvious. The mid-dorsal stripe is more irregular in shape in this species, and is obviously the interconnection of lateral white stripes. The subocular bar is more obvious in northern pike than in muskellunge or their hybrids.

Figure 2. Typical color patterns of muskellunge, muskellunge × northern pike hybrid, and northern pike. (A) Muskellunge: (i) Clear, or immaculate--83.9 cm total length (TL), male, from Lake of Two Mountains, St. Eustache, Quebec; (ii) Spotted--93.0 cm TL, male, from St. Lawrence River, Verdun, Quebec; (iii) Barred--79.8 cm TL, female, from Little Vermillion Lake, Ontario. (B) Muskellunge × northern pike hybrid--97.5 cm TL, male, from Little Vermillion Lake, Ontario. (C) Northern pike--87.6 cm TL, female, from the upper St. Lawrence River, Ontario.

Hybrids

The general impression given by the color and pattern of the hybrid is an extreme or bizarre variation of the muskellunge (Fig. 2B). It is not greatly different from that seen in some true muskellunge in Wisconsin and Minnesota. However, virtually no northern pike characteristics are visible.

Color pattern is extremely variable, but some generalizations are possible. The pattern is dark pigment on a lighter ground color. It consists of narrow, dark, vertical or oblique bars running downward from the dark back. These bars are narrower and more irregular in shape than those in muskellunge, and extend farther towards the ventral surface. There is very little pale area on the flanks. Very often the narrow bands appear like chain links, are forked, and alternate with vertical rows of small, dark spots. The median fins are often more densely covered with small, brown to black spots than in the muskellunge. The lateral regions of the head, especially the cheeks and opercular area, show a similar spotted coloration that sometimes is very exaggerated and not seen in muskellunge. The paired fins are often very darkly colored.

In large hybrids, the head seems massive (Fig. 2B), and is longer and heavier than that of muskellunge. The tail appears equally large. The body of the hybrid is deeper and thicker for its length and is generally more robust than that of northern pike (Fig. 2C), and much more so than that of muskellunge (Fig. 2A).

The pattern of young hybrids varies depending upon the sex of the parent used in the cross (illustrated in Crossman and Buss 1965). Those described here result from the most frequently used combination of female muskellunge × male northern pike. At 10 cm total length, the pattern of hybrids is more like that of muskellunge at approximately 20 cm. The earlier pattern of the muskellunge does not seem to appear regularly in the hybrid. At approximately 20 cm total length, the body is silvery to golden with 10 or 12 wide, mostly vertical, darkly pigmented bars or saddles, with the paler intervening spaces narrower than the bars. The bars are rounded or only broadly pointed at the vertical end, and some bars are shorter with a large, round spot below. The head is sometimes spotted, darkly pigmented above the level of the eye, with no suborbital bar, but often the snout has dark horizontal "sight line" bars. The dorsal and caudal fins have fewer large spots or blotches, and the anal fin is most often without pigmentation. Pectoral and pelvic fins are pale and without pigmentation. The mid-dorsal stripe is absent, or consists of a very zigzag interconnection of the pale areas. The young could be confused only with the young of muskellunge.

External Morphometrics

The morphometric study was conducted to quantify morphometric differences between muskellunge and northern pike, to determine the statistical significance of such differences and the intermediacy of hybrids between the parental types, and to provide a means by which field biologists can quickly and correctly classify wild-caught specimens to either parental type or to the putative hybrid class (hybrid of unknown parentage; i.e., wild-caught individuals).

Morphometric samples

Ninety specimens of northern pike were used—30 each from mid-west Canada (drainage to Hudson Bay), southern Ontario (Great Lakes drainage), and mid-west United States (Mississippi drainage). One hundred specimens of muskellunge were used. The muskellunge samples were obtained from all areas of the distribution (for details on localities see Reist 1983) so that there would be considerable geographic overlap of the samples for each species. This ensured that any ecophenotypic variation induced by environmental effects was likely equivalent for both species. Furthermore, the samples contained individuals from Dalrymple Lake, Ontario, in which both species are sympatric and hybridize; thus the efficiency of the separation between the species could be checked directly. All these specimens represented non-hybrid, "pure" members of the appropriate species. Additional samples of five of each parental type, ten hybrids of known parentage, and two putative hybrids from Dalrymple Lake, were used to examine the efficiency of a discriminant function that separated the parental types.

Morphometric characters

Ten measurements were chosen (Fig. 3). All were parallel to a major body axis, between topographical features, and were made to the nearest 0.1 mm with dial calipers. The distance from the tip of the lower jaw to the end of the fleshy part of the caudal peduncle was used as a standard length measurement (STL) for each specimen.

Statistical approach

Growth of fish is considered to be continuous, so comparisons of body morphometry must be conducted in terms of the shapes (relative sizes) of body

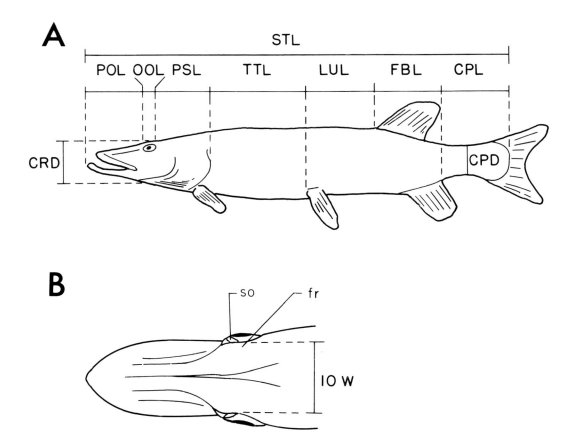

Figure 3. Morphometric variables used in the study, depicted on a typical esocid. (A) Lateral view: CPD--caudal peduncle depth; CPL--caudal peduncle length; CRD--cranial depth; FBL--fin base length; LUL--lumbar length; OOL--orbital length; POL--preorbital length; PSL--postorbital length; STL--standard length; and TTL--trunk length. (B) Dorsal view of head: fr--frontal bone; IOW--interorbital width; and so--supraorbital bone.

parts. To achieve such comparison, the raw data must be transformed to dissociate size and shape. This transformation was achieved by generating the ratio of a body part (POL to CPD inclusive) to standard length, although this method is theoretically incorrect (Simpson et al. 1960; Atchley et al. 1976). These ratios were then used for subsequent analyses. Ratios were used to provide the simplest method possible so that field biologists could readily use the technique on unidentified specimens without applying elaborate computations. Although not reported here, all the results based upon ratios were substantiated by parallel analyses conducted upon residuals from the pooled, common within-groups regression line for each variable (POL to CPD inclusive) and standard length (Reist 1983). No substantial differences were found between the results herein and those based on residuals.

Univariate descriptive statistics, univariate analysis of variance, and multivariate discriminant analysis were each conducted on the ratios to determine the presence and degree of differences between species. Discriminant analysis is a powerful multivariate technique that consists of three parts. First, the existence of a significant difference between group centroids (multivariate means) is determined. Second, if such a difference exists, the groups are separated in terms of the original variables. This is achieved by mathematically combining these variables into a linear discriminant function. Third, this function with the coefficients in an unstandardized form can be used with the raw data (ratios) from an individual to generate a score for that individual. Proximity of this score to one or the other of the group centroids can then be used to hypothesize group membership of the individual. These scores represent mathematically proper "hybrid scores" based upon several variables, and as such are substantially better than so-called "hybrid indices" generated by the mathematically unsound practice of combining univariate differences. The classification process can be applied to specimens used to construct the discriminant function; the percentage of correct classifications is a measure of the efficiency of the function. Correct classification of other individuals known to be "pure" members of either group ensures that this discriminant function is reasonable for classifying unknown specimens from a particular water body. Finally, the classification process can also be applied to true unknown specimens or putative hybrids. If the unknowns belong to one or the other of the species, assignment should be unequivocal. In the case of hybrids, the scores should be intermediate between the two species, but the exact positioning depends upon the degree of parental genetic background inherent in each character and the importance of these variables in distinguishing between the species.

Univariate analysis

Some body parts of northern pike are, on average, relatively larger than those of muskellunge (preorbital length, orbital length, postorbital length, cranial depth, lumbar length, and fin base length). Northern pike have relatively smaller parts for the remaining variables (Table 1). These differences were statistically significant ($P<0.01$ to $P<0.001$) for all variables except caudal peduncle depth. In terms of the explained variance (r^2), the differences due to trunk length were greatest, those due to cranial depth next greatest, and so on for orbital length, postorbital length, lumbar length, preorbital length, fin base length, caudal peduncle length, and interorbital width.

Mean values of the hybrid samples were intermediate between, or similar to, those for the parental species for all variables except fin base length and caudal peduncle length (Table 1). For preorbital length, postorbital length, cranial depth, and lumbar length, means for the hybrids were most similar to those for muskellunge. For orbital length and trunk length, mean values for hybrids were most similar to those for northern pike. For interorbital width and caudal peduncle depth, mean values of all samples were quite similar. For most variables a univariate estimate of variation, the coefficient of variation (CV), was smaller for northern pike than for muskellunge. The CV for the hybrids was smaller than that for either parental type (Table 1).

Multivariate analysis

The difference between the centroids representing the species was significant (P approaching 0), thus a discriminant analysis was conducted. The standardized discriminant function was:

Table 1. Descriptive statistics of the morphometric ratios (variable ÷ STL, expressed as percent) and univariate analysis of variance (ANOVA) for muskellunge, northern pike, and their hybrids.

Variable	Northern pike		Muskellunge		Female northern pike × male muskellunge		Female muskellunge × male northern pike		ANOVA	
	$\bar{x}$	CV	$\bar{x}$	CV	$\bar{x}$	CV	$\bar{x}$	CV	Sign.[a]	r^2
POL	13.4	5.94	12.8	8.97	12.8	2.27	13.1	2.82	***	0.080
OOL	4.7	14.02	3.6	13.69	4.4	5.68	4.4	3.41	***	0.447
PSL	12.5	4.55	11.7	8.79	11.6	1.81	11.3	8.76	***	0.185
CRD	10.1	6.02	8.5	7.33	8.8	3.07	8.9	4.27	***	0.622
IOW	5.1	6.24	5.2	10.97	5.1	0.98	5.2	6.54	***	0.013
TTL	24.0	5.66	28.6	4.54	25.9	2.51	25.6	2.93	***	0.753
LUL	19.3	6.40	18.3	7.92	18.0	2.78	18.6	3.76	***	0.135
FBL	12.9	7.60	12.4	7.71	13.6	3.38	13.3	7.37	**	0.055
CPL	10.3	8.46	12.8	8.56	13.4	4.70	13.9	4.46	**	0.049
CPD	6.1	7.29	6.2	13.46	6.0	2.83	5.9	4.41	ns	0.004

[a]Significance levels
    *** = $P<0.001$
    ** = $P<0.01$
    ns = non-significant

Score = − 0.35POL + 0.16OOL + 0.29PSL
      + 0.77CRD − 0.52IOW − 0.65TTL
      + 0.46LUL + 0.25FBL + 0.11CPL
      − 0.25CPD

The *a posteriori* classification accuracy was 99% overall, with two specimens of northern pike being misclassified. The difference between the species is reflected in the histogram of scores on the discriminant function (Fig. 4).

The absolute magnitudes of the above coefficients indicate the power of individual variables in distinguishing between the species. Since hybrids of known parentage fall intermediate to the parents, the variables with the largest absolute coefficients also indicate which regions of the body yield the best information for discrimination and identification purposes. These regions of the body are, in decreasing order of contribution, cranial depth, trunk length, interorbital width, lumbar length, and preorbital length. The remaining variables contribute progressively less information.

The discriminant scores for muskellunge were generally less variable (single mode, range from 0 to −4.5) than were those for northern pike (multiple modes, range from 0 to +6.0) (Fig. 4), in contrast to the observations for univariate variation (Table 1).

Classification of individuals

Three specimens of muskellunge and five of northern pike from Dalrymple Lake were included in the analyses. The discriminant scores for these specimens of muskellunge (symbol A, Fig. 4A) were close to the muskellunge centroid. Scores from northern pike from Dalrymple Lake (symbol B, Fig. 4A) ranged from the northern pike centroid towards the muskellunge end of the function. One of the two misclassified specimens of northern pike was from Dalrymple Lake, the other from Heming Lake, Manitoba.

To obtain the discriminant score of an unknown individual, 1) compute the appropriate ratios as proportions rather than percentages, and 2) insert the appropriate ratio at the proper place in the following equation and calculate the score. The unstandardized discriminant function was:

Score = − 35.37POL + 27.30OOL + 34.69PSL
      + 125.6CRD − 110.99IOW
      − 49.06TTL + 33.83LUL
      + 25.43FBL + 10.39CPL
      − 36.56CPD − 2.27

The resulting score can then be situated relative to the group centroids (northern pike, 2.77; muskellunge, −2.50) or the group limits of Figure 4A, and an identity hypothesized. For example, for one specimen known to be muskellunge, but not included in the analysis to generate the function, the appropriate ratios are: POL--0.114, OOL--0.024, PSL--0.121, CRD--0.087, IOW--0.058, TTL--0.310, LUL--0.186, FBL--0.124, CPL--0.121, and CPD--0.060. Substituting these values in the above function, a discriminant score of −3.682 is obtained. Positioning this individual indicated that it was indeed closest to the muskellunge centroid (symbol A, Fig. 4B).

This classification process was conducted with 22 specimens extrinsic to the analysis used to construct the function. Of these, five were muskellunge, five were northern pike, two were wild-caught putative hybrids, and ten were hybrids of known parentage. By this technique, specimens of pure parentage were classified with their parental types; specimens of known hybrid parentage were classified intermediate to both parental types; and specimens of suspected hybrid origin, but unknown parental type, generally were classified intermediate to the parental types (Fig. 4B). The scores of many of the known hybrids (7 of 10) were closer to the muskellunge centroid. The two putative hybrids of unknown parentage from Dalrymple Lake scored well into the region of the function occupied by specimens of muskellunge (symbol E, Fig. 4B).

Discussion of morphometrics

The high degree of accuracy of *a posteriori* classification of the individuals to their respective species, the classification to correct species of specimens that were extrinsic to the samples used to construct the function, and the positioning of hybrids of known parentage to the expected, intermediate region of the discriminant function, all indicate that external morphometry in general and the particular discriminant function enumerated herein are valid for hybrid identification. The classification to correct species of specimens extrinsic to the function suggests that the discriminant function used here can be generalized to all parts of the geographical distribution of these fish where hybridization could occur. Since the scores of hybrids overlapped those of some individuals of each species, the presence of an intermediate score on the function should be taken as an indication of a potential hybrid rather than absolute proof. The hypothesis of the hybrid nature of an

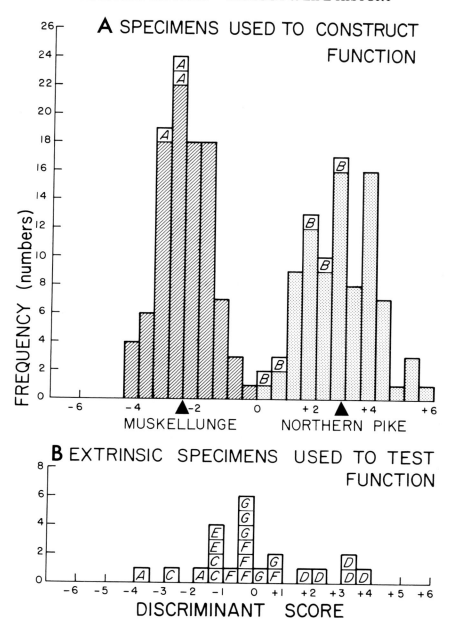

Figure 4. Histograms of discriminant scores for individuals. (A) Distribution of the scores for 100 muskellunge and 90 northern pike used to construct the discriminant function. Group centroids for each species are indicated by closed triangles. (B) Classification of specimens extrinsic to those used to construct the discriminant function. Symbols: *A*--muskellunge from Dalrymple Lake, Ontario, Royal Ontario Museum (ROM) 22599; *B*--northern pike from Dalrymple Lake, Ontario; *C*--muskellunge from Niagara River, Fort Erie, Ontario, ROM 37093; *D*--northern pike from Knowlton Lake, Ontario, ROM 21063; *E*--two putative hybrids (ROM 21834 and ROM 22562) from Dalrymple Lake, Ontario; *F*--cultured hatchery hybrids of known parentage (female muskellunge × male northern pike, ROM 21962); *G*--cultured hatchery hybrids of known parentage (female northern pike × male muskellunge, ROM 21961).

indivdual could then be tested using other data such as color pattern, meristics, etc.

Salient differences between species were noted in two groups of variables--those of the head (cranial depth, interorbital width, and preorbital length) and those of the anterior portion of the body (trunk and lumbar lengths). The importance of these groups of variables most likely has a functional explanation. Those of the former group are likely associated with the prey type, while those of the body reflect numbers of abdominal vertebrae. Both the total count and counts of abdominal vertebrae differ substantially between northern pike and muskellunge (Crossman and Buss 1965; Cavender 1969). Univariate variation is low, and the strength of the relationship separating northern pike from muskellunge (r^2 in ANOVA) is high for the two variables measuring this region of the body. Therefore additional variables derived from these areas of the body should be most effective in identifying hybrids. It is likely that these variables (TTL, LUL, and vertebral counts) reflect changes in morphometry to accommodate prey items of different sizes (Cavender 1969).

Variables from the posterior region of the body (FBL, CPL, CPD) contributed minimally to univariate or multivariate differences between the species. These variables estimate the large expanse of fin at the rear of the body that is used for maximal acceleration of these predators during the specifics of their predatory act (Webb 1978). Since all esocids prey in the same manner, these regions of the body presumably cannot vary outside of certain functional limits common to all species.

The general intermediacy of known hybrids indicates that the regions of the body described herein are under substantial genetic control, and can provide useful information about the proximity of relationships of esocid fishes. The association of most hybrids of known parentage, more so with muskellunge than northern pike, suggests differing parental contribution of genetic information. This may result from unequal dominance of parental alleles. A similar situation occurs for the color pattern (Crossman and Buss 1965). These observations question the validity of the general assumption of hybrid intermediacy between the parental types, thus many hybrids may be unrecognizable by cursory examination.

Meristics

Although numerous types of meristics have been used to distinguish among fishes, and early studies were conducted on muskellunge, northern pike, and their hybrids (e.g., Eddy 1944), the meristics considered here are those that have been shown to be most valuable in distinguishing among these fish. The numbers of rays in the various fins overlap, and do not provide unequivocal separation. The pattern of the sensory pores and canals on the head is the same for both parents (Fig. 1A) and the hybrids.

Branchiostegal rays

An enumeration of branchiostegal rays is given as the range in the total number for each side and for the "divided count"; that is, the number of rays associated with both the anterohyal (ceratohyal) and posterohyal (epihyal) bones (Crossman 1960). The divided count of the rays on one side (e.g., Fig. 1B) is often diagnostic, even if the totals are not. The total number of branchiostegal rays separates the two parent species, but not their hybrid (Table 2). The modal combination of the divided count in muskellunge is 8 on the anterohyal bone and 10 on the posterohyal, with 8/9 the next most frequent combination. In northern pike, 7/8 is the modal combination, and 6/8 the next most frequent. A rather high degree of bilateral variation in esocids results in a range of possible combinations in both parents. However, the range is greater in the hybrid, in which the total number on one side ranges from 12 to 18. The combinations vary from 3/9 to 9/9, and the modal and next most frequent combinations are 6/9 and 8/9. The divided branchiostegal ray count should be diagnostic for all three fish, but alone may not perfectly diagnose a hybrid, although it is useful if combined with values for other characteristics.

Mandibular pores

The mandibular pores, sometimes referred to as submandibular pores, are sensory canal pores, or holes, located on the ventral surface of the lower jaw (Fig. 1A). The terminology used here is that of Nelson (1972), who provided the best details on these pores and canals. The number of obvious pores on the ventral side of the mandible readily separates the two parents, and the hybrids from the muskellunge; however, the ranges of the total numbers overlap in the northern pike and the hybrid (Table 2). In the hybrid the most frequent combination of the number of pores on both sides of the jaw is also that characteristic of the northern pike. The next most frequent count, 7/6 (left side/right side), is a muskellunge-like characteristic, and the total range of the two-side combinations varies from a

Table 2. Comparison of counts of branchiostegal rays, mandibular pores, lateral line scales, and vertebrae for muskellunge, northern pike, and muskellunge × northern pike hybrids.

Species	Branchiostegal rays		Mandibular pores		Lateral line scales		Vertebrae
	Total one side	Divided count[a]	Total both sides	One side[b]	Total	Notched	
Muskellunge	16-20	8/10 (8/9)	12-20	6-10	140-176	55-77	63-67
Northern pike	13-16	7/8 (6/8)	9-11	4-5	105-148	42-55	57-64
Muskellunge × northern pike hybrid	12-18	3/9-9/9 6/9 (8/9)	9-13	5/4-7/6 5/5 (7/6)	126-145	45-56	61-65

[a] Values in this column represent range of combinations for the divided count (anterohyal/posterohyal), the modal combination, and the next most frequent count in parentheses.
[b] Values in this column represent the range in numbers of pores on one side and, for hybrids, the range of combinations of one side counts (left/right), the modal combination, and the next most frequent combination in parentheses.

northern pike-like value of 5/4 to a muskellunge-like value of 7/6.

Hence, any individual with a muskellunge-like pigmentation pattern and a 5/4, 5/5, or 5/6 mandibular pore combination will be a hybrid.

Lateral line scales

The degree of overlap in the total numbers of lateral line scales and the range of individual variability render the count of a single unidentified fish of little use. The total count should separate the two parents, but will not separate the hybrids from many northern pike (Table 2). A count of the less numerous cardioid, or notched, lateral line scales is less laborious and much better (Table 2). That count separates most individuals of the parent species, but will not alone separate the hybrid from northern pike. The notch in these scales partially encircles sensory structures (Merrilees and Crossman 1974).

If an individual has a color pattern similar to that of a muskellunge, but has fewer than 52 notched scales in the lateral line, it is probably a hybrid.

Vertebrae

The ranges of numbers of vertebrae of the parent species barely overlap (Table 2). However, the vertebral counts for most individuals in either species fall near the center of the range, so the vertebral number normally separates the two species: muskellunge--65 and northern pike--60. The range of the numbers of vertebrae in the hybrids more closely resembles that of muskellunge than that of northern pike. The number of vertebrae in most individual hybrids is in the middle of the range (62-63), and these values are intermediate to the modal values of the parent species.

The intermediacy of the vertebral number is a useful character, and any individual that resembles a muskellunge and has 60 to 62 vertebrae is most likely a hybrid.

Scalation on the Cheeks and Opercula

The extent to which both the cheeks and opercula are covered with scales (Fig. 1C) has long been used as a diagnostic characteristic to separate muskellunge, northern pike, and their hybrids (Fig. 5). The degree of scalation is a useful character, but has too often been thought to be precise. The extent of individual variation within a species has not been well described. As with any one characteristic, it should not be used alone to separate these three fish. A rule of thumb for muskellunge is that only the upper half of the cheek and of the operculum is scaled (Fig. 5A); for northern pike the entire cheek, but only the upper half of the operculum, is scaled (Fig. 5C). The degree of scalation on the cheek can be so variable that some individuals in each species can approach the typical condition of the other species.

Black and Williamson (1947) provided information on 23 hatchery-reared hybrids, approximately 20 cm total length, and on individuals of the parental species from Wisconsin. For two of 42 muskellunge, scale coverage on the cheek was 6/10 and 8/10, within the variation typical for northern pike. However, in the remaining 40 the variation was in the typical range, 1/10 to 5/10. In northern pike the variation was narrower, 8/10 to 10/10, and 15 of the 23 specimens examined had 10/10, or complete coverage of the cheek. Of the 23 hybrids, the range of coverage of the cheek was 3/10 to 10/10. However, all but four individuals were in the 7/10 to 10/

A MUSKELLUNGE

B MUSKELLUNGE x NORTHERN PIKE HYBRID

C NORTHERN PIKE

Figure 5. A side view of the head of (A) muskellunge, (B) hybrid, and (C) northern pike, illustrating the extent and location of the scalation (shaded area) on the cheeks and opercula (0.4×). The muskellunge was 38.0 cm TL, male, from White Lake, Ontario. The hybrid was 36.8 cm TL, sterile, from Daggett Lake, Michigan. The northern pike was 38.2 cm TL, female, from Dalrymple Lake, Ontario.

10 range.

It is difficult to evaluate this characteristic in fish less than approximately 20 cm since the scales are small and appear in these locations late in development. In 16 larger hybrids (28 to 80 cm total length) from Pennsylvania and Michigan collected for the present study, scalation on the cheeks varied from 7/10 to 9/10 (Fig. 5B), the intermediate between the two species, but somewhat characteristic of northern pike. Scales on the operculum varied from 3/10 to 6/10; well within the condition of the parent species. Although scalation on the cheeks of hybrids usually is intermediate between the parents, they often exhibit conditions approaching northern pike,

but with increased variability. Hence, any individual that has the color pattern of a muskellunge, but has 7/10 or more of the cheek covered with scales, is a hybrid or "tiger" muskellunge.

Scale Shape, Size, and Configuration of Circuli

It is not difficult to accept that scale features can be species-specific when it has already been shown that groups of fish of the same species (stocks) that live sympatrically can be separated by scale shape (Casselman et al. 1981; Ihssen et al. 1981). The present study was conducted to establish scale criteria that could be used to separate muskellunge, northern pike, and their hybrids.

The scale samples used in this part of the study came from fish from diverse sources. The muskellunge samples came mainly from the St. Lawrence River, but some came from fish from seven other Ontario populations (Nogies Creek, Niagara River, and Eagle, Wabigoon, Dalrymple, White, and Young lakes). Similarly, most of the northern pike samples came from the St. Lawrence River, but a few came from two other Ontario populations (Dalrymple and Theatre lakes). The hybrid samples were obtained much more diversely. Samples from natural hybrids came from The Cleithrum Project (Casselman and Crossman, 1986) and the St. Lawrence River. Artificial hybrids came from Illinois, Michigan, and Pennsylvania.

Scale shape

Very sophisticated techniques are available for quantifying scale shape (Casselman et al. 1981). However, only very simple radial measurements were used in this comparison because specialized equipment and complex programs necessary for Fourier series analysis often are not readily available when tests are needed.

Scales came from the midlateral region of the trunk. In esocids these scales are typically tri-lobed in the anterior field and moderately symmetrical (e.g., Fig. 6A). For this comparison, measurements were made from photographic prints of scale impressions using a digitizer interfaced with an IBM-PC microcomputer. Only scales that were tri-lobed, laterally symmetrical, and of average size for the sample were used. Two or three scales from each fish were measured, and the results were averaged. All radial measurements were made about the focus, or origin. Mean anterior radius was obtained by determining the average length of the maximum

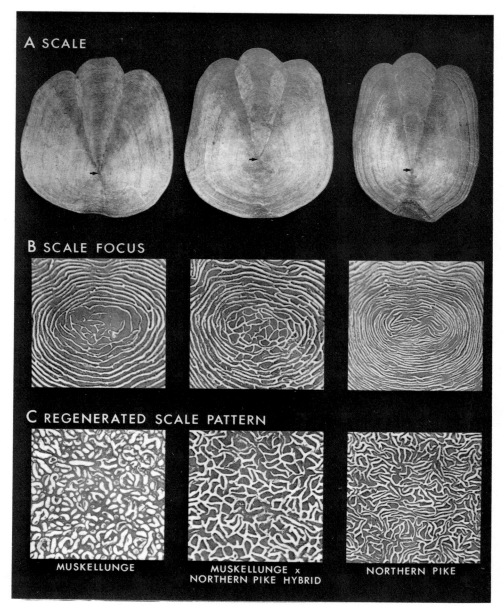

Figure 6. Scale characteristics: (A) Typical scales from the midlateral region of the body (5.7×): left--muskellunge, 91.7 cm TL, male, from the St. Lawrence River, Ontario; center--artificial hybrid, 83.8 cm TL, sterile, from Clinton Lake, Illinois; right--northern pike, 84.8 cm TL, female, from the St. Lawrence River, Ontario. The arrow indicates the position of the focus. (B) Typical foci of scales from the midlateral region of the body (96×): left--muskellunge, 41.9 cm TL, female, from the St. Lawrence River, Ontario; center--artificial hybrid, 36.8 cm TL, sterile, Daggett lake, Michigan; right--northern pike, 60.7 cm TL, male, from the St. Lawrence River, Ontario. (C) Typical configurations of the circuli on regenerated scales (96×): left--muskellunge, 105.1 cm TL, male, from the St. Lawrence River, Ontario; center--hybrid, 100.3 cm TL, sterile, from Eagle Lake, Ontario; right--northern pike, 60.0 cm TL, male, from the St. Lawrence River, Ontario.

radius of each of the three lobes. Posterior radius was the distance from the focus to the posterior edge of the scale on a line that was an extension of the anterior radius of the central lobe (the anterior-posterior axis). This anterior radius usually passed through the middle of the central lobe. Lateral radii were measured from the focus to the edge of the scale, at right angles to the anterior-posterior axis. Mean lateral radius was the average of the dorsal-lateral and the ventral-lateral radii. Scale length was the combined length of the mean anterior scale radius and the posterior scale radius. Scale width was the combined length of the dorsal-lateral radius and the ventral-lateral radius. Scale shape was examined by averaging the proportions of the various radii for fish in each 10-cm body length-class (Table 3).

Scale shape changes with age, tending to grow more oval with increasing age. However, the 95% confidence limits (CL) indicated that scales of muskellunge (Fig. 6A, left) are significantly rounder than are scales of northern pike (Fig. 6A, right), or the hybrids (Fig. 6A, center), both of which are relatively oval (Table 3). Scales of muskellunge are 10% longer than wide, whereas scales of the other two forms are 25% longer than wide.

The focus is slightly more posterior on scales of northern pike than on scales of the other two forms (Table 3). However, the relative location of the focus changes with age, affecting the comparison so that location is not significant over all size-classes (Table 3). The focus is more centrally located on scales of hybrids and muskellunge than on scales of northern pike. The average lateral radius is approximately 30% greater than the posterior radius on scales of muskellunge. The difference is not so great, though similar, in hybrids (Table 3).

Scale annulus

Checks, or breaks, in the configuration of the circuli that are associated with annuli are quite different in these fish. Annuli on northern pike scales are the most distinct. A prominent hyaline check (a band devoid of circuli) is deposited around the scale in older northern pike (Fig. 6A, right). This type of annulus is not found on the scales of muskellunge or hybrids. In these two forms there are subtle, yet distinct, differences in the type of annuli laid down on the scales, but to differentiate these would require a description too detailed to be considered here.

Scale focus

There are significant differences in the patterns of the circuli first laid down on the scale platelets, and subsequently apparent about the focus. In muskellunge the focus is relatively devoid of circuli (Fig. 6B, left). When circuli are present, they are usually thin and fragmented, with some interconnection. In northern pike the characteristics of the focus are less variable and more distinct (Fig. 6B, right). Often the focus on the scales of this species appears to have paired nuclei, or two centers. The circuli are thin but very numerous, prominent, coalescent, and form a unique parallel type of pairing. The circuli in the focus on scales of hybrids are more uniformly and widely spaced, and appear to interconnect, forming a network (Fig. 6B, center).

Regenerated scale pattern

The pattern of the circuli is quite distinct on scales that have regenerated after the original scales have been lost. Regenerated muskellunge scales contain circuli that are fragmented, extremely thick and short; some are curved and occasionally branched, but generally not interconnected (Fig. 6C, left). Short, oval, and even rounded circuli are common. In northern pike, the circuli are interconnected, and often appear paired (Fig. 6C, right). On regenerated scales, the circuli are generally thinner in northern pike than in the other two forms, but have charac-

Table 3. Scale morphometrics for muskellunge, northern pike, and muskellunge × northern pike hybrids, comparing the ratio of scale length to scale width, mean anterior scale radius to posterior scale radius, and mean lateral scale radius to posterior scale radius.

Species	No. of samples	No. of 10-cm length classes	$\bar{x}$	Scale length / Scale width		Mean anterior scale radius / Posterior scale radius			Mean lateral scale radius / Posterior scale radius		
				Lower 95% CL	Upper 95% CL	$\bar{x}$	Lower 95% CL	Upper 95% CL	$\bar{x}$	Lower 95% CL	Upper 95% CL
Muskellunge	71	12	1.10	1.08	1.12	2.02	1.76	2.28	1.38	1.27	1.49
Northern pike	83	11	1.26	1.23	1.29	2.14	1.96	2.32	1.26	1.19	1.33
Muskellunge × northern pike hybrid	71	9	1.25	1.18	1.32	2.00	1.78	2.23	1.22	1.06	1.38

teristics similar to the focus in normal scales, although they are more irregular in shape. In hybrids, circuli are thicker, but not as robust as in muskellunge. They are strongly interconnected and often form an open network (Fig. 6C, center).

The patterns of the circuli laid down in the focus and on regenerated scales are quite distinct. The extreme differences in the configurations of the circuli deposited on scale platelets during extremely rapid growth (either first formation or regeneration) may indicate conditions that are more indicative of genetic differences than when growth rate is moderate or slow.

Scale size

To examine scale size relative to body size, mean anterior scale radius was averaged for 10-cm total length intervals. Two linear relations were used to fit the data for both muskellunge and northern pike because the data were linear on each side of an inflection point of approximately 90 cm and 50 cm respectively (Fig. 7). Although the relations for both species are undoubtedly curvilinear, two linear relations provided excellent fits of the data, and permitted easy comparisons among the hybrids and the two species. Up to a total length of approximately 75 cm, scales of muskellunge are shorter than are scales of northern pike and hybrids, which are very similar in size (Fig. 7). Since muskellunge scales are shorter than those of the other two forms, as would be expected muskellunge have more scales along the lateral line (Table 2). For body lengths greater than 90 cm, scales of hybrids are significantly longer. Body-scale relations for muskellunge and northern pike indicate two growth phases, whereas the relationship for the hybrid is essentially the same throughout the entire size range. Since body size at the point of inflection coincides rather well with size

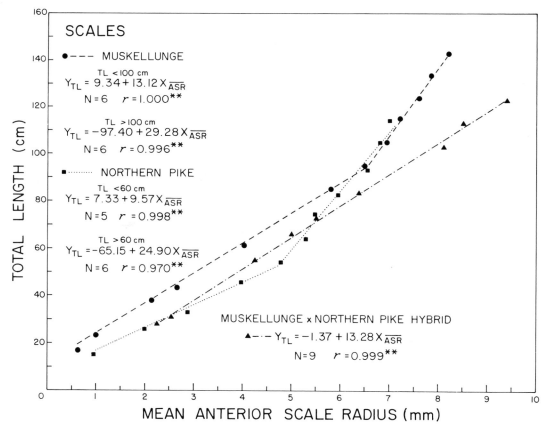

Figure 7. Relation between total length (cm) and mean anterior scale radius ($\overline{ASR}$, mm) for muskellunge, hybrids, and northern pike. The relations were constructed from data averaged for 10-cm length intervals.

at first maturity for these two species in the St. Lawrence River, and since the hybrid is sterile and shows no inflection, these differences in relative size of the scales are probably related to growth differences effected by the onset of maturity.

Osteological Characteristics

There are many osteological characters that distinguish the two parental species and may, in fact, be represented in the hybrid by intermediate conditions. For example, it has been shown that the epipleurals (intramuscular or "Y" bones) are distinctly different in muskellunge and northern pike (Crossman and Casselman 1969). However, we have limited ourselves here to a few osteological features that can be seen externally or in the body cavity of an eviscerated fish. One exception is the cleithrum, a superficial bone now replacing scales as the medium for the collection and analysis of age, growth, and of other biological data of esocid fishes (Casselman 1974, 1979).

Dentition on vomer and palatine bones

The vomer and palatine are tooth-bearing bones of the roof of the mouth. The vomer, a central bone, bears a single club-shaped patch of teeth. Each of the paired palatines that borders the vomer bears a somewhat trapezoidal patch of teeth.

In muskellunge, the vomerine tooth patch is greatly constricted at the neck, the point that can be thought of as the connection between a short triangular head, and a narrow, long shaft (Fig. 8). In muskellunge, there are rarely more than two larger teeth in a cross row on the neck. In northern pike, the neck is less obvious, much thicker, and usually the cross row at this point bears four smaller teeth (Fig. 8). The anterior edge of the head of the vomer in muskellunge bears a pair of lateral, larger, canine-like

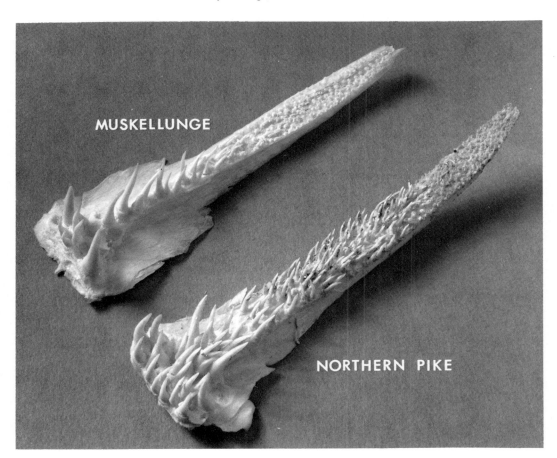

Figure 8. Ventral view of the vomer of muskellunge and northern pike.

teeth with much shorter teeth between these. The teeth on the head of the vomer of northern pike are more uniformly graded in size, with no obvious canine-like teeth. The shaft of the vomer of muskellunge is a long oval in outline, partly as a result of the restricted neck, and does not extend backward as far as the posterior edge of the palatine patches. In northern pike, the outer edges of the shaft of the vomer are straight and parallel, the shaft is wider, and does extend back as far as the posterior edge of the palatine patches.

In muskellunge, the anterior, inner (medial) edge of each palatine tooth patch bears three or four canine-like teeth that are larger and much longer than adjacent teeth. These larger palatine teeth are adjacent to the larger canine-like teeth on each side of the vomer. In northern pike, the teeth of the palatine patch are much more uniformly graded in length and stoutness, and none stands out as canine-like. These differences in the dentition of the vomerine and palatine tooth patches are obvious if the mouth is opened and examined, or if the open jaws are viewed head-on (Fig. 9). Unfortunately, the hybrid is intermediate in these characters, and more resembles the generalized condition of the northern pike than the unique condition of the muskellunge. The difference is one of degree, but can be detected with experience. An individual with the color and pattern of a muskellunge is a hybrid if the vomer tooth patch has very little constriction at the neck, three or four teeth in each cross row, parallel sides, and is without prominent canine-like teeth on the head of the tooth patch.

Dentary teeth

Often all that is available for identification (e.g., from archaeological sites) are large, bony elements such as the dentary and its large teeth. Hence, diagnostic characteristics that use these large dentary teeth (Fig. 10A) would be especially useful because they could also be used in live fish.

The large dentary teeth of muskellunge are robust, uniformly tapered, and slightly recurved (Fig. 10B, left). These teeth in northern pike are more laterally compressed, with prominent blade-like edges near the base (Fig. 10B, right). These "compressed lateral edges" of northern pike teeth have been observed by Crossman and Harington (1970). In northern pike the dentary teeth are somewhat sigmoid or "S" shaped over their length, a feature that makes them quite distinct from muskellunge, and can easily be observed in the open mouth (Fig. 9). The dentary teeth of the hybrid are intermediate in character, but are generally more like the teeth of northern pike. Lateral compression is not so obvious, but the tip is usually slightly curved (Fig. 10B, center).

If cross-sections near the base of the tooth are compared, the muskellunge tooth is more oval (Fig. 10C, left), whereas the northern pike tooth is round, with more prominent anterior-posterior blade-like protuberances. The cross-section of the hybrid tooth is intermediate in character, but generally more similar to that of northern pike.

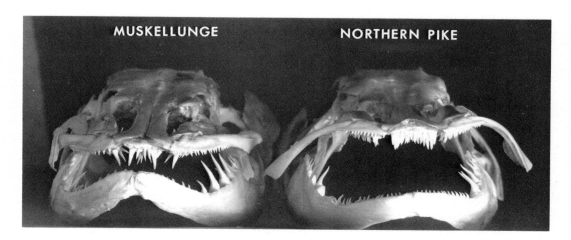

Figure 9. Anterior view of the head of muskellunge and northern pike, indicating the differences in canine-like teeth on the vomer and palatines, as well as the differences in the typical shape of the large, centrally located teeth on the dentaries.

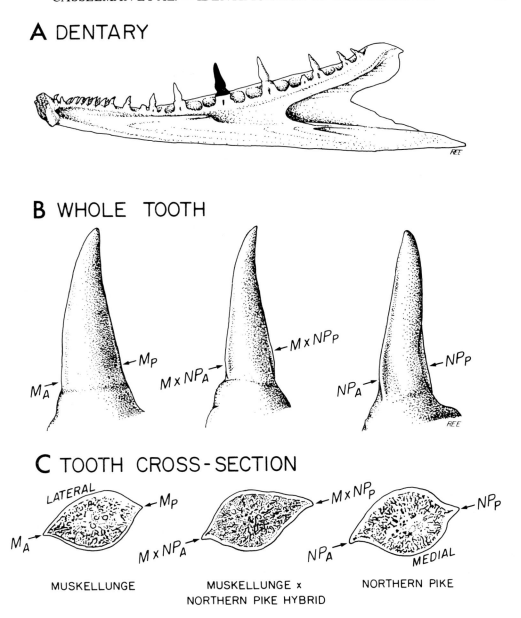

Figure 10. Characteristics of the large teeth centrally located on the dentary. (A) Medial view of right dentary of a northern pike, 114.2 cm TL, female, from Dalrymple Lake, Ontario (0.7×). The darkened tooth is the largest, usually the most centrally located tooth in the dentary, and the one used in these comparisons. (B) Medial view (slightly rotated) of typical whole, large, dentary teeth: left--muskellunge, 78.7 cm TL, female, from Nogies Creek, Ontario (5.8×); center--artificial hybrid, 82.7 cm TL, sterile, from Daggett Lake, Michigan (4.9×); right--northern pike, 114.2 cm TL, female, from Dalrymple Lake, Ontario (4.9×). Arrows with species abbreviations and subscripts indicate the precise anterior-posterior orientation of the cross-sections illustrated in (C). (C) Cross-sections of teeth in (B) illustrating anterior and posterior carinae. The magnification and orientation of the cross-sections are indicated by arrows, and are identical to (B).

Cleithra

Since cleithra are now often collected for assessing age, features of these structures could be useful in confirming species identification and detecting hybrids. The specific terminology used to describe cleithrum morphology is available in Casselman (1978).

Shape and appearance of cleithra of muskellunge and northern pike are quite distinct and easily recognized (Casselman 1980). The salient differences are illustrated in Fig. 11 as follows: 1--curvature of the medial costa, or inner rib (delineated by dark line), essentially straight in muskellunge but curved in northern pike; 2--the shape of the anterior edge of the blade (the blade is the portion on which annuli are depicted) is uniformly rounded in muskellunge but indented in northern pike; 3--translucent zones or annuli are narrow and distinct in muskellunge but broad and rather diffuse in northern pike; 4--shape of the heel is relatively flat or straight in muskellunge but strongly curved in northern pike; 5--shape of the dorsal spine is straight in muskellunge, but curved, especially on the posterior edge, in northern pike; 6--angle subtended by the medial costa and dorsal spine is approximately 90° in muskellunge but usually much greater than 90° in northern pike.

The shape of the cleithra of hybrids is distinctly different from that of the parental species. Cleithra from hybrids sometimes have characteristics that are intermediate but usually appear abnormal, more variable, and sometimes grossly different. Almost all cleithra from hybrids can be classified as one of four types. Type A (Fig. 12) appears malformed and contains a prominent rib, keel, or fold on a rather thickened blade. Of the cleithra from hybrids that were examined (N = 129), the frequency of occurrence of type A was 16%. In this type of cleithrum the annuli are indistinct, and age usually cannot be interpreted. In type B (occurrence 29%) the blade is uniformly narrow throughout and abnormally long, the outer edge of the blade forms a uniform curve, and the heel is slender. In type C (occurrence 8%) the anterior of the blade appears abnormally slender, but the posterior of the blade is robust and thick. Because the blade of this type of cleithrum is malformed and thickened, the annuli are indistinct, and often it is impossible to interpret age. In type D (occurrence 45%) the cleithrum, especially the anterior blade, is extremely robust and the mid-part of the blade is very broad.

To examine size of the cleithrum relative to size of the body, anterior cleithral radius (Casselman 1979) was averaged for 10-cm total length intervals. Cleithral samples used in this comparison came from the same sources as the scale samples described above. Size of the cleithrum relative to size of the body is very similar in muskellunge and northern pike, but slightly smaller in the former (Fig. 13). Cleithra from hybrids are not intermediate in size relative to the parental species, but are significantly larger. Hybrid vigor is obvious when relative sizes of bony elements such as the cleithrum are compared. Anterior cleithral radius of the hybrid is almost exactly 1/10 of the total length of the fish. These results are similar to those reported by Casselman and Crossman (1986), even though the data are different.

Vertebral column and centra

Over most of the length of the body cavity the dorsal aorta of muskellunge has a distinctive ventral-lateral location on the right side of the vertebral column. This provides a useful tool for recognizing muskellunge and separating them from northern pike (Crossman and Casselman 1969). The asymmetrical location of the dorsal aorta in muskellunge coincides with the asymmetrical location of the dorsal aortic groove in the centra. In muskellunge, the dorsal aortic groove (Fig. 14) is located on the right side of the vertebral column after approximately the sixth vertebra, but runs along the entire length of the mid-ventral in northern pike. It is intermediate in location in the hybrid. If the hybrid is to be recognized, it is necessary to locate precisely the dorsal aortic groove. This can be done if measurements are made to determine the position of the radial plane of the dorsal aortic groove relative to the radial plane of the right and left basiventral fossae (Fig. 15). This involves measuring the outer rim of the anterior face of the centrum and calculating and comparing the angles subtended by these radial planes (J. Casselman and R. Elliott, unpublished data).

Although the difference in the location of the dorsal aorta and the aortic groove applies only to vertebrae in the central portion of the body cavity, throughout the vertebral column the centra of these three forms are sculptured quite differently. The centra of muskellunge are smooth with only slight striations, and the dorsal aortic groove is shallow (Fig. 14). However, the centra of northern pike are strongly striated, and the dorsal aortic groove is very deep. The centra of hybrids are intermediate in character. Muskellunge, northern pike, and their hybrids can be distinguished by using the relative location of the dorsal aorta in the eviscerated car-

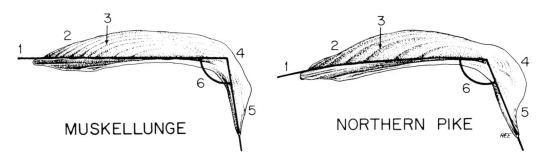

Figure 11. Typical shape and appearance of cleithra from muskellunge and northern pike. Illustrations depict the medial view of the left cleithrum, anterior to the left. Cleithra are from a muskellunge, 117.3 cm TL, male, from St. Lawrence River, Ontario (0.5×), and a northern pike, 89.0 cm TL, female, from North Channel, Lake Huron, Ontario (0.6×). Numbers indicate salient differences: 1) curvature of the medial costa; 2) shape of the anterior edge of the blade; 3) type of translucent zonation, depicted as viewed in incident light; 4) shape of the heel; 5) shape of the dorsal spine; 6) angle subtended by the dorsal and ventral rami.

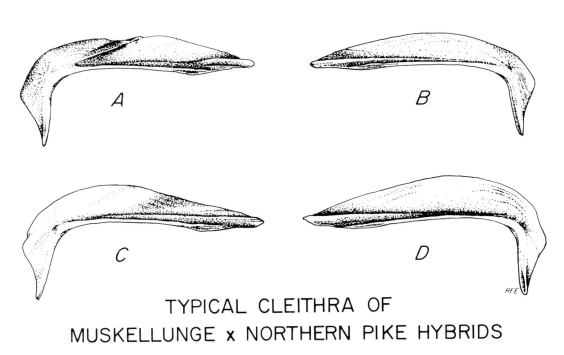

Figure 12. Typical shape and appearance of cleithra from natural hybrids. Four types of cleithra have been identified, and are represented by medial views of right cleithra (*A* and *C*) and left cleithra (*B* and *D*). The cleithra are from the following fish: *A*--119.4 cm TL, sterile, from Dalrymple Lake, Ontario (0.55×); *B*--123.5 cm TL, sterile, from Eagle Lake, Ontario (0.38×); *C*--122.2 cm TL, sterile, from Lake Simcoe, Ontario (0.43×); *D*--109.2 cm TL, female, from Indian Lake, Michigan (0.50×).

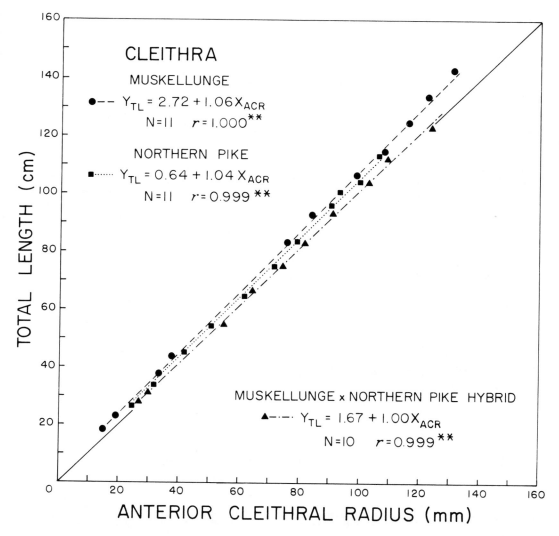

Figure 13. Relation between total length (cm) and mean anterior cleithral radius (ACR, mm) for muskellunge from the St. Lawrence River, Ontario (N = 59), northern pike from the St. Lawrence River, Ontario (N = 83), and hybrids from several localities (N = 129). The relations were constructed from data averaged for 10-cm length intervals. The solid line on either end of the regression lines indicates the diagonal where total length = 10× anterior cleithral radius.

cass, the degree of sculpturing on the centra, and the relative location and depth of the dorsal aortic groove in the vertebral column.

Karyotypes

Chromosome number and morphology of the karyotype have been reported for northern pike and muskellunge (Beamish et al. 1971; Davisson 1972) as containing 50 acrocentric chromosomes. Similar results had been reported by McGregor (1970) for muskellunge, and by Nygren et al. (1968) for northern pike. Svärdson and Wickbom (1939), Prakken et al. (1955), and Lieder (1959) have reported different diploid numbers for northern pike in Europe which were 18, 48, and 46 respectively. Svärdson and Wickbom (1939) also reported 5 metacentric, 1 submetacentric, and 3 acrocentric chromosomes within the northern pike karyotype while Prakken et

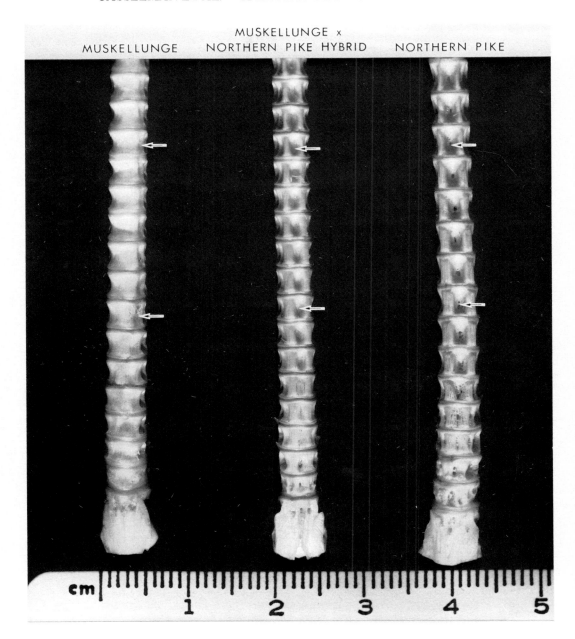

Figure 14. Ventral view of the anterior portion of vertebral columns with a small portion of the cranium intact (bottom). Vertebrae are from muskellunge, 38.6 cm TL, female, from Young Lake, Ontario; hybrid, 36.8 cm TL, sterile, from Daggett Lake, Michigan; and northern pike, 38.2 cm TL, female, from Dalrymple Lake, Ontario. Arrows indicate the center of the ventral groove in the centra associated with the dorsal aorta.

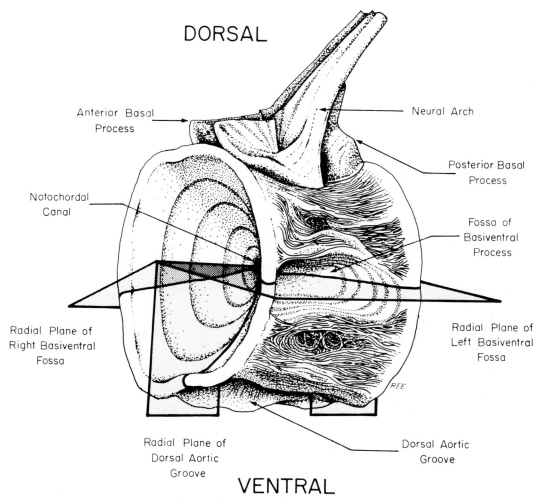

Figure 15. Centrum and neural arch of a trunk vertebra (13th) from a northern pike (10.2×). The vertebra is from a fish 53.8 cm TL, female, from Dalrymple Lake, Ontario. The various fossae and planes, used to determine the precise position of the groove in the ventral of the centrum that is associated with the dorsal aorta, are illustrated.

al. (1955) and Lieder (1959) found all acrocentric chromosomes. The karyotype determining methods used in these studies varied from embryonic smears (Svärdson and Wickbom 1939) to tissue air drying--squash techniques (McGregor 1970; Beamish et al. 1971, Davisson 1972).

Muskellunge × northern pike hybrids have a diploid number of 50; all chromosomes are acrocentric, but the karyotype did exhibit some secondary arms on five pair of chromosomes (Davisson 1972). Beamish et al. (1971) and Davisson (1972) found no chromosomes in northern pike and muskellunge that exhibited any morphological distinction indicative of heteromorphic sex chromosomes.

Karyotypes of the species show no notable differences that can be used to distinguish one species from the other whether it be these two or any other esocid. However, some of the early chromosome methods may not have permitted investigators to distinguish subtle differences in chromosome morphology as was done more recently (Booke 1968). The hybrid karyotype indicates that the centromere on some of the presumed acrocentric chromosomes is located such that two-armed chromosomes may exist in the karyotype of either parent (Davisson 1972).

Electrophoresis

Electrophoresis of soft tissue samples such as skeletal muscle and liver was used to examine the interspecific electrophoretic differentiation between northern pike, muskellunge, and their hybrids. Also, we were interested in determining which electrophoretic loci are the most useful in distinguishing among these two species and their hybrids. Lens protein differences have been reported for esocids by Eckroat (1974). The three pickerels were found by Eckroat to be electrophoretically different from muskellunge or northern pike, but no differences were found between muskellunge and northern pike.

Electrophoretic samples

We were primarily interested in interspecific rather than intraspecific variation; therefore, tissue from only a few hybrids and individuals from each species was assayed. The hybrids were artificial and from two sources: seven from Daggett Lake, Michigan, and 21 from Pleasant Mount Fish Culture Station, Pennsylvania. The eight muskellunge came from the St. Lawrence River and three lakes in central Ontario--White, Young, and Dalrymple. The 13 northern pike analysed came from Dalrymple

Table 4. Enzyme-protein systems analyzed electrophoretically to compare muskellunge, northern pike, and muskellunge × northern pike hybrids.

Enzyme-protein	Loci	Tissue	Buffer[a]	Mobility--interspecific difference
Aspartate aminotransferase	Aat-1	Liver	A	No
Aspartate aminotransferase	Aat-2,3	Liver	A	No
Alcohol dehydrogenase	Adh	Liver	A	Yes
Aldolase	Ald-1	Heart	B	No
Aldolase	Ald-2	Eye	B	No
Creatine kinase	Ck-1	Muscle	A	No[b]
Creatine kinase	Ck-2	Heart	A	No
Creatine kinase	Ck-3	Eye	A	No
Cathodal muscle protein	Cmp	Heart	A	Yes[c]
Diaphroase	Dia	Liver	B	Yes
Glycerol-3-phosphate dehydrogenase	G3p-1,2	Muscle	D	Yes[d]
Glycerol-3-phosphate dehydrogenase	G3p-3	Liver	D	Yes
General muscle protein	Gmp	Muscle	A	Yes
Glucosephosphate isomerase	Gpi-1	Muscle, heart	A	Yes[e]
Glucosephosphate isomerase	Gpi-2	Muscle, liver	A	Yes
Isocitrate dehydrogenase	Idh-1	Muscle	D	No
Isocitrate dehydrogenase	Idh-2	Liver	D	Yes[e]
Lactate dehydrogenase	Ldh-1	Muscle	A,D	Yes
Lactate dehydrogenase	Ldh-2	Muscle	A,D	No
Lactate dehydrogenase	Ldh-3	Liver	A,D	No
Lactate dehydrogenase	Ldh-4	Heart	A	No
Lactate dehydrogenase	Ldh-5	Eye	A,D	No
Malate dehydrogenase	Mdh-1,2	Liver, muscle	B	Yes
Malate dehydrogenase	Mdh-3	Muscle	B	No
Malate dehydrogenase (mitochondrial)	m-Mdh-1,2	Muscle	B	No
Malic enzyme	Me-1	Muscle	D	No
Malic enzyme	Me-2	Muscle	D	Yes
Mannosephosphate dehydrogenase	Mpi	Muscle	B	No
4-Methylubelliferyl phosphate	Mup	Liver	B	Yes
Peptidase (leucyl-alanine)	Pep	Muscle	B	Yes
Phosphogluconate dehydrogenase	Pgd	Muscle	B	Yes[e]
Phosphoglucomutase	Pgm-1	Muscle	B	No
Phosphoglucomutase	Pgm-2	Liver	B	Yes
Sorbitol dehydrogenase	Sdh	Liver	A	Yes[f]
Superoxide dismutase	Sod-1	Heart	A	No
Superoxide dismutase	Sod-2	Liver	D	Yes
Xanthine dehydrogenase	Xdh	Liver	A	No

[a]Buffer systems are described in detail in Table 3 in Casselman et al. 1981; A, discontinuous Tris-citric acid (gel pH 8.5) lithium hydroxide--boric acid (tray pH 8.1); B, citric acid-N-(3-aminopropyl)-morpholine (pH 6.1); D, Tris-citric acid (pH 8.0).

[b]Single bands of same mobility in northern pike and muskellunge, three bands in hybrids.

[c]Very poor resolution.

[d]Polymorphic in northern pike, see Fig. 17.

[e]Polymorphic in muskellunge, see Fig. 17.

[f]Deduced to be polymorphic in northern pike because of extra isozymes in the hybrid.

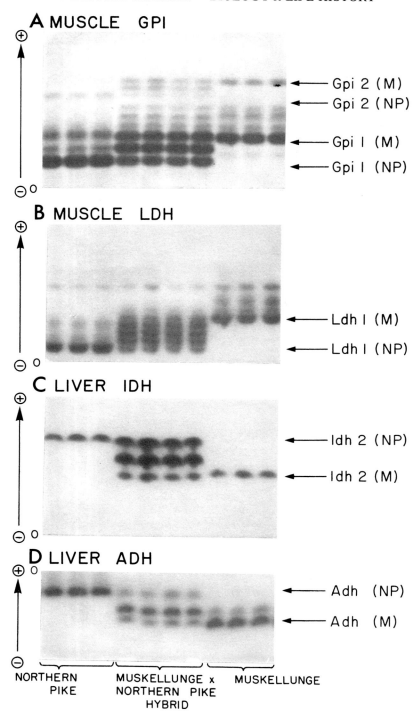

Figure 16. Typical starch gels for four enzyme systems for muskellunge, hybrid, and northern pike: (A) muscle glycosephosphate isomerase; (B) muscle lactate dehydrogenase; (C) liver isocitrate dehydrogenase; (D) liver alcohol dehydrogenase.

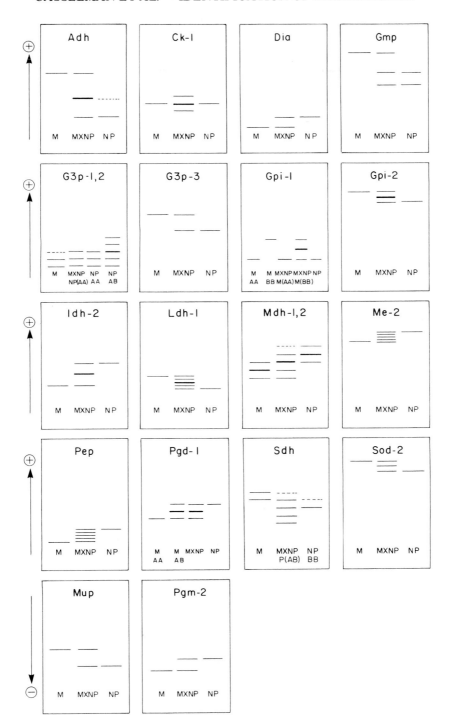

Figure 17. Schematic representation of the allelic products for esocid loci with interspecific and intraspecific polymorphism in electrophoretic mobility.

Lake, Ontario.

Electrophoretic methods

Three buffer systems resolved 75 allelic products encoded by 41 presumptive loci (Table 4). Of these 41 loci, 18 differentiated between northern pike and muskellunge. Species-specific isozymes of different mobility were clearly resolved on starch gels from both liver and muscle tissue (Fig. 16). Three additional loci (Ck-2, Ald-1, Mdh-3) produced isoenzymes that differed consistently in activity for the two species. But without breeding data we cannot be certain whether this is a genetic effect.

Electrophoretic differences

For all but two of the 18 loci, for which the two species differed in electrophoretic mobility, the hybrids had the expected intermediate phenotypes; i.e., two, three, and five isozyme bands for monomeric, dimeric, and tetrameric enzymes, respectively (Fig. 17). The allelic products for the tetrameric enzymes (Ldh, Me, Pep) could not always be resolved in the hybrids into all of their five heterotetramers. Ck-1 and Sdh produced isozymes in the hybrids with mobility outside the range of both parents (Fig. 17). Our interpretation of this result is that at least one of the parents of these hybrid individuals was polymorphic at these loci.

Three intraspecific polymorphisms were observed in addition to the two deduced from the hybrids discussed above. G3p-1,2 was polymorphic in northern pike, and Gpi-1 and Pgd-1 in muskellunge (Fig. 17). Gpi-1 polymorphism was also observed in the hybrids. The hybrids from Pennsylvania had the St. Lawrence River muskellunge Gpi-1 allele predominantly, whereas the hybrids from Michigan had the muskellunge allele found predominantly in samples from three central Ontario lakes. From ecological considerations, we suspect that there are several genetic stocks of muskellunge in Ontario. We have just commenced a study to examine this genetic variability in Ontario muskellunge stocks.

Any one of the 18 loci that differentiated between northern pike and muskellunge could be used to identify unknown specimens or even tissue samples of unknown origin. However, for practical considerations some of these loci are preferred over others. If a locus is polymorphic, and especially if the polymorphism is for an allele forming a product with identical migration in both species, there is a possibility of misidentification if no other locus is used. Hence, for positive identification, loci such as Gpi-1, Pgd-1, and Sdh should not be used alone. For other loci, the enzymes did not seem to store well under ordinary conditions ($-20C$). Still others give poor resolution except under ideal electrophoretic conditions, hence are not recommended for routine application. The tissue in which the loci are expressed may also be a consideration. If live specimens are to be identified, the preferred tissue is skeletal muscle from which biopsy samples can be taken. For enforcement purposes, only fish fillets may be available, and again loci expressed in muscle tissue must be assayed. These considerations suggest Gmp, Mdh-1,2, and Ldh-1 (Fig. 17) as the ideal loci for identification. All three seem to store well, have not yet been found to be polymorphic in either muskellunge or northern pike, have good resolution, and are available in skeletal muscle tissue.

CONCLUSIONS

A broad range of techniques that are useful for identifying and distinguishing muskellunge, northern pike, and their hybrids has been considered. Several of these techniques were examined not only on fish from the same lake (e.g., Dalrymple), but also on the same specimens (e.g., hybrids from Michigan and Pennsylvania), thus reducing variability and permitting a better comparison of techniques. It is obvious that the techniques vary in ease of application and resolving power, but to list or summarize them would be redundant.

No one criterion should be used in isolation, especially if the interpretations are at all subjective. Where characteristics are intermediate in nature, as is often the case for morphometrics and meristics of the hybrid, special effort should be made to examine those criteria that demonstrate heterosis (e.g., relative size of the cleithrum).

We have reviewed, developed, and assembled a range of techniques, varying from powerful forensic tools, such as electrophoresis, to simple yet equally useful technqiues, involving osteological characters such as type and degree of dentition, that can be used easily by fishermen. These can be used to help protect muskellunge, recognize hybrids, and assist in the management of the large esocids.

ACKNOWLEDGEMENTS

We thank the many muskellunge fishermen and taxidermists who have provided both information and scales and cleithra from their muskellunge and muskellunge × northern pike hybrids. Cultured hybrids were supplied by G. Beyerle and J. Capezynski of the Michigan Department of Natural Resources, by S. Hood and Z. Bean of the Pennsylvania Fish Commission. M. Pike and P. Taylor working for the Ontario Ministry of Natural Resources (OMNR) on a Fisheries Special Employment Program sponsored by the Ontario Federation of Anglers and Hunters (OFAH), helped obtain the samples. S. Campbell, C. Horkey, J. Ross, M. Burridge-Smith, and M. Walters of the Department of Ichthyology and Herpetology, Royal Ontario Museum, assisted with numerous aspects of the development of the manuscript. We acknowledge the technical assistance of W. Martin of OMNR, Maple, with the electrophoretic phase of the study.

We are especially indebted to the late G. Vanston, a commercial fisherman on the St. Lawrence River at Mallorytown Landing, Ontario, who over the years has provided JMC with samples of northern pike and muskellunge used here, as well as the northern pike used in Figure 2. We wish to thank R.E. Elliott who, working on an OMNR-OFAH Fisheries Special Employment Program, assisted with the osteological study and whose artistic talents are obvious in his fine illustrations.

This joint study was coordinated by JMC, and was conducted as follows: general description, key, and meristics by EJC; morphometrics by JDR; scales and cleithra by JMC; other osteological characteristics by EJC and JMC; electrophoresis by PEI; and karyotype review by HEB.

We thank the following referees for their helpful reviews of the manuscript: R. Bailey, B. Coad, D. McAllister, and J. Ramsey.

REFERENCES

Atchley, W.R., C.T. Gaskins, and D. Anderson. 1976. Statistical properties of ratios, I. Empirical results. Systematic Zoology 25:137-148.

Beamish, R.J., M.J. Merrilees, and E.J. Crossman. 1971. Karyotypes and DNA values for members of the suborder *Esocoidei* (Osteichthyes: Salmoniformes). Chromosoma (Berlin) 34:436-447.

Berg, L.S. 1948. Freshwater fishes of the USSR and adjacent countries. Volume 1. Academy of Sciences of the USSR. Translation published by the National Science Foundation and the Smithsonian Institute, Washington, DC, USA.

Black, J.D., and L.O. Williamson. 1947. Artificial hybrids between muskellunge and northern pike. Transactions of the Wisconsin Academy of Science, Arts and Letters 38:299-314. 1946.

Booke, H.E. 1968. Cytotaxonomic studies of the coregonine fishes of the Great Lakes, USA: DNA and karyotype analysis. Journal of the Fisheries Research Board of Canada 25:1667-1687.

Casselman, J.M. 1974. Analysis of hard tissue of pike *Esox lucius* L. with special reference to age and growth. Pages 13-27 *in* T. B. Bagenal, editor. Ageing of fish--Proceedings of an international symposium. Reading, England, July 19-20, 1973. Unwin Brothers Ltd., England.

Casselman, J.M. 1978. Calcified tissue and body growth of northern pike, *Esox lucius* Linnaeus. Ph.D. thesis, University of Toronto, Toronto, Ontario, Canada. 782 pages (2 volumes).

Casselman, J.M. 1979. The esocid cleithrum as an indicator calcified structure. Pages 249-272 *in* J. Dubé and Y. Gravel, editors. Proceedings of the 10th (1977) Warm Water Workshop. Ministère du Loisir, de la Chasse et de la Pêche du Québec, Direction de la recherche faunique. Montréal, Québec, Canada.

Casselman, J.M. 1980. Identification of pike and muskellunge from cleithra. Muskies, Inc. (The official publication of Muskies, Inc.), May 1980, pages 10-12.

Casselman, J.M., J.J. Collins, E.J. Crossman, P.E. Ihssen, and G.R. Spangler. 1981. Lake whitefish *(Coregonus clupeaformis)* stocks of the Ontario waters of Lake Huron. Canadian Journal of Fisheries and Aquatic Sciences 38:1772-1789.

Casselman, J.M., and E.J. Crossman. 1986. Size, age, and growth of trophy muskellunge and muskellunge-northern pike hybrids. American Fisheries Society Special Publication 15:93-110.

Cavender, T. 1969. An Oligocene mudminnow (Family Umbridae) from Oregon with remarks on relationships within the Esocoidei. Occasional Papers of the Museum of Zoology, University of Michigan 660:1-33.

Crossman, E.J. 1960. Variation in number and asymmetry in branchiostegal rays in the family Esocidae. Canadian Journal of Zoology 38:363-375.

Crossman, E.J. 1978. Taxonomy and distribution of North American esocids. American Fisheries Society Special Publication 11:13-26.

Crossman, E.J., and K. Buss. 1965. Hybridization in the family Esocidae. Journal of the Fisheries Research Board of Canada 22:1261-1292.

Crossman, E.J., and J.M. Casselman. 1969. Identification of northern pike and muskellunge from axial skeletons, scales, and epipleurals. Journal of the Fisheries Research Board of Canada 26:175-178.

Crossman, E.J., and C.R. Harington. 1970. Pleistocene pike, *Esox lucius*, and *Esox* sp., from the Yukon Territory and Ontario. Canadian Journal of Earth Sciences 7:1130-1138.

Davisson, M.T. 1972. Karyotypes of the teleost family Esocidae. Journal of the Fisheries Research Board of Canada 29:579-582.

Eckroat, L.R. 1974. Interspecific comparisons of lens proteins of Esocidae. Copeia 1974:977-978.

Eddy, S. 1944. Hybridization between northern pike (*Esox lucius*) and muskellunge (*Esox masquinongy*). Proceedings of the Minnesota Academy of Science 12:38-43.

Hourston, A.S. 1955. A study of the variation in the maskinonge from three regions in Canada. Royal Ontario Museum of Zoology and Palaeontology, Contribution 40:13 pages.

Ihssen, P.E., H.E. Booke, J.M. Casselman, J.M. McGlade, N.R. Payne, and F.M. Utter. 1981. Stock identification: materials and methods. Canadian Journal of Fisheries and Aquatic Sciences 38:1838-1855.

Legendre, V. 1952 (MS). The pike of Quebec: the ogres of our waters. Unpublished manuscript, Department of Fish and Game, Office of Biology, Province of Quebec, Montreal, Quebec, Canada. 42 pages + 27 figures.

Lieder, U. 1959. Die Chromosomenverhaltnisse von *Esox lucius* L., *Perca fluviatilis* L. and *Lucioperca lucioperca* L. (Vertebrata, Pisces). Arkiv für Zoologi 12:123-126.

McGregor, J.R. 1970. The chromosomes of the maskinonge (*Esox masquinongy*). Canadian Journal of Genetics and Cytology 12:224-229.

Merrilees, M.J., and E.J. Crossman. 1974. Species and geographic variability of surface pits in the Esocidae. Copeia 1974:893-909.

Nelson, G.J. 1972. Cephalic sensory canals, pitlines, and the classification of esocid fishes, with notes on galaxiids and other teleosts. American Museum Novitates 2492:49 pages.

Nygren, A., P. Edlund, U. Hirsch, and L. Ahsgren. 1968. Cytological studies in perch (*Perca fluviatilis* L.), pike (*Esox lucius* L.) and ruff (*Acerina cernua* L.). Hereditas 59:518-524.

Prakken, R., J. Bekendam, and G. Peters. 1955. The chromosomes of *Esox lucius* L. Genetica 27:484-488.

Reist, J.D. 1983. Studies on the systematic significance of the external body morphometry of esocid fishes. Ph.D. dissertation. Department of Zoology, University of Toronto, Toronto, Ontario, Canada.

Scott, W.B., and E.J. Crossman. 1973. Freshwater fishes of Canada. Bulletin 184. Fisheries Research Board of Canada, Ottawa, Ontario, Canada.

Simpson, G.G., A. Roe, and R.C. Lewontin. 1960. Quantitative zoology. Harcourt, Brace and Co., New York, New York, USA.

Svärdson, G., and T. Wickbom. 1939. Notes on the chromosomes of some teleosts, *Esox lucius* L., *Lucioperca lucioperca* L. and *Perca fluviatilis* L. Hereditas 25:472-476.

Webb, P.W. 1978. Fast start performance and body form in seven species of teleost fish. Journal of Experimental Biology 74:211-226.

A Review of Lymphosarcoma of Muskellunge and Northern Pike

R.A. SONSTEGARD AND T.T. CHEN

Department of Biology-Pathology
McMaster University
Hamilton, Ontario L8S 4L8

ABSTRACT

Feral muskellunge and northern pike in widely separated geographical regions in North America and northern pike in Europe suffer epizootics of lymphosarcoma. The disease is highly contagious and virus associated. The tumor is transmitted fish to fish via physical contact during spawning activities. To prevent spread of the disease, it is recommended that stocking be limited to the progeny of disinfected eggs propagated under quarantine hatchery conditions.

During the past 100 years, North American freshwater fishery resources have been heavily impacted by man's activities. Furthermore, the rate of losses resulting from these impacts (i.e., sports fishing, commercial fishing, habitat loss, discharge of toxic chemicals, etc.) continued to escalate. Looking ahead to muskellunge management in the year 2050, it is apparent that we may have to increasingly rely on fish culture-stocking programs to maintain this highly valuable sports fishery.

Whereas infectious diseases are important modifiers of feral fish populations, they have the potential of disastrous consequences in highly crowded hatchery situations where transmission of infectious agents is greatly enhanced. Today, control of infectious diseases has become of strategic importance in fishery management.

In this paper, the biology and epizootiology of lymphosarcoma (malignant cancer of lymphoid tissue) in muskellunge and northern pike are reviewed with particular emphasis as to strategies for disease control.

HISTORY AND GEOGRAPHIC DISTRIBUTION

The occurence of lymphosarcoma in muskellunge and northern pike has been known for some time. Olhmacher (1898) diagnosed the disease in northern pike nearly 100 years ago, while Sonstegard (1976) documented the disease in muskellunge collected in 1909.

The disease occurs in widely separated geographical regions in feral muskellunge and northern pike. To date, the disease in muskellunge has been diagnosed in the St. Lawrence River, the Kawartha Lake region of Ontario, Georgian Bay of Lake Huron and several tributaries of the same, and in several lakes in the Rainy Lake region of Ontario. The tumor in northern pike has been diagnosed in Ireland (Mulcahy and O'Rourke, 1964), Sweden (Ljunberg, 1976) and in Finland, Netherlands, France, Quebec, Ontario, Manitoba, Saskatchewan, Alberta, Northwest Territories, New York and Michigan.

Fish of the genus *Esox* are believed to have evolved more than 60 million years ago (Crossman and Harington, 1970). The evolutionary and geographic history of the species, and the occurrence of the same or closely related disease in both the Old and the New World, might suggest a common evolution of the disease rather than separate pathogenic events.

The frequency of occurrence of the disease is the highest known for malignant neoplasm in any free-living vertebrate. For example, 16% of muskellunge and 20.9% of northern pike populations have been found with the disease (Sonstegard 1976). Similarly, Mulcahy and O'Rourke (1964) and Ljunberg (1976) reported epizootics of the disease in European pike.

Muskellunge Lymphosarcoma

Sonstegard (1970, 1975, 1976a,b) and Sonstegard and Hnath (1978), in a series of papers, reported investigations of the etiology and epizootiology of lymphosarcoma in muskellunge. These studies are summarized below:

In field epizootiological studies, no muskellunge had lymphosarcoma without involvement of skin. The tumors appear as "red sores" composed of blister-like growths or large, ulcerated pinkish-white lesions. The external lesions were found (listed in descending order of occurrence) on the sides, fins, and head. The clinical course of the disease in muskellunge is as follows: The tumor is unicentric arising in the cutaneous tissues and forming a nodular tumor. The neoplasm spreads from cutaneous foci via intercellular spaces to involve the flank musculature. In the late stages of the disease, the kidney and spleen become infiltrated, diffusely enlarged, and may display solid sacromatous masses. The liver is the only other organ commonly involved, being infiltrated only after kidney and spleen infiltration. Occasionally the superficial skin lesions heal, however, in most cases, the tumor progresses to the death of the animal (less than 1% survive).

The disease in the muskellunge exhibits marked seasonal periodicity in both frequency and progression of the lesions. During mid-summer, virtually no muskellunge exhibit the disease (tumor frequency 0.5%). In the fall there is an abrupt increase in the percentage of fish affected (tumor frequency 10%), the majority of these having early skin lesions. In early spring, the disease is found at about the same frequency as in the fall; however, a higher percentage of lesions are in an advanced stage (muscle and organ involvement) and a lower percentage have only superficial skin lesions. In late spring (May and June), the percentage of the population exhibiting lesions decreases, while the percentage of lesions in an advanced stage (muscle, organ, and peripheral blood involvement) increases. During this period, there is almost a complete die-off of muskellunge with lymphosarcoma. The seasonal periodicity in muskellunge is attributed to the progression of small superficial cutaneous tumors which are found in the late summer and fall. The growths are rare during the summer months because their rapid growth during the late spring and early summer causes death of the majority of the affected fish.

The above progression of the muskellunge lymphosarcoma is the clinical course of the disease as documented in field tagging studies and laboratory studies of muskellunge from Lake Scugog in the Kawartha region of Ontario. In field collections at other sites in the Kawartha Lakes (i.e. Stoney Lake, Buckhorn Lake), muskellunge with healed lesions and lesions which appeared to be in remission were common. It is unknown which factor(s) (i.e., genetic variation in virus, selection for resistance in the gene pool of the population, etc.) may contribute to the same.

It is hypothesized that the mode of transmission is horizontal since the stocking of thousands of muskellunge fingerlings, raised under hatchery conditions from spawn collected in endemic watersheds over several decades, have failed to introduce the disease into an isolated watershed. This, of course, constitutes strong circumstantial evidence against vertical transmission. The clinical picture of progression of the disease from a localized tumor in muskellunge dermis suggests that the disease is introduced percutaneously.

In considering the possible modes of horizontal transmission, the behavioural patterns of muskellunge must be considered. Muskellunge is a solitary, nongregarious animal except in spring at spawning time. The constant involvement of the skin, plus the observation that only sexually mature muskellunge develop the disease (4-5 year-old fish) suggests that the disease is transmitted by contact during spawning. By virtue of the solitary nature of the animal, this is the only time of the year that contact is likely to occur, and certainly the only time that contact might be frequent enough to account for the prevalence recorded. This is particularly strong evidence against a vector hypothesis, and one would have to consider a vector that would affect only sexually mature fish, an unlikely possibility. The fact that only sexually mature fish exhibit the neoplasm must also be considered in relation to the possible role of hormones in expression and transmission of the disease, as found in virus diseases in other animals. However, since transmission trials in fingerling fish were successful, the role of hormones in expression of the disease does not appear to be of paramount importance (Sonstegard, 1976).

The muskellunge lymphosarcoma is highly contagious (Sonstegard, 1970, 1973, 1976a). For example, normal muskellunge held in tanks with tumor-bearing fish develop the neoplasm. When the lesion is rubbed on an abraded surface of a normal muskellunge, the fish develops the neoplasm at the site.

Evidence for a viral etiology of the neoplasm in muskellunge is equivocal. Reverse transcriptase enzyme activity has been demonstrated in postmitochondrial preparations of tumor samples (enzyme specifically associated with retroviruses). However, cell-free transmission trials have been unsuccessful and/or equivocal (Sonstegard, 1976a).

Northern Pike Lymphosarcoma

The biology of lymphosarcoma in northern pike was reviewed by Sonstegard (1976a, b) and Hnath and Sonstegard (1978). The disease is summarized as follows:

All feral pike with lymphosarcoma have cutaneous lesions (i.e., sides, fins, head). The anatomical distribution of the lesions is not random, 83% being located on the posterior one-half of the fish.

Horizontal contact transmission during the act of spawning is hypothesized as the mode of transmission of the pike lymphosarcoma based on the following considerations. All pike with the disease have involvement of skin. Normal pike held in the same tanks as tumor-bearing pike develop skin tumors. The anatomical predilection for the neoplasm for the posterior one-half of the fishes' body, a prime site of physical contact during spawning-courting activities, suggests transmission by contact during the act of spawning. Furthermore, there is a dramatic increase in frequency of occurrence one year post-sexual-maturity in both sexes. Several tumors were recorded in two-year-old male pike which were captured on spawning marshes which argues against the horizontal spawning contact hypothesis. These fish may, however, have been inadvertently exposed by confinement in fish trapping operations together with tumor-bearing fish.

In field epizootiological studies of the pike lymphosarcoma, pike with healed lesions and/or lesions in remission were documented. Confirmation as to the healed lesions as being lymphosarcoma was by recapture of tagged fish which had active lesions. Histopathological examination of the healed lesions also revealed that they had clusters or islets of neoplastic cells in the dermis. Although evidence of spontaneous remission was not seen in laboratory-exposed fish and/or in other geographical locations monitored, the lake environment in which the field studies were made support the hypothesis that most of the pike do not die of the neoplasm (Sonstegard, 1976).

The microscopic characteristics of the lymphosarcoma in the northern pike are characterized by myriads of isomorphic cells. The cytological characteristics of the neoplastic cells appear uniform in fish within a given watershed, however, variation has been recorded in different geographical areas. The progenitor cell giving rise to the neoplasm is unknown.

Electron microscopic studies of post-mitochondrial preparations made from frozen northern pike tumors revealed virus-like particles banding at 1.15-1.17 g/ml in density gradients (Papas et al., 1976, 1977). Viral enzyme (reverse transcriptase) was detected in association with the banded preparations; suggesting a viral etiology of the tumor (Sonstegard and Papas, 1977; Papas et al., 1976, 1977). The viral enzyme activity dropped rapidly at temperatures above 20°C (Papas et al., 1976, 1977). It is hypothesized that the elevated water temperatures of the summer months are non-permissive temperatures for virus expression and may be the mechanism of spontaneous regression. Although virus has been found associated with the tumor, a viral etiology has not been confirmed.

Karyotype studies (direct) made from tumor-bearing pike revealed the presence of markers (metacentric and minute chromosome; diploid number of pike is 50; all acrocentric) (Whang-Peng et al., 1976). The etiological significance of the chromosome marker is unknown.

Springtime field collections of northern pike and muskellunge included fish with firm, slightly elevated whitish-grey epithelial plaques on the flank, fins, and head. The frequencies of these plaque-like proliferations were about 2% and 6% in pike and muskellunge, respectively. They occasionally involved the epidermis directly overlying active lesions of lymphosarcoma. The condition has been found only in the springtime and it is hypothesized that the condition regresses during the warm water periods of summer.

Electron microscopic studies of the epithelial plaque-like lesions in both muskellunge and northern pike revealed the presence of C-type virus particles (similar to those found in lymphosarcoma lesions). The etiological significance of the C-type virus particles associated with plaque-like lesions in relation to the lymphosarcoma in pike and muskellunge are unknown. Similar observations of virus associated with epidermal hyperplastic lesions have also been reported in Swedish pike (Winquist et al., 1968).

RELATION TO HUMAN HEALTH

In recent years, there has been increasing concern as to the association of neoplasms in fishes with environmental toxicants. To date, no association of the frequency of occurrence of the *Esox* lymphosarcoma with environmental toxicants has been made. Furthermore, the widespread geographical distribution of the disease in *Esox* suggests that the disease is not a "new" phenomenon, and likely has persisted in nature for thousands of years. Considerable inter-lake differences in frequency of occurrence do, however, occur. It is hypothesized (based on field and laboratory studies) that physical characteristics of spawning habitat(s) which increase the opportunity for epidermal abrasions during courtship and spawning increases the risk for transmitting the disease. Reports from field biologists and anglers occasionally describe occurrences of epizootics of the disease in watersheds with no previous history of the disease. These reports are biased by awareness of personnel to recognize the disease, age of animals captured, etc. Annual fluctuations in frequency of occurrence likely occur reflecting environmental conditions facilitating transmission, factors influencing clinical course of the disease, population density, etc.

Although the disease is highly infectious, there is no evidence to suggest that the disease presents a human health hazard. The viral enzyme (reverse transcriptase) associated with the pike and muskellunge lymphosarcoma loses activity at 37°C (body temperature), strongly suggesting that viral infection could not occur. Pragmatically, no known animal cancer has been found associated with an infectious risk to man. Furthermore, there is increasing evidence in the medical literature that most human cancers are not of an infectious nature.

DISEASE CONTROL

Field and laboratory investigations of both the muskellunge and northern pike lymphosarcoma support the hypothesis that the disease is infectious and is horizontally transmitted by physical contact between spawning fish. This gives a "window" for the development of guidelines for disease control. Specifically, it is recommended that the stocking-introductions be limited to eggs, fry, fingerlings, or sub-adults which have been propagated under hatchery conditions which exclude contact and/or exposure to adult fish. Introduction via wild capture fish should not be made. In addition, egg disinfection techniques should be incorporated into hatchery practices to decrease the potential of disease introduction.

Lymphosarcoma as a disease has persisted in fish of the genus *Esox* for thousands of years. As such, disease introduction(s) are the ultimate form of pollution into the environment. Whereas this might seem pragmatic today, it may well be paradoxical tomorrow.

REFERENCES

Crossman, E.J. and Harington, C.R. 1970. Pleistocene pike, *Esox lucius* and *Esox* sp. from the Yukon Territory and Ontario. Canadian Journal of Earth Sciences 7: 1130-1138.

Ljunberg, O. 1976. Epizootiological and experimental studies of skin tumors in northern pike *(Esox lucius)* in the Baltic Sea. Progress in Experimental Tumor Research 20: 156-165.

Mulcahy, M.F. and O'Rouke, F.J. 1964. Cancerous pike in Ireland. Irish Naturalists Journal 34: 312-315.

Ohlmacher, H.P. 1898. Several Examples illustrating the comparative pathology of tumors. Bulletin of the Ohio Hospital for Epileptics 1: 223-233.

Papas, T.S., Dahlberg, J.E., and Sonstegard, R.A. 1976. The presence of Type C virus in lymphosarcoma in fish (northern pike, *Esox lucius*). Nature 261: 506-508.

Papas, R.S., Shaffer, M., Pry, T., and Sonstegard, R.A. 1977. DNA polymerase in lymphosarcoma in northern pike *(Esox lucius)*. Cancer Research 37: 3214-3217.

Sonstegard, R.A. 1970. Descriptive and epizootiological studies of infectious pancreatic necrosis virus of salmonids and lymphosarcoma of *Esox masquinongy*. Ph.D. Dissertation, University of Guelph, Guelph, Ontario.

Sonstegard, R.A. 1975. Lymphosarcoma in muskellunge *(Esox masquinongy)*. *In* W.E. Riblein and G. Migaki, editors. The Pathology of Fishes. University of Wisconsin Press.

Sonstegard, R.A. 1976a. Studies of the etiology and epizootiology of lymphosarcoma in northern pike *(Esox lucius)* and muskellunge *(Esox masquinongy)*. Bibliotheca Haematologica 43: 242-244.

Sonstegard, R.A. 1976b. Studies of the etiology and epizootiology of lymphosarcoma in Esox *(Esox lucius* L. and *Esox masquinongy)*. Progress in Experimental Tumor Research 20:141-145.

Sonstegard, R.A. and Hnath, J.G. 1978. Lymphosarcoma in muskellunge and northern pike: Guidelines for disease control. American Fishery Society Special Publication 11: 235-237.

Sonstegard, R.A. and Papas, R.S. 1977. Descriptive and comparative studies of C-type virus DNA polymerase of fish (northern pike, *Esox lucius*). American Association Cancer Research. Proceedings of 1977 Meeting, p. 806.

Sonstegard, R.A. and Sonstegard, K.S. 1973. Establishment of a cell line from lymphosarcoma of muskellunge *(Esox masquinongy)*. In Vitro 8: 410-411.

Whang-Peng, J., Sonstegard, R.A., and Dawe, C.J. 1976. Chromosomal characteristics of malignant lymphoma in northern pike *(Esox lucius)* from the U.S.A. Cancer Research 36: 3665-3560.

Winquist, G., Ljunberg, O., and Hellstrom, B. 1968. Skin tumors of northern pike *(Esox lucius* L.). II. Viral particles in epidermal proliferations. Bulletin de l'Office International des Epizooties 64: 1023-1031.

Movement, Activity, and Habitat Use Patterns Of Muskellunge In West Okoboji Lake, Iowa[1]

MARLYN L. MILLER[2] AND BRUCE W. MENZEL

Department of Animal Ecology
Iowa State University
Ames, Iowa 50011

ABSTRACT

Ultrasonic telemetry was used to determine habitat use and activity patterns of nine adult muskellunge in a 1540-hectare natural lake from spring 1978 to fall 1979. About 80% of summer and early fall contacts (n=1292) occurred in three large, relatively shallow, vegetated bays. Study fish exhibited significant summertime changes in depth of water occupied, association with vegetation, and general activity. Fish activity was negatively associated with surface water temperatures; activity was greatest over the 11-17 C range. There was a positive association of fish activity with water transparency (Secchi disk depth). Significant diel behavioral variations occurred during June, when there were crepuscular peaks in fish movement rates, activity, and water depth use. There were no identifiable sex- or size-dependent differences in any of the studied behaviors.

The study fish appeared to adjust their basic foraging patterns to maintain an optimal feeding strategy in response to seasonally changing environmental factors. After the spawning period and through midsummer, the fish behaved as searching predators, as evidenced by relatively high levels of activity, extensive movements, use of a variety of water depths and habitat types, and pronounced crepuscular activity. By late summer, the fish exhibited behavioral characteristics of a sedentary ambush predator, e.g. reduced activity, strong allegiance to activity centers associated with vegetation, and little diel variation in activity. These seasonal changes in muskellunge habitat use and behavior may contribute to the distinct seasonality of the fishery for this species in West Okoboji Lake, maximum catch rates occurring in late summer and early fall.

The muskellunge occurs naturally in a variety of both lotic and lentic habitats is east-central North America and has been frequently introduced beyond its native range (Crossman 1978; Carlander et al. 1978). It is generally considered a shallow-water fish, associated with areas of submergent vegetation, stumps and fallen timber, or other cover (Oehmcke et al. 1958; Scott and Crossman 1973). Large individuals, however, may also be found in deep waters where vegetation is sparse or absent. The species is commonly described as a solitary, sedentary predator.

Knowledge of muskellunge behavior and habitat use has increased recently through application of biotelemetric techniques. Crossman (1977) and Minor and Crossman (1978), using ultrasonic and radio telemetry respectively, conducted investigations on adult fish in a small flowage and in a large lake in central Ontario. Dombeck (1979) used radio telemetry to study seasonal movement of adults in two northern Wisconsin lakes. Additionally, this volume contains reports on two radio telemetry studies; Gilbertson discusses winter movements of juveniles in a Minnesota lake, and Strand describes movement and spawning of adults in Leech Lake, Minnesota.

The purpose of our study was to examine, through the use of ultrasonic telemetry, the behavior of adult muskellunge in a large natural lake, and to relate the observed behaviors to environmental features. Elsewhere, we have reported patterns of home range and homing behavior (Miller and Menzel, in press). The present report focuses on seasonal movement, activity and habitat use patterns of the tagged fish.

STUDY AREA

West Okoboji Lake is one of a chain of recreationally-important natural lakes located in Dickinson Co., northwestern Iowa. The physical characteristics of this deep (mean depth = 1.10; max. depth = 42.7 m) 1540-ha eutrophic lake were described by Bachmann et al. (1966). The deep, central channel is bordered by several large, relatively shallow bays (Fig. 1). Sand and small gravel substrates are common in the broad littoral zone whereas muck bottom is dominant in the midlake area.

---

[1] Journal Paper No. J-11327 of the Iowa Agriculture and Home Economics Experiment Station, Ames, Iowa. Project No. 2236.

[2] Present address: Ohio Department of Natural Resources, Division of Wildlife, 360 E. State Street, Athens, Ohio 45701.

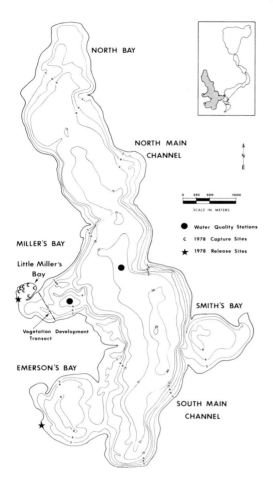

Figure 1. Contour map of West Okoboji Lake showing water sampling sites, 1978 muskellunge capture and release sites, and lake's location relative to other area lakes. Contour intervals are in meters.

Prominent coarse gravel and boulder areas occur in most bays as well as the central channel.

Most of the littoral area supports dense growths of aquatic macrophytes during the summer to a depth of about 6 m. The most widespread and abundant species are *Potamogeton* spp., *Vallisneria americana, Heteranthera dubia, Ceratophyllum demersum,* and *Myriophyllum exalbescens* (Crum and Bachmann 1973).

The lake's thermal and chemical characteristics were reported by Bachmann and Jones (1974). There is late summer thermal stratification, and hypolimnetic oxygen depletion below 15 to 20 m. The upper 10 m of the water column generally remain homothermous and saturated with dissolved oxygen. Summer surface temperatures can exceed 25 C.

The lake supports a large and diverse fish community. The sport fishery, in order of importance to the creel, includes yellow perch, bluegill and pumpkinseed, walleye, northern pike, smallmouth bass, black and white crappie, and black bullhead (Christianson 1978). The nongame fish community is dominated by carp, freshwater drum, bigmouth buffalo, and golden shiner.

The Iowa Conservation Commission introduced muskellunge into the lake in 1960. Early stockings were limited to a few hundred fish, but from 1970 to 1979 approximately 16,700 fingerlings produced by brood fish from West and East Okoboji lakes were stocked. Although only West Okoboji has been stocked, legal-sized fish (76 cm, total length) have been caught throughout the five-lake chain. It is not known if natural reproduction contributes to the maintenance of the population.

METHODS

Adult muskellunge were captured by gill netting during the 1978 spring spawning period in late April. Six males (designated A-E, H) and one female (G) were caught in Little Miller's Bay, a small (10 ha) shallow bay on the west side of the lake (Table 1, Fig. 1). Two other females (F, J) were selected from wild-caught brood fish held at nearby Spirit Lake Hatchery; their exact capture sites were unknown. Males were fitted with transmitters immediately after capture, whereas females were held at the hatchery for 1 to 10 days to allow for stripping of eggs before tagging. Ultrasonic transmitters were implanted in the fish's body cavities after anaesthetization with quinaldine (25 ppm). Surgical procedures followed methods of Hart and Summerfelt (1975) and Crossman (1977), except that there was no postoperative injection of antibiotics. The entire procedure required 20 to 30 min. Fish were returned to the lake shortly after they regained equilibrium. Fish known to have originated from Little Miller's Bay were released in the vicinity of their capture sites. The other two females were released into Emerson's Bay (Fig. 1).

Ultrasonic transmitters and receiving equipment were manufactured by Donald L. Brumbaugh, Tucson, Arizona. Transmitters (16 x 60 mm) weigh approximately 8 g in water. Six transmitters were temperature-sensing, producing pulse rates that

Table 1. Study fish characteristics and observation histories.

| | | | | | Number of contacts | | |
| | | | Total | Weight | 1978 | | 1979 |
Fish	Sex	Age	length (cm)	(kg)	Summer	Autumn	
E	M	?[a]	69	2.5	144	13	13
A	M	III	71	2.6	130	13	12
B	M	IV	79	3.3	128	14	13
D	M	IV	82	4.1	109	13	36
H	M	IV	85	4.0	99	14	13
C	M	VII	90	6.0	151	15	29
G	F	IV	85	3.9[b]	138	13	23
J	F	IV	86	4.5[b]	131	15	32
F	F	VII	101	6.9[b]	148	4	—[c]

[a] Scales were unsuitable for aging.
[b] Weight after being stripped of eggs.
[c] Caught and removed by angler on September 23, 1978.

varied predictably with changes in a fish's body temperature. Three others had unique fixed pulse rates and provided only location. Transmitters were individually identified through a combination of their specific frequencies (74 to 92 kHz) and pulse rates (56 to 103 pulses/min. at 23 C). The portable receiving equipment consisted of a battery-powered receiver, headphones, and a directional, hand-held hydrophone. Under optimum conditions, transmissions were received up to 1.5 km.

Preliminary observations of tagged muskellunge in late April and May 1978 confirmed that all transmitters were functioning and that the fish had survived the tagging procedure. From June 1 to August 31, fish movements were monitored daily, as weather conditions and recreational boat traffic permitted. Tracking operations were conducted from a motorboat. On routine tracking days, starting points and the order in which lake areas were searched were randomized. Within lake areas, contacts with individual fish were made opportunistically. All fish were located, or the entire lake was searched, at least once before a subsequent monitoring series was begun. Periodically, 24 hour tracking efforts were conducted on individual fish, with locations recorded every 1 to 3 hrs. From September through November 1978, monitoring was conducted at 2- or 3-week intervals. No tracking was attempted during the period of ice cover. From late April through May 1979, observations were made daily, or as conditions allowed. During June, July, and August 1979, tracking was limited to once per week.

Upon locating a study fish, the observer positioned the boat in the immediate vicinity of the fish and recorded its identity, time of day, weather conditions, and identities of other study fish within range of reception. Approximate fish location was noted on a lake map, and precise position was determined by sextant triangulation on lakeshore landmarks. The behavioral parameter, fish activity, was recorded as either "active", when the fish was judged to be moving, or "stationary." Water temperature profiles were taken in the vicinity of fish fitted with temperature-sensing tags. An electronic depth recorder-grapher (Lowrance Model LRG-605) was used to measure water depth (depth of water column at fish's location) and vegetation association (presence or absence of submergent vegetation in immediate vicinity of fish). Occasionally, images of study fish were identifiable on the depth recorder graph, permitting measurement of actual depth occupied. Often the presence of other fish was noted on the graph recordings as well.

Measurements of water transparency (Secchi disk depth), water temperature, and dissolved oxygen were taken regularly at midlake and midbay locations (Fig. 1). Development of aquatic vegetation was monitored in fall 1978 and summer 1979 by making biweekly transects across Miller's Bay with the depth recorder (Fig. 1). From the graph images, average height of the submergent vegetation was determined along a prominent reef and at the openwater edge of the littoral zone, hereinafter termed the "weedline".

Fish locations were plotted on a 1:7920 scale map of the lake. At this scale, a 1-mm diameter pencil dot marking a fish location represents a 7.9-m diameter circle. This is an indication of the minimum possible error associated with distance measurements; actual error was probably several times greater (Hart and Summerfelt 1973). The X-Y coordinates from a superimposed grid-square system (1 grid square=0.63 ha) were assigned to each location. These coordinates and associated grid-square assignments were used in analyses of habitat use. Distance between consecutive locations of fish were measured on the map and represent minima that fish must have traveled. Movement rate, expressing minimum movement rate (m/hr), was calculated as the distance between consecutive locations divided by the elapsed time between contacts. Only observations where elapsed time was at least 1.0 hr., but not more than 6.5 hr, were used in analyses of movement rate (Miller 1982). For analyses of diel behavior patterns, the average movement rate for a given hourly period was calculated by averaging the minimum movement rates from all pairs of observations which included that period (Miller 1982).

Potential muskellunge habitat availability was determined from a detailed lake map and from published reports. Proportions of the lake's total surface area encompassed within 2-m depth contour intervals were calculated from Bachmann et al. (1966). Waters shallower than 6 m (approximate weedline depth in 1978) were classified as vegetated. This somewhat overestimates the proportion of the lake actually vegetated because some shallows are virtually devoid of rooted plants, e.g. beaches, rocky shoals, and shorelines subjected to intense wave action. Based on map information and personal observations, substrates of map grid-squares were classified as either hard (gravel and/or rock) or soft (sand and/or muck).

Data on the 1978 sport fishing catch was obtained from voluntary report records kept by four local bait and tackle dealers and from a mail survey of selected anglers.

STATISTICAL ANALYSIS

Preliminary analyses, using Wilcoxon's Two-sample test and Spearman's Rank Correlation Test (Gibbons 1976), showed no differences relative to sex and size for the behavioral variables. Therefore, data from all fish were pooled for subsequent statistical tests.

Muskellunge habitat use was evaluated by a chi-square goodness-of-fit test (Gibbons 1976) on the fish's observed frequency of occurrence in a given habitat with respect to that habitat's relative availability in the lake. Chi-square independence tests were used to evaluate the interactions of these variables with seasonal and diel time periods. These analyses were performed on data for individual fish, and on pooled monthly and summer data. Spearman's Rank Correlation Test (Gibbons 1976) was used to examine relationships between behavioral and environmental variables (water transparency and water temperature). Behavioral data were reduced to means of individual fish's weekly means for continuous variables, or to means of individuals' weekly percentages for dichotomous variables (fish activity and vegetation association). Seasonal and diurnal differences in behavior were evaluated by two-way analyses of variance (ANOVA) performed on individual fish weekly and hourly means. Arcsin transformation of means was applied to dichotomous variables. For all statistics, null hypotheses were rejected at $P < 0.05$.

RESULTS

During the two field seasons, 1463 fish locations (126 to 195 contacts/fish) were obtained; the intensive tracking effort of summer 1978 produced 1178 contacts (Table 1). From the second week of June through August 1978, there was an average of 10 contacts/fish/week (range 3 to 22). All transmitters functioned flawlessly and had no evident adverse physiological effects, as determined by superficial examination of seven fish recaptured after periods of 4 to 66 weeks at large. Four fish (C, E, G, and H), recaptured by gill netting in spring 1979, were in normal spawning condition and were used as hatchery brood fish before their return to the lake. Three fish were captured by anglers; one (B) was released and the others (F, J) kept. All recaptured fish exhibited well-healed incisions and normal growth and weight gain for Okoboji muskellunge (Miller 1982).

After tagging and release, the fish dispersed throughout the lake. By mid-June, as surface water temperatures rose above 17 C, most moved to bays where they established summer residency (Fig. 2). Miller's, Emerson's, and Smith's bays accounted for about 80% of the summer locations. All fish maintained well-defined summer home ranges in one or two of the major bays or the central channel. The home ranges were 39 to 443 ha in surface area (ave. 146 ha) and thus encompassed large portions of the bays or lake regions of residence. In spring 1979, after presumably overwintering elsewhere, several fish returned to their 1978 capture and release site. By mid-summer, all eight surviving fish returned to the home ranges maintained during the previous year. Miller and Menzel (in press) provide a detailed account of spring movements, summer home ranges, and homing behavior.

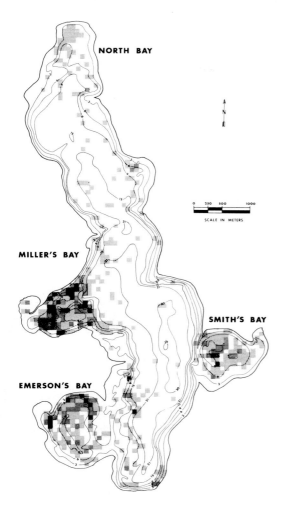

Figure 2. Muskellunge distribution in West Okoboji Lake during summer and autumn 1978. Shaded grid-squares express relative frequency of use: 1 or 2 observations in lightest squares; 3 to 5 in medium density squares; 6 or more in darkest squares. Contour intervals are in meters.

SEASONAL BEHAVIOR PATTERNS

Movement rates, activity, association with vegetation, and depth of the fish changed seasonally in 1978 (Figs. 3 and 4). During the early weeks of the summer, the fish moved from littoral areas into pelagic water, and average weekly water depth increased accordingly, from 4.0 m in early June to 8.8. m in early July. They gradually returned to more intermediate depths, averaging 4.8 to 6.0 m from mid-August to early November. Concomitant with shifts in water depth, vegetation association values decreased in early summer to a minimum weekly average of 28% by mid-July, then increased as the fish returned to the littoral area, and remained

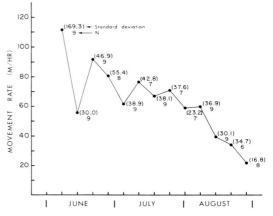

Figure 3. Seasonal patterns of muskellunge movement rate during summer 1978. Values are grand means of individual weekly means.

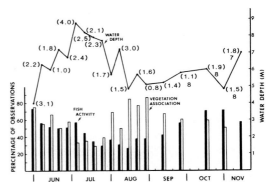

Figure 4. Seasonal patterns of water depth, fish activity, and vegetation association exhibited by muskellunge during summer and autumn 1978. Water depth values are grand means of individual weekly means; standard deviations are noted in parentheses. Fish activity and vegetation association values are means of individual weekly percentages. Data are based on 9 fish from June through September, and 7 or 8 fish in October and November.

Table 2. Association of behavioral and environmental variables during summer 1978 as determined by Spearman's Rank Correlation (R); N=13.

	Water depth	Vegetation association	Fish activity	Movement rate	Body temperature	Water temperature
Water depth	-					
Vegetation association	-0.94**	-				
Fish activity	0.31	-0.14	-			
Movement rate	0.64*	-0.60*	0.37	-		
Body temperature	-0.21	0.14	-0.68**	-0.62*	-	
Water temperature	0.25	-0.36	-0.52	-0.35	0.74**	-
Water transparency	0.33	-0.17	0.90**	0.41	-0.65*	-0.56*

* probability of greater R is less than 0.05 ($p<0.05$)
** $p<0.01$

above 60% from early August to mid-October. As expected, this inverse relationship between water depth and vegetation association is statistically significant (Table 2).

During their early summer pelagic residency, the fish were active; weekly averages for fish activity exceeded 50% and those for movement rate typically exceeded 60 m/hr (Figs. 3 and 4). Both indices of fish activity progressively declined as the fish returned to littoral areas, reaching seasonal low values in mid- to late-August. Movement rates could not be calculated from the limited autumn data set, but an increase in fish activity comparable to late spring levels was recorded for the period. Over the summer, movement rate was directly associated with water depth and inversely related to vegetation association. Although relationships between fish activity and the other behavioral parameters followed predictable directions, there were no significant associations (Table 2).

Two-way ANOVA of the 13 weeks of summer behavioral data revealed highly significant weekly differences ($P < 0.001$, d.f. = 12,96) for water depth ($F = 4.44$), vegetation association ($F = 5.12$) and fish activity ($F = 3.29$), despite significant individual variation for each parameter ($F = 7.64, 2.85$, and 5.20, respectively; $P < 0.01$, df = 8, 96).

Midlake surface water temperature increased steadily from 17 C in early June, fluctuated between 23 and 25 C from early July to mid-September, and decreased rapidly in autumn to 5 C by late November (Fig. 5). Body temperatures of six monitored fish closely paralleled this pattern (Table 2). Fish temperatures tended to average 1 C higher than mid-lake temperature during the warmest period, when the fish chiefly occupied the shallower and warmer bays. There was a significant inverse association of fish activity and water temperature. Water transparency, measured by Secchi disk depth, changed seasonally from maxima of > 4 plus m in June and November to minima of 2 m in August. This resulted in a strong direct association of fish activity and water transparency (Table 2).

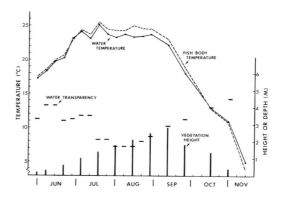

Figure 5. Seasonal patterns of muskellunge body temperature, mid-lake water temperature, and Secchi disk depth in West Okoboji Lake during summer and autumn 1978, and height of submerged aquatic vegetation during autumn 1978 and summer 1979.

Macrophyte growth, based on a composite of observations in 1978 and 1979, reached greatest height from mid-August to mid-September (Fig. 5; details in Miller 1982).

Evidence from depth recorder images suggests that fish activity was associated with quantity of locally available potential prey. From 595 depth recordings taken at open-water fish locations, 416 revealed the presence of small fishes, and in 51% of these cases, muskellunge were active. In contrast, the fish were active in only 37% of the cases where no potential forage was detected (pooled data $x^2 = 10.32$, $P = 0.001$, d.f. = 1).

Summer Habitat Use

The implied selection by muskellunge for shallow vegetated areas during summer 1978 was tested by comparing actual habitat use with distribution patterns expected on the basis of random occupation of available habitat in the entire lake. Muskellunge occurred in shallows (less than 6 m deep) far more frequently than could be accounted for by random associations with the available range of depths ($x^2 = 1393.06$, $P < 0.001$), viz. 78% of fish locations were in water 2 to 8 m deep, while only 37% of the lake's area is within these contours. Although individual fish differed significantly relative to water depth (Miller 1982), all occurred primarily in shallow water. Similarly, at 57% of the locations, muskellunge were associated with vegetation, which was significantly greater than the expected 33% based on random associations ($x^2 = 303.59$, $P < 0.001$, d.f.= 1). All but one fish exhibited this positive association with vegetation (Miller 1982). Selection of shallow, vegetated habitat was pronounced in late summer, when most fish were located in the narrow depth contour range of 4 to 6 m. At this time, depth-recorder images or transmitter signals frequently indicated that muskellunge were at the weedline, either suspended above the vegetation or between discrete vegetation patches. The fish occurred over gravel and rock substrates somewhat more frequently (15%) than expected (12%) by the distribution of hard bottom area ($x^2 = 3.47$, $P = 0.06$, d.f. = 1). Three fish (C, F, H) were frequently observed on or near prominent vegetated reefs, and all others were occasionally found near such habitat features. In July, two fish (F and J) regularly appeared in water above unvegetated rock reefs, but there was little additional evidence to suggest that substrate was an important factor in habitat selection or that the fish were closely associated with the bottom. To the contrary, 115 depth recorder images depicting study fish position indicated that fish which were not associated with vegetation tended to be suspended in the middle and upper portions of the water column, regardless of water depth (Miller 1982).

Throughout the summer, fish occurring in deeper (>6 m) open-water tended to be more active and exhibited greater movement rates than when they were in vegetated shallows. These behavior-habitat associations applied to all fish (Miller 1982).

Summer Diel Behavior Patterns

Significant diel differences in behavior occurred during early summer (June) only (Fig. 6). Then, muskellunge tended to be most active in late evening and early- to mid-morning, both in terms of fish activity ($x^2 = 20.94$, $P < 0.001$, d.f. = 5) and movement rate (F = 1.69, P = 0.03, d.f. = 23, 184). Despite this general trend, there was considerable variation, and individuals could be categorized as morning-active, evening-active, or crepuscular (Miller 1982). Diel variations in water depth were suggested for June (F = 1.85, P = 0.058, d.f. = 11, 85). Fish tended to be found in deeper water (7.5 m) during early to mid-morning and late evening, and on the average occurred in water about 6 m deep during the remainder of the day. Throughout the summer months, there was no significant diel variation in vegetation association (Figs. 6, 7, and 8). Although other behavioral elements shown in Figs. 6 to 8 are suggestive of diel patterns, none were declared statistically significant because of high within-group variance.

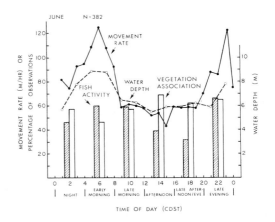

Figure 6. Diel behavior patterns of average movement rate, water depth, fish activity, and vegetation association exhibited by tagged muskellunge during June 1978.

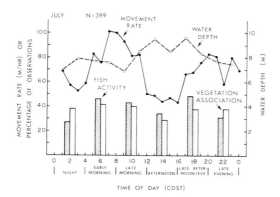

Figure 7. Diel behavior patterns of average movement rate, water depth, fish activity, and vegetation association exhibited by tagged muskellunge during July 1978.

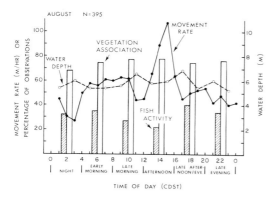

Figure 8. Diel behavior patterns of average movement rate, water depth, fish activity, and vegetation association exhibited by tagged muskellunge during August 1978.

Seasonality of the Sport Fishery

As determined from angler reports, muskellunge catches were greatest in August and September, and least in June and July 1978 (Fig. 9). Catch/release affidavits of a local fishing club for the previous 7 years revealed a similar pattern (Miller 1982).

DISCUSSION

The results describe a pattern of seasonal environmental change in West Okoboji Lake that may influence, or at least reflect, a dynamic behavioral strategy among resident muskellunge. For analytical purposes, these environmental and behavioral patterns may be organized into three temporal phases. In early summer, the water gradually warms, aquatic vegetation increases in height and density, and water transparency is relatively high. At this time, muskellunge move from shallow, nearshore areas into deeper, open water, where they exhibit a high level of activity, presumably associated with foraging.

During mid- to late-summer, water temperature is at a relatively constant seasonal maximum, and aquatic macrophytes and phytoplankton achieve greatest biomass. Although macrophyte growth is gradual, phytoplankton blooms often develop rapidly and drastically reduce water transparency, as in mid-July 1978. Muskellunge behavior during this phase involved a gradual return to the littoral zone, where the fish became relatively sedentary close to the weedline. In 1978, this behavior was most characteristic from late July through late August when water transparency was seasonally low and macrophytes had achieved nearly full growth. Finally, during autumn vegetation rapidly deteriorates, water temperature decreases, and transparency increases. At this time, the fish exhibit a general increase in activity and ultimately abandon their summer home ranges for overwintering areas, probably located at greater depths offshore (Miller and Menzel, in press).

Muskellunge thermal ecology and its behavioral correlates have received considerable research attention. In spring, mature fish initiate movement to spawning areas as water temperature increases above 5 C, and spawning typically occurs over the range, 9 to 15 C (Dombeck 1979, Gammon 1986). In West Okoboji Lake, transition from spring to summer behavior patterns occurs as the central channel surface temperature increases above approximately 16 C (Miller and Menzel, in press).

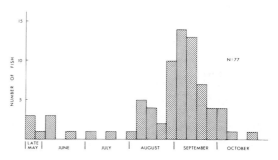

Figure 9. Reported muskellunge catch in West Okoboji Lake during 1978.

Greater fish activity occurred in spring and autumn, at 11 to 17 C, whereas fish were least active in late summer, at temperatures of 23 to 25 C. With minor variations, this pattern has been observed for muskellunge in Ontario and Wisconsin waters as well (Minor and Crossman 1978; Dombeck 1979).

Interestingly, muskellunge exhibit thermal preferenda over the range of temperatures at which they seem to be least active in nature. The experimentally-determined preferendum of 25.1 C for juveniles (Reynolds and Casterlin 1979) is similar to field observations of preferenda for adults in the range 25 to 28 plus C (Oehmcke et al. 1958; Minor and Crossman 1978; Dombeck 1979). Additionally, Okoboji muskellunge tended to occupy littoral habitats in late summer where temperatures slightly exceeded midlake surface temperatures. In contrast, under laboratory conditions, both juvenile muskellunge and northern pike exhibited greatest spontaneous activity when held at or near their thermal preferenda (Fry 1971; Casselman 1978). Possibly, these seemingly conflicting results are attributable to methodological differences in activity measurement, environmental differences, or differences in preferenda between juveniles and adults. In the laboratory, activity may be measured with precision, but only small individuals can be readily tested, the fish's scope for activity is physically restricted, and other environmental elements are similarly artificial. In nature, measurement of activity is crude, but the animals enjoy freedom of movement, presumably in response to a wide variety of environmental stimuli.

For a sight-feeding predator such as muskellunge, environmental factors that affect the field of vision influence foraging behavior. Typical elements in nature include light intensity, water transparency, and obstructions within the visual field. Decreases in the former two factors tend to lower the predator's reactive distance to prey and thus reduce its frequency of attack. This relationship is predicted by theoretical models of feeding strategies (Schoener 1971) and has been demonstrated for a number of sight-feeding fishes (Moore and Moore 1976; Gardner 1981).

Relative to muskellunge, Oehmcke et al. (1958) stated that feeding was hampered by turbid water, but provided no documentation. In our study, the strong direct association (Spearman's $R = 0.90^{**}$) between water transparency and fish activity might be interpreted as presumptive evidence for this relationship. A turbidity-reduced visual field might also dampen reactions to congeners, other larger predators, and human disturbances. Manteifel et al. (1978) suggested that reduced water transparency can suppress diel and, especially, crepuscular behavior patterns. Alternatively, changes in transparency might affect prey behavior and, thus, muskellunge reactions to prey activity. It seems plausible, therefore, that the seasonal change in Okoboji muskellunge diel behavior is at least partly a function of transparency changes. The significant inverse association of water transparency and water temperature ($R = -0.56^*$) confounds this interpretation, however.

The affinity of muskellunge for areas of submergent and emergent vegetation is generally known (Oehmcke et al. 1958; Scott and Crossman 1973). Muskellunge in Nogies Creek, Ontario, occupied winter home ranges that were 30% to 50% covered in submergent vegetation and selected home ranges that were 75% macrophyte covered (Minor and Crossman 1978). In West Okoboji Lake, vegetated areas are large and well defined. Thus, the strong association of muskellunge with the weedline in late summer when plant height was greatest implies active selection for the vegetated areas. There may be a threshold value of vegetation height or biomass required before the fish are attracted to macrophyte patches.

In 1978, the muskellunge were closely associated with the weedline when about two-thirds of maximum vegetation height had been achieved. This occurred in early August, about 2 weeks after the rapid decrease in water transparency and a month after the water temperature maximum had been reached (compare Figs. 4 and 5). Alternatively, association with the weedline might also reflect seasonal changes in the availability of forage fishes in vegetated and nonvegetated areas. Investigations by Hall and Werner (1977), Emery (1978), Keast (1978) and others have elucidated the importance of macrophytes to the distribution and interactions of species in littoral zone fish communities. Assuming that our measures of fish activity and movement rate are primarily indices of foraging behavior during the growth season, the observed behavioral-environmental associations may be spurious, or at least only reflections of unmeasured, seasonally-dynamic predator-prey relationships.

Intra- and interspecific competition for food, space, and other resources also represent potential influences on muskellunge behavior. Several observations seem pertinent to this point. Home ranges of the fish were broadly overlapping, especially those located in Emerson's, Miller's and Smith's bays. Occasionally, two or more of the fish were found in proximity to each other, conceivably within mutual view. These observations suggest an absence of territorial behavior. Another top predator in the lake, walleye, exhibits similar intraspecific spatial relationships and, moreover, seems to be spatially segregated from muskellunge. At the time when the muskellunge are concentrated at the weedline, at the edges and tops of the vegetation, walleye are either in deep open-water at considerable depth or near the bottom within the littoral vegetation (Pitlo 1978). Also, Okoboji walleye typically move to shoreline shallows at night to forage, but muskellunge do not.

Assuming that feeding dominates muskellunge activity in summer and autumn, the observed behaviors may be related to Schoener's (1971) basic foraging types. After the spawning period and through midsummer, the fish seem to be searching predators (Schoener's Type II predators). During this time, they exhibit relatively high levels of activity, long displacements, distribution across a variety of water depths and habitat types, and crepuscular behavior. Schoener (1969) suggested that such a predator requires a discrete time period each day devoted exclusively to feeding. The crepuscular peaks in activity among Okoboji muskellunge are probably intensive feeding periods. By late summer, however, the fish exhibit a complex of behaviors more characteristic of a "sit and wait" predator (Schoener's Type I): reduced activity, shorter displacements, high degree of association with potential ambush sites, and no significant crepuscular behavior. This pattern continues until mid- to late-fall, when vegetation height and water temperature are reduced to early season levels. This shift in behavior patterns, therefore, may reflect a dynamic optimal feeding strategy in response to seasonal changes in environment and prey availability.

Our angling survey results imply that Okoboji muskellunge are most vulnerable to angling during late summer and early fall. The relatively small areas of preferred habitat into which the adult population is seemingly concentrated at this time are readily located by fishermen, either through visual sighting of submergent vegetation or with electronic depth-recorders. Furthermore, the weedline and vegetation patch areas with which the fish are associated are effectively fished by the locally popular techniques: casting lures toward the vegetation and speed-trolling. While muskellunge behavior and distribution vary considerably during the warm season, our survey indicates that angler behavior and distribution do not. Thus, harvest of the muskellunge population is seasonally maximized when the paths of fishermen and fish intersect in time and space.

ACKNOWLEDGEMENTS

Project funding was provided by the Iowa State Agriculture and Home Economics Experiment Station, the Iowa State University Graduate College, and the Keith McNurlen Scholarship of the Iowa Chapter of the Izaak Walton League. Equipment and facilities were provided by the Iowa Conservation Commission and Iowa Lakeside Laboratory. We are grateful to Richard McWilliams, Wallace Jorgenson, and James Christianson of the Iowa Conservation Commission, and to Dr. Richard Bovbjerg and Robert and Tanya Benson of Iowa Lakeside Laboratory for their special assistance. We also appreciated the excellent work of our field and laboratory assistants; David Moffitt, Thomas Chapman, and Steven Brayton.

REFERENCES

Bachmann, R.W., R.V. Bovbjerg, and J.D. Hall. 1966. The morphometry of West Lake Okoboji. Proceedings of the Iowa Academy of Science 73:165-168.

Bachmann, R.W. and J.R. Jones. 1974. Water quality in the Iowa Great Lakes. Iowa Great Lakes Water Quality Control Plan Report. 573 pp.

Carlander, K.D., J.S. Campbell, and R.J. Muncy. 1978. Inventory of percid and esocid habitat in North America. American Fisheries Society Special Publication 11:27-38.

Casselman, J.M. 1978. Effects of environmental factors on growth, survival, activity, and exploitation of northern pike. American Fisheries Society Special Publication 11:114-128.

Christianson, J. 1978. Fishing and boating survey, East and West Okoboji Lakes. Project No. 77-I-P-7. 1977 Job Completion Reports, 1978 Management Plan, Fisheries Management Section, Iowa Conservation Commission 1978:44-67.

Crossman, E.J. 1977. Displacement, and home range movements of muskellunge determined by ultrasonic tracking. Environmental Biology of Fishes 1:145-158.

Crossman, E.J. 1978. Taxonomy and distribution of North American esocids. American Fisheries Society Special Publication 11:13-26.

Crum, G.H. and R.W. Bachmann. 1973. Submersed aquatic macrophytes of the Iowa Great Lakes Region. Iowa State Journal of Research 48:147-173.

Dombeck, M.P. 1979. Movement and behavior of the muskellunge determined by radio-telemetry. Wisconsin Department of Natural Resources Technical Bulletin Number 113. 19 pp.

Emery, A.R. 1978. The basis of fish community structure: marine and freshwater comparisons. Environmental Biology of Fishes 3:33-47.

Fry, F.E.J. 1971. The effect of environmental factors on the physiology of fish. Pages 1-98 *in* W. S. Hoar and D.J. Randall, editors. Fish physiology, vol VI. Academic Press, New York.

Gammon, J.R. 1986. Sperm production and spawning success for muskellunge and northern pike. American Fisheries Society Special Publication 15:87-92.

Gardner, M.B. 1981. Effects of turbidity on feeding rates and selectivity of bluegills. Transactions of the American Fisheries Society 110:446-450.

Gibbons, J.D. 1976. Nonparametric methods for quantitative analysis. Holt, Rinehart, and Winston, New York. 463 pp.

Gilbertson, B.A. 1986. Winter rescue of juvenile muskellunge from shallow winterkill lakes. American Fisheries Society Special Publication 15:74-78.

Hall, D.J., and E.E. Werner. 1977. Seasonal distribution and abundance of fishes in the littoral zone of a Michigan lake. Transactions of the American Fisheries Society 106:545-555.

Hart, L.G., and R.C. Summerfelt. 1973. Homing behavior of flathead catfish *Pylodictus olivaris* (Rafinesque) tagged with ultrasonic transmitters. Proceedings of the 27th. Annual Conference of the Southeastern Association of Game and Fish Commissioners 27:520-531.

Hart, L.G., and R.C. Summerfelt. 1975. Surgical procedures for implanting ultrasonic transmitters in flathead catfish (*Pylodictus olivaris*). Transactions of the American Fisheries Society 104:56-59.

Keast, A. 1978. Trophic and spatial interrelationships in the fish species of an Ontario temperate lake. Environmental Biology of Fishes 3: 7-31.

Manteifel, B.P., I.I. Girsa, and D.S. Pavlov. 1978. On rhythms of fish behavior, pp. 215-224. in: J.E. Thorpe (ed.), Rhythmic activity of fishes. Academic Press, New York.

Miller, M.L. 1982. Behavior of muskellunge (*Esox masquinongy*) in West Okoboji Lake, Iowa as determined by ultrasonic telemetry. Master of Science thesis, Iowa State University, Ames, 174 pp.

Miller, M.L. and B.W. Menzel. (in press) Movements, homing, and home range of muskellunge (*Esox masquinongy*) in West Okoboji Lake, Iowa. Environmental Biology of Fishes.

Minor, J.D., and E.J. Crossman. 1978. Home range and seasonal movements of muskellunge as determined by radio-telemetry. American Fisheries Society Special Publication 11:146-153.

Moore, J.W., and I.A. Moore. 1976. The basis of food selection in flounders, *Platichthyes flesus* (L.) in the Severn estuary. Journal of Fish Biology 9:139-156.

Oehmcke, A.A., L.D. Johnson, J. Klingbiel, and C. Wistrom. 1958. The Wisconsin muskellunge. Its life history, ecology, and management. Wisconsin Conservation Department Publication 225, 12 pp.

Pitlo, J.M. 1978. Walleye movement and behavior in West Lake Okoboji, Iowa. Master of Science thesis. Iowa State University. Ames, 86 pp.

Reynolds, W.W., and M.E. Casterlin. 1979. Thermoregulatory rhythm in juvenile muskellunge (*Esox masquinongy*): evidence of a diel shift in the lower set-point. Comparative Biochemistry and Physiology 63A:523-525.

Schoener, T. W. 1969. Models of optimal size for solitary predators. American Naturalist. 103:277-313.

Schoener, T.W. 1971. Theory of feeding strategies. Annual Revue of Ecology and Systematics 2:369-404.

Scott, W.B., and E.J. Crossman. 1973. Freshwater fishes of Canada. Bulletin 184. Fisheries Research Board of Canada, Ottawa. 966 pp.

Strand, R. 1986. Identification of principal spawning areas and seasonal distribution and movements of muskellunge in Leech Lake, Minnesota. American Fisheries Society Special Publication 15:62-73.

Identification Of Principal Spawning Areas And Seasonal Distribution And Movements Of Muskellunge In Leech Lake Minnesota[1]

ROBERT F. STRAND

Minnesota Department of Natural Resources
2114 Bemidji Avenue
Bemidji, MN 56601

ABSTRACT

Six specific muskellunge spawning locations were documented and preferred spawning habitat was determined by surgically implanting radio transmitters in 14 adult muskellunge captured by angling in summer 1979. Adult fish from each of three areas on the lake were radio-tagged. Spawning sites were distinctly offshore in depths of 1 to 2 m, over a soft calcareous substrate with *Chara* spp. as the dominant vegetation. Physical characteristics of the six examined areas were very similar even though widely separated geographically. Selection for these sites occurred even though a wide variety of habitat types was available. Mean tracking time for 12 fish was 396 days which included the 1980 spawning period. Significant differences relating to home range and movement were observed for muskellunge inhabiting disparate environments within the lake. Muskellunge residing in the main basin occupied total home ranges which were five times larger than those of fish inhabiting Walker Bay. The main lake fish also tended to have winter home ranges distinctly separate from summer ranges, and winter home ranges were larger than summer ones by a factor of nearly two. Walker Bay fish had winter home ranges which were smaller than those of summer by a factor of over six and were contained within the summer ranges. Survival of all angler-caught fish, additionally subjected to implant surgery, strongly suggests that catch and release of muskellunge is a realistic management option.

Leech Lake supports the most important muskellunge fishery in Minnesota. The sport catch from this naturally-reproducing population approximates one-half the annual catch in the state. Prior efforts by Minnesota Department of Natural Resources, Section of Fisheries (MDNR) over a period of 15 years were unsuccessful in attempts to determine spawning areas and to obtain muskellunge eggs for propagation.

A cooperative muskellunge tagging program was undertaken in 1976 by MDNR and Muskies, Inc. to provide information on distribution, movement and survival of captured and released muskies. From the data on catch and recapture, it was apparent that muskellunge were most abundant in three general areas of the lake. There was also some evidence of different seasonal distribution and movement but there was not adequate data to define specific patterns (Scidmore 1978). The percentage of recaptured tagged fish was low (11.7% in three years) compared with that for most other species. This could be due to several causes, one of which was the possibility that there was substantial mortality among caught, tagged and released fish. With the available data, these factors could not be accurately assessed (Scidmore 1978).

The objective of this study was to determine spawning habitat type and specific spawning areas used by muskellunge. This information would enable MDNR to respond properly to permit applications for alteration of the shoreline or lake bed of Leech Lake. Identification of the spawning areas would allow adequate protection to be given them, and data on seasonal distribution and movements of muskellunge within the lake would enhance management for this important species.

STUDY SITES

Leech Lake, in north-central Minnesota, is an irregularly shaped lake of 45,134 ha that was formed by a moraine dam (Fig. 1). Since 1884, the lake has been part of a system of reservoirs retaining water for navigation on the upper Mississippi River. Four major tributaries and many smaller streams drain the 301,000 ha watershed. The 306 km of shoreline are about equally divided between high or intermediate shore bordered by upland cover and low shore bordered by muskeg and sedges interspersed with stands of tamarack or aspen. Trophic conditions within the lake range from eutrophic to

---

[1]This study was supported by Dingell-Johnson project F26-R, Project Number 123, Minnesota.

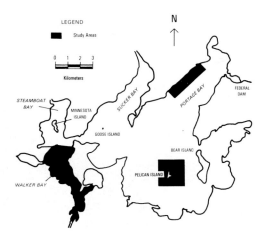

Figure 1. Leech Lake and the three study areas within the lake.

morphometrically-oligotrophic but the main basin is mesotrophic. For a more detailed lake description see Schupp (1978).

The fish community is dominated by percids and esocids. Walleye, northern pike and muskellunge are the principal piscivores and yellow perch are abundant throughout the lake. White sucker, cisco, lake whitefish and rock bass are common in the mesotrophic areas. The weedy, eutrophic bays also support populations of brown and black bullhead, black crappie, pumpkinseed and bluegill. Study fish were captured from three areas on the lake where muskellunge are most abundant. These were Walker Bay, and two areas in the main basin, Portage Bay and Pelican Island (Fig. 1).

The morphometry of Walker Bay is similar to that of many oligotrophic lakes on the Precambrian Shield but the chemical fertility of the water is high, and thus the basin is classified as mesotrophic. A thermocline forms at about 10 m during mid-summer and an oxygenated hypolimnion is present. Maximum depth is 54 m. In most years, the bay is ice-covered from early December to early May. Productive muskellunge angling areas are scattered throughout Walker Bay and consist mainly of steep shoreline drop-offs associated with weed beds (*Potamogeton* spp.), and to a lesser extent, rock.

The main basin areas differ in the relative amounts of substrate materials and the regularity of bottom contours. Neither of the main basin areas thermally stratify and both are ice-covered from late November to late April each year. The area about Pelican Island is characterized by rocky reefs to depths of 3-4 m radiating in all directions from the island. Scattered *Potamogeton* beds are present in the area and maximum depth is 8.2 m. Portage Bay has very gradual contours and a maximum depth of 9.2 m. Muskellunge habitat consists entirely of *Potamogeton* beds in depths of 2-4 m extending for approximately 4.8 km along the western shore and to a distance of up to 0.8 km into the bay.

METHODS

All study fish were captured by angling with artificial lures and were landed with a cradle which prevented violent thrashing (Fig. 2). Struggle time from hooking to landing varied from 3 to 4 minutes per fish. Hooks were removed either before or after anesthetization, depending on circumstances. Lithium battery-powered radio transmitters in the 53 MHz band, with trailing external whip antennae, were implanted during the period 29 June to 29 August 1979. Each transmitter frequency was separated by at least 10 kHz to permit identification of individual fish. The transmitters weighed 28 g in water. Study fish ranged from 85 to 122 cm total length. Mean surface temperature at the time of surgery was 21.4 C and ranged from 18.9 to 23.9 C (Table 1). Radio tags were implanted shortly after capture and all study fish were immediately returned to the water at the capture site.

Table 1. Data on 14 radio-tagged muskellunge in Leech Lake.

Fish Freq. Number	Capture Date (1979)	Total length (cm)	Minutes anesthesia	Surface temp. (C)	Capture location
503	29 June	85	15	19.4	Portage Bay
461	2 July	91	20	20.6	Portage Bay
521	12 July	113	23	20.6	Portage Bay
401	16 July	89	19	22.8	Portage Bay
481	17 July	110	18	23.3	Portage Bay
379	24 July	91	20	22.8	Pelican Island
302[a]	24 July	89	18	22.8	Pelican Island
580	16 Aug.	104	21	18.9	Pelican Island
421	16 Aug.	91	18	18.9	Pelican Island
440	16 Aug.	102	19	18.9	Pelican Island
319	2 Aug.	86	16	23.9	Walker Bay
361	2 Aug.	122	18	23.9	Walker Bay
538	7 Aug.	94	18	22.8	Walker Bay
302[a]	29 Aug.	101	18	20.0	Walker Bay

[a]Transmitter recovered from angler-killed fish and reused.

Figure 2. Cradle used to land and handle muskellunge. Constructed of knotless nylon webbing, mesh size 6.3 mm, length = 127 cm, width = 46 cm and depth = 41 cm.

After 5-6 minutes in lake water containing 80 mg/l MS-222, fish were adequately sedated for surgery. They were then placed in a surgical tube mounted on an incline on top of a live tank containing half strength anesthetic solution (Fig. 3). The inclined mount permitted the head and gill region to remain submersed while the abdominal implant area was dry and exposed when the fish was positioned on its back. Full strength anesthetic solution or fresh lake water was applied directly to gills by spray bottle as needed. The time of anesthetization ranged from 15 to 23 minutes per fish and averaged 18.5 minutes.

Insertion of antenna and placement of transmitter was by the shielded-needle technique described by Ross and Kleiner (1982). The transmitter was positioned above the pelvic girdle and did not rest on the incision. The incision was closed with nonabsorbable surgical suture (3/0) with one suture placed around the antenna at the point of exit from the body for additional support. After closure, the fish was placed in an upright position and a numbered plastic dart tag was attached at the base of the dorsal fin so that the anchor penetrated the interneurals. This provided an external identification mark visible from above. Study fish were held in a recovery tank containing antiseptic and supplied with oxygen for 12 minutes.

The three major limiting factors to locating radio-tagged fish were: deep water; extensive movement; and ice conditions in mid to late winter. The use of deep water by radio-tagged muskellunge occurred frequently in Walker Bay but was not a factor in the main basin. Equipment was not capable of determining depth of study fish in the water column but for fish frequenting areas deeper than 10 m there was often a greatly reduced range of detection accompanied by a loss of directional capability. This was a greater problem during the capture-implant period of 1979 when tracking was done with a hand-held directional loop antenna which had limited range. This was also true for marked fish making

Figure 3. Boat mounted two compartment live tank complete with oxygen supply and surgical tube mounted on an incline. Tank dimensions: length = 122 cm, width = 61 cm and depth = 61 cm. Surgical tube (plastic) dimensions: length = 132 cm, inside diameter = 20 cm.

extensive moves. Those problem were partially overcome by using a yagi mast antenna after capture was completed.

A yagi mast antenna, hand-held directional loop, whip antennae of several lengths and twin directional loop antennae were used during tracking operations from boat, snowmobile and airplane. When in close proximity to tagged fish, whip antennae of decreasing length were used to verify the actual location of study fish. Triangulation with visible landmarks was used to plot locations on a lake contour map. A depth sounder was used to aid in plotting locations as well as to provide details at each fish location. Depth, presence of weed growth, bottom structure and changes in contour were recorded each time a fish was located. Dissolved oxygen and temperature were recorded at tagged fish locations during mid-summer.

Home ranges were determined by the convex polygon method described by Winter and Ross (1981). All known locations for each fish were plotted on a map and the appropriate minimum perimeter polygons, excluding land mass areas, were drawn. Areas within these polygons were determined using a digitizing pad interfaced with a microcomputer. This procedure was repeated for each fish for total, winter and summer home ranges. Home range differences between basins and differences between winter and summer home ranges were compared by t-test. Distance traveled by study fish was determined by direct measurement between plotted locations of each fish on a lake map. The actual course traveled by the fish was unknown.

RESULTS

Study fish regained equilibrium and exhibited swimming behavior within 3 to 5 minutes in the recovery tank. Response upon release was of two types. Six of 14 sounded immediately while the remaining 8 fish circled slowly on the surface, orienting themselves into the wind for 0.5 to 2.5 minutes before sounding.

Behavior during the first 24 hours after release varied dramatically. All fish receiving implants were captured from water depths of 2 to 4 m and were associated with submerged weed beds or rock reefs. Of 5 fish released in Portage Bay, 3 remained in the immediate vicinity, one moved 1.2 km to deeper water and one moved 2.4 km to shallower water. Similarly, of 5 fish released in the Pelican Island area, 3 remained in the immediate vicinity, one moved 3.2 km during 24 hours and one was not located on the day following release but was located on the second day at a distance of 4.8 km from release site. Presumably, the long distance movement occurred shortly after release and is the reason the fish was not located. In Walker Bay, 3 of 4 fish moved to deep water during the first 24 hours and in at least one instance this occurred within 7.5 hours of release. One fish was not located until 4 days after release and when found was also in deep water. Presumably, the move to deeper water occurred shortly after release.

All 14 muskellunge survived capture by angling and the surgery required for tag implants. A single study fish (302) was lost to sportfishing 20 days after release but the transmitter was recovered and reused. Movement ceased on one fish (503) approximately 172 days after release and the fish was assumed to be dead. Recovery was not attempted because of ice cover and a water depth of nearly 7 m.

Mean tracking time for the 12 remaining fish was 396 days with a range of 291 to 469 days (Table 2). A total of 405 radio locations was determined. Most locations were determined by boat (293) during the open water period of late April to November. Eleven aircraft searches resulted in 102 locations. No signals were heard on the 12th and final flight on 11 March 1981 and it was assumed that all transmitters were dead. Only 10 radio locations were determined by snowmobile search although greater effort was expended than suggested by this result. The maximum range of signal detection during ice-free periods was approximately 0.4 km with a hand-held loop antenna, 1.6 km with a yagi mast and 4.8 km when using aircraft.

Spawning

Spawning sites were determined for 10 of 12 study fish in spring 1980. They are represented by six separate areas numbered one through six (Fig. 4). The locations were distinctly offshore in depths of 1 to 2 m over a soft calcareous substrate where *Chara* spp. was the dominant vegetation. Physical characteristics of the six areas were similar, even though widely separated geographically. Selection for these sites occurred even though a wide variety of habitat types were available and thus they were presumed to be representative of preferred spawning habitat. The identified spawning sites were not frequented by the tagged muskellunge during the rest of the year.

Table 2. Tracking information for tagged muskellunge in Leech Lake.

Fish frequency number	Period tracked	Number of locations	Total days
503	29 June 79 to 18 Dec 79	28	172
461	2 July 79 to 2 June 80	39	335
521	12 July 79 to 24 Sept 80	36	440
401	16 July 79 to 2 Sept 80	30	414
481	17 July 79 to 23 Sept 80	36	434
379	24 July 79 to 15 Aug 80	32	388
302a	24 July 79 to 13 Aug 79	6	20
580	16 Aug 79 to 24 Sept 80	27	405
421	16 Aug 79 to 3 June 80	26	291
440	16 Aug 79 to 24 Sept 80	28	405
319	2 Aug 79 to 14 Nov 80	32	469
361	2 Aug 79 to 30 Sept 80	42	425
538	7 Aug 79 to 30 Sept 80	36	420
302a	29 Aug 79 to 25 Sept 80	41	330

[a]Transmitter recovered from angler-killed fish and reused.

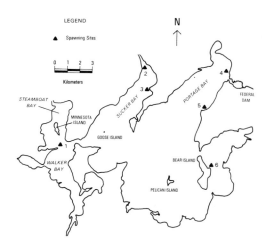

Figure 4. Leech Lake with six known muskellunge spawning areas. Indicated spawning areas are not drawn to scale.

Two of the spawning sites identified in 1980 (Areas 1 and 3) were subsequently described by Dombeck et al. (1984). They reported that the sites were characterized by nearly pure stands of *Chara* spp. interspersed with bare patches resulting in mean vegetative cover of 67%. The plants attained a maximum and median height of 25 and 15 cm, respectively. The substrate was flocculent marl of low organic content ($\bar{x} = 15\%$) and composed of 67% carbonates. Dissolved oxygen (DO) concentration was monitored both within the vegetative growth and on the bare patches at two levels in the water column above the substrate-water interface at each of 12 sample sites. No DO microstratification and depletion at the substrate-water interface occurred in the spawning areas during the time of egg incubation. Temperature was quite uniform throughout the water column and was comparable to temperatures in adjacent areas of similar depth due to mixing by wind-induced currents.

Two of four fish from Portage Bay spawned in Sucker Bay while two used areas within Portage Bay (Table 3). Spawning locations were determined for only two of four Pelican Island fish, one of which spawned in Sucker Bay, the other on Area 6 to the east of Bear Island. All four Walker Bay fish used Area 1 located on the south end of Steamboat Bay.

Two individuals (521 and 580) frequented more than one spawning site. In both cases, the fish utilized Areas 2 and 3 in Sucker Bay which are approximately 3 km apart. Fish 521, known to be a female from later recapture, moved onto Area 2 on 1 May and remained there for 5 days before moving to Area 3 where she remained for 4 days. Fish 580, thought to be a male based on length-age at capture, moved onto Area 2 on 26 April. The following day it had moved to Area 3 where it spent 8 days before returning to Area 2 which it frequented for 11 days.

A minimal estimate of size for Area 1 in Steamboat Bay, the largest of the 6 identified spawning areas, approximated 135 ha based on observation and net catches of spawning fish. The others, although smaller, represent extensive areas as well, resulting in relatively low densities of muskellunge, even on the spawning grounds.

Movement to the spawning areas began shortly after ice-out when mid-lake surface temperatures reached 8.3 C. These moves were represented by distances of 5 to 21 km for 8 of 10 fish and by shorter moves for two fish which were frequenting areas close to spawning sites during the late winter and early spring period. Movements were direct and frequently included crossing of large expanses of open water rather than irregular courses of travel along depth contours or shoreline. Radio signals during this period were very strong, suggesting the traveling fish were high in the water column.

Study fish 521 and 461 exhibited typical movements to spawning sites (Fig. 5). Fish 521 was located on a rock reef northwest of Pelican Island on 9 April. Next contact was on 26 April when the fish was found to the east of Goose Island, a distance of 9.6 km. Radio signals on 26 April indicated the fish was moving in a westerly direction and was remaining under the receding ice pack. The 9.6 km move probably occurred over a short period of time, as evidenced by other tagged fish movements. On 28 April, the fish was 1.1 km further west and still under the receding ice pack. The fish then moved 11.4 km to the north end of Sucker Bay where it was located on Spawning Area 2 on 1 May. Fish 461 was located in Uram Bay on the south end of the lake from 25 to 27 April; it then moved 3.5 km north where it was located on 1 May and remained until at least the afternoon of 6 May. On 8 May, the fish could not be located but was found on 9 May on Spawning Area 3 in Sucker Bay. This represents a minimum travel distance of nearly 18 km in less than 72 hours.

Emphasis was placed on monitoring Spawning Areas 1, 2 and 3 because of numbers of muskellunge using those sites and accessibility. Muskellunge using Spawning Areas 2 through 6 moved onto or

Table 3. **Spawning areas used by 10 muskellunge relative to tagging origin in Leech Lake.**

Fish frequency and tagging origin	Spawning areas used					
	Steamboat Bay 1	Sucker Bay 2	Sucker Bay 3	Portage Bay 4	Portage Bay 5	Bear Island 6
Portage Bay						
461			X			
521		X	X			
401				X		
481					X	
Pelican Island						
379						X
580		X	X			
Walker Bay						
319	X					
361	X					
538	X					
302[a]	X					

[a] Second use of this transmitter.

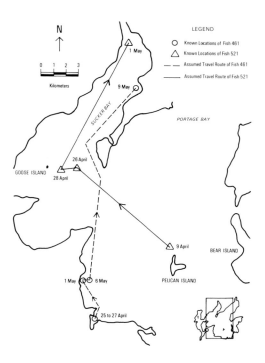

Figure 5. Leech Lake with known locations and assumed travel routes relative to spawning of fish 521 and 461.

immediately adjacent to spawning sites within 7 to 14 days after complete ice-out at surface temperatures of 8.9 to 10.5C. Three of 4 Walker Bay fish moved into or adjacent to area 1 within 3-4 days of complete ice-out on Walker Bay, and one moved on the same day as complete ice-out at surface temperatures of 10.5 to 11.6 C. Spawning Area 1 in Steamboat Bay actually becomes ice-free from 5 to 10 days earlier than Walker Bay.

During the initial period on the spawning sites, the fish were extremely wary. They could not be approached by boat and observed even though water clarity would allow it from a considerable distance. A marked change in fish behavior was noted several days later when water temperatures increased to 11.6 C or higher and remained that high for up to 24 hours. Muskellunge became unafraid and could be approached and observed, both during daylight hours and at night. Non-tagged muskellunge as well as tagged fish were observed.

Large individuals, presumed to be females, were frequently observed lying at or near the surface during periods of calm sunny weather. On one sample night, a single muskellunge approximately 125 cm in length was observed by spotlight just under the surface in an area 1.5 m in depth. The fish was attracted by splashing an oar in a manner similar to that described by Dombeck (1979). The fish was brought up to the boat three different times by this process. On another occasion during calm sunny conditions, while following fish 580 in Area 2 in Sucker Bay, another muskellunge (non-tagged) of large size was observed lying just under the surface. This fish was approached quite closely by paddling the boat; it then swam slowly off and positioned itself just under the surface 30 m away. The fish was approached six times before it left the area. Later when some tagged fish were leaving the spawning sites, muskellunge again became difficult to approach; if one was observed, it would swim off rapidly and could not be approached again.

The actual spawning act was not observed on these sites, and attempts to sample eggs by siphon pump for verification of use were unsuccessful. Negative egg-sampling results were attributed to the large size of these areas, which precluded intensive sampling, and to relatively low densities of spawning muskellunge.

The areas described as spawning sites were based on intensive monitoring of tagged fish throughout the time period when spawning would occur. Study fish traveled directly to these areas and remained there, except when severe cold weather might cause them to temporarily move off the sites to deeper water. They returned to the same areas with the recurrence of warm weather. It is also based on the observation of muskellunge other than study fish on these sites and the behavioral changes previously described. Ripe muskellunge have been successfully captured in nets on these locations for three consecutive years (1981-83) following the telemetry study. From the combined data, muskellunge spawning in Leech Lake occurs from 15 to 35 days after complete ice-out at water temperatures of 11.6 to 14.0 C, although short term fluctuations in temperature outside of that range frequently occur.

Study fish were not sexed at the time of tag implantation during mid-summer. Later recaptures of six of these fish by anglers allowed examination and sex determination. While this could not be done for all study fish, it did permit limited observations on spawning differences between sexes. Males in these observations moved onto the spawning grounds first, penetrated farther into the areas and remained on the sites more consistently and later than females.

Females tended to frequent the periphery of spawning sites to a greater extent than males. Both sexes were sensitive to drastic weather changes and females were more sensitive than males. During the spawning period of 1980, a severe cold front caused a 11 C drop in surface temperature and caused females to move temporarily off the spawning areas to deeper water for distances of up to 2 km.

The internal radio transmitters apparently did not hamper the reproductive process. Eight fish (seven post-telemetry study) were captured by anglers and one was taken in a gillnet in the period of 20 days to just over 3 years after release. Of eight angler captures, six were killed and two were measured and released again. Study fish apparently resumed feeding shortly after release as evidenced by the capture of one individual by angling only 20 days after release. Examination of three angler-caught fish which carried transmitters for 23, 25, and 37 months showed that they were in good condition and of normal length-weight relationships. Internal examination revealed that in two specimens the transmitter remained where placed with very little scar tissue or adhesion evident. In one, the transmitter had moved anterior to the pelvic girdle with a moderate amount of tissue adhesion. Incisions were so well healed they were difficult to detect. In all specimens a slight cratering was evident where the external antenna exited the body.

Home Range and Movement

Return movements from spawning grounds to summer ranges were also characterized by directed movements, frequently crossing large open stretches of lake but at a slower rate of travel. Five of 6 fish from the main basin had returned within the bounds of their summer home range within 17 to 33 days ($\bar{x} = 23.8$) of vacating the spawning sites. Contact with fish 580 was lost between 21 May, when it moved off the spawning area to deeper water, and 2 July when it was again located within its summer home range near Pelican Island. All 4 fish from Walker Bay returned to summer ranges within 7 to 9 days. This is similar to what occurred in the main basin, considering distances traveled.

Home-range area calculations included movements associated with spawning, which were frequently extensive and substantially increased the size of total home range. The areas varied from 9.5 to over 140 km^2 (Table 4). With spawning movements excluded from total home range, the values ranged from 4.9 to over 61 km^2 with mean values of

Table 4. Home range (km^2) of 12 radio-tagged muskellunge in Leech Lake.

Fish frequency and location	Home range			
	Total with spawning	Total without spawning	Summer	Winter
Main basin				
Portage Bay				
461	121.7	58.3	2.0	5.8
521	140.3	61.6	2.6	5.0
401	46.5	41.9	34.0[a]	NA[b]
481	49.9	23.9	6.4	NA
Pelican Island				
379	41.7	24.0	3.5	7.2
580	71.1	28.7	5.2	NA
421	31.3	16.6	NA	2.1
440	50.6	40.2	5.4	18.4
	$\bar{x}$ 69.1	$\bar{x}$ 36.9	$\bar{x}$ 4.3	$\bar{x}$ 7.7
Walker Bay				
319	19.4	8.9	5.8	0.5
361	9.5	7.0	6.5	2.0
538	11.8	9.6	9.4	0.8
302	14.1	4.9	4.6	0.8
	$\bar{x}$ 13.7	$\bar{x}$ 7.6	$\bar{x}$ 6.6	$\bar{x}$ 1.0

[a]Excluded from mean.
[b]Insufficient radio locations for determination.

36.9 km^2 for fish residing in the main basin and 7.6 km^2 for Walker Bay fish. The size of the summer home range was determined for 11 fish with values ranging from 2.6 to 33.9 km^2 with a mean of 4.3 km^2 for fish in the main basin and 6.6 km^2 in Walker Bay. Fish 401 was excluded from the mean calculation because of too few radio locations.

Winter ranges of 9 individuals varied from 0.5 to 18.4 km^2 with a mean of 7.7 km^2 in the main basin and 1.0 km^2 in Walker Bay. Winter-summer home range estimations were not calculated for individuals with less then four locations during the appropriate time period or when plotted locations were linear indicating travel rather than habitation. Winter home ranges of fish 319 and 461 were based on four radio-locations each. All but two of the values expressed in Table 4 are based on data exceeding the minimum standard.

Marked differences in home range and movement were observed for muskellunge inhabiting disparate environments within the lake. Those residing in the main basin occupied total home ranges (exclusive of spawning) which were nearly 5 times larger than those of fish inhabiting Walker Bay; mean values of 36.9 and 7.6 km^2, respectively, were significantly different by t-test (t = 4.928, df = 7, p<0.0001). Muskellunge in the main lake basin tended to have winter home ranges distinctly separate from summer home ranges; winter ranges were larger by a factor of nearly two (Fig. 6), but the results were not statistically significant (t = -1.167, df = 4, P = 0.308). All fish originating in Portage Bay, for which winter ranges were determined, wintered to the south of Portage Bay at distances of 11 to 18 km from the approximate center of summer range to winter range. The opposite was noted for Pelican Island fish which wintered to the north, and one individual to the east, of their summer range. Individuals of both groups for which summer or winter ranges were not determined exhibited the same directional patterns.

Walker Bay fish had winter home ranges which were smaller than summer by a factor of over six (P = 0.014), with winter ranges essentially contained within the summer ranges (Fig. 7). Summer home ranges of Walker Bay fish ($\bar{x} = 6.6$ km^2) were larger than those of fish from the main basin ($\bar{x} = 4.2$ km^2), but the results were not statistically significant (P = 0.121).

The summer home range of fish 521 differed from other tagged fish in that it utilized three widely sepa-

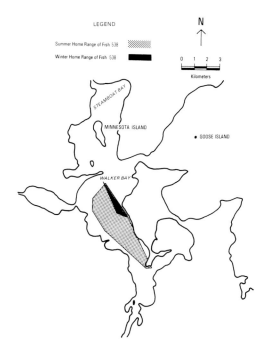

Figure 7. Leech Lake with representative summer-winter home range of fish 538 in Walker Bay.

rated areas (Fig. 8). A minimum perimeter polygon incorporating these areas resulted in a summer home range estimation of 26 km^2. The fish apparently traveled directly to the indicated areas and did not inhabit the region between; therefore, each area was calculated separately and summed for the summer home range value (Table 4). Use of the three areas was not characterized by seasonal progression, but the fish used the same three areas both years of investigation (1979-1980).

Winter movements appear greater than summer movements in the main basin (Table 5), but the results were not statistically significant (t = 1.024, df = 12, P = 0.326). Winter contacts were fewer and with greater time lapses ($\bar{x} = 24.9$ days) between contacts than in summer ($\bar{x} = 7.7$ days). This would tend to increase values for winter movement although all individuals should be influenced equally by this factor. Winter movements in the main basin were greater than summer with the two exceptions previously noted. Fish 521 traveled frequently between three widely separated summer areas and fish 401 began traveling extensively by August. This resulted in summer movement greater than winter

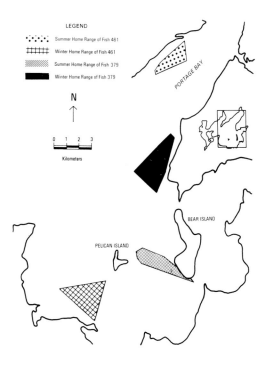

Figure 6. Leech Lake with representative summer-winter home ranges of fish 461 and 379 in the main basin.

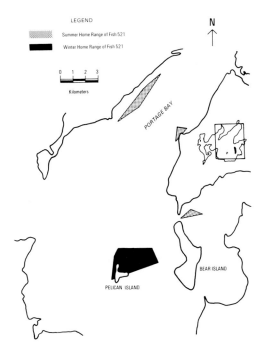

Figure 8. Leech Lake with three separate summer home ranges of fish 521 and the winter home range of fish 521.

Table 5. Mean distances (km) between telemetry-determined locations of 12 muskellunge in Leech Lake.

Fish frequency and location	Distance traveled			
	Total with spawning	Total without spawning	Summer	Winter
Main basin				
Portage Bay				
461	3.78	2.36	2.30	2.64
521	4.47	4.03	4.60	1.75
401	2.95	2.69	2.76	2.09
481	2.49	2.01	1.53	4.60
Pelican Island				
379	2.43	2.02	1.28	3.74
580	4.28	2.74	1.98	6.16
421	3.07	1.99	NA	1.46
440	4.29	4.55	4.16	5.25
	$\bar{x}$ 3.47	$\bar{x}$ 2.80	$\bar{x}$ 2.66	$\bar{x}$ 3.46
Walker Bay				
319	1.83	1.92	2.22	0.93
361	1.47	1.56	1.45	1.97
538	1.79	1.80	1.76	1.89
302	2.34	2.12	1.89	2.57
	$\bar{x}$ 1.86	$\bar{x}$ 1.85	$\bar{x}$ 1.83	$\bar{x}$ 1.84

movement for these two individuals. No difference was apparent between summer and winter distances traveled for Walker Bay fish even though this group exhibited winter home ranges significantly smaller than summer.

Dissolved oxygen and temperature were not decisive factors in muskellunge location in Leech Lake during mid-summer. Fish location was strongly associated with physical structure and was a factor in 146 of 185 observations (79%). The relative amounts of structure available in Leech Lake were not quantified but from extensive work on the lake it is apparent that they represent a small portion of the available habitat. Association with vegetation ranked highest and occurred 86 times, 78 of which were submergent and 8 emergent vegetation. This was followed by changes in bottom contour (38) and rock reefs (22). No discernable physical association was apparent in 39 of 185 observations (21%).

DISCUSSION

The physical characteristics of muskellunge spawning areas in Leech Lake differ markedly from the normal muskellunge spawning habitat described by Scott and Crossman (1973) and Eddy and Underhill (1974). They more closely resemble spawning habitat described by Haas (1978) for muskellunge in Lake St. Clair, Michigan. Haas (1978) reported that the spawning area in Lake St. Clair was in open water with a minimum depth of 3 m and little or no vegetation. The Leech Lake spawning areas are similar to Lake St. Clair with respect to open water location but Leech Lake spawning areas are shallower (1-2 m). Also, the Leech Lake spawning areas are over dense beds of *Chara* spp. with a flocculent marl substrate. This type habitat does not become anoxic at the water-mud interface.

Scott and Crossman (1973) stated that where muskellunge and earlier hatching northern pike utilize the same spawning area, muskellunge fry are utilized as prey by northern pike. They felt that this predation was responsible for lowered muskellunge populations. Selection for the off-shore areas by spawning muskellunge in Leech Lake results in spatial separation, in addition to temporal differences, from spawning areas used by northern pike. Since the spawning areas of the two species are distinctly separated, it is reasonable to assume that their progeny also occupy separate nursery areas for some period of time.

Effective range of radio detection decreased significantly during mid to late winter due to the non-uniformity of the ice pack. In most years on Leech

Lake the ice reaches a thickness of about 76 cm, and this ice sheet is composed of various layers resulting from pressure flooding and freezing and the accumulation of slush ice. Each layer, differing in physical properties, potentially represents an additional surface to attenuate the signal. Winter tracking was also hampered by study fish avoiding the disturbance created by the snowmobile. This often required an approach on foot, after a signal was detected, which decreased the avoidance reaction of the fish. Ground search by snowmobile was not an efficient means of locating tagged fish on this large lake, and aircraft proved to be the most reliable method of detection during the period of ice cover.

Knowledge of muskellunge home ranges is limited. Minor and Crossman (1978) found that home range size of muskellunge of both sexes increased linearly with fish size for all fish during summer months. No relationship was evident between fish size and home range size in the main basin of Leech Lake. For Walker Bay, no relationship was evident between fish size and total home range or summer home range size, but winter home range size did increase with fish size ($r = 0.909$, $df = 2$, $p = 0.09$).

Home range values determined for Leech Lake muskellunge are considerably greater than values reported by Minor and Crossman (1978), Dombeck (1979) and Miller and Menzel (1982). These investigations involved bodies of water from 35 to 3,725 ha as compared to 45,134 ha Leech Lake. Haas (1978) indicated that muskellunge have an extensive seasonal movement pattern (40 km) on Lake St. Clair (surface area 111,000 ha). These data suggest that home range size may be a function of lake size. Prey abundance and movement may be a factor influencing muskellunge movement in Leech Lake but those considerations were beyond the scope of this study.

Investigations of fish movement and behavior based on telemetry often describe extensive fish movement, or movement outside the norm or activity center, as exploratory or transient movement. The Leech Lake data, with the exception of the 1980 spawning period, represents gross movement data. The size of the lake and number of tagged fish being monitored resulted in contacts with individual fish as infrequently as once every 7 to 14 days during the open water period. The extensive time period over which the fish were monitored revealed that exploratory movements were repeated by individual fish. For this reason, the movements were included in total home range even though the amount of time spent in the areas was of shorter duration. More frequent contact with tagged fish would probably have revealed more of this type of behavior. Observed differences among individual fish locations (habitat) for fish frequenting the same general locale during the same time period, in addition to the variations in movement, suggests some degree of individuality for the tagged muskellunge.

The 100% survival of tagged fish is attributed to careful handling methods, cool water temperatures during implantation and favorable transmitter-to-fish ratios. The minimum acceptable size for study fish was set at 85 cm to assure a transmitter weight in water less than 1% of the fish's weight out of water, which was slightly below the ratio of 1 to 1.25% recommended by Winter et al. (1978). This weight relationship allows a fish to regain equilibrium within minutes after tagging and to keep total acclimation time to a minimum.

The high survival is comparable to the results of Minor and Crossman (1978), who also used internal implants but their study fish were captured by trap nets. Dombeck (1979) reported no mortality for 18 fyke net-captured muskellunge receiving external radio transmitters over a 14 month period. Miller and Menzel (1982) also reported no mortality for nine muskellunge which carried ultrasonic transmitters.

The Leech Lake results are not in agreement with the 30% angling mortality (8 of 25) reported by Beggs et al. (1980) in their evaluation of physiological consequences of angling stress in muskellunge. The size of study fish used by Beggs et al. (1980) was smaller than those used at Leech Lake. The basic difference between the two studies were that Leech Lake fish received surgically-implanted transmitters and were released immediately at the capture site, while in Beggs et al. (1980) their fish were transported to the laboratory, cannulated in the ventral aorta and retained in plexiglass respirometers supplied with creek water for observation and recovery periods of up to 84 hours. A comparison of methods and results of the two studies suggests that cannulation and retention, employed by Beggs et al. (1980), may have induced stress in addition to capture by angling which contributed to the 30% mortality.

The high survival of Leech Lake study fish, captured by angling and additionally subjected to surgery for radio implantation, strongly suggests that catch and release of muskellunge is a realistic option.

ACKNOWLEDGEMENTS

I thank J.D. Connell for his able assistance with field work for the duration of the study as well as the capture of six muskellunge used in the study. I am very grateful for the voluntary assistance of Dr. C. Cotton and his wife, Betty, for the capture of study fish as well as sharing their muskellunge expertise on Leech Lake. They contributed immeasurably to the success of the investigation. B. Collins and R. Rosenburg each provided a muskellunge for our use and R. Halvorson contributed generously of his time and knowledge. The New Leech Lake Campground provided boat access and harbor space. My special thanks to the many anglers who expressed a desire to cooperate in the capture of fish and expended considerable effort in doing so. M.J. Ross provided technical guidance on implant techniques and on site training in tracking procedures. Cedar Creek Bio-Electronics Lab, Bethel, Minnesota built the transmitters, receivers and attennae and they also field tested the equipment. D.H. Schupp, P.J. Wingate and J. Hoenig critically reviewed the manuscript.

REFERENCES

Beggs, G.L., G.F. Holeton, and E.J. Crossman. 1980. Some physiological consequences of angling stress in muskellunge, *Esox masquinongy*, Mitchell. Journal of Fish Biology 17:649-659.

Dombeck, M.P. 1979. Movement and behavior of the muskellunge determined by radio-telemetry. Wisconsin Department of Natural Resources, Technical Bulletin No. 113. Madison, Wisconsin. 19 pp.

Dombeck, M.P., B.W. Menzel, and P.N. Hinz. 1984. Muskellunge spawning habitat and reproductive success. Transactions of The American Fisheries Society. 113:205-216.

Eddy, S., and J.C. Underhill. 1974. Northern Fishes. 3rd edition. University of Minnesota Press. Minneapolis, Minnesota. 414 pp.

Haas, R.C. 1978. The muskellunge in Lake St. Clair. American Fisheries Society Special Publication Number 11:334-339.

Miller, M.L., and B.W. Menzel. 1982. Behavior of muskellunge in West Okoboji Lake. Iowa Cooperative Fishery Research Unit. Annual Report, Vol. 47. Ames, Iowa.

Minor, J.D., and E.J. Crossman. 1978. Home range and seasonal movements of muskellunge as determined by radiotelemetry. American Fisheries Society Special Publication Number 11:146-153.

Ross, M.J., and C.F. Kleiner. 1982. Shielded-needle technique for surgically implanting radio-frequency transmitters in fish. Progressive Fish-Culturist 44:41-43.

Schupp, D.H. 1978. Walleye abundance, growth, movement, and yield in disparate environments within a Minnesota lake. American Fisheries Society Special Publication Number 11:58-65.

Scidmore, W.J. 1978. Cooperative musky tagging project, Minnesota Department of Natural Resources-Muskies, Incorporated. Minnesota Department of Natural Resources, Section of Fisheries. Staff Report. St. Paul, Minnesota. 7 pp.

Scott, W.B., and E.J. Crossman. 1973. Freshwater Fishes of Canada. Bulletin 184. Fisheries Research Board of Canada, Ottawa. 966 pp.

Winter, J.D., V.B. Kuechle, D.B. Siniff, and J.R. Tester. 1978. Radio tracking freshwater fish. University of Minnesota, Agricultural Experiment Station. Miscellaneous Report 152. Minneapolis, Minnesota. 18 pp.

Winter, J.D., and M.J. Ross. 1982. Methods in analyzing fish habitat utilization from telemetry data. Pages 273-279 *in* N. Armantrout, ed. Proceedings of the symposium of the acquisition and utilization of aquatic habitat inventory information. American Fisheries Society Western Division. 376 pp.

Winter Rescue of Juvenile Muskellunge from Shallow Winterkill Lakes

BRUCE GILBERTSON

Minnesota Department of Natural Resources
Section of Fisheries
1200 Warner Road, St. Paul, MN 55106

ABSTRACT

Fall harvesting with fyke nets and winter rescue of juvenile muskellunge from shallow winterkill lakes were compared to determine relative efficiency of each harvest method. Fall fyke-net harvesting was more efficient in terms of muskellunge harvested per hour of labor expended, but winter rescue was shown to be a viable alternative. A modification of winter rescue methods that involved monitoring of dissolved oxygen concentrations and installation of a portable pumping system when dissolved oxygen diminished to near critical levels was tested on young-of-the-year and yearling muskellunge. Fish were attracted to the flow of oxygenated water created under the ice where they were captured in hardware cloth-mesh traps. Movements of juvenile muskellunge, in relation to dissolved oxygen concentrations, were studied by observing radio-tagged fish. Muskellunge were shown to avoid anoxic waters and to concentrate in areas of higher oxygen as dissolved oxygen diminished in the lake. Recommendations for winter rescue of juvenile muskellunge include increased monitoring of oxygen levels, placing of winter-rescue equipment in the area with the highest oxygen content, initiation of rescue operations until oxygen levels diminish to 0.5 mg/l and no muskellunge are caught for 4 days.

Small, shallow winterkill lakes are used by the Minnesota Department of Natural Resources to rear young-of-the-year muskellunge. These lakes are less than 40 ha in area and less than 3 m in maximum depth. Waters are selected which have a high probability of becoming anoxic during the winter ice-cover period to eliminate large predatory fish before the next rearing season. In autumn, the fish are removed by fyke netting and stocked elsewhere. If the autumn harvest is not complete, remaining muskellunge often die and become a wasted resource. The high cost of muskellunge production has led to the use of winter rescue as a method of harvesting fish when sizable numbers remain in rearing areas after traditional harvest methods have been used.

Winter rescue operations have been used successfully for a number of years to capture northern pike in similar situations (Johnson and Moyle 1969). These workers found that catch rates of northern pike at a permanent trapping location increased dramatically when lake oxygen levels under the ice dropped to 1.0 mg/l but catches were erratic. They speculated that when oxygen levels dropped below 2.0 mg/l northern pike became more active and attracted to water currents generated by pumps in constructed channels. Petrosky and Magnuson (1973) noted that the activity of northern pike increased as oxygen levels decline, with the greatest increase occurring between 0.5-0.25 mg/l. This activity was accompanied by an upward movement toward the ice.

The present paper describes a modified, winter-rescue methodology and results of preliminary trials on young-of-the-year and yearling muskellunge. The procedure involved monitoring dissolved oxygen concentrations during ice cover. When oxygen concentrations diminished to near critical levels, a portable pumping system was installed to create a flow of oxygenated water under the ice. Fish were attracted to this flow and captured in trap nets. If movement patterns could be related to an easily-measured quantity, such as dissolved oxygen, it would be possible to increase the efficiency of winter rescue procedure by optimizing timing and location of trapping operations. For this reason, the movements of juvenile muskellunge were studied by radio telemetry before and during a winterkill event.

METHODS

Autumn trapping and winter rescue operations in east-central Minnesota were used to recover stocked muskellunge from four small, shallow lakes from 1978 to 1981. The lakes ranged in size from 1.5 to 33.1 ha and had maximum depths from 1.3 to 2.3 m (Table 1). Transplant fingerlings (approximately 38 mm total length) were stocked in the spring, and the population size was estimated in October and November by the Petersen and Schnabel mark-recapture method (Ricker 1975). Yearling muskellunge were recovered from Laddie Lake in 1981 because the lake did not experience winterkill conditions the previous year.

Table 1. Comparison of autumn trapping and winter rescue of muskellunge in four Minnesota lakes.

	LAKE					
	Knucklehead	Hanlo's	Hanlo's	Laddie	McGroarty	McGroarty
Year	1978-79	1978-79	1979-80	1981-82	1979	1980
Area (ha)	1.5	3.1	3.1	33.1	6.1	6.1
Maximum depth (m)	2.1	2.3	2.3	1.3	2.1	2.1
Population estimate (Oct. - Nov.)	300	367	227	747	1,102	283
Harvest (%)						
Autumn trapping	---	---	---	642(86)	869(79)	144(51)
Winter rescue	142(47)	150(41)	155(68)	38(36)[a]	---	---
Cost						
Man-hours	142	96	130	72.110[b]	164	141
Trips	47	9	14	12.20[b]	18	20
Fish/man-hour	0.5	1.6	1.2	8.9, 0.4[b]	5.3	1.0

[a] percent harvest is of the remaining population (=Oct. - Nov. population estimate minus autumn harvest).

[b] first number refers to autumn trapping; second, to winter rescue.

Autumn trapping operations consisted of setting standard 1 x 1.5-m x 9.5-mm mesh fyke nets, with a 12-m lead perpendicular to the shore. Netting was conducted on three occasions in Laddie and McGroarty lakes and lasted from 12 to 20 days.

Winter rescue equipment, modified from Johnson and Moyle (1969), consisted of a pump and two portable box traps measuring 1 x 1 x 2-m and covered with 64-mm mesh hardware cloth. Each trap was fitted with an 8-m hardware cloth lead. One trap enclosed the pump-intake pipe and the other trap enclosed the outflow pipe. Heavy plastic sheets (3 mil) were placed around both sides, one end and the bottom of the outflow trap to direct the flow of water through the throat of the trap. When power was available, an electric sump pump with maximum capacity of 0.01 m^3/sec was used to run the system. The electric pump was less expensive and easier to use than a gasoline-powered pump, since the latter required construction of protective housing. Tests of the system were performed in Knucklehead, Hanlo's and Laddie Lakes.

Movements of muskellunge in relation to dissolved oxygen were studied in Laddie Lake. Fish were collected by fyke-netting or electrofishing prior to ice cover. Two fish of age I and one of age II were tagged with radio transmitters on 12 December 1981. The fish measured 435, 437, and 595 mm TL with weights of 403, 403, and 1,316 g, respectively. Transmitters were surgically implanted after the muskellunge were anesthetized (Ross and Kleiner 1982; Strand 1986). The fish were held overnight to monitor any stress from surgical complications. Transmitters weighed 7 g in water and had a life expectancy of 180 days. Each transmitter was set to a separate frequency to enable indentification of individual fish.

Receiver and antenna design and construction for the 53 MHz tracking system were similar to that described by Winter et al. (1978). Three temporary tracking towers with yagi antennae and compass rows were constructed at Laddie Lake for locating fish. A hand-held loop antenna was used for tracking fish when observers walked on the ice, and a whip antenna was used for pinpointing fish to within a 1.5 m radius.

Dissolved oxygen concentrations were monitored with a YSI oxygen meter and probe, 0.3 m below the ice and 0.3 m above the bottom. Weekly measurements were taken. On the three smaller lakes, at least two sampling locations were monitored; on Laddie Lake, a minimum of 3 stations were monitored, and on four occasions a grid of 20 stations were visited to enable construction of oxygen isopleths. In addition, oxygen levels were monitored at the locations determined for the telemetered fish.

RESULTS

Winter rescue recovered 0.4 to 1.6 fish/man-hour of labor compared to 1.0 to 8.9 fish/man-hour in the autumn trapping (Table 1). Man-hours of labor were the total labor hours expended for conducting trapping operations. The man-hour was used to provide comparisons because wage variations existed between laborers and between years. Autumn trapping captured 51-86% of the estimated October-November population of Laddie and McGroarty Lakes, whereas winter rescue in Knucklehead and Hanlo's Lakes recovered 41-68% (Table 1). Winter rescue at Laddie Lake recovered an additional 36% of the muskellunge estimated to remain after fall netting. Catches in the winter rescue operations varied greatly and were erratic (Figs.

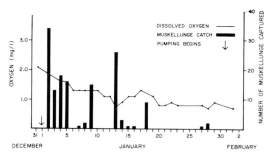

Figure 1. Relationship of dissolved oxygen to catch of juvenile muskellunge by winter rescue in Knucklehead Lake, Minnesota, winter 1978-79.

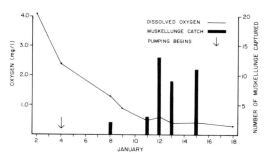

Figure 2. Relationship of dissolved oxygen to catch of juvenile muskellunge by winter rescue in Hanlo's Lake, Minnesota, winter 1979-80.

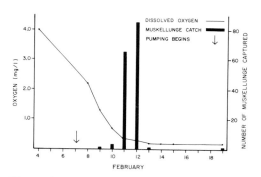

Figure 3. Relationship of dissolved oxygen to catch of juvenile muskellunge by winter rescue in Laddie Lake, Minnesota, winter 1981-82.

1-3) with respect to pumping operations and dissolved oxygen levels.

Oxygen levels were high (9.0 mg/l) throughout Laddie Lake at the beginning of the study, and tagged muskellunge ranged over the entire lake. Most of the southern half of the lake became anoxic by 5 January 1982, but oxygen levels greater than 3.0 mg/l were found in the northern half of the lake until 8 January 1982 (Fig. 4). After 4 January 1982, none of the 36 locations of tagged muskellunge were found in the southern half of the lake. In the northern half, the muskellunge were generally found in water with oxygen levels close to the highest in the lake. On 5 January, the highest oxygen level measured on the sampling grid was 3.5 mg/l, while oxygen levels measured at 3 fish locations were 2.5, 3.2 and 3.6 mg/l. Oxygen levels measured at fish locations on 8 January ranged from 2.2 to 2.5 while the highest level measured in the pond was 3.1 mg/l (Fig. 4).

Winter rescue was conducted on Laddie Lake from 4 to 18 January 1982. The pump and traps were set in an area where oxygen levels had been the highest the previous year and where muskellunge had been rescued in the past. This area did not have the highest oxygen levels as shown by the isopleths (Fig. 4). The three tagged fish remained in water with higher oxygen levels rather than around the rescue traps until anoxia occurred in the northern half of the lake. The tagged muskellunge survived for 2 to 3 days when oxygen concentrations at the sampling stations and at the fish locations were 0.3 to 0.5 mg/l, but succumbed to the anoxic conditions on 12-13 January 1982.

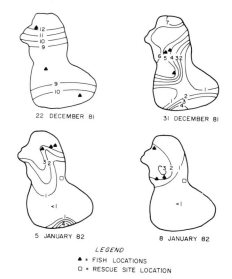

Figure 4. Oxygen isopleths on Laddie Lake, Minnesota, winter 1981-82. Isopleths represent oxygen concentration in mg/l at 0.3 m under ice.

DISCUSSION

Fall trap netting was more efficient than winter rescue for harvesting juvenile muskellunge from shallow rearing lakes. Nevertheless, winter rescue is a suitable alternative for harvesting populations when autumn trapping cannot be completed or results in a poor harvest. The two methods can be used in tandem to increase the total harvest.

Johnson and Moyle (1969) reported that most of their catch of northern pike was from traps through which aerated water was pumped when lake oxygen levels were below 2.0 mg/l. The best catches of juvenile muskellunge in winter rescue operations on Hanlo's Lake (1979-80) and on Laddie Lake (1981-82) were observed when dissolved oxygen ranged from 0.3-0.5 mg/l. Muskellunge exhibited trap avoidance, and none were captured during the study when lake oxygen levels were above 2.0 mg/l. The tagged muskellunge were observed to survive for 2 to 3 days at these oxygen levels.

Knucklehead Lake (Winter 1978-79) catch rates of muskellunge were highest at dissolved oxygen concentrations between 1.0 and 2.0 ppm. This catch rate appears to be related to the density of muskellunge in the lake. There was no fall harvest of muskellunge, and densities at winter-rescue time were 2.7 and 9.1 times greater than in Hanlo's (Winter 1979-80) and Laddie Lakes (Winter 1981-82) respectively. In addition, oxygen concentration never declined as rapidly as in the other lakes because of climatic conditions during the winter of 1978-79.

In the four trials with the rescue equipment, catches in the trap surrounding the discharge pipe averaged 75, 84, 100, and 100% of the total catch. These findings indicate that muskellunge are capable of avoiding low oxygen when an oxygen gradient exists, and they are attracted to flowing water, which is similar to behavior of northern pike under low oxygen concentrations (Johnson and Moyle 1969; Casselman 1978). A large catch in the intake trap may be indicative of a tight water circulation pattern caused by poor separation of the intake and outflow pipes. If catches in the intake trap are consistently low, one may dispense with this trap altogether.

The speculations by Johnson and Moyle (1969) that esocids may be attracted by flowing water under low oxygen conditions is supported by evidence from Hanlo's Lake (1979-80). Large numbers of muskellunge were captured in the rescue traps even when oxygen levels in the traps were the same as in the lake (0.4 mg/l), with most captured in the discharge trap. At these low oxygen concentrations, 23% of the captured muskellunge were dead when the trap was opened. Casselman (1978) reported that northern pike can survive minimum oxygen concentrations of 0.3 mg/l, and that some northern pike were captured alive in stationary gear at oxygen concentrations of 0.04 mg/l, when the maximum oxygen concentration in the lake was 0.8 mg/l; no mortalities were associated with those conditions. He proposed that the ability of northern pike to withstand critically-low winter oxygen concentrations may be genetic since the critical levels varies between lakes.

Since juvenile muskellunge were found to concentrate in waters of highest oxygen concentration when oxygen levels decline, it is recommended that the winter rescue equipment be placed in the area with the highest oxygen content. Pumping operations should begin when oxygen levels have declined to approximately 2.0 mg/l; above that level trap avoidance is a problem. Pumping should continue until levels are less than 0.5 mg/l throughout the lake and no fish have been caught for 4 days.

It is important to establish an adequate grid of sampling stations, since oxygen levels are not necessarily uniform throughout a lake and water masses can periodically shift location. Close monitoring of oxygen levels should enable the fisheries manager to judge where and when to set up rescue equipment and when to cease operations, thus increasing harvest and reducing labor expenditure.

ACKNOWLEDGEMENTS

I wish to thank the Minnesota Chapter of Muskies, Incorporated for purchasing the transmitters, M.J. Ross of the University of Minnesota for technical advice and assistance and the residents of Laddie Lake on whose property tracking towers were constructed. Minnesota Fisheries personnel generously shared their comparative data. P.J. Wingate and J.M. Hoenig provided editorial assistance.

REFERENCES

Casselman, J.M. 1978. Effects of environmental factors on growth, survival, activity, and exploitation of northern pike. American Fisheries Society Special Publication 11:114-128.

Douderoff, P., and D.L. Shumway. 1970. Dissolved oxygen requirements of freshwater fishes. FAO Fisheries Technical Paper 86. 291p.

Johnson, F.H., and J.B. Moyle. 1969. Management of a large shallow winterkill lake in Minnesota for the production of pike (*Esox lucius*). Transactions of the American Fisheries Society 98:691-697.

Petrosky, B.R., and J.J. Magnuson. 1973. Behavioral responses of northern pike, yellow perch, and bluegill to oxygen concentrations under simulated winterkill conditions. Copeia 1:124-133.

Ricker, W.E. 1975. Computation and interpretation of biological statistics of fish populations. Bulletin 191. Fisheries Research Board of Canada, Ottawa. 382 p.

Ross, M.J., and C.F. Kleiner. 1982. Shielded-needle technique for surgically implanting radio-frequency transmitters in fish. Progressive Fish Culturist 44:41-43.

Strand, R.F. 1986. Identification of principal spawning areas and seasonal distribution and movements of muskellunge in Leech Lake, Minnesota. American Fisheries Society Special Publication 15:62-73.

Winter, J.D., V.B. Kuechle, D.B. Siniff, and J.R. Tester. 1978. Equipment and methods for radio tracking freshwater fish. University of Minnesota, Institute of Agriculture Miscellaneous Report 152. 45p.

Nursery Habitat of Muskellunge in Southern Georgian Bay, Lake Huron, Canada

ROBIN E. CRAIG AND RONALD M. BLACK

Ministry of Natural Resources
Midhurst, Ontario L0L 1X0

ABSTRACT

Summer nursery habitats of Great Lakes muskellunge were studied during July and August, 1981. The "typical" habitat was occupied by young-of-year (YOY) muskellunge about 50 mm long and by five other fish species, the most abundant being largemouth bass, pumpkinseed and yellow perch. Each habitat was comprised of eight families of emergent and floating vegetation and nine species of submergent vegetation. Sedge (Cyperaceae) was the most abundant emergent family while bushy pondweed (*Najas flexilis*), muskgrass (*Chara* sp.) and variable pondweed (*Potamogeton gramineus*) were the predominant submergent species. Emergent vegetation was dense in a narrow band along the shore where submergent vegetation was sparse. Off shore habitat consisted of a wider band of less dense emergent and floating vegetation and increased bottom cover of submergent vegetation. Shoreline residents often alter littoral areas and their activities may restrict muskellunge production by reducing macrophytes in critical spawning and nursery habitats. Fishery managers of southern Georgian Bay can use the "typical" nursery area description to identify potentially important habitats and possibly mitigate the impacts of proposed alterations.

Muskellunge provide an important trophy fishery in Georgian Bay, Lake Huron, Canada. This bay, located within 200 km of major population centers in southern Ontario, has long been an important source of recreation for anglers, boaters, and summer residents. Although resorts, cottages, and marinas are numerous, many shores are still natural and unaltered, but demand to develop more waterfront land is continuing. Shoreline residents eventually wish to "improve" their waterfronts by dredging for boat access, filling marshy areas to build up property, removing aquatic vegetation for swimming and aesthetics, and dumping sand for beach creation. Fishery managers must mitigate these proposals to protect important littoral areas and accommodate the needs of various recreationists, developers, and property owners. Littoral habitat requirements of muskellunge, however, are unknown, making the managers' task difficult.

Spawning habitats of muskellunge and other esocids and nursery habitats of YOY northern pike have been described (Dombeck 1979; Dombeck et al. 1984; Forney 1968; Holland and Huston 1984; McCarraher and Thomas 1972). Newly-hatched muskellunge remain among aquatic plants before becoming active and feeding (Scott and Crossman 1973), but information on species and densities of vegetation and other parameters of fry nursery habitat is lacking. Brewer (1980) found several species of centrarchids in muskellunge habitats, and Parsons (1959) and Oehmcke, et al. (1958) reported that young muskellunge require forage fish shortly after hatching, but further details are unavailable. The objectives of this study were to describe the nursery requirements and fish community associations of YOY muskellunge in southern Georgian Bay of Lake Huron, Canada.

STUDY SITE

Although most of Georgian Bay is deep and oligotrophic, muskellunge are found along the shallow, eutrophic eastern shore. This investigation was conducted in Severn Sound at the southeastern extremity of Georgian Bay (Fig. 1). The sound is located between the Precambrian Shield on the north and the Paleozoic land mass of southern Ontario on the south. The areas studied were all situated on the north shore, among the many islands, rock outcrops, and protected bays of the Shield. The waters are productive and aquatic vegetation is extensive because the sound receives nutrients from rivers flowing through agricultural lands to the south.

METHODS

Potential muskellunge spawning areas were initially identified by local fishing guides. As muskellunge fry remain close to spawning sites (John Casselman, Ontario MNR, personal communica-

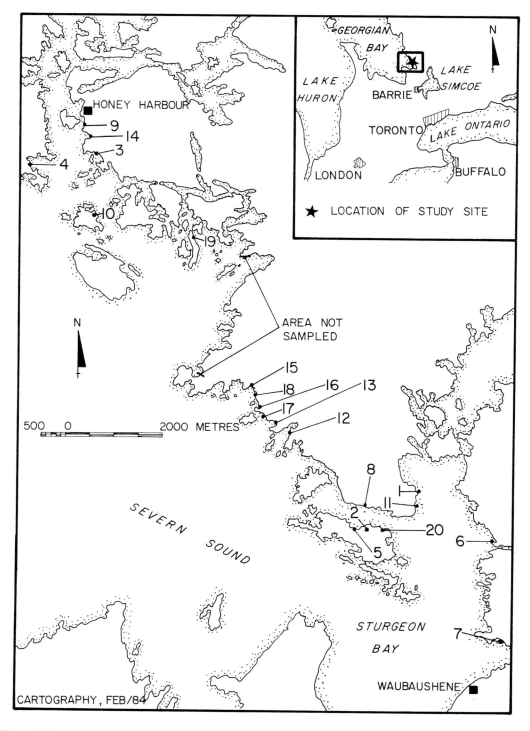

Figure 1. Location of Severn Sound, Georgian Bay, Lake Huron, Canada and muskellunge nursery habitats (number 1 - 20)

tion), potential nursery areas were considered to be near spawning areas. Additional sampling areas were selected visually, based upon their physical and vegetative similarities to these other sites. Fishes were collected during the day from potential nursery areas between June 1 and July 17, 1981, by a single haul of a seine (6.4 mm mesh x 15 m x 1 m). A seine haul consisted of running the seine out full length perpendicular to the shore and drawing it back, describing a quarter circle. In a few sites, where bottom slope was very gentle, the seine was carried out to 1 m depth of water and drawn straight in, parallel to the shore. The abundance of each fish species was recorded for each seine haul. The total lengths of all muskellunge were measured and those <305 mm were considered YOY (Scott and Crossman 1973). Because the dimensions of the seine limited the water depth for fish capture, all potential habitat data were collected from water ≤1 m deep. The physical characteristics, water temperature, distance from shore to 1 m water depth (transect length), and bottom type were recorded for each sample area.

Aquatic vegetation was surveyed between July 20 to August 18, 1981 in areas where muskellunge YOY were captured. Vegetation was sampled at each site, along three transects, 10 m apart, running perpendicular from shore to a depth of 1 m. In small bays, three transects were run equal distances apart along the shore, to a centre point in the bay usually at a depth of 1 m. Transect lengths were constant for each sample site but varied between sites depending upon bottom slope. Sample quadrats (0.5 m x 0.5 m = 0.25 m^2) were established along each transect. The first quadrat was at the shore - water interface and the remainder were at 2m intervals along the transect.

Families and numbers of emergent and floating vegetation were identified and recorded for each quadrat. Percent composition was determined by comparing total counts for all families from all quadrats at each sample site and their means were calculated for all sites. Percent of the bottom covered by submergent vegetation was estimated for each quadrat. To determine relative abundance, submergent species were listed in order of decreasing abundance for each quadrat. Data from all quadrats at each sample site were compared and then species were listed in order of decreasing abundance for each sample site. Each species was then assigned a value from seven for the most abundant to one for the least abundant. When more than seven species occurred at a sample site, additional species, equally low in abundance, were each given the value of one. Means were calculated by comparing relative abundance values for each species from all sample sites. Frequencies of occurrence were expressed as percentages of the number of sites each family or species of vegetation occurred.

To facilitate descriptive analysis, sample quadrats were divided into four depth groups. Those within 2 m of shore formed one group as did those within 2 m of transect ends; 21.8 m ($\pm$2.1) (95% confidence interval)—23.8 m ($\pm$2.1) from shore. The remaining quadrats were separated into two equal groups by the mid-point of the transects; 2 m—11.9 m ($\pm$1.1) and 11.9 m ($\pm$1.1) - 21—8 m ($\pm$2.1) from shore.

RESULTS

Fish were netted at 103 potential nursery areas. Thirty-four muskellunge YOY were caught at 20 sites (1.4/site). Nineteen other fishes were captured and an average of 4.9 ($\pm$0.9) species cohabited each muskellunge nursery area (Table 1). Pumpkinseed, yellow perch and largemouth bass occurred in 75% or more of nursery areas; they were also the most abundant species. Most centrarchids and percids were YOY or yearlings; the few cyprinids captured were generally adults. The densities of YOY muskellunge found by seining here were much lower than densities determined by electrofishing in the upper St. Lawrence River (pers. comm. Casselman). Possibly catches of muskellunge in southern Georgian Bay would have been greater with electrofishing. The mean total length of 34 muskellunge YOY captured at nursery sites from June 15 to July 14 was 49.7 mm ($\pm$6.3). The shortest was 16.0 mm, caught June 23 and the longest was 97.5 mm, caught July 14.

The mean water temperature of 20 nursery areas between June 15 and August 18 was 24°C ($\pm$1.1). The most common bottom types were sand (9) and muck (7), but silt-detritus (4) was also found. The mean transect length to 1 m depth contour for 18 areas was 23.8 m ($\pm$2.1).

Aquatic vegetation was analysed at 18 of the 20 nursery areas. A total of 15 families of emergent and floating vegetation were encountered with an average of 7.9 ($\pm$0.7) families found at the various sites (Table 2). Arrowhead and grasses occurred at all sites while sedges, pickerel weed, and water lilies were present at 89% or more. The most abundant family was sedge at 52% ($\pm$10.7); the remaining 14 families each comprised 10% or less of the

Table 1. Frequency of occurrence (%) and abundance (No./site) of fish species co-habiting 20 muskellunge nursery areas in southern Georgian Bay, Lake Huron, Canada.

Species	Frequency of Occurrence (%)	Mean No./Site	Confidence Interval (95%)
muskellunge	100	1.4	±0.3
pumpkinseed	90	14.1	±5.5
yellow perch	80	10.9	±4.7
largemouth bass	75	25.7	±10.7
black crappie	40	1.7	±1.5
bluntnose minnow	40	4.9	±3.3
blackchin shiner	25	4.5	±4.4
other species (14)[a]	75	16.9	±16.1

[a]Other species, each occurring at <20% of the sites, in order of abundance: golden shiner, brown bullhead, smallmouth bass, brook silversides Johnny darter, mottled sculpin, rock bass, longnose gar, Iowa darter, logperch, tadpole madtom, central mudminnow and bowfin.

Table 2. Frequency of occurrence (%) and mean composition of emergent and floating vegetation growing in 18 muskellunge nursery areas in southern Georgian Bay, Lake Huron, Canada.

Common Name (Family)	Frequency of Occurrence (%)	Mean Percent Composition	Confidence Interval (95%)
grasses (Gramineae)	100	7	± 3.5
arrowhead (Alismaceae)	100	9	± 9.3
sedges (Cyperaceae)	94	52	±10.7
pickerelweed (Pontederiaceae)	83	10	± 4.9
water lilies (Nymphaeaceae)	83	7	± 3.3
pondweeds (Potamogetonaceae)	78	5	± 3.1
rushes (Juncaceae)	72	4	± 3.5
burreeds (Sparganiaceae)	67	2	± 1.7
other families (7)[a]	72	2[b]	± 2.0

[a]Other families include: irises (Iridaceae), cattails (Typhaceae), pipeworts (Eriocaulaceae), horsetails (Equisetaceae), arums (Araceae), parsley (Umbelliferae), smartweeds (Polygonaceae).
[b]Due to rounding, total does not equal 100%

Table 3. Frequency of occurrence (%) and relative abundance ratings of submergent vegetation in 18 muskellunge areas in southern Georgian Bay, Lake Huron, Canada.

Species	Frequency of Occurrence (%)	Mean Relative Abundance Rating	Confidence Interval 95%
bushy pondweed (*Najas flexilis*)	94	5.6	±4.7
muskgrass (*Chara sp.*)	89	4.9	±5.0
variable pondweed (*Potamogeton gramineus*)	72	2.9	±4.4
wild celery (*Vallisneria americana*)	67	2.6	±5.0
Richardson's pondweed (*P. Richardsonii*)	61	1.0	±2.5
waterweed (*Anacharis canadensis*)	50	1.9	±4.5
coontail (*Ceratophyllum demersum*)	50	1.1	±3.5
water milfoil (*Myriophyllum sp.*)	50	0.8	±2.3
Other species (14)[a]	100	---	---

[a]Other species, each occurring at < 50% of the sites, in order of abundance: arrowheads (2) (*Sagittaria spp.*), pondweeds (7) (*Potamogeton spp.*), pipewort (*Eriocaulon septangulare*), bladderwort (*Utricularia sp.*), water marigold (*Megalodonta beckii*), quillwort (*Isoetes sp.*), pickerelweed (*Pontederia cordata*).

vegetation. An average of 8.9 (±1.7) species of submergent vegetation was found at each site (Table 3). Bushy pondweed, muskgrass, and variable pondweed occurred at 70% or more of the sites and were also the most abundant species recorded. The remaining 19 species occurred at 67% or less of the sites.

Emergent and floating vegetation were most

Table 4. Densities (No. stems and petioles/m²) of emergent and floating vegetation and coverage (%) of submergent vegetation by depth zone in 18 muskellunge nursery areas in southern Georgian Bay, Lake Huron, Canada.

Depth Zone	Distance from Shore	Mean No. of Stems and Petioles/m²	Mean percent Coverage of the bottom by Submergent Vegetation
1	0.0 - 2.0 m	293.6	19
		(±54.8)[a]	(±8)
2	2.0 m - 11.9 m	108.4	40
	(±1:1)	(±13.8)	(±6)
3	11.9 m - 21.8 m	42.8	51
	(±1.1) (±2.1)	(±6.8)	(±6)
4	21.8 m - 23.8 m	24.4	50
	(±2.1) (±2.1)	(±6.4)	(±8)

[a] 95% confidence interval

dense within the near shore depth zone with a mean density of 293.6 stems and petioles /m² (Table 4). Sedges were the most abundant emergent in this habitat, and common spikerush and common three square were the predominant species. Mean densities decreased considerably in the respective depth zones farthest from shore. Sedges remained predominant, but the most abundant species was soft stem bulrush. Submergent vegetation covered an average of 19% of the bottom near shore and about 50% off shore.

DISCUSSION

The "typical" southern Georgian Bay muskellunge nursery habitat consisted of dense emergent vegetation near shore adjacent to a wider expanse (approx. 10x) of less dense emergents off shore (Fig. 2). Aquatic vegetation plays an impor-

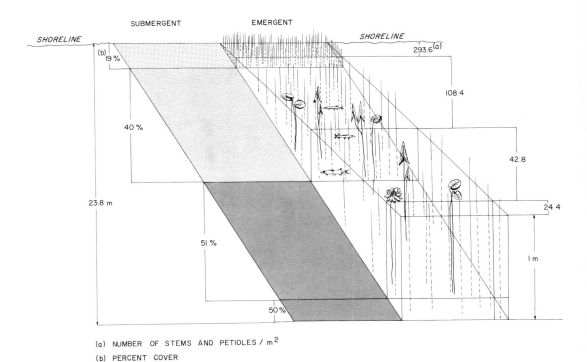

Figure 2. A "typical" Georgian Bay, Lake Huron, Canada muskellunge nursery habitat with emergent vegetation densities and submergent vegetation coverage.

tant role during the early life of muskellunge. Eggs are deposited and develop on the previous season's growth of emergent plants (Dombeck 1979); newly-hatched fry find protection from predators in new-growth emergents; and young muskellunge ambush prey, such as yellow perch and various cyprinids (Scott and Crossman 1973; Hourston 1952), from within aquatic vegetation (Parsons 1959). Forage fish availability is an important determinant of esocid stock densities (Snow 1978), but predator-prey systems may be influenced by macrophyte densities. Lake management studies in Wisconsin indicated that high densities of macrophytes reduced fish production (Dunst et al. 1974). Research on bluegill and pumpkinseed in Iowa revealed that populations were reduced when submergent vegetation was sparse (DiCostanzo 1957). Possibly high vegetation densities reduced production by reducing predator-feeding efficiency while low densities provided few prey, thus also leading to reduced production (Cooper and Crowder 1974). Optimum fish production may be best achieved in habitats containing intermediate densities of aquatic vegetation (Eipper and Regier 1962; Glass 1971), where populations of preferred prey can be easily hunted but not decimated. The southern Georgian Bay study areas provided all the elements of vegetation and forage that would seem necessary for optimum muskellunge production and growth.

Muskellunge fry are difficult to net (Brewer 1980), but data about their habitats are relatively easy to acquire. Knowledge of the fishes, vegetation and physical characteristics in the "typical" southern Georgian Bay muskellunge nursery habitats will enable fishery managers to better identify potential nursery areas and more easily assess and mitigate the impacts of proposed shoreline perturbations.

HABITAT ALTERATION - A MANAGEMENT CONSIDERATION

Muskellunge nursery habitats are not generally suited to the needs or wants of shoreline property owners - they are usually "too weedy", "too shallow", or "too mucky" - and changes desired by most waterfront residents would considerably alter these habitats. Herbicides applied at prescribed rates to restrict submergent plant growth may have minimal impact on fish eggs and fry (Hiltibran 1967), but invertebrate food supplies can be reduced temporarily (Gilderhus 1967). Herbicide toxicity varies between macrophytes (Yeo 1967) and extended use could shift plant communities to mono cultures of tolerant species (Hestand and Carter 1977). Mechanical harvesting, an alternative to chemical treatment of emergent vegetation, can kill small fish directly (Wile 1978), and cutting operations can disrupt fish distribution (Swales 1982). The impacts of mechanical cutting, however, vary directly with the extent of harvesting operations (Carpenter and Gasith 1978; Haller et al 1980). Dredging and filling can kill aquatic plants and change littoral area depths so that emergents can not re-establish. Suspended silt may restrict light penetration and reduce submergent vegetation (Edwards 1969), or, during esocid spawning, can increase egg mortality substantially (Hassler 1970).

Alterations that reduce abundance of macrophytes may effect muskellunge populations by eliminating or reducing spawning and nursery habitats. Disruptions to growth of aquatic flora by herbicide treatments and mechanical harvesting have been implicated in muskellunge declines in Chautauqua Lake, New York (Bimber and Nicholson 1981). Northern pike and chain pickerel populations have also decreased in other waters because of modifications to aquatic plant communities (Threinen 1969; Bidgood 1973; Baumann et al. 1974; Clady 1976; Willemsen 1980; Benson 1980).

Future management of muskellunge should include not only harvest regulations and control of poaching but should also recognize the importance of spawning and nursery habitats. Fishery biologists need to assess the impacts of proposals which will seriously alter these habitats and should recommend restricting such alterations in muskellunge production areas. Environmentally-sensitive shorelines should be designated for limited development in official planning and zoning documents. Biologists will need to educate local governments, planning agencies, and existing property owners about the need to protect important habitats along already-developed shores. The continued production and survival of the highly valuable, trophy muskellunge in southern Georgian Bay will necessitate the identification, protection, and management of critical habitats in concert with onshore and inwater activities of the local human population.

ACKNOWLEDGEMENTS

The authors wish to thank Dr. John Casselman for providing valuable direction early in the project and for critically reviewing the manuscript.

We are also indebted to Doug Gilmore and Al Hovingh for their technical assistance throughout the field season. Appreciation is extended to the fishing guides of Port Severn who assisted in locating muskellunge spawning habitats initially; Norm White, Morley Woods and the Miller family: Jane, Ian, Bruce, and Richard. Cathy Martin provided skilled assistance in preparing figures while Sheila Dobson helped summarize the data.

REFERENCES

Baumann, P.C., J.F. Kitchell, J.J. Magnuson, and T.B. Kayes. 1974. Lake Wingra, 1873-1973: a case history of human impact, Wisconsin Academy of Sciences, Arts and Letters 62:57-94.

Benson, N.G. 1980. Effects of post-impoundment shore modifications of fish populations in Missouri River reservoirs. Research report - U.S. Fish and Wildlife Service, Washington, District of Columbia, USA.

Bigood, B.F. 1973. Divergent growth in two lake whitefish (*Coregonus clupeaformis*) populations. Journal of the Fisheries Research Board of Canada 30:1683-1696.

Bimber, D.L. And S.A. Nicholson. 1981. Fluctuations in muskellunge (*Esox masquinongy* Mitchill) populations of Chautauqua Lake, New York. Environmental Biology of Fish 6:207-211.

Brewer, D.L. 1980. A study of native muskellunge populations in eastern Kentucky streams. Fisheries Bulletin 64 Kentucky Department of Fish and Wildlife Resources, Frankfort, Kentucky, USA.

Carpenter, S.R. and A. Gasith. 1978. Mechanical cutting of submersed macrophytes: immediate effects on littoral chemistry and metabolism. Water Research 12:55-57.

Clady, M.D. 1976. Change in abundance of inshore fishes in Oneida Lake, 1916 to 1970. New York Fish and Game Journal 23:73-81.

Cooper, W.E. and L.B. Crowder. 1979. Patterns of predation in simple and complex environments. Pages 257-267 in R.H. Stroud and H. Clepper (editors). Predator-prey systems in fisheries management. Sport Fishing Institute, Washington, District of Columbia, USA.

DiConstanzo, C.J. 1957. Growth of bluegill (*Lepomis macrochirus*) and pumpkinseed (*Lepomis gibbosus*) of Clear Lake, Iowa. Iowa State College Journal of Science 32:19-34.

Dombeck, M.P. 1979. Movement and behaviour of the muskellunge determined by radio-telemetry. Technical Bulletin of the Wisconsin Department of Natural Resources, Madison, Wisconsin, USA.

Dombeck, M.P., B.W. Menzel, and P.M. Hinz. 1984. Muskellunge spawning habitat and reproductive success. Transactions of the American Fisheries Society 113:205-216.

Dunst, R.C., S.M. Born, P.D. Uttomark, S.A. Smith, S.A. Nichols, J.O. Peterson, D.R. Knauer, S.L. Serns, D.R. Winter and T.L. Wirth. 1974. Survey of lake rehabilitation techniques and experiences. Technical Bulletin of the Wisconsin Department of Natural Resources, Madison, Wisconsin, USA.

Edwards, D. 1969. Some effects of siltation upon aquatic macrophyte vegetation in rivers. Hydrobiologia 34:29-36.

Eipper, A.W. and H.A. Reiger. 1962. Fish management in New York farm ponds. New York State College of Agriculture. Cornell Extension Bulletin, Ithaca, New York, USA.

Forney, J.L. 1968. Production of young northern pike in a regulated marsh. New York Fish and Game Journal 15:143-154.

Gilderhus, P.A. 1967. Effects of Diquat on bluegills and their food organisms. The Progressive Fish-Culturist 29:67-73.

Glass, N.R. 1971. Computer analysis of predation energetics in the largemouth bass. Pages 325-363, in B.C. Patten (editor) Systems analysis and simulation in ecology. Volume 1. Academic Press, New York, USA.

Haller, W.T., J.V. Shireman, and D.F. DuRant. 1980. Fish harvest resulting from mechanical control of hydrilla. Transactions of the American Fisheries Society 109:517-520.

Hassler, T.J. 1970. Environmental influences on early development and year-class strength of northern pike in Lakes Oahe and Sharpe, South Dakota. Transactions of the American Fisheries Society 99:369-375.

Hestand, R.S. and C.C. Carter. 1977. Succession of various aquatic plants after treatment with four herbicides. Journal of Aquatic Plant Management 15:60-64.

Hiltibran, R.C. 1967. Effects of some herbicides on fertilized fish eggs and fry. Transactions of the American Fisheries Society 96:414-416.

Holland, L.E. and M.L. Huston. 1984. Relationship of young-of-the-year northern pike to aquatic vegetation types in backwaters of the upper Mississippi River. North American Journal of Fisheries Management 4:514-522.

Hourston, A.S. 1952. The food and growth of the maskinonge (*Esox masquinongy* Mitchill) in Canadian waters. Journal of the Fisheries Research Board of Canada 8:347-368.

McCarraher, D.B. and R.E. Thomas. 1972. Ecological significance of vegetation to northern pike, *Esox lucius*, sqawning. Transactions of the American Fisheries Society 101:560-563.

Oehmcke, A.A., L. Johnson, J. Klingbiel, and C. Wistrom. 1958. The Wisconsin muskellunge its life history, ecology, and management. Publication of the Wisconsin Conservation Department, Madison, Wisconsin, USA.

Parsons, J.W. 1959. Muskellunge in Tennessee streams. Transactions of the American Fisheries Society 88:136-140.

Scott, W.B. and E.J. Crossman. 1973. Freshwater Fishes of Canada. Bulletin 184. Fisheries Research Board of Canada, Ottawa.

Snow, H.E. 1978. Responses of northern pike to exploitation in Murphy Flowage, Wisconsin. American Fisheries Society Special Publication 11:320-327.

Swales, S. 1982. Impacts of weed-cutting on fisheries: an experimental study in a small lowland river. Fisheries Management 13:125-127.

Threinen, C.W. 1969. An evaluation of the effect and extent of habitat loss on northern pike populations and means of prevention of losses. Management Report of the Wisconsin Department of Natural Resources, Madison, Wisconsin, USA.

Wile, I. 1978. Environmental effects of mechanical harvesting. Journal of Aquatic Plant Management 16:14-20.

Willemsen, J. 1980. Fishery-aspects of eutrophication. Hydrological Bulletin 14:12-21.

Yeo, R.R. 1967. Dissipation of Diquat and Paraquat, and effects on aquatic weeds and fish. Journal of the Weed Society of America 15:42-46.

Sperm Production and Spawning Success for Muskellunge and Northern Pike

J.R. GAMMON

Department of Zoology
DePauw University
Greencastle, IN 46135

ABSTRACT

Sperm production and success of natural fertilization were examined for both muskellunge and northern pike to assess their potential importance in natural reproduction. The annual cycle of spermatogenesis, the relative sizes of the testes, and the production of spermatozoa were determined for both species. Mature testes in muskellunge weigh less than half those of northern pike of the same size. The density of spermatozoa within the testes of muskellunge was about 7% less than for northern pike.

The viability of naturally spawned eggs of both species was determined by collecting eggs with trays and handnets in 1960 and 1964. Fertlization rates of 47.2% and 70.6% were determined for two collections of northern pike eggs on trays. Fertilization rates of newly-spawned muskellunge eggs ranged from 30.0% to 88.4% for three tray collections. Slightly less than 50% of these eggs remained alive after 3-4 days laboratory incubation in aerated water. The viability of eggs collected with hand nets from *Anacharis* beds was low (20.7 to 32.8%). Predation by minnows caused significant egg mortality at one site.

Muskellunge are rarely found in abundance and in most Wisconsin waters it has been necessary to supplement the natural population by stocking young fish (Klingbiel 1966; Johnson 1978). In attempting to examine the success of natural reproduction in muskellunge very few young-of-the-year fish were found even in some northern Wisconsin lakes known to support good populations of adults. Males of both northern pike and muskellunge produce small volumes of milt, a characteristic which presents hatchery workers with difficulties in the artificial culture of both species (Huet 1953; Johnson 1958; De Montalembert et al. 1978).

These two seemingly-unrelated observations prompted this study to assess the importance of spermatozoa production as a controlling factor of reproductive success in natural populations of muskellunge. When possible the results were compared to those for the northern pike.

This study included (1) a description of the annual cycle of spermatogenesis, (2) a determination of the quantity of sperm produced by the males, and (3) the collection of naturally-spawned muskellunge and northern pike eggs and and an assessment of their viability. The first two components are described in detail elsewhere (Gammon 1961) and will only be summarized here.

METHODS

Cycle of Spermatogenesis

Testes of muskellunge were obtained in 1960, in all months except June, November, and December. From January through March residents of the Chippewa Indian Reservation provided testes of fish speared in lakes near Reserve, Wisconsin; others were obtained from anglers at Resort of the Woods, Boulder Junction, Wisconsin during August to October.

Portions of 28 muskellunge testes were fixed in Bouin's fixative, embedded in paraffin, sectioned at 8 μm and stained with Harris' hematoxylin and eosin. Sections of northern pike testes were examined when available, but the descriptions of Turner (1919) and Lofts and Marshall (1957) provided the primary information for this species.

Spermatozoa Production

Fifty-two mature male northern pike and 23 mature male muskellunge were collected from the Rock River (Walworth Co.), and several lakes in Sawyer and Vilas Counties, Wisconsin between January 19 and April 26, 1960. All had completed spermatogenesis, but had not spawned.

Individual fish were weighed and measured. Testes were removed and weighed to the nearest gram. Initially, one testis was examined for longitudinal differences in spermatozoa density. Because no differences were apparent from anterior to posterior, a 0.5 to 1.0 g section removed from the middle portion of one testis was then used to determine sperm density. The sample was macerated in a Potter-Elvehjem homogenizer containing normal saline solution and diluted to 100 ml. Ten ml of this suspension was then diluted to 100 ml using a 0.75% sodium chloride solution. Spermatozoa concentration in the final dilution was estimated using a hemacytometer. Sperm concentration was also determined for lobular fluid by lacerating the testis, expressing the fluid, measuring a quantity into a micropipette, and then processing as previously described for whole testis.

Collection of Naturally Spawned Eggs

Naturally-spawned eggs of both species were collected using egg trays and handnets. Egg trays consisted of a 0.61 m x 0.91 m pine frame with fiberglass window screening stapled to it. Longitudinal half-round ribs were secured midway between the edge and center ribs to minimize sagging when the trays are lifted. A 2.25 kg piece of lead was placed in the center to sink the tray. At one location a top screen of 0.63 cm hardware cloth was needed to adequately protect the eggs from minnow predation. The trays were placed in likely spawning areas, usually in water less than 1 m deep, and tended daily. About 2 hours were required to tend 20 trays.

Once eggs were found on the trays it was relatively simple to obtain more in the vicinity of the trays by dredging vegetation or other bottom substrate with a handnet. Eggs were hand picked from this material. The trays eased the task of collecting eggs and locating primary spawning areas. When tended daily they also provided recently-spawned eggs of known age. The eggs obtained with the handnet included a mixture of day-old eggs and those of unknown age. Following collection eggs were examined immediately for viability, were held in the laboratory for 24 hours or more and re-examined.

RESULTS

Annual Cycle of Spermatogenesis in Muskellunge

Muskellunge and northern pike testes are similar in many respects. They consist of two separate, elongated bodies, triangular in cross section. They are suspended ventrally from the kidney and extend the entire length of the body cavity on either side of the swim bladder. At maximum size a northern pike testis tapers only slightly throughout its length while the anterior portion of a muskellunge testis is much more attenuated. Within the fibrous connective tissue sheath which envelops the testis is a complex mass of interbranched, blind lobules which ultimately connect with several large channels located in the ventral edge of the posterior third of each testis, the ductus efferens. These channels, in turn, lead to the ductus deferens which opens a few millimeters posterior to the anus.

During the summer months the testes of muskellunge are very small. The walls of the lobules are lined with nests of spermatogonia called cysts. Small quantities of mature spermatozoa remaining from the spawning period were found in the lumen of the lobules even in July. This retention of spermatozoa was also noted for the northern pike by Zaitzev (1955) and Shikhshabekov (1979), as well as for *Cottus* (Foley 1926) and *Umbra* (Hann 1927).

By early August most mature spermatozoa have disappeared and each lobule is somewhat enlarged and composed of numerous thin-walled cysts. By late August spermatogenesis is proceeding rapidly. Cysts in different stages of development are abundant. Because all of the cells within each cyst divide at nearly the same time, each cyst contains cells at the same stage of development. There was no evidence of an anterioposterior gradient of spermatogenesis. Some cysts have already produced mature spermatozoa and have ruptured to release their contents into the lumen of the lobules.

Throughout September the testes increase in size and more and more lobules fill with mature spermatozoa. Secondary spermatocytes at this time are more scarce than either primary spermatocytes or spermatids (Fig. 1a).

By October most of the lobules are filled with mature spermatozoa and cysts containing intermediate stages of cells are less common than before. Spermatogonia are common as are 'lobule boundary cells'. Lofts and Marshall (1957) found that for northern pike these latter cells accumulate cholesterol-positive lipids beginning in September and release them into the lumen of the lobule at spawning. They believe that these cells are homologous to Leydig cells of higher vertebrates.

No testes were obtained in November or December, but it seems likely that the intensity of spermatogenesis diminishes rapidly in response to declining water temperatures.

By January all of the cysts in northern pike testes have matured and the lobules are filled exclusively with mature spermatozoa (Lofts and Marshall 1957; Hoffmann et al. 1980); this is not so in muskellunge testes where many cysts are incompletely developed in January. Most of the immature cells are primary spermatocytes although some spermatids are also present. These cells appear to be undergoing degeneration judging from their appearance and shape and the staining qualities of the nuclei. Cysts of these degenerating cells comprise 3.0-7.0% of the total volume of testes as determined by linear optical measurements.

After January the degenerating cells appear to be

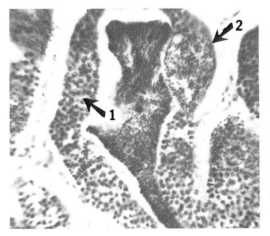

Figure 1. (a) Lobule of muskellunge testis in late September. Maturing sperm in lumen, surrounded by cysts of primary spermatocytes (1) and spermatids (2). H and E × 300.

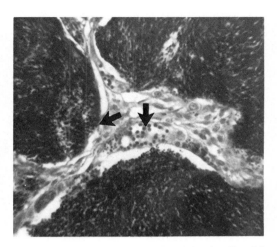

(b) Muskellunge testis in late April with masses of mature spermatozoa and small cysts of undeveloped intermediate stages (<−). H and E × 300.

reabsorbed, gradually diminish in frequency, and make up only about 1.5% of the testes volume at the time of spawning (Fig. 1b). Spermatogonia are abundant throughout this period and are located in a thin layer of connective tissues of the lobule walls that protudes into the lumen. Many of these cells have divided to form two separate cells, some of which appear to have 2-4 nuclei. Most of the volume of the testis consists of packed masses of mature spermatozoa.

This cycle of spermatogenesis in muskellunge is quite similar to that described for the northern pike in Europe and Asia (Zaitzev 1955; Lofts and Marshall 1957, Shikhshabekov 1979, Hoffman et al. 1980) except in two respects. According to these authors all spermatogonia of northern pike disappear by November and do not reappear until after spawning in the spring, whereas in muskellunge they appear to be present at all times of the year. The other marked difference for muskellunge is the in-

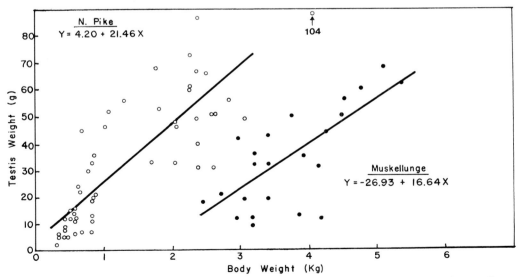

Figure 2. Regressions of mature testes weight (g) on total weight (kg) for northern pike and muskellunge.

complete transformation of intermediate stages and subsequent degeneration.

Spermatozoa Production in Northern Pike and Muskellunge

The relationship of testes weight to body weight for both northern pike and muskellunge during the winter months, when the testes are full of mature spermatozoa, is shown in Fig. 2. For muskellunge the size of the testes in grams relative to body weight in kilograms was described by the equation $Y = 4.20 + 21.46X$ ($r = .826$).

The testes of a 2 kg northern pike and a 5 kg muskellunge weigh about the same. The mature testes of northern pike constitute about 2.25% of the body weight while those of muskellunge are only about 0.8-1.4%. By comparison bluegill testes comprise 1.0-1.5% at maximum development (James 1946), white bass 1.7-5.7% (Kilpatrick 1959), yellow perch 4.6-5.9% (Turner 1919), and the Pacific herring nearly 25% (Wynne-Edwards 1929).

The density of mature spermatozoa in whole testes and in the lobular fluid was determined for nine male muskellunge taken between Feb. 6 and April 22, 1960 (Table 1). The mean density of sperm in whole muskellunge testes is about 7% less than in northern pike testes, a statistically nonsignificant difference. The connective tissue walls surrounding the lobules of muskellunge testes appeared to be somewhat thicker than in northern pike. The mean concentration of sperm in the lobular fluid is about 11% higher in muskellunge than in northern pike, but this difference is also not statistically significant.

Collection of Naturally Spawned Eggs

Numbers of live and dead northern pike and muskellunge eggs collected from natural spawning sites in Wisconsin Lakes in 1960 and 1964 are summarized in Table 2. Northern pike were observed actively spawning on April 14, 1960 when a total of 1056 eggs were collected. Spawning fish were not directly observed during the period of April 24 - May 2, 1960 when 1547 eggs were obtained because of windy, cloudy weather, but it is believed the eggs were from northern pike since water temperatures did not exceed 9 C. Two spent female northern pike were taken in nets on April 25, 1960.

Muskellunge were observed spawning from May 12 through May 17, 1960 when water temperatures increased to as much as 14C. However, spawning activity was widely dispersed and only 133 eggs were collected on trays, 30.8% of which were viable.

The water temperature in the spawning bay in Trout Lake was 14-16 C during May 13-21, 1960 and muskellunge were actively spawning. As many as eight pairs of muskellunge were observed, spawning over beds of *Anarcharis* and also near shore over dead maple leaves, sticks, and other debris. Despite intense spawning activity, only 121 eggs were collected on the trays because of minnows actively feeding on the deposited eggs. An additional 1225 eggs were recovered from vegetation dredged with a handnet, of which 402 were live and in various stages of development.

Muskellunge apparently regarded the trays as a rather poor substrate for egg deposition since eggs were usually found in a corner or near the edges rather than scattered over the tray surface. However, there was no strong tendency to avoid the trays altogether.

In 1964 no spawning activity was observed until May 4. On May 7 a total of 15 trays with protective lids were distributed in the Trout Lake spawning site and tended daily for 5 days. The water temperature during this period ranged from 7.5 C to 10 C in early morning to about 13 C in midafternoon. A total of 206 eggs were collected on trays during this period.

Table 1. Estimates of mature spermatozoa density ($\times 10/g$) in testes of northern pike and muskellunge following completion of spermatogenesis prior to spawning.

Species	No.	Whole Testis Mean	(1S.E.)	Lobular Fluid Mean	(1S.E.)
Muskellunge	9	3.165	(0.082)	5.131	(0.240)
Northern pike	22	3.386	(0.089)	4.565	(0.205)

Table 2. Summary of collections of muskellunge and northern pike eggs and their viability at the time of collection.

Lake (County)	Dates	Live Eggs No.	(%)	Dead Eggs No.	(%)	D[a]
NORTHERN PIKE						
Wingra (Dane)	4-14 to 4-15-60	498	(47.2)	558	(52.8)	T
Allequash (Vilas)	4-24 to 5-2-60	1092	(70.6)	455	(29.4)	T
MUSKELLUNGE						
Allequash (Vilas)	5-13 to 5-17-60	41	(30.8)	91	(69.2)	T
Trout (Vilas)	5-14 to 5-21-60	86	(71.1)	35	(28.9)	T
Trout (Vilas)	5-14 to 5-21-60	402	32.8	823	(67.2)	N
Trout (Vilas)	5-7 to 5-12-64	182	(88.4)	24	(11.6)	T
Trout (Vilas)	5-7 to 5-12-64	170	(20.7)	653	(79.3)	N

[a]D = collecting device: T = Trays N = hand net

A total of 153 eggs were deposited the first 2 days and incubated in aerated water at 11-13 C for 3 more days. Viability was 95% initially and on successive days of incubation was 89.1%, 67.5%, and 47.6%. Only a few eggs were found on trays on days 3 and 4, but 40 more were found on day 5, 80% of which were viable.

Many more eggs were recovered from vegetation, but only 20.7% were viable at the time of collection. Most of these were at the mid-epiboly stage of embryologic development, indicating they were about 2 days old.

DISCUSSION

The potential production of sperm is lower for muskellunge than for northern pike. Not only do muskellunge mature much later, in about their fourth year of life, but also their testes are less than half as large as northern pike of the same size. In addition, the density of sperm within muskellunge testes is about 7% less, probably because of a greater proportion of connective tissues and the failure of intermediate stages of sperm cells to fully complete their development.

The fertilization rate of eggs collected on trays at different times suggests that temporal differences may exist. The highest fertility rate for muskellunge (95%) was for eggs known to have been spawned during the early part of the spawning period in 1964. However, for other samples the time of collection in relation to the entire spawning period was not determined. The newly spawned eggs taken on trays in Allequash Lake, only 30.8% of which were fertile, may have been spawned toward the end of the spawning period, but there is insufficient data to support or reject that possibility.

The data do indicate that under some conditions the fertility rate of muskellunge eggs is high. This finding differs from that of E.L. Schneberger (unpublished data, Wis. Dept. Conservation) who in 1936 found that only 47 of 166 (34%) naturally spawned muskellunge eggs were fertile. Those eggs were apparently collected during the peak spawning period on May 3 and 4, incubated for a few days, and then examined. He also found that 3% of the eggs died after being fertilized.

Of the two batches of eggs collected on trays in the present study, approximately 95% had been fertilized, but less than half were viable after 3 to 4 days of incubation (44% and 48%). Thus it appears that considerable mortality occurs during early development. This may have been an important factor in Schneberger's study also. Also, a marked disparity was noted between the viability of incubated eggs (44% and 48%) and eggs extracted at the same time from vegetation (21% to 33%). These dredged eggs were of mixed ages and had developed to different stages, but their survival rate was lower than newly-spawned, incubated, and aerated eggs from trays. Nearly all dredged eggs were found in direct contact with *Anacharis* leaves. Dombeck, et al. (1984) found very high mortality rates (99% and 76%) of muskellunge eggs incubated for only 3 days on willow leaves and coontail, and attributed this high mortality rate to low dissolved oxygen concentrations (0-0.1 mg/l). Dissolved oxygen concentrations were not determined in this study.

Some mortality observed here was from predation by minnows and other small fish. In 1960 egg trays deployed in the spawning bay had only one side protected by quarter-inch hardware cloth. Eggs were found only on the screen-protected side of the trays. Undoubtedly, eggs on the flat surface of the unprotected trays were more vulnerable to predation than those spawned in vegetaion.

Future research should include coordinated studies in both the field and laboratory in order to assess the relative importances of spawning substrate, dissolved oxygen concentration, temperature, and predation as causes of egg mortality.

ACKNOWLEDGEMENTS

This research was supported in part by the Rahr Foundation and the Danforth Foundation. The assistance of Dr. Philip Doepke was especially helpful in 1964. Mr. Steve Taylor and Mr. Herb Sahar provided invaluable assistance in securing testes samples. This contribution is dedicated to two people who have had an unusually large influence on my life; Mr. C.L. O'Beirne, a good friend and avid muskellunge fisherman and Dr. Arthur D. Hasler, University of Wisconsin.

REFERENCES

De Montalembert, G., C. Bry, and R. Billard. 1978. Control of reproduction in northern pike. American Fisheries Society Special Publication 11:217-225.

Dombeck, M.P., B.W. Menzel, and P.N. Hinz. 1984. Muskellunge spawning habitat and reproductive success. Transactions American Fisheries Society 113: 205-216.

Foley, J.O. 1926. The spermatogenesis of *Umbra limi* with special reference to the behavior of the spermatogonial chromosomes and the first maturation division. Biological Bulletin 50:117-147.

Gammon, J.R. 1961. Contributions to the biology of the muskellunge. Ph. D. Thesis, University of Wisconsin. 144 pp.

Hann, H.W. 1927. The history of the germ cells of *Cottus bairdii* (Girard). Journal of Morphology 443: 427-498.

Hoffmann, R., P. Wondrak, and W. Groth. 1980. Seasonal anatomical variations in the testes of European pike, *Esox lucius* L. Journal of Fish Biology 16:475-482.

Huet, M. 1953. Traité de Pisciculture, 2nd ed. Bruxelles: Editions La Vie Rustique.

James. M.F. 1946. Histology of gonadal changes in the bluegill, *Lepomis macrochirus* Rafinesque. and the largemouth bass, *Huro salmoides* (Lacépède). Journal of Morphology 79:63-92.

Johnson, L.D. 1958. Pond culture of muskellunge in Wisconsin. Wisconsin Conservation Department Technical Bulletin Number 17:1-54.

Johnson, L.D. 1978. Evaluation of esocid stocking program in Wisconsin. American Fisheries Society Special Publication 11:298-301.

Kilpatrick, E.B. 1959. Seasonal cycle in the gonads of the white bass (*Roccus crysops*), in Lake Texoma, Oklahoma. Ph. D. Thesis, University of Oklahoma. 44 pp.

Klingbiel, J. 1966. An evaluation of stocking large muskellunge fingerling. Wisconsin Department of Natural Resources, Fisheries Management Division, Report 3:11 pp.

Lofts, B., and A.J. Marshall. 1957. Cyclical changes in the distribution of the testis lipids of a teleost fish, *Esox lucius*. Quarter Journal of Microscopical Science 98:79-88.

Oehmcke, A.A., L. Johnson, J. Klingbiel, and C. Wistrom. 1958. The Wisconsin Muskellunge, its life history, ecology, and management. Wisconsin Conservation Department Publication 225:1-12.

Shikhshabekov, M.M. 1979. The sexual cycles of the catfish *Silurus glanis,* the pike, *Esox lucius,* the perch, *Perca fluviatilis,* and the pike-perch, *Lucioperca lucioperca*. Journal of Ichthyology 18:457-468.

Svardson, G. 1947. Gaddlekstudier. Sondra Sveriges Fiskeriforening. Skrifter: 34-59.

Turner, C.L. 1919. The seasonal cycle in the spermary of the perch. Journal of Morphology 32:681-711.

Wynne-Edwards, V.C. 1929. The reproductive organs of the herring in relation to growth. Journal Marine Biological Association 16:49-65.

Zaitzev, A.V. 1955. The annual cycle of testes in pike. Akademii Nauk SSSR (Doklady) 101:185-187.

Size, Age, and Growth of Trophy Muskellunge and Muskellunge-Northern Pike Hybrids -- The Cleithrum Project, 1979-1983[1]

J.M. CASSELMAN

Ontario Ministry of Natural Resources
Box 50, Maple, Ontario L0J 1E0

E.J. CROSSMAN

Department of Ichthyology and Herpetology
Royal Ontario Museum
100 Queen's Park, Toronto, Ontario M5S 2C6

ABSTRACT

In 1979 a joint study, called "The Cleithrum Project," was initiated by the Ontario Ministry of Natural Resources and the Royal Ontario Museum. This involved the establishment of a repository for the collection and subsequent dissemination of basic biological data on angled, trophy muskellunge and muskellunge-northern pike hybrids. Basic data on length, weight, and sex are provided voluntarily by cooperating taxidermists and anglers. The cleithrum, a flat bone from the pectoral girdle, is also supplied. It is an essential part of the project, and is used to obtain biological data and an accurate age assessment (Casselman 1974, 1979), which is provided to the cooperator.

Data on 506 trophy muskellunge up to 30 years of age and 25 kg in weight were received up to August 1983, and were analyzed separately for Canada (N = 190) and United States (N = 316). Trophy muskellunge from Canadian waters are on the average significantly older, longer, and heavier (12.7 yr, 114.0 cm, 10.9 kg) than those from United States waters (9.9 yr, 104.4 cm, 8.6 kg). Biological data from across the range of the species are used to describe averages, confidence limits, extremes, and scope for growth as well as age-length, age-weight, length × girth-weight, length-weight, and cleithrum-body relations. Von Bertalanffy growth models and estimates of ultimate lengths (average 139.3 cm) and weights (average 20.5 kg) are presented.

Data are similarly described for 60 natural muskellunge-northern pike hybrids from 3 to 18 years of age, averaging 7.2 yr and 6.8 kg, but weighing up to 15.4 kg. Although the hybrids were slightly faster growing than were muskellunge up to approximately age 8, they did not appear to attain the maximum age or size of muskellunge.

The data are used to construct growth standards for the species and a current, quantitative definition of "trophy" muskellunge and muskellunge-northern pike hybrids.

In 1979 a joint study, The Cleithrum Project, was initiated by the Department of Ichthyology and Herpetology, Royal Ontario Museum, and the Age-Growth and Environmental Studies Fisheries Research Unit, Ontario Ministry of Natural Resources. The purpose was to collect and disseminate biological data on trophy muskellunge. The repository, which was intended to assemble information independent of geographic and political boundaries, resides at the Royal Ontario Museum. The project focuses on the cleithrum, the major bone of the pectoral girdle of the fish, which is an important tool in age and growth assessment of northern pike and muskellunge (Casselman 1974, 1979).

We were concerned that potentially valuable data, which could be collected on trophy muskellunge, were being overlooked and lost. With increasing fishing pressure, large trophy size muskellunge were being rapidly "fished up", and valuable biological information, such as growth potential, longevity, and ultimate size, needed to be documented for future comparisons. In addition, the increasing practice of releasing even very large muskellunge will probably result in fewer of these data in the future.

The study was also launched because fisheries workers often find it difficult, using conventional sampling techniques, to collect adequate quantities of data on muskellunge, especially those of trophy

---

[1]Contribution No. 85-08 of the Ontario Ministry of Natural Resources, Fisheries Branch, Research Section, Box 50, Maple, Ontario L0J 1E0.

size. In most Canadian populations, muskellunge are rarely seen during routine creel surveys. They are not caught consistently; indeed, this is part of the "musky mystique". Under these conditions it is almost impossible to monitor rate of exploitation or to collect even scanty amounts of biological data on angled muskellunge of any size. Once a trophy is caught, the angler immediately returns to port to share his success. Even if he is intercepted by a biologist, the angler is usually reluctant to allow his prize to be "tampered with or mutilated", even for the sake of science, but wants it safe in the hands of his taxidermist and mounted for the wall.

Through this project muskellunge anglers, taxidermists, and biologists are being alerted to the importance of collecting all the biological information possible from every muskellunge that is killed, especially those of trophy size. The Cleithrum Project not only provides a mechanism for intercepting and using these data, but also stimulates interaction and builds rapport between muskellunge anglers and fisheries workers.

Many well-known muskellunge populations are fished by anglers from both Canada and the United States, and the international transport of angled muskellunge for mounting is common. The collection of biological data must not be influenced by man-made boundaries, and necessitates an international effort.

This paper describes the type of information that is being derived from The Cleithrum Project, discusses its limitations, and provides a summary of the data received and compiled up to August 1983 on: 1) age-length, age-weight, age distribution, scope for growth, length-weight, length × girth-weight, and cleithrum-body relations; 2) growth standards that can be used to examine changes in growth rate relative to exploitation; 3) estimated ultimate size; and 4) a current, general description of "trophy" muskellunge and muskellunge-northern pike hybrids.

MATERIALS AND METHODS

The data are exclusively from angler-caught trophy muskellunge and muskellunge-northern pike hybrids that were provided voluntarily by cooperative taxidermists and anglers. In this study a "trophy" fish is defined simply as one that is mounted or that the successful angler considers worthy of mounting.

From 1979 until August 1983, The Cleithrum Project has received data on over 600 trophy esocids from across the North American range. The majority of the samples came from Wisconsin and Ontario. The initial request for data was outlined in Crossman and Casselman (1979a, b). In return for specific information such as length, weight, girth, date and origin of capture, sex determination, as well as the cleithrum, the cooperator is given an accurate age determination of the fish.

The cleithrum was chosen as a basic data collection unit because it is a useful indicator calcified structure (Casselman 1979), which records the environmental physiology of the fish, and provides an accurate age assessment and other important biological information (e.g., growth rate). The cleithrum is vital to the project because it provides the only known valid method of determining age of old muskellunge. The scale method is invalid for old or slow-growing muskellunge (Casselman 1983).

Body length of the fish was most often reported as total length, and if necessary was converted to the metric equivalent and expressed to the nearest 0.1 cm. When only fork length (FL) was provided, it was converted to total length (TL) using the following equation:

$$Y_{FL(cm)} = -2.54 + 0.9617 X_{TL(cm)}$$
$$N = 45 \qquad r = 0.999$$

The conversion from total length to fork length is:

$$Y_{TL(cm)} = 2.89 + 1.0376 X_{FL(cm)}$$

These conversion equations are from trophy muskellunge (>80 cm) from the St. Lawrence River (Casselman, unpublished data). Girth measurement was if necessary converted to metric and expressed to the nearest 0.1 cm. Total weights provided in pounds were converted to the nearest 0.1 kg. Metric values were used in subsequent analyses.

Data for fish from Canadian and United States waters were analyzed first together, then separately. Samples from international waters were few, and were included with the Canadian samples. It was apparent from preliminary examination of the data that length, weight, and age distributions of fish from Canadian and United States waters were significantly different, and should be presented separately. Although biology could be influenced by latitude, differences in rate of exploitation are probably a major factor. Almost twice as many samples (1.7×) were submitted from United States waters; this is probably indicative of the difference in rate of exploitation.

Sex of the fish was provided for 43% of the samples from Canadian waters, and for 76% of the samples from United States waters. Females dominated in both samples: Canada—6.3:1 and United States—5.1:1. Analyses were not separated by sex because it is not known if this extreme sex ratio is typical of angled trophy muskellunge, or is an artifact of mistaken sex determination, since sex was determined by untrained people, and could not be verified. The sex ratio of a similar type of sample of trophy muskellunge from the upper St. Lawrence River was slightly in favour of females, 2.1:1 (Casselman, unpublished data). This indicates that anglers harvest disproportionately more female than male muskellunge. However, the sex ratio of the samples submitted to The Cleithrum Project appears extreme, and is presented with reservation until it can be substantiated.

Most of the 181 samples for which date of capture was reported were from fish that were caught between June and September (86%): June—17%, July—24%, August—25%, September—20%, October—9%, November—4%. Specific regional differences in date of capture were noticed. In Great Lakes populations, the majority of the trophy muskellunge were taken from late September to mid October. However, in most other populations fish were caught most frequently in July and August, and to a lesser extent in June.

Since length, weight, and girth data were supplied by numerous cooperators, the measurements could be somewhat variable. The measurements were made not only by different people using slightly different procedures, but also under different conditions and at different times after capture. However, the cleithral bone reduces this variability, because if undamaged this hard structure can provide precise information about the fish from which it was extracted. Cleithral measurements were made in a consistent fashion by trained technicians on a hard structure relatively unaltered by preservation and handling.

When cleithra were received, they were cleaned of muscle and connective tissue by repeatedly immersing them for a few seconds in simmering water, with a small amount of detergent added, each time wiping away the loosened soft tissue with a soft toothbrush and paper or cotton towelling. They were then allowed to air-dry.

Age was assessed by enumerating the number of translucent zones associated with annuli (Casselman 1983). Because the majority of the cleithra were taken from fish during the middle of the growing season, opaque bony tissue was often observed outside the last annulus on the edge of the cleithrum ("+" condition). Under these conditions the assessed cleithral age would be equal to the number of annuli, which equals the calendar age expressed in years. However, approximately 10% of the cleithra in the sample contained a translucent zone on the edge of the bone ("o" condition). This translucent zone is considered to be associated with conditions that would produce an additional annulus on the edge if it were demarcated with opaque bony tissue (if growth resumed or increased). If this "o" condition occurs at the beginning of the growing season before growth resumes, the calendar age at time of capture would be one more than the number of annuli, whereas if the condition occurs at the end of the growing season the calendar age at time of capture would be equal to the number of annuli.

Anterior cleithral radius (ACR) is the distance from the origin to the anterior tip of the blade, and is considered the best linear measurement of the cleithrum for size calculations (Casselman 1978). The anterior radius of the cleithrum was measured to the nearest 0.1 mm on cleithra that were complete and undamaged.

Relations for size at age were constructed for total length, total weight, and anterior cleithral radius, using averages, and are presented with 95% confidence limits and extremes. Average values were used to calculate the von Bertalanffy growth curve using the iterative method (Ricker 1975), and to determine the growth parameters (k) and asymptotic values, or ultimate sizes (TL_∞, TW_∞, ACR_∞). Average values for each age group were used to examine the relation between total length, girth, and weight and the length-weight and cleithrum-body relations for muskellunge and muskellunge-northern pike hybrids.

The von Bertalanffy growth models for length and weight were used to create growth standards for muskellunge, and the 95% confidence limits and extremes at each age are presented as a percent of this standard. The length and weight growth standards are the von Bertalanffy growth curve estimates of length and weight at cleithral age. The calculated growth indices are the length or weight expressed as a percent of this growth standard. Growth-length indices were calculated for data from other muskellunge studies to ascertain the applicability of the linear growth standard and the growth indices.

RESULTS AND DISCUSSION

Length

Length-frequency distributions of trophy mus-

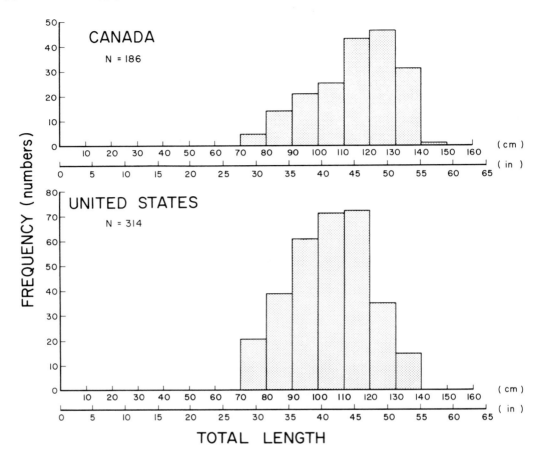

Figure 1. Length-frequency distributions with a class interval of 10 cm total length for 500 trophy muskellunge, separated according to origin.

kellunge from Canadian and United States waters (Fig. 1) were significantly different ($X^2 = 48.6$, 6 df P<0.001). Trophy muskellunge from Canada were on the average 10 cm longer (114.0 ± 2.07 cm-- average ± 95% confidence interval), than those from the United States (104.4 ± 1.47 cm). The modal length class for the Canadian sample was 120 to 130 cm and for the United States sample 110 to 120 cm. In the sample from Canadian waters one fish in six exceeded 130 cm, as compared to one in 21 from United States waters. The longest fish in the Canadian sample was 141 cm, and 137 cm in the United States sample.

Weight

Weight frequency distributions of trophy muskellunge from Canadian and United States waters (Fig. 2) were also significantly different ($X^2 = 50.6$, 7df, P<0.001). Trophy muskellunge from Canada were on the average 2.3 kg heavier (10.9 ± 0.55 kg) than were those from the United States (8.6 ± 0.37 kg). The modal weight class for the Canadian sample was 14 to 16 kg and for the United States sample 8 to 10 kg. In the sample from Canadian waters, one fish in eight exceeded 16 kg, as compared to one in 35 from United States waters. The heaviest fish in the Canadian sample was 25.0 kg, as compared to 19.1 kg in the United States sample.

Age

Age frequency distributions of trophy muskellunge from Canadian and United States waters (Fig. 3) were highly significantly different ($X^2 = 46.2$, 16 df, P<0.001). Trophy muskellunge from Canada were on the average 2.8 years older (12.7 ± 0.77 years) than those from the United States (9.9

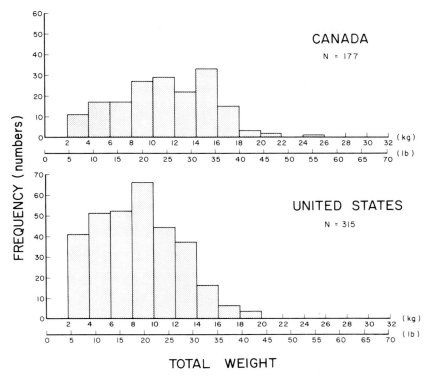

Figure 2. Weight-frequency distributions with a class interval of 2 kg total weight for 492 trophy muskellunge, separated according to origin.

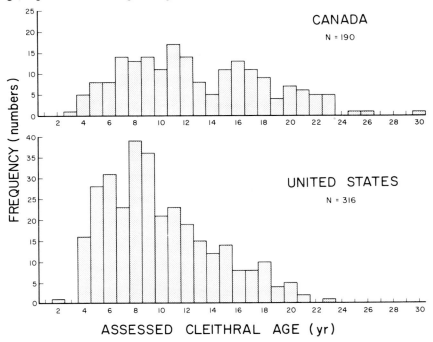

Figure 3. Age-frequency distributions of 506 trophy muskellunge, separated according to origin.

± 0.50 years). The sample from Canadian waters included 24 age groups with a modal age of 11 years, whereas the United States sample included 20 age groups with a modal age of 8 years. In the sample from Canadian waters, 10% of the fish exceeded 20 years of age, as compared with only 1% from United States waters. The oldest fish from Canadian waters was assessed at 30 years, as compared with 23 from United States waters. The age distribution of this sample of fish is unique because it is rare in fisheries work that a sample of fish encompasses so many age groups and contains so many old individuals.

Length at Age and Linear Growth Rate

There was no significant difference between the lengths at age for the samples from Canadian and United States waters, so samples were combined, and results were obtained from data for 500 trophy muskellunge representing 25 age groups (Fig. 4). It was apparent that angler selectivity occurred in the smaller, younger fish. This selectivity occurs when anglers retain only the largest of their small fish, and results in abnormally large length values for the youngest age groups. This bias is usually greatest at, or slightly above, minimum length limits. Hence, the youngest individuals in the sample appear to be abnormally fast growing. This type of selectivity no doubt influenced results for size at age for the younger age groups in the sample, especially at ages 2 and 3, and possibly even at age 4.

The 95% confidence limits were, on the average, 4.5% above and below the average length at each age (Table 1). The longest individuals at each age averaged 11.5% above average, and the shortest individuals 14.3% below average. These extremes and limits delineated the boundaries (Table 1) of the scope for linear growth of angled trophy muskellunge from across the range (illustrated in Fig. 4).

The von Bertalanffy growth model, calculated using average total length at assessed cleithral age for ages 4 to 23 (Fig. 4), estimates the average ultimate total length (TL_∞) of trophy muskellunge to be 139.3 cm. This is a good approximation because muskellunge caught in recent years have rarely exceeded 140 cm total length. Using the length-weight relation for trophy muskellunge (presented below), it is estimated that a fish of this length would weigh 19.8 kg.

Table 1. Averages, 95% confidence limits, extremes, number of samples, linear growth standard, and growth-length indices (percent of growth standard) for 495 muskellunge from across the range.

Assessed cleithral age (yr)	Average total length (cm)	N	Growth standard (von Bertalanffy growth model) length (cm)	95% confidence limits				Range of lengths				Assessed cleithral age (yr)
				Lower (cm)	% of growth standard	Upper (cm)	% of growth standard	Smallest (cm)	% of growth standard	Largest (cm)	% of growth standard	
4	81.4[a]	21	84.0	78.9	93.9	84.0	100.0	71.9	85.6	94.0	111.9	4
5	82.7	36	89.0	80.4	90.3	85.0	95.5	72.4	81.3	101.6	114.2	5
6	92.8	39	93.8	90.8	96.8	94.8	101.1	81.3	86.7	106.7	113.8	6
7	96.1	37	98.0	93.6	95.5	98.6	100.6	75.6	77.1	111.8	114.1	7
8	100.8	50	101.9	98.4	96.6	103.2	101.3	76.2	74.8	116.8	114.6	8
9	106.3	49	105.3	103.8	98.6	108.8	103.3	85.1	80.8	123.2	117.0	9
10	108.3	31	108.5	105.0	96.8	111.5	102.8	91.4	84.2	121.9	112.4	10
11	113.9	40	111.4	111.3	99.9	116.5	104.6	90.5	81.2	129.5	116.2	11
12	114.3	31	114.0	111.4	97.7	117.3	102.9	97.8	85.8	127.0	111.4	12
13	116.5	23	116.3	113.1	97.2	120.0	103.2	96.5	83.0	127.0	109.2	13
14	118.6	17	118.4	114.6	96.8	122.6	103.5	102.9	86.9	127.0	107.3	14
15	121.6	25	120.4	117.8	97.8	125.3	104.1	104.1	86.5	134.6	111.8	15
16	125.9	21	122.1	121.7	99.7	130.0	106.5	101.6	83.2	141.0	115.5	16
17	122.5	19	123.7	118.8	96.0	126.3	102.1	109.2	88.3	134.6	108.8	17
18	127.8	19	125.2	124.5	99.4	131.1	104.7	109.2	87.2	137.2	109.6	18
19	125.3	8	126.5	117.1	92.6	133.5	105.5	111.8	88.4	139.7	110.4	19
20	129.9	12	127.7	125.8	98.5	134.0	104.9	116.8	91.5	137.2	107.4	20
21	127.0	8	128.8	119.7	92.9	134.3	104.3	118.1	91.7	139.7	108.5	21
22	129.9	3	129.7	117.9	90.0	141.9	109.4	125.7	96.9	134.5	103.7	22
23	127.4	6	130.6	118.5	90.7	136.4	104.4	111.8	85.6	132.1	101.1	23
Average					95.93		103.24		85.34		110.95	

[a] Angler selectivity, created by a minimum size limit, has resulted in a value that is abnormally large for 4-year-old fish and should not be used in individual comparisons.

If the von Bertalanffy growth equation is calculated using the longest individuals at each age, the estimated maximum ultimate length is 173 cm. However, this is a rather artificial projection because it is based on estimations from several individuals from different populations growing at the maximum rate for the life span of the species. It is unlikely that any one fish could maintain this excessively rapid growth rate for its entire life. Nevertheless, this theoretical ultimate length is actually 5.6% longer than the current world weight record muskellunge (TL--163.8 cm; TW--31.7 kg) caught

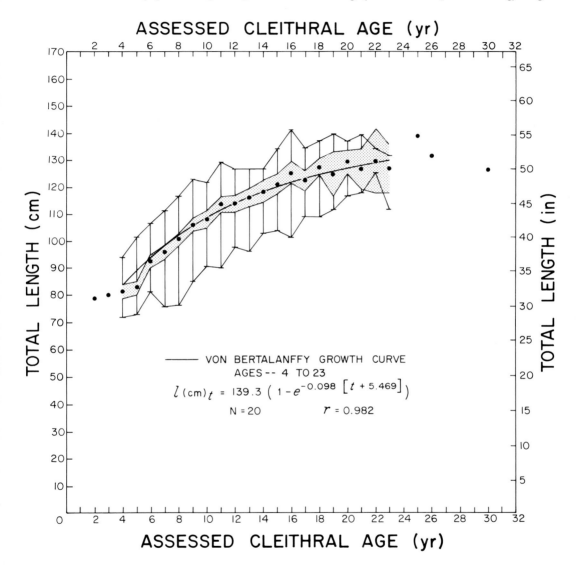

Figure 4. Average total length at assessed cleithral age, growth rate, and scope for linear growth for 500 trophy muskellunge from across the North American range. The smallest and largest at each age are illustrated by crossbars on the vertical range line. The 95% confidence interval (shaded area) about the average for each age (closed circle) is illustrated for ages containing 6 or more individuals. Closed circles at the extremes indicate single values. The von Bertalanffy growth equation for average length is given, and the curve is illustrated over the range of ages used in the calculation.

in the upper St. Lawrence River in 1957 (Ramsell 1982).

To examine the applicability of the linear growth standard and the growth index, size at age data for 18 studies were examined (Table 2). An average growth-length index was calculated for muskellunge of ages 5 to 10. Eight data sets gave indices significantly less than the standard, six were approximately equal, and four were greater. It is apparent from these indices that populations of muskellunge in streams and relatively unexploited small lakes are significantly slower growing than the standard. Populations in lakes at the southern part of the range are faster growing. Growth of muskellunge in large rivers is very similar (St. Lawrence River--95%, 98%, and 99%; Niagara River--97%).

Table 2. Growth-length indices for 18 studies providing size at age data for muskellunge. The growth index is an average calculated for ages 5 to 10.

Growth index (%)	Origin	Authority
78	Nogies Creek, Ontario	Muir 1960
81	Chautauqua Lake, New York	Bimber 1982
84	Bone Lake, Lac Court Oreilles, Big Spider Lake, Wisconsin	Johnson 1971
87	streams in Tennessee	Parsons 1959
87	streams in western Virginia	Miles 1978
89	Chautauqua Lake, New York	Greeley 1938 (from Carlander 1969)
92	Eagle and Wabigoon lakes, Ontario	Casselman and Gibbard, unpublished data
94	streams in Kentucky	Brewer 1980
95	upper St. Lawrence River	Casselman, unpublished data
97	waters in Michigan	Williams 1959
97	Niagara River	Harrison and Hadley 1979
98	lakes St. Francis and St. Louis	Hourston 1952
99	lakes St. Francis and St. Louis	Simard and Magnin 1968
100	Cave Run Lake, Kentucky	Axon 1978
101	lakes and drainage areas in northern Wisconsin	Schloemer 1936
103	Lake St. Clair	Haas 1978
105	waters of Pennsylvania	Buss and Miller 1961
109	Lake Pomme de Terre, Missouri	Belusz 1978

Although this growth standard is somewhat biased, it provides a useful means for comparing growth of individual fish, samples, and populations. The standard is biased because it is constructed from a model using lengths of fish caught throughout the growing season (the majority in midsummer). Hence indices for fish taken at the beginning of the growing season, or lengths obtained by "back calculation", are smaller, and fish appear to be slower growing. From these indices (Table 2) it is estimated that back-calculated lengths would underestimate growth by only 2 to 4% when compared with this growth standard constructed from midsummer lengths. Also, the standard could be somewhat biased because it contains disproportionately more female fish, which are faster growing than male fish. Nevertheless, this may be typical of a sample of angled muskellunge. These types of biases do not reduce the utility of the standard, but, if understood, may help to explain some of the slight differences that may be observed.

This growth standard was constructed using age assessment from cleithra. If scale ages are used in comparisons with this standard, sizes at scale ages greater than 10 should be excluded from the analysis. In most populations muskellunge older than age 10 cannot be accurately aged by the scale method (Casselman 1983).

Weight at Age and Growth Rate in Weight

There was no significant difference between weights at age for samples from Canadian and United States waters, so samples were combined, and results are presented from data for 495 trophy muskellunge representing 25 age groups (Fig. 5). Young fish, ages 2 to 4, were heavier than expected for their age, indicating that angler selectivity was greater in weight than in length. Weight data were considerably more variable than were length data. The 95% confidence limits averaged 11.3% above and below the average weight at age (Table 3). The heaviest individuals averaged 46.7% above average, and the lightest individuals 40.0% below average. These extremes and limits delineate the boundaries (Table 3) of the scope for growth in weight for angled trophy muskellunge from across the range (illustrated in Fig. 5).

The von Bertalanffy growth model was calculated using average weight at assessed cleithral age for ages 4 to 23 (Fig. 5). The average ultimate weight (TW_∞) for trophy muskellunge, as estimated by this model, is 20.49 kg. This is not large, but is a reasonable approximation of the average ultimate weight of muskellunge because muskellunge caught in recent years have not exceeded 25 kg total weight. If the von Bertalanffy growth equation is calculated

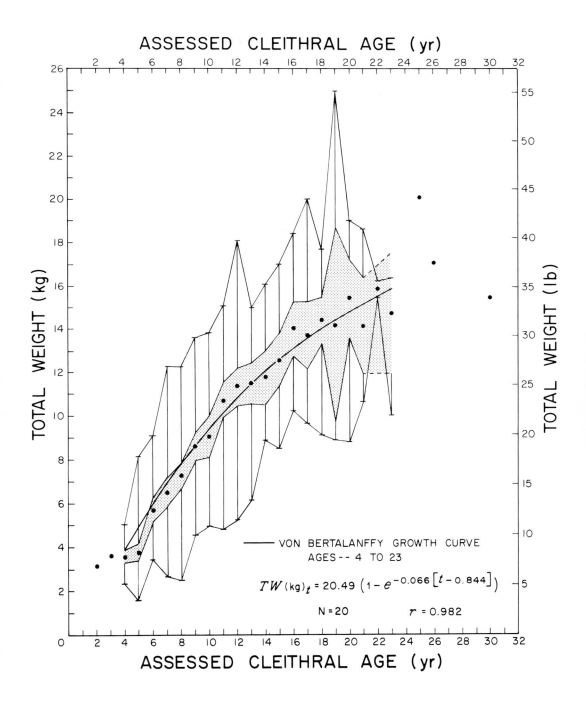

Figure 5. Average weight at assessed cleithral age, growth rate, and scope for growth in weight for 495 trophy muskellunge from across the North American range. The methods used to illustrate the results are the same as in Figure 4. Only two individuals were 22 years old.

using the heaviest individuals at each age, the estimated maximum ultimate weight would be 33.3 kg. This theoretical maximum ultimate weight is 5.0% heavier than the weight of the current world record. If a muskellunge grew at an excessively rapid rate for its entire life and lived to a very old age, it would be theoretically possible for it to attain a weight equal to, or even slightly heavier than, the present world record.

The von Bertalanffy growth model, calculated using average total weight at assessed cleithral age, was used to construct a weight growth standard for muskellunge by estimating weight at each cleithral age (Table 3).

Length and Weight Relation

There was no significant difference between the length-weight relations calculated for samples collected from Canadian and United States waters, so samples were combined and results were obtained from data for 495 trophy muskellunge representing 25 age groups ranging from 2 to 30 (Fig. 6). The length-weight equation for trophy muskellunge (Fig. 6) is very similar to those presented by Krumholz (1947) for muskellunge from Lake St. Clair, Michigan, and Chautauqua Lake, New York.

Girth

Girth is often measured on large fish to provide some verification of weight. There was a highly significant direct correlation between girth and age for 18 age groups ranging from 4 to 21; the logarithmic relation is presented in Fig. 7. The scope for girth in relation to age is indicated; however, the youngest fish, which were selected by size, had a larger girth than would be expected from an extrapolation of the relation. At the other extreme, fish over age 20 had lower girth values than might be expected. Senescence, with an accompanying loss in condition and girth, may have occurred in these older fish.

There was a highly significant correlation between length × girth and total weight for trophy muskellunge (Fig. 8). This relation was constructed

Table 3. Averages, 95% confidence limits, extremes, number of samples, weight growth standard, and growth-weight indices (percent of growth standard) for 490 muskellunge from across the range.

Assessed cleithral age (yr)	Average total weight (kg)	N	Growth standard (von Bertalanffy growth model) weight (kg)	95% confidence limits				Range of weights				Assessed cleithral age (yr)
				Lower (kg)	% of growth standard	Upper (kg)	% of growth standard	Smallest (kg)	% of growth standard	Largest (kg)	% of growth standard	
4	3.6[a]	21	3.9	3.3	84.6	3.9	100.0	2.4	61.5	5.0	128.2	4
5	3.8	36	4.9	3.4	69.5	4.2	85.7	1.5	30.6	8.2	167.3	5
6	5.7	38	5.9	5.2	88.1	6.2	105.1	3.4	57.6	9.1	154.2	6
7	6.5	36	6.9	5.8	84.1	7.2	104.3	2.6	37.7	12.3	178.3	7
8	7.2	52	7.7	6.7	87.0	7.8	101.3	2.5	32.5	12.3	159.7	8
9	8.6	49	8.6	8.0	93.0	9.2	107.0	4.5	52.3	13.6	158.1	9
10	9.0	31	9.3	8.1	87.1	10.0	107.5	4.9	52.7	13.8	148.4	10
11	10.7	39	10.0	9.9	99.0	11.5	115.0	4.8	48.0	15.1	151.0	11
12	11.3	33	10.7	10.4	97.2	12.2	114.0	5.2	48.6	18.1	169.2	12
13	11.5	23	11.3	10.6	93.8	12.4	109.7	6.1	54.0	14.9	131.9	13
14	11.7	16	11.9	10.5	88.2	13.0	109.2	8.8	73.9	16.1	135.3	14
15	12.6	23	12.5	11.4	91.2	13.8	110.4	8.5	68.0	17.0	136.0	15
16	14.0	19	13.0	12.8	98.5	15.3	117.7	10.3	79.2	18.4	141.5	16
17	13.7	19	13.5	12.1	89.6	15.2	112.6	9.6	71.1	20.0	148.1	17
18	14.4	19	13.9	13.3	95.7	15.4	110.8	9.1	65.5	17.7	127.3	18
19	14.2	8	14.3	9.8	68.5	18.7	130.8	8.9	62.2	25.0	174.8	19
20	15.4	12	14.7	13.6	92.5	17.2	117.0	8.9	60.5	19.1	129.9	20
21	14.1	8	15.1	11.9	78.8	16.3	107.9	10.6	70.2	18.6	123.2	21
22	15.8	2	15.5					15.4	99.4	16.2	104.5	22
23	14.7	6	15.8	11.9	75.3	17.5	110.8	10.0	63.3	16.3	103.2	23
Average					87.45		109.31		59.69		143.51	

[a]Angler selectivity, created when anglers retain only the largest of their small fish, has resulted in a value that is abnormally large for 4-year-old fish and should not be used in individual comparisons.

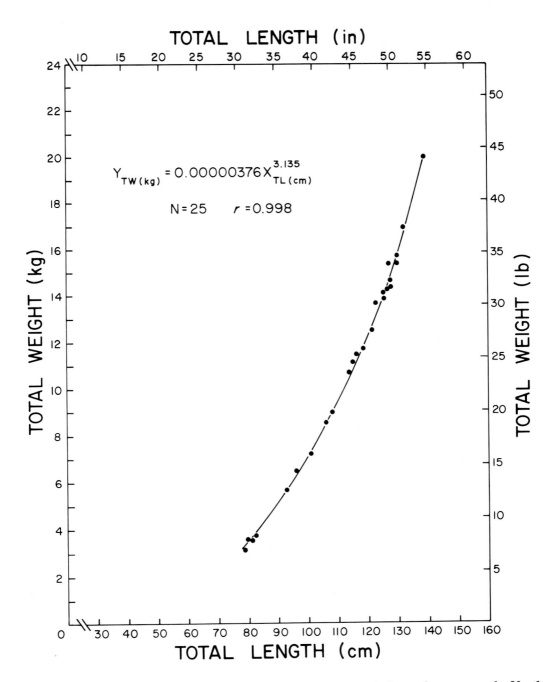

Figure 6. Relation between length and weight for 495 trophy muskellunge from across the North American range, calculated by using average size at age for 25 age groups ranging from 2 to 30.

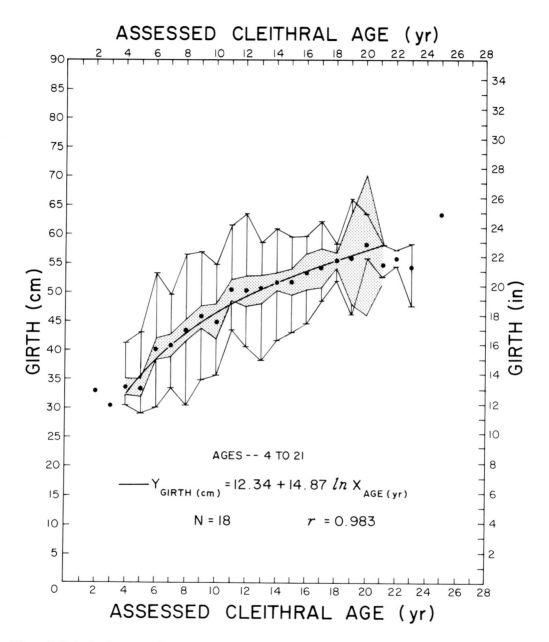

Figure 7. Relation between girth and assessed cleithral age for 341 trophy muskellunge from across the North American range. The extremes, 95% confidence interval, and average (closed circle) are illustrated for each age containing 6 or more individuals. The equation for average girth is given, and the curve is illustrated over the range of ages used in the calculation.

using mean values for 23 age-groups for fish ranging in age from 2 to 25. Although this relation is slightly curvilinear, it can be adequately fitted linearly (Fig. 8). Using these equations, weight of trophy muskellunge can be estimated with a high degree of confidence when only linear measurements are available.

Cleithrum Size at Any Age and Growth Rate

The size of the anterior cleithral radius at age did not differ significantly between samples from Canadian and United States waters, so samples were combined and results were obtained from data for 495 trophy muskellunge representing 25 age groups

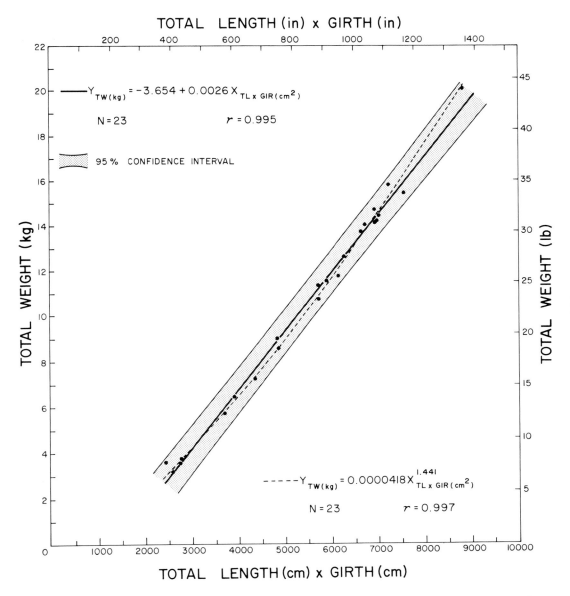

Figure 8. Relation between total length × girth and total weight for 341 trophy muskellunge from across the North American range. The linear and power curve equations are shown, calculated by using average values at age for 23 age groups ranging from 2 to 25.

(Fig. 9). The 95% confidence limits averaged 4.7% above and below the average anterior cleithral radius at age. The largest cleithral bones averaged 13.0% above average, and the smallest 12.9% below average. These extremes and limits delineated the boundaries for the scope for growth of cleithra of angled trophy muskellunge.

The von Bertalanffy growth equation, calculated using average anterior cleithral radii at assessed cleithral age for ages 4 to 23, is also provided in Fig. 9. The average ultimate anterior cleithral radius (ACR∞) for trophy muskellunge is estimated from this equation to be 130.0 mm. Using the cleithrum-body relation for trophy muskellunge (presented below), it is estimated that a bone of this size would come from a fish with a total length of 141 cm. This estimate is confirmed because muskellunge now rarely exceed 140 cm total length.

Size, Age, and Growth of Muskellunge-Northern Pike Hybrids

The muskellunge-northern pike hybrids analysed in this study were indigenous fish. The average length of the individuals in the combined sample

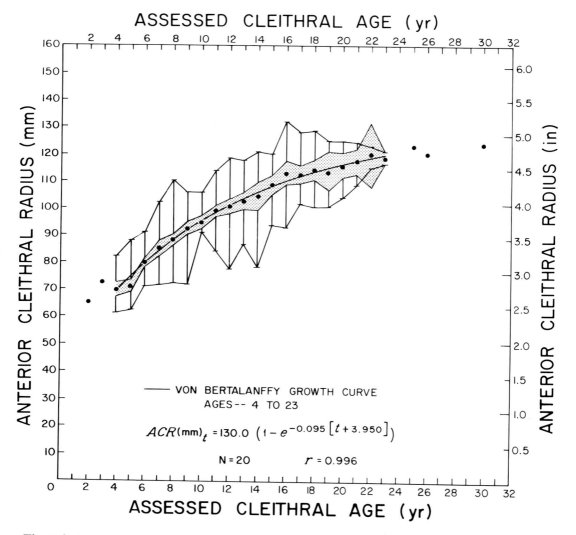

Figure 9. Average anterior cleithral radius at assessed cleithral age, cleithral growth rate, and scope for linear growth of the cleithrum for 495 trophy muskellunge from across the North American range. The methods used to illustrate the results are the same as in Figure 4.

was 98.6 ± 3.17 cm, and the longest individual was 123.5 cm. The average weight was 6.8 ± 0.6 kg, and the heaviest individual was 15.4 kg. The average age was 7.2 ± 0.91 years, and the oldest individual was 18. The average size at age and scope for growth for angled trophy muskellunge-northern pike hybrids are illustrated in Fig. 10. Angler selectivity of the largest, fastest growing small fish was even more apparent in hybrids than in muskellunge, as indicated by the disparity between the sizes of the hybrids at ages 3 and 4. Although hybrids were slightly faster growing than muskellunge up to approximately age 8, they did not appear to attain the age or size of muskellunge. The von Bertalanffy growth model, calculated using average total length at assessed cleithral age for ages 3 to 14, is also provided in Fig. 10. The average ultimate length for trophy hybrids estimated by this equation is 145.9 cm. Using the length-weight relation, it is calculated that a hybrid of this length would weigh 20.9 kg.

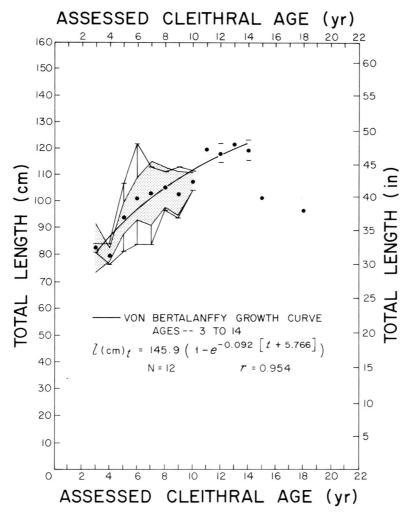

Figure 10. Average total length at assessed cleithral age, growth rate, and scope for linear growth for 60 trophy muskellunge-northern pike hybrids from across the North American range. The methods used to illustrate the results are the same as in Figure 4. Hybrids older than age 10 are represented by single individuals, except fish with assessed ages of 12 and 14, which were represented by two individuals each. The range in size of these individuals is indicated by crossbars.

$$l(cm)_t = 145.9 \left(1 - e^{-0.092 [t + 5.766]} \right)$$

N = 12   r = 0.954

Cleithrum Size and Body Length for Muskellunge and Muskellunge-Northern Pike Hybrids

Anterior cleithral radius relative to total length of the body of muskellunge was not significantly different between samples from Canadian and United States waters, so samples were combined and results were obtained from data for 495 trophy fish representing 25 age groups ranging from 2 to 30 years. The cleithrum-body relation was linear and highly significantly correlated (Fig. 11). Hence, size of the cleithrum can be used to estimate precisely length of the body.

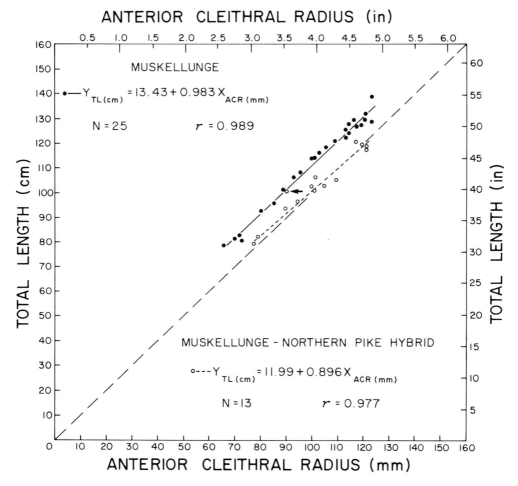

Figure 11. Relation between total length and anterior cleithral radius for 495 trophy muskellunge and 60 trophy muskellunge-northern pike hybrids from across the North American range. The relation for muskellunge was calculated by using average length at age (closed circle) for 25 age groups, ranging from 2 to 30. The relation for muskellunge-northern pike hybrids was calculated by using average length (open circle) at age for 13 age groups, ranging from 3 to 18. The open circle indicated by the arrow represents the values for one specimen that was submitted as a hybrid. However, by the appearance of the cleithrum and the size of the bone in relation to the body it was subsequently identified as a muskellunge and was excluded from the analysis. The diagonal line (dashed) represents total body length = 10 × the anterior cleithral radius.

The cleithrum-body relation for muskellunge-northern pike hybrids is similarly correlated (Fig. 11). However, the cleithral bones from the hybrid, when compared to muskellunge, are significantly larger relative to body size. The open circle indicated by an arrow in Fig. 11 represents one specimen that was submitted as a hybrid, but specific characteristics in the cleithrum (Casselman 1980; Casselman et al. 1986 and the size of its cleithrum relative to body size indicated that it was from a muskellunge. Therefore, this specimen was excluded from subsequent analyses. Muskellunge-northern pike hybrids have significantly larger cleithra than do muskellunge, a characteristic that can be used to differentiate these two fish.

Indeed, size of the cleithrum relative to size of the body provides a unique thumb rule for estimating body length from the anterior radius of the cleithral bone (Casselman 1979), because the length of the fish is very close to ten times the length of the anterior radius of the cleithral bone (see diagonal, Fig. 11). This is especially applicable for trophy muskellunge-northern pike hybrids, but also is useful for this sample of trophy muskellunge. For these slower-growing fish, when the ten times thumb rule is applied, the length of the cleithrum estimates fork length better than total length.

CONCLUSIONS

Large trophy muskellunge are often very old fish. It takes 25 to 30 years to produce a trophy fish of record size that would approach the average ultimate length of 139 cm and weight of 21 kg.

The weight distributions can be used to estimate mathematically the chance than a trophy fish will exceed a particular weight. If a trophy fish is heavier than the modal weight class (Fig. 2), it is three times more likely to exceed a specific weight if it is taken from Canadian waters. For example, for a trophy muskellunge that exceeds the modal weight class, if taken from Canadian waters there is one chance in 25, as compared with one chance in 81 from United States waters, that it will exceed 18 kg.

The size, age, and growth of angler-harvested trophy muskellunge and muskellunge-northern pike hybrids have been generally described and defined. For the present, we have separated the results, where possible, only for Canadian and United States waters because there are obvious differences that are probably related to level of exploitation. When enough data have been acquired, we will attempt to conduct these analyses for specific bodies of water. Calculation of the growth model by body of water would indicate ultimate size, and define precisely what constitutes a "trophy" for each body of water. To accomplish this, fishermen must be willing to supply specific, accurate information about the source of their specimens.

Before this degree of resolution can be attempted, the analyses must be separated by sex; therefore, taxidermists and fishermen must be provided with an easy, reliable method of determining the sex of their fish. Such studies are underway, as are attempts to develop a method for determining sex of muskellunge from characteristics in the cleithra. This has been accomplished with cleithra of northern pike (Casselman 1979).

The Cleithrum Project is an example of a system whereby fishermen and taxidermists can help acquire important fisheries data that might otherwise not be collected. Once the sample size becomes large enough that a meaningful subsample can be drawn for specific bodies of water, specific analyses will be attempted. These data can be used to develop and refine indices (e.g., changes in age structure and growth rate) for measuring and monitoring changes in the rate of exploitation of trophy muskellunge and for documenting and detecting changes in the quality of trophy fisheries.

ACKNOWLEDGEMENTS

We thank the many taxidermists and muskellunge fishermen who supplied us with basic data and cleithra from over 600 muskellunge and muskellunge-northern pike hybrids. Larry Ramsell supported our project and encouraged wide participation; for this we are indebted. We especially thank Ron Lax, a taxidermist from Wisconsin, who provided us with large numbers of samples from fish from United States waters. We thank Adrienne London, Elizabeth Hay-Chmielewski, Mary Burridge-Smith, Margaret Pike-Zur, and Fielding Sherwood for their technical assistance in preparing the cleithra for age assessment, cataloguing the samples, and conducting preliminary analyses of the data. We appreciate the comments of Peter Colby, Ken Carlander, and one anonymous reviewer. We acknowledge with thanks the funds provided by Muskies Canada for the purchase of specially printed envelopes used in The Cleithrum Project.

REFERENCES

Axon, J.R. 1978. An evaluation of the muskellunge fishery in Cave Run Lake, Kentucky. American Fisheries Society Special Publication 11:328-333.

Belusz, L.C. 1978. An evaluation of the muskellunge fishery of Lake Pomme de Terre and efforts to improve stocking success. American Fisheries Society Special Publication 11:292-297.

Bimber, D.L. 1982. Longevity, growth and mortality of muskellunge in Chautauqua Lake, New York. New York Fish and Game Journal 29:134-141.

Brewer, D.L. 1980. A study of native muskellunge populations in eastern Kentucky streams. Kentucky Department of Fish and Wildlife Resources, Fisheries Bulletin 107 pages.

Buss, K, and J. Miller. 1961. The age and growth of the muskellunge in Pennsylvania. Pennsylvania Angler, April 1961.

Carlander, K.D. 1969. Handbook of freshwater fishery biology, Volume 1. Iowa State University Press, Ames, Iowa, USA.

Casselman, J.M. 1974. Analysis of hard tissue of pike *Esox lucius* L. with special reference to age and growth. Pages 13-27 *in* T.B. Bagenal, editor. Ageing of fish--Proceedings of an international symposium. Reading, England, July 19-20, 1973. Unwin Brothers, England. 234 pages.

Casselman, J.M. 1978. Calcified tissue and body growth of northern pike, *Esox lucius* Linnaeus. Ph.D. thesis, University of Toronto, Toronto, Ontario, Canada. 782 pages (2 volumes).

Casselman, J.M. 1979. The esocid cleithrum as an indicator calcified structure. Pages 249-272 *in* J. Dubé and Y. Gravel, editors. Proceedings of the 10th Warmwater Workshop, Special Publication, Northeast Division, American Fisheries Society. Published by Québec Ministère du Loisir, de la Chasse et de la Pêche, Direction de la recherche faunique, Montréal, Québec. 285 pages.

Casselman, J.M. 1980. Identification of pike and muskellunge from cleithra. Muskies, Inc. (the official publication of Muskies, Inc.), May 1980, pages 10-12.

Casselman, J.M. 1983. Age and growth assessment of fish from their calcified structures--techniques and tools. Pages 1-17 *in* E.D. Prince and L.M. Pulos, editors. Proceedings of the international workshop on age determination of oceanic pelagic fishes: Tunas, billfishes, and sharks. NOAA Technical Report NMFS 8:211 pages

Casselman, J.M., E.J. Crossman, P.E. Ihssen, J.D. Reist, and H.E. Booke, 1986. Identification of muskellunge, northern pike, and their hybrids. American Fisheries Society Special Publication 15:14-46.

Crossman, E.J., and J.M. Casselman. 1979a. The cleithrum aging project. Muskies Canada Inc. Newsletter 1:3-8.

Crossman, E.J., and J.M. Casselman. 1979b. The cleithrum aging project. Muskies, Inc. Newsletter 13:12-17.

Haas, R.C. 1978. The muskellunge in Lake St. Clair. American Fisheries Society Special Publication 11:334-339.

Harrison, E.J., and W.F. Hadley. 1979. Biology of muskellunge *(Esox masquinongy)* in the upper Niagara River. Transactions of the American Fisheries Society 108:444-451.

Hourston, A.S. 1952. The food and growth of the maskinonge *(Esox masquinongy* Mitchill) in Canadian waters. Journal of the Fisheries Research Board of Canada 8:347-368.

Johnson, L.D. 1971. Growth of known-age muskellunge in Wisconsin: and validation of age and growth determination methods. Wisconsin Department of Natural Resources, Technical Bulletin 24 pages.

Krumholz, L.A. 1947. Length-weight relationship of the muskellunge, *Esox m. masquinongy*, in Lake St. Clair. Transactions of the American Fisheries Society 77:42-48.

Miles, R.L. 1978. A life history study of the muskellunge in West Virginia. American Fisheries Society Special Publication 11:140-145.

Muir, B.S. 1960. Comparison of growth rates for native and hatchery-stocked populations of *Esox masquinongy* in Nogies Creek, Ontario. Journal of the Fisheries Research Board of Canada 17:919-927.

Parsons, J.W. 1959. Muskellunge in Tennessee streams. Transactions of the American Fisheries Society 88:136-140.

Ramsell, L. 1982. A "compendium" of muskie angling history. Echo Printing, Alexandria, Minnesota, USA. 420 pages.

Ricker, W.E. 1975. Computation and interpretation of biological statistics of fish populations. Bulletin 191. Fisheries Research Board of Canada, Ottawa 382 pp.

Schloemer, C.L. 1936. The growth of the muskellunge, *Esox masquinongy immaculatus* (Garrard), in various lakes and drainage areas of northern Wisconsin. Copeia 1936:185-193.

Simard, A., and E. Magnin. 1968. Age et croissance du masquinonge, *Esox masquinongy*, dans quatre lacs du Québec. Journal of the Fisheries Research Board of Canada 25:1831-1842.

Williams, J.E. 1959. The muskellunge in Michigan. Michigan Department of Conservation, Fish Division Pamphlet 5 pages.

Genetic Applications in Muskellunge Management

JEFFREY B. KOPPELMAN AND DAVID P. PHILIPP

Aquatic Biology Section
Illinois Natural History Survey
607 East Peabody Drive
Champaign, Illinois 61820

ABSTRACT

Several subspecies of muskellunge have, at one time, been recognized by biologists as distinct. Although these distinctions have been based on morphological and geographic distributional differences, they do indicate the basis for potentially significant genetic variation among populations of muskellunge. Current muskellunge management practices include hatchery propagation for subsequent stocking programs within and outside of its natural range. The purpose of this study was to assess the level of genetic variability existing among muskellunge populations from a variety of regions within the range of this species. Up to 26 individuals were collected from populations in each of nine different geographic areas and genetically analyzed using vertical starch gel electrophoresis couples with histochemical staining procedures. Specifically, for each individual, liver, eye and white skeletal muscle extracts were used to determine the genotypes at 57 loci. The resulting allele frequencies demonstrated that the level of genetic variability within and among muskellunge populations examined in this study is substantial and supports the hypothesis that distinct genetic stocks of muskellunge exist in the United States.

Recent genetic studies of natural populations indicate that programs designed to manage certain fish species must incorporate principles of genetic conservation (Philipp et al. 1981, 1983; Ryman 1981; Smith and Chesser 1981). Management programs need to protect the genetic resources of these species by preventing the loss of genetic variability within individual populations or stocks and within the species as a whole. Through natural selection, a population becomes genetically tailored for the specific environment it inhabits. During this process, specific allele combinations and frequencies which confer a selective advantage are promulgated within each population. Therefore, to protect the integrity of specific genetic stocks, management practices which disrupt these advantageous allele combinations or which introduce new, maladaptive alleles must be avoided.

For these protective measures to be met, management programs must be based upon the stock concept (see Altukhov 1981; Booke 1981; Kutkuhn 1981). The conservation aims of these programs should not be solely at the level of the species as a whole, but at the level of the unit genetic stock, as well. Three stages of development are required for implementation of this concept. First, the individual genetic stocks which comprise the species must be identified. Next, the various distinctive characteristics of these stocks must be defined and their interrelationships understood. Finally, management efforts need to be structured so that the separate stocks are individually managed in an optimal manner. Programs of this nature would not only protect the genetic integrity of the individual stocks, but also the genetic resources of the species as a whole (Hynes et al. 1981; Krueger et al. 1981; MacLean and Evans 1981).

The natural range of the muskellunge extends in Canada from Lake of the Woods through the Hudson Bay drainage to the St. Lawrence River, then southward into the United States along portions of the Great Lakes, Mississippi River, and Ohio River drainages (Carlander 1969; McAllister and Crossman 1973; Phillips et al. 1982). In the past, three distinct subspecies have been described on the basis of morphological and geographic characteristics (Hubbs and Lagler 1958; McClane 1974; Smith 1979). Although fisheries biologists and systematists generally do not currently recognize these subspecific classifications, their past description suggests that detectable levels of genetic variation may exist among muskellunge populations.

Due to an increased demand for this popular sportfish, muskellunge are often propagated in hatcheries and introduced into bodies of water for supplementation of existing populations or for the establishment of new populations. The cost-to-benefit ratio of managing muskellunge stocks in this manner can be extremely high if the ability of the chosen donor stock to survive and reproduce is relatively poor compared to that of the natural recipient population. Assessing the amount of genetic variation which exists within and among muskellunge populations from different geographic locations serves as a beginning for determining how genetic differences among stocks may affect their performance in different environments.

The purpose of this investigation was to use electrophoretic techniques to identify polymorphic loci within the muskellunge and to use the allelic variation at these polymorphic loci to assess the relative amount of genetic differentiation among several populations. Although such information does not reveal the quality of performance for any particular stock in a given environment, the information concerning the degree of interpopulational genetic differentiation may reflect the amount of adaptive specificity among populations. In any case, this initial assessment of muskellunge genetic variation represents an essential first step in the design and implementation of truly effective muskellunge management programs.

Figure 1. Location of populations sampled. Numbers correspond to populations listed in methods.

METHODS

Muskellunge samples taken from populations across the natural range were obtained from various state agencies, taxidermists, and fishermen. Collections were made from the nine hatchery and wild populations listed below and shown in Figure 1: (1) Linesville Fish Cultural Station, Linesville, PA (Pennsylvania Fish Commission); (2) Chautauqua Fish Hatchery, Mayville, NY (New York State Department of Environmental Conservation); (3) St. Lawrence River, NY (Chippewa Bay); (4) Wolf Lake State Hatchery, Mattawan, MI (Michigan Department of Natural Resources); (5) Tribal Fish Hatchery, Lac Du Flambeau, WI (6) Lac Vieux Desert, upper Wisconsin River drainage, northeast Wisconsin; (7) Pioneer Group (Pioneer, N. Twin, S. Twin, Buckatabon, Snipe, Sand, Big St. Germain, Tomahawk, Bass, and Boot lakes), upper Wisconsin River drainage, northeast Wisconsin; (8) Palmer Group (Palmer, Tenderfoot, One Thousand Island, Big, Mamie, Cisco, W. Bay, and Spring lakes), Ontonogon River drainage, Lake Superior basin, northeast Wisconsin and Upper Peninsula, Michigan; and (9) Flambeau Group (Flambeau, Big Muskellunge, and Long lakes), Flambeau River drainage, northeast Wisconsin. Groupings were determined on the basis of drainage for purposes of comparing populations of muskellunge within a traditionally productive area of the natural range. Lac Vieux Desert was treated as a separate population even though located in the Wisconsin River drainage due to the large sample size from this lake. Sample sizes ranged from 2 to 26 individuals per population. Difficulty in obtaining samples of this valuable sportfish resulted in the less than desired minimum of 20 individuals in most populations analyzed.

Once collected, fish were shipped frozen to laboratory facilities at the Illinois Natural History Survey in Champaign, Illinois, and were kept frozen until the required tissue samples (white skeletal muscle, eye, and liver) were excised from each individual. Storage temperature of -20C was found to be adequate for the preservation of enzyme activity for the short period of time prior to analysis (less than one month). Tissue extracts were prepared and subjected to vertical starch gel electrophoresis and histochemical staining essentially as described by Philipp et al. (1979). In this study, eleven additional stains were modified from Siciliano and Shaw (1976) and used as described below:

1. Malic enzyme (EC 1.1.1.40)
   Staining solution: 50 ml 0.25 M Tris-HCl, pH 8.0, 5.0 ml 1.0 M sodium malate, pH 7.0, 50.0 mg $MgCl_2$, 10.0 mg NADP, 3.5 mg NBT, 3.5 mg PMS.
2. Xanthine dehydrogenase (EC 1.2.3.2)
   Staining solution: 50 ml 0.25 M Tris-HCl pH 8.0, 100 mg hypoxanthine, 7.5 mg NAD, 3.5 mg NBT, 3.5 mg PMS.
3. Fructose-1,6-diphosphatase (EC 2.7.1.11)
   Staining solution: 2 ml 0.25 M Tris-HCl, pH 8.0, 100 mg fructose-1, 6-diphosphate, 100 mg $MgCl_2$, 300 units glucosephosphate isomerase, 100 units glucose-6-phosphate dehydrogenase,

10 mg NADP, 10 mg NBT, 1 mg PMS, 16 ml 1% agar solution.
4. Pyruvate kinase (EC 2.7.1.40)
Staining solution: 8 ml 0.25 M Tris-HCl, pH 8.0, 30 mg ADP, 40 mg MgSO$_4$, 40 mg KCl, 15 mg fructose-1, 6-diphosphate, 160 units lactate dehydrogenase, 10 ml 1% agar solution, 10 mg NADH$^+$-add last. View under ultraviolet light.
5. Phosphoglycerate kinase (EC 2.7.2.3)
Staining solution: 2 ml 0.25 M Tris-HCl, pH 8.0, 4 ml deionized H$_2$O, 25 mg ATP, 40 mg MgCl$_2$, 1 mg EDTA, 15 mg 3-phosphoglyceric acid, 280 units glyceraldehyde-3-phosphate dehydrogenase, 5 ml 1% agar solution, 15 mg NADH$^+$-add last. View under ultraviolet light.
6. Adenosine deaminase (EC 3.5.4.4)
Staining solution: 4 ml 0.25 M Tris-HCl, pH 8.0, 7 ml deionized H$_2$O, 80 mg adenosine, 1.7 units xanthine oxidase, 5.5 units nucleoside phosphorylase, 50 mg sodium arsenate, 3.5 mg NBT, 3.5 mg PMS, 10 ml 1% agar solution.
7. Fumarase (EC 4.2.1.2)
Staining solution: 50 ml 0.25 M Tris-HCl, pH 8.0, 50 mg fumaric acid, 75 units malate dehydrogenase, 7.5 mg NAD, 3.5 mg NBT, 3.5 mg PMS.
8. Aconitase (EC 4.2.1.3)
Staining solution: 2 ml of solution of 0.1 M cis-aconitic acid, pH 8.0, 4 ml 0.25 ml Tris-HCl, pH 8.0, 2.3 units isocitric dehydrogenase, 100 mg MgCl$_2$, 12 mg NAD, 15 mg NBT, 1 mg PMS, 16 ml 1% agar solution.
9. Triosephosphate isomerase (EC 5.3.1.1)
Staining solution: 1 ml substrate solution (16 g Dowex and 5 ml deionized H$_2$O, swirled 30 seconds and filtered, add deionized H$_2$O to 25 ml and 80 mg dihydroxyacetone, incubated 4 hours at 38-40°C and adjusted to pH 4.5 with KHCO$_3$), 10 ml 0.25 M Tris-HCl, pH 8.0, 5 ml deionized H$_2$O, 820 units glyceraldehyde-3-phosphate dehydrogenase, 100 mg sodium arsenate, 25 mg NAD, 15 mg NBT, 1 mg PMS, 10 ml 1% agar solution.
10. Mannose-6-phosphate isomerase (EC 5.3.1.8)
Staining solution: 50 ml 0.25 M Tris-HCl, pH 8.0, 50 mg mannose-6-phosphate, 100 units glucosephosphate isomerase, 80 units glucose-6-phosphate dehydrogenase, 100 mg MgCl$_2$, 10 mg NADP, 10 mg NBT, 1 mg PMS, 15 ml 1% agar solution.
11. Esterase (EC N/A)
Staining solution: 50 ml 0.25 ml Tris-HCl, pH 8.0, 100 mg fast blue BB, 3 ml 1% naphthylacetate in 50% acetone.

The specific tissue extracts and electrophoretic conditions used to assess the phenotypes at each of the 57 loci studied are shown in Table 1. All enzyme loci are designated by an abbreviation of the enzyme followed by a capital letter (See Philipp et al. 1983). In addition, small letter superscripts designate duplicate (most likely tandem duplicates) loci: Mdh-B^a and Mdh-B^b; Me-A^a and Me-A^b; Aat-B^a and Aat-B^b; Aat-M^a and Aat-M^b, respectively. Alleles for each polymorphic locus are designated by numerals, with the numeral 1 representing the allelic form migrating the least anodal distance from the origin (Table 2). A locus was considered polymorphic if more than one allele was detected in any population studied. The mean number of alleles per locus, the percentage of loci polymorphic, and the mean heterozygosity levels were calculated for each population. Unbiased estimates were used to determine expected amounts of heterozygosity (see Nei 1978).

To evaluate the degree of heterogeneity among populations, contingency tables were computed for each polymorphic locus and the probability that population samples were drawn from a single homogeneous population was evaluated using a likelihood ratio test (G-test; Sokal and Rohlf 1981). Population subdivision was examined using F-statistics (Wright 1978), expressing the absolute amount of genetic variance (σ^2DT) among demes (populations) relative to the limiting variance under complete fixation at the same allele frequencies. These quantities may be expressed equivalently in terms of Nei's (1973, 1977) gene diversity analysis, in which the total gene diversity (H_T=Wright's limiting variance) is partitioned into within-deme ($H_D/H_T = 1-F_{DT}$) and among-deme ($D_{DT}/H_T = F_{DT}$) components (Swofford, 1982). Genetic distance and similarity coefficients were calculated using Nei (1978) methods. Cluster analysis by the unweighted pair-group method (UPGMA) was performed using the Nei (1978) unbiased genetic distance matrix to produce a dendrogram delineating clustered populations. All statistical analyses were performed using the BIOSYS-1 program developed by Swofford and Selander (1981).

RESULTS

Among the populations studied, allelic variation was detected at ten loci: Ldh-C, Idh-A, Idh-B, 6-Pgdh-A, Xdh-A, Aat-M^b, Ak-A, Gpi-B, Est-A, and Est-B (Table 2). At two of these loci variant alleles

Table 1. Enzyme loci, tissue, and electrophoretic conditions under which muskellunge isozymes were analyzed.

Enzyme Council number	Enzyme name	Locus	Tissue analyzed	Electrophoretic system[a]
1.1.1.1	Alcohol dehydrogenase	Adh-A	Liver	EBT
1.1.1.8	Glycerol-3-phosphate dehydrogenase	Gpdh-A	Muscle	TC
		Gpdh-B	Liver	TC
1.1.1.14	Sorbitol dehydrogenase	Sdh-A	Liver	TC
1.1.1.27	Lactate dehydrogenase	Ldh-A	Eye	TC
		Ldh-B	Eye	TC
		Ldh-C	Eye	TC
1.1.1.37	Malate dehydrogenase	Mdh-A	Muscle	TC
		Mdh-B[a]	Muscle	TC
		Mdh-B[b]	Muscle	TC
1.1.1.40	Malic enzyme	Me-A[a]	Muscle	EBT
		Me-A[b]	Muscle	EBT
1.1.1.42	Isocitrate dehydrogenase	Idh-A	Eye	TC
		Idh-B	Liver	TC
1.1.1.43	6-phosphogluconate dehydrogenase	6-Pgdh-A	Liver	TC
1.2.1.12	Glyceraldehyde-3-phosphate dehydrogenase	Gapdh-A	Muscle	TC
		Gapdh-C	Eye	TC
1.2.3.2	Xanthine dehydrogenase	Xdh-A	Liver	EBT
1.15.1.1	Superoxide dismutase	Sod-A	Liver	TC
2.6.1.1	Aspartate aminotransferase	Aat-A	Muscle	TC
		Aat-B[a]	Eye	TC
		Aat-B[b]	Eye	TC
		Aat-M[a]	Eye	TC
		Aat-M[b]	Eye	TC
2.7.1.11	Fructose-1,6-diphosphatase	F-1,6-dp-A	Muscle	TC
2.7.1.40	Pyruvate kinase	Pk-A	Muscle	TC
		Pk-B	Eye	TC
2.7.2.3	Phosphoglycerate kinase	Pgk-A	Muscle	TC
2.7.3.2	Creatine kinase	Ck-A	Muscle	TC
		Ck-B	Eye	EBT
		Ck-C	Eye	EBT
2.7.4.3	Adenylate kinase	Ak-A	Eye	HC
2.7.5.1	Phosphoglucomutase	Pgm-A	Muscle	TC
3.5.4.4	Adenosine deaminase	Ada-A	Muscle	TC
4.1.2.13	Aldolase	Ald-A	Muscle	TC
		Ald-C	Eye	TC
4.2.1.2	Fumarase	Fum-A	Muscle	TC
		Fum-B	Muscle	TC
4.2.1.3	Aconitase	Acon-A	Muscle	TC
		Acon-B	Eye	TC
5.3.1.1	Triosephosphate isomerase	Tpi-A	Muscle	TC
5.3.1.8	Mannose-6-phosphate isomerase	Mpi-A	Eye	HC
5.3.1.9	Glucosephosphate isomerase	Gpi-A	Muscle	TC
		Gpi-B	Eye	TC
	Esterase	Est-A	Muscle	TC
		Est-B	Liver	EBT
		Est-C	Liver	EBT
		Est-D	Muscle	TC
	Calcium binding protein	Cbp-A	Muscle	TC
	General protein	Prot-A	Eye	EBT
		Prot-B	Eye	EBT
		Prot-C	Eye	EBT
		Prot-D	Eye	EBT
		Prot-E	Eye	EBT
		Prot-F	Eye	EBT
		Prot-G	Muscle	EBT
		Prot-H	Muscle	EBT

[a]EBT: gel = 15.1% starch (w/v), 0.042 M Tris, 0.023 M borate, 0.00089 M EDTA, pH 8.6; electrode chambers = (top) 0.18 M Tris, 0.100 M borate, 0.0036 M EDTA, pH 8.6 and (bottom) 0.123 M Tris, 0.07 M borate, 0.0026 M EDTA, pH 8.6; electrophoresis = 17 h at 225 V, 4 C. HC: gel = 17.4% starch (w/v), 0.00125 M histidine-NaOH, pH 8.4; electrode chambers = 0.10 M citrate, pH 8.2; electrophoresis = 4 h at 225V, 4 C. TC: gel = 18.6% starch (w/v), 0.0175 M Tris, 0.0058 M citrate, pH 7.0; electrode chambers = 0.075 M Tris, 0.025 M citrate, pH 7.0; electrophoresis = 18 h at 225 V, 4 C.

Table 2. Allele frequencies at each polymorphic enzyme locus surveyed.

Locus	Population[a]								
	1	2	3	4	5	6	7	8	9
N	20	20	4	2	13	13	26	12	4
Ldh-C									
1	0.600	0.550	0.125	0.000	0.269	0.000	0.000	0.000	0.000
2	0.400	0.450	0.875	1.000	0.731	1.000	1.000	1.000	1.000
Idh-A									
1	0.600	0.750	1.000	0.500	0.231	0.538	0.481	0.458	0.500
2	0.400	0.250	0.000	0.500	0.769	0.462	0.519	0.542	0.500
Idh-B									
1	1.000	1.000	0.750	1.000	1.000	1.000	1.000	1.000	1.000
2	0.000	0.000	0.250	0.000	0.000	0.000	0.000	0.000	0.000
6-Pgdh-A									
1	0.625	0.350	1.000	1.000	0.692	0.731	0.673	0.792	0.625
2	0.375	0.650	0.000	0.000	0.308	0.269	0.327	0.208	0.375
Xdh-A									
1	0.975	1.000	0.250	1.000	0.885	1.000	1.000	1.000	1.000
2	0.025	0.000	0.750	0.000	0.115	0.000	0.000	0.000	0.000
Aat-M[b]									
1	0.400	0.500	0.500	0.500	0.462	0.500	0.519	0.500	0.500
2	0.600	0.500	0.500	0.500	0.538	0.500	0.481	0.500	0.500
Ak-A									
1	0.075	0.000	0.000	0.000	0.038	0.077	0.019	0.000	0.000
2	0.925	0.425	1.000	1.000	0.962	0.923	0.942	1.000	1.000
3	0.000	0.575	0.000	0.000	0.000	0.000	0.038	0.000	0.000
Gpi-B									
1	1.000	0.725	0.500	0.000	0.538	0.000	0.000	0.042	0.000
2	0.000	0.275	0.500	1.000	0.462	1.000	1.000	0.958	1.000
Est-A									
1	0.150	0.000	0.000	0.000	0.000	0.000	0.000	0.000	0.000
2	0.850	1.000	1.000	1.000	1.000	1.000	1.000	1.000	1.000
Est-B									
1	0.950	1.000	0.625	1.000	0.846	1.000	1.000	1.000	1.000
2	0.050	0.000	0.375	0.000	0.154	0.000	0.000	0.000	0.000

[a]Population numbers correspond to those on map in Figure 1 and those described in Methods.

were discovered in only a single population each, Est-A[1] in the Linesville Hatchery, and ldh-B[2] in the St. Lawrence River population. Phenotypic variation at five loci, Ldh-C, ldh-A, 6-Pgdh-A, Xdh-A, and Gpi-B (shown in the zymograms presented in Figs. 2-6, respectively), contributed substantially to the amount of genetic differentiation revealed among the populations. Table 3 summarizes the genetic variability (mean number of alleles per locus, percentage of loci polymorphic, mean heterozygosity) observed among the nine populations studied. Offspring of the Linesville and Lac Du Flambeau hatchery broodstocks had the highest observed percentage of polymorphic loci (14.0), followed by the offspring from the Chautauqua hatchery (10.5) and the St. Lawrence River sample (10.5) (Table 3). The Linesville, Chautauqua, and Lac Du Flambeau hatchery samples also contained the highest mean number of heterozygous individuals over all loci, 0.052, 0.054, and 0.053, respectively (Table 3). The greatest differences be-

Table 3. Genetic variability at 57 loci in all populations.

Population	Mean sample size per locus	Mean no. of alleles per locus	Percentage of loci polymorphic	Mean heterozygosity Direct-count (S.E.)	Expected (S.E.)
1. Linesville Hatchery	20.0	1.1	14.0	0.052 (0.022)	0.044 (0.017)
2. Chautauqua Hatchery	20.0	1.1	10.5	0.054 (0.024)	0.049 (0.019)
3. St. Lawrence River	4.0	1.1	10.5	0.044 (0.022)	0.049 (0.020)
4. Mattawan Hatchery	2.0	1.0	3.5	0.018 (0.018)	0.023 (0.016)
5. Lac Du Flambeau Hatchery	13.0	1.1	14.0	0.053 (0.021)	0.049 (0.018)
6. Lac Vieux Desert	13.0	1.1	7.0	0.032 (0.019)	0.028 (0.015)
7. Pioneer Group	26.0	1.1	7.0	0.039 (0.023)	0.028 (0.015)
8. Palmer Group	12.0	1.1	7.0	0.039 (0.024)	0.026 (0.014)
9. Flambeau Group	4.0	1.1	5.3	0.048 (0.028)	0.029 (0.017)

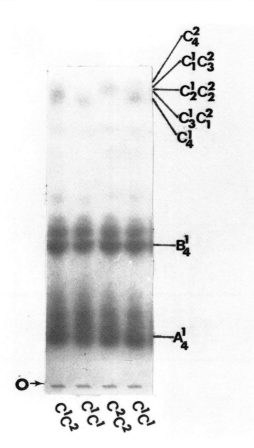

Table 4. Summary of contingency table analyses at all polymorphic loci.

Locus	No. of alleles	D.F.	G	P
Ldh-C	2	8	98.080	0.00000
Idh-A	2	8	28.529	0.00038
Idh-B	2	8	5.310	0.72398
6-Pgdh-A	2	8	27.399	0.00060
Xdh-A	2	8	34.063	0.00004
Aat-M^b	2	8	1.497	0.99276
Ak-A	3	16	71.941	0.00000
Gpi-B	2	8	192.920	0.00000
Est-A	2	8	14.062	0.08017
Est-B	2	8	19.854	0.01090
Totals		88	493.656	0.00000

heterozygote was not observed for any individual. Additionally, the electrophoretic resolution at this locus was not well enough defined to quantify individual band staining intensities to the point of distinguishing the single versus duplicate locus condition (Magee and Philipp 1982).

Figure 2. Isozyme phenotypes encoded by the Ldh-C locus in muskellunge.

tween the observed and the Hardy-Weinberg expected levels of heterozygosity (although not statistically significant) were observed for the three population samples formed by grouping individuals from adjacent lakes in northeastern Wisconsin (Pioneer, Palmer, and Flambeau groups).

A summary of the contingency table analyses is presented in Table 4. Assuming independence of loci, the probability that these samples represent a single, homogeneous population is extremely small ($P < 10^{-5}$). Most differentiation levels for the muskellunge populations (Table 5) are high (mean F_{DT} = 0.280) compared to reported values for other organisms (Lewontin 1974, Felley and Avise 1979), with the exception of the Aat-M^b locus (0.055). It is likely that the Aat-M^b locus represents not a single locus but a pair of duplicated loci sharing common alleles. However, a three-allele double

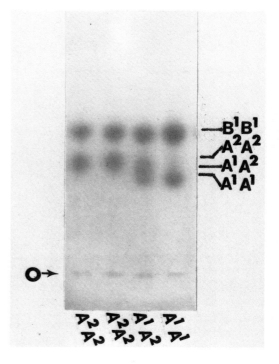

Figure 3. Isozyme phenotypes encoded by the Idh-A locus in muskellunge.

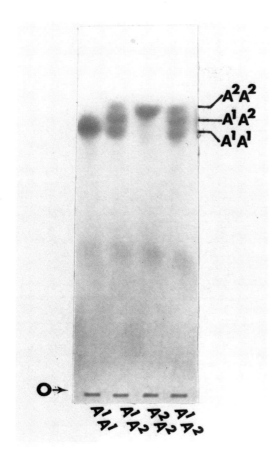

Figure 4. Isozyme phenotypes encoded by the 6-Pgdh-A locus in muskellunge.

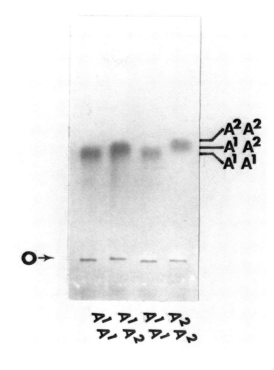

Figure 5. Isozyme phenotypes encoded by the Xdh-A locus in muskellunge.

A summary of F-statistics and gene diversity indices are shown in Table 5. Two loci, Xdh-A and Gpi-B, had relatively high F_{DT} values of 0.609 and 0.618., respectively, indicating that these two loci were closer to differential fixation than the other polymorphic loci. In terms of Nei's gene diversity analysis, 72.0% of the total gene diversity is in-

Table 5. Summary of F-statistics and gene diversity indices at all polymorphic loci.

Locus	Limiting Variance (H_T)	σ^2_{DT}	F_{ID}	F_{IT}	$F_{DT} = \dfrac{D_{DT}}{H_T}$	$\dfrac{H_D}{H_T}$
Ldh-C	0.284	0.108	-0.093	0.322	0.380	0.620
Idh-A	0.492	0.081	-0.236	-0.033	0.164	0.836
Idh-B	0.054	0.012	-0.333	-0.029	0.229	0.771
6-Pgdh-A	0.402	0.071	0.077	0.113	0.177	0.823
Xdh-A	0.178	0.109	0.798	0.921	0.609	0.391
Aat-M[b]	0.500	0.002	-0.803	-0.794	0.005	0.995
Ak-A	0.169	0.063	0.169	0.480	0.374	0.626
Gpi-B	0.429	0.265	-0.190	0.545	0.618	0.382
Est-A	0.033	0.004	-0.176	-0.017	0.136	0.864
Est-B	0.120	0.029	0.323	0.485	0.239	0.761
Total	0.266	0.074	-0.253	0.097	0.280	0.720

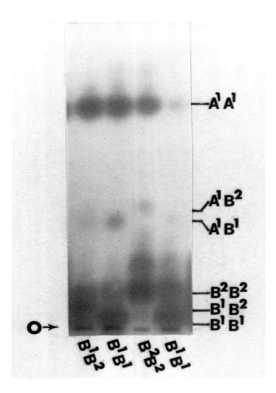

Figure 6. Isozyme phenotypes encoded by the Gpi-B locus in muskellunge.

cluded in the within-deme component; the among-deme diversity represents 28.0% of the total.

All Wisconsin collections appear very similar to each other (D = 0.000) with the exception of the Lac Du Flambeau hatchery sample (D = 0.006-0.008). Genetic distance values were used to present phenetic relationships in a dendrogram (Fig. 7). This dendrogram reflects the close similarity among the Wisconsin populations studied. The dendrogram also reveals that the Linesville and Chautauqua hatchery collections were most similar to each other while the St. Lawrence River collection does not cluster closely with any other collection.

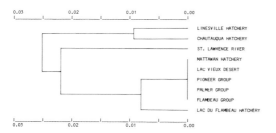

Figure 7. Unweighted pair group cluster analysis of unbiased genetic distance values (Nei 1978) based on all loci surveyed.

INTERPRETATIONS AND IMPLICATIONS

According to Selander and Kaufman (1973), the percentage of monomorphic loci for most vertebrate populations ranges from 80 to 90%. Among the muskellunge populations examined the percentage of loci which possessed at least a single variant allele was 17.7% (Table 3), thus placing this species within the commonly found level of genetic polymorphism. Our study also revealed that the hatchery populations possessed a greater amount of genetic variation than the samples taken from natural populations. This is interesting since decreased genetic variability from generations of inbreeding is often the problem among hatchery offspring.

There are several possible explanations for the increased genetic variation among hatchery stocks observed in this study. First, the "natural" populations may have been influenced by strong selectional pressures or may have undergone one or more "bottleneck" situations. Both of these circumstances could serve to reduce the level of genetic variablity in these populations. Secondly, hatchery criteria for selecting broodstock may have been biased towards artificially enhancing the amount of polymorphic loci in the offspring produced. Thirdly, the samples taken as representative from the hatchery may have in fact represented only a small fraction of the offspring produced there and may not closely resemble other broods produced. Finally, if fish from regions geographically quite distant from a given hatchery were obtained to mix with the local broodstock, the introduction of new alleles from this distant population would result in an increased heterozygosity level in the hatchery offspring. Although the affect that increased parental heterogeneity has on the performance of hatchery offspring in the wild can not be predicted, it should be the subject of future studies.

Heterozygosity levels within populations did not differ significantly from Hardy-Weinberg expectations (Table 3) although a slight excess of heterozygotes was apparent (F_{IS} = -0.253), particularly in northeast Wisconsin. This may represent an artificial enhancement caused by grouping individuals from several adjacent lakes, or merely sampling error.

Allele frequency estimates (Table 2) revealed two loci (Gpi-B and Xdh-A) for which there was a high degree of fixation for alternative alleles when comparing different populations. The eastern populations, Linesville and Chautauqua, displayed allele frequencies of 1.000 and 0.725, respectively, for the Gpi-B^1 allele; in contrast, the western populations (Lac Vieux Desert, Pioneer group, Palmer group, and Flambeau group) exhibited Gpi-B^1 allele frequencies of 0.000, 0.000, 0.042, and 0.000, respectively. Interestingly, the St. Lawrence River (eastern) and Lac Du Flambeau hatchery (western) populations were exceptional to this pattern and possessed relatively equal amounts of both alleles (Gpi-B^1 = 0.500 and 0.538, respectively). These exceptions may not be surprising when one considers the extent of the introductions which have been made in the past, using fish sometimes transported over great distances.

The St. Lawrence River population exhibited a distinctly different Xdh-A^1 allele frequency, 0.250, than did all of the other eight populations (0.885-1.000). Allele frequencies at several other loci (Ldh-C, Idh-A and Ak-A) also showed differences among the populations but did not contribute the same level of genetic differentiation as did the Gpi-B and Xdh-A loci. Contingency table analyses revealed a very low probability ($P < 10^{-5}$) that the muskellunge collections represent subsamples from a homogeneous population (Table 4). Six loci, Ldh-C, Idh-A, 6-Pgdh-A, Xdh-A, Ak-A, and Gpi-B displayed significant heterogeneity among the populations, although only Gpi-B and possibly Xdh-A seemed to show fixation according to geographic location. The F-statistics (Table 5) reiterate these conclusions and serve to illustrate the substantial amount of differentiation which exists among the populations, particularly at the Xdh-A and Gpi-B loci.

According to the dendrogram, (Fig. 7), the populations surveyed branch off into three groups or clusters. The four Wisconsin natural lake groups and the Mattawan hatchery progeny were virtually indistinguishable. A larger sample size of offspring from the Mattawan hatchery could possibly change this assessment of their relatedness, although small samples can provide a source of information regarding population differences. The Lac Du Flambeau hatchery sample (population 5), was somewhat more differentiated from the natural Wisconsin populations, undoubtedly for one or more previously discussed reasons. Although the three clusters appear to represent different genetic stocks based on the amount of genetic differentiation present, a more detailed genetic analysis incorporating samples from a much larger number of natural populations across the entire range of this species (see Philipp et al. 1983) is needed to more accurately define the stock structure of the muskellunge.

In a specific environment, the performance of one stock of muskellunge relative to another depends upon the relative fitnesses of those stocks in that environment. Furthermore, the combination of genotypes present among the individuals comprising a stock determines its relative fitness. Stocks of muskellunge have evolved as a result of different selection pressures encountered in different local environments. In a given environment some specific genes in specific combinations, which have proven to be selectively advantageous, have been retained. Through generations of selection, the resultant stocks have then become tailored for that environment. The introduction of a new stock into an established population of muskellunge would most likely disrupt selectively advantageous allele combinations and genic arrangements present in the recipient population. This would probably result in a reduction in the fitness of the recipient population. Once specific allele combinations have been altered or lost, they will be difficult, if not impossible, to reconstruct (Ryman 1981; Philipp et al. 1983).

We have demonstrated the existence of a substantial amount of genetic variation among populations of muskellunge across its range. Furthermore, our preliminary data suggest the existence of several distinct stocks of this species. These results demonstrate the need for a more extensive genetic analysis, involving many more populations and covering the entire natural range in depth, in order to fully describe the stock structure of this species. Until the geographic distribution of the various muskellunge stocks, their physiological characteristics, and their biotic and abiotic interrelationships are understood, management programs based upon supplementing natural populations with hatchery-produced individuals may very well be compromising the genetic integrity of those populations they were designed to enhance. Great care should be exercised that stocking practices for short-term benefits do not result in long-term detrimental effects. Natural resource managers and administrators should seriously consider the genetic implications of all fishery management programs. Genetic conservation must become a central principle in the design and implementation of management programs for muskellunge and other sport fishes.

ACKNOWLEDGEMENTS

We thank these people and others at the following organizations who supplied the samples of muskellunge: Ron Lax Taxidermy Studio, Conover, Wisconsin; James Harvey and Shyrl Hood, Pennsylvania Fish Commission; John Bubnack, New York State Department of Environmental Conservation; Jim Copeland and Harry Westers, Michigan Department of Natural Resources; Ken Clodfelter, Southern Illinois University Research Laboratory, LaSalle, Illinois; Larry Ramsell, President, Muskies, Inc.; and Steve Pallo, Illinois Power Company, Decatur, Illinois, who was instrumental in coordinating efforts for sample collections through Muskies, Inc. We also thank David L. Swofford for assistance in the statistical analyses of the data; Christine Kaminski for technical and editorial assistance; William W.M. Steiner, David L. Swofford and William D. Harvey for critically reviewing the manuscript; and Suzanne Peratt and Beverley Hubert for technical preparation of the manuscript.

REFERENCES

Altukhov, Y.P. 1981. The stock concept from the viewpoint of population genetics. Canadian Journal of Fisheries and Aquatic Sciences 38:1523-1538.

Booke, H.E. 1981. The conundrum of the stock concept--are nature and nurture definable in fishery science? Canadian Journal of Fisheries and Aquatic Sciences 38:1479-1480.

Carlander, K.D. 1969. Handbook of freshwater fishery biology, Volume 1. The Iowa State University Press, Ames, Iowa, USA.

Felley, J.D., and J.C. Avise. 1980. Genetic and morphological variation of bluegill populations in Florida lakes. Transactions of the American Fisheries Society 109:108-115.

Hubbs, C.L., and K.F. Lagler. 1958. Fishes of the Great Lakes region. Cranbrook Institute of Science Bulletin 26.

Hynes, J.B., E.H. Brown Jr., J.H. Helle, N. Ryman, and D.A. Webster. 1981. Guidelines for the culture of fish stocks for resource management. Canadian Journal of Fisheries and Aquatic Sciences 38:1867-1876.

Krueger, C.C., A.J. Gharrett, T.R. Dehring, and F.W. Allendorf. 1981. Genetic aspects of fisheries rehabilitation programs. Canadian Journal of Fisheries and Aquatic Sciences 38:1877-1881.

Kutkuhn, J.H. 1981. Stock definition as a necessary basis for cooperative management of Great Lakes fish resources. Canadian Journal of Fisheries and Aquatic Sciences 38:1476-1478.

Lewontin, R.C. 1974. The genetic basis of evolutionary change. Columbia University Press, London, England.

MacLean, J.A., and D.O. Evans. 1981. The stock concept, discreteness of fish stocks, and fisheries management. Canadian Journal of Fisheries and Aquatic Sciences 38:1889-1898.

Magee, S.M., and D.P. Philipp. 1982. Biochemical genetic analyses of the grass carp ♀ x bighead carp ♂ F_1 hybrid and the parental species. Transactions of the American Fisheries Society 111:593-602.

McAllister, D.E., and E.J. Crossman. 1973. A guide to the freshwater sport fishes of Canada. National Museums of Canada, Ottawa, Canada.

McClane, A.J. 1974. McClane's field guide to freshwater fishes of North America. Holt, Rinehart and Winston, New York, U.S.A.

Nei, M. 1973. Analysis of gene diversity in subdivided populations. Proceedings of the National Academy of Science 70:3321-3323.

Nei, M. 1977. F-statistics and analysis of gene diversity in subdivided populations. Annals of Human Genetics 41:225-233.

Nei, M. 1978. Estimation of average heterozygosity and genetic distance from a small number of individuals. Genetics 89:583-590.

Philipp, D.P., W.F. Childers, and G.S. Whitt. 1979. Evolution of patterns of differential gene expression: a comparison of the temporal and spatial patterns of isozyme locus expression in two closely related fish species (northern largemouth bass, *Micropterus salmoides salmoides*, and smallmouth bass, *Micropterus dolomieui*). Journal of Experimental Zoology 175:283-296.

Philipp, D.P., W.F. Childers, and G.S. Whitt. 1981. Management implications for different genetic stocks of largemouth bass (*Micropterus salmoides*) in the United States. Canadian Journal of Fisheries and Aquatic Sciences 38:1715-1723.

Philipp, D.P., W.F. Childers, and G.S. Whitt. 1983. A biochemical genetic evaluation of the northern and Florida subspecies of largemouth bass. Transactions of the American Fisheries Society 112:1-20.

Phillips, G.L., W.D. Schmid, and J.C. Underhill. 1982. Fishes of the Minnesota region. University of Minnesota Press, Minneapolis, Minnesota, USA.

Ryman, N. 1981. Conservation of genetic resources: Experiences from the brown trout (*Salmo trutta*). In Ryman, N. (ed.) Fish Gene Pools. Ecological Bulletins, Stockholm, Sweden 34:61-74.

Selander, R.K., and D.W. Kaufman. 1973. Genic variability and strategies of adaptation in animals. Proceedings of the National Academy of Science 70:1875-1877.

Siciliano, M.J. and C.R. Shaw. 1976. Separation and visualization of enzymes on gels. In Smith, I. (ed.) Chromatographic and Electrophoretic Techniques, Vol. 2, 4th edition. Wm. Heinemann Medical Books, Ltd., London, England.

Smith, M.H., and R.K. Chesser. 1981. Rationale for conserving genetic variation of fish gene pools. *In* Ryman, N. (ed.) Fish Gene Pools. Ecological Bulletins, Stockholm, Sweden 34:13-20.

Smith, P.W. 1979. The fishes of Illinois. University of Illinois Press, Urbana, Illinois, USA.

Sokal, R.R., and F.J. Rohlf. 1981. Biometry, 2nd edition. W.H. Freeman, San Francisco, California, USA.

Swofford, D.L. 1982. Genetic variability, population differentiation, and biochemical relationships in the family Amblyopsidae. Master's thesis, Eastern Kentucky University, Richmond, Kentucky, USA.

Swofford, D.L., and R.B. Selander. 1981. BIOSYS-1: a FORTRAN program for the comprehensive analysis of electrophoretic data in population genetics and systematics. Journal of Heredity 72:281-283.

Wright, S. 1978. Evolution and genetics of populations. Volume 4. Variability within and among natural populations. University of Chicago Press, Chicago, Illinois, USA.

COMMUNITY INTERACTIONS

Natural Muskellunge Reproduction In Midwestern Lakes[1]

MICHAEL P. DOMBECK[2]

United States Department of Agriculture, Forest Service
Park Falls, Wisconsin 54552

BRUCE W. MENZEL

Department of Animal Ecology

PAUL N. HINZ

Department of Statistics
Iowa State University
Ames, Iowa 50011

ABSTRACT

Throughout the native range of the muskellunge, self-sustaining lake populations of the species are declining. Because little is known of muskellunge reproductive requirements, this study had the objective of statistically identifying ecological variables that may influence natural reproduction in lakes and, further, to develop a procedure for estimating muskellunge reproductive potential in individual lakes by using readily accessible ecological data. The Natural Resources departments of Michigan, Minnesota, and Wisconsin provided file data on 117 selected lakes. Information on water chemistry, hydrology, watershed characteristics, fish communities, and cultural perturbations comprised 94 variables.

Nine variables accounted for 57% of the variability in reproduction. These included northern pike abundance, inflow water source (drainage or seepage lake), conductivity, spring water-level condition, discharge volume, shoreline development factor, cultural development of adjacent lands, and alkalinity. Conditions identified as most strongly promoting reproduction were limited northern pike abundance, rising springtime water level, high alkalinity, and high shoreline development factor, and drainage-lake systems. Complete data sets for these 5 variables were available for 89 muskellunge lakes. When organized by discriminant function analysis, 58% of the lakes were classified identically to manager-estimated reproductive level, and 91% were classified within ±1 reproductive level. Application of the discriminant function produced from this analysis may be useful for determining lake-stocking and habitat-improvement strategies directed toward creating or maintaining self-sustaining muskellunge populations.

The muskellunge is a prized gamefish in northern states and Canada. In many midwestern lakes that formerly supported self-sustaining populations, natural reproduction of the species has greatly declined or ceased. Ecological reasons accounting for this decline include competition with the congeneric northern pike (Inskip and Magnuson 1983; Oehmcke et al. 1974) and reproductive failure due to loss or modification of habitat (Dombeck et al. 1984).

Muskellunge management strategies commonly involve harvest regulations and stocking of hatchery-reared fingerlings (Miller 1983). To date, little effort has been made to manage muskellunge habitats. In large part, this is attributable to a paucity of detailed ecological knowledge, especially that relating to muskellunge habitat requirements and fish species interactions.

One approach to gaining needed ecological information might be through long-term case studies. Such research, however, is costly, labor intensive, and often only of site-specific utility. Because of these disadvantages, management agencies tend to give such studies low priority. An alternative method is to gain ecological insights through statistical analysis of an extensive data base, such as that readily available in management agency files. This approach requires no additional field work, is relatively inexpensive, can cover a broad geographical area, and is quite possible with presently available computer technology. Various multivariate statistical techniques can be used to identify general ecological correlations, which also may have site-specific applications.

---

[1] Journal Paper J-11373 of the Iowa Agriculture and Home Economics Experiment Station, Ames, Iowa. Project 2236.

[2] Present address: USDA - Forest Service, Fisheries and Wildlife Management, 630 Sansome Street, San Francisco, CA 94111.

This study employed such techniques to (1) identify ecological factors that influence natural muskellunge reproduction in midwestern lakes; and (2) develop a quantitative procedure for evaluation of muskellunge reproductive potential in individual lakes by using readily-accessible ecological data.

METHODS

Collection of Data and Location of Lakes

Ecological data were obtained from Department of Natural Resources (DNR) survey files on 117 selected lakes within USDA-Forest Service Region 9, in Michigan, Minnesota, and Wisconsin. Lakes were selected on the basis of the availability and completeness of ecological information and their occurrence within the muskellunge's native range. All were surveyed between 1961 and 1981, but primarily after 1972. They occur within six major drainages: Upper Mississippi, Chippewa, St. Croix, and Wisconsin rivers, and lakes Michigan and Superior (Figure 1).

Muskellunge Reproduction Estimates

For each lake, estimates of the proportion of the muskellunge population recruited from natural reproduction were obtained from DNR fisheries biologists. Twenty-seven lakes support self-sustaining muskellunge populations, 67 are non-self-sustaining (i.e., populations maintained in part or wholly by stocking), and 23 were either former muskellunge lakes or non-muskellunge lakes randomly selected (excluding dystrophic lakes) from the same geographic range as, and of similar size to, the muskellunge lakes.

For statistical purposes, each lake was placed into one of five discrete groups, each group representing a level of natural muskellunge reproduction: 0 (none), 1 (poor), 2 (low), 3 (moderate), and 4 (high or self-sustaining). Categories 0 and 4 were based on knowledge that either no muskellunge were recruited from natural reproduction or no stocking was necessary to maintain the population while categories 1, 2, and 3 were based upon manager estimates. The total number of lakes representing each reproductive category were: none-23, poor-24, low-17, moderate-19, and high or self-sustaining-34 (Table 1). In the subsequent statistical analysis, the discrete level of estimated natural reproduction was the dependent variable.

Ecological Variables

Ecological data represented five general categories: fish communities, water quality, lake morphometry, natural watershed features, and cultural development of watersheds, totaling 94 variables. For a complete list of these variables see Dombeck (1984). Terminology and survey methods followed the procedures of Wisconsin Department of Natural Resources (1982). Where necessary, Michigan and Minnesota data were standardized to conform with Wisconsin's terminology and methods. Selected ecological features of study lakes that were used are given in Table 1.

Data Analysis

Analysis of variance (ANOVA) was used to test for significant relationships between level of muskellunge reproduction and each ecological variable. Because of the broad geographical distribution of the 117 lakes, there was concern that ecological distinctions exist between the six major drainage basins. Therefore, ANOVA was first used to remove significant interdrainage variance and then to determine those parameters most useful as predictors of muskellunge reproduction. For this analysis, $P < 0.1$ was used as the level of significance, because the hazard in analysing a data set compiled from several sources is, not that false relationships will be identified, but that existing relationships will be masked or weakened due to large variances (Type I error).

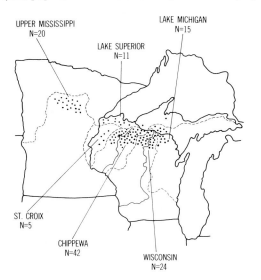

Figure 1. Geographic distribution of study lakes showing drainage boundaries (---) and sample size (N). Each dot represents one lake.

Table 1. Ecological features of the 117 study lakes by drainage basin showing the number (*) of lakes in each category or the mean value and range ( ) for the parameter.

Parameter	Drainage					
	Lake Michigan	Lake Superior	Wisconsin River	Chippewa River	St. Croix River	Mississippi River
Total lakes	15	11	24	42	5	20
Muskellunge reproduction (*)						
High	5	6	1	9	0	13
Moderate	0	1	3	9	1	5
Low	0	2	9	6	0	0
Poor	1	0	6	15	2	0
None	9	2	5	3	2	2
Northern pike abundance (*)						
Abundant	0	1	1	3	2	15
Common	6	0	8	10	2	3
Rare	5	4	10	5	1	2
Absent	4	6	5	24	0	0
Lake types (*)						
Natural	12	8	14	17	4	7
Impoundment	3	3	10	23	1	13
Reservoir	0	0	0	2	0	0
Area	324	150	331	521	465	4675
(hectares)	(65-806)	(21-259)	(55-1373)	(38-6192)	(62-1044)	(141-45135)
Alkalinity	65	37	40	44	41	150
mg/l	(18-145)	(17-86)	(1-175)	(10-97)	(11-84)	(121-171)
Shoreline development factor	2	2	2	3	2	2
Cultural development index	0.01	0.01	0.03	0.02	0.02	0.001

After an array of nine predictive variables was developed, simple linear correlation analysis was employed to determine the direction of the relationship. Data for all nine variables were available for 68 lakes. The remaining lakes had missing data; therefore, sample sizes varied from parameter to parameter. Stepwise multiple linear regression, utilizing the maximum R^2 improvement technique (Helwig and Council 1979), was performed to order the variables in a fashion that would explain the largest proportion of total variance in level of reproduction.

Because a strong relationship between muskellunge reproduction and northern pike emerged from these analyses, special attention was given to this relationship. Simple correlation analysis of muskellunge reproductive level versus northern pike abundance was first performed for each drainage basin. Resulting interbasin correlations suggested that confounding factors were involved; therefore, to determine factors influencing the muskellunge-pike association, lakes were identified as either: (1) self-sustaining muskellunge/ abundant pike populations; (2) self-sustaining muskellunge/ pike rare or absent; (3) non-self-sustaining muskellunge/abundant pike; (4) non-self-sustaining muskellunge/pike rare or absent. ANOVA procedures then were used to identify variables significantly related ($P < 0.1$) to the association (statistical interaction) of the two species in the four groups of lakes. The means of the variables identified by ANOVA were next examined to determine the nature of the association between the species.

Discriminant analysis was used to derive a predicted muskellunge reproductive level for individual lakes, based upon a limited array of ecological variables previously demonstrated as correlates of muskellunge reproductive success. Lakes lacking natural muskellunge reproduction presented a special problem because we could not always distinguish between those that had never supported natural muskellunge populations and those from which natural populations have been extirpated. We resolved this issue by deleting all lakes with a reproductive level of zero and, thus, employed only lakes with poor, low, moderate, or high reproduction. Ecological variables considered for use in the discriminant analysis included nine variables identified as being significantly ($P < 0.1$) related to level of reproduction by ANOVA. From this group, four variables were subjectively eliminated after consideration of their redundancy with other parameters, difficulty and accuracy of measurement, and completeness of the data set. Thus, the discriminant analysis employed five variables determined to best represent four of the five ecological data categories. These were: abundance of northern pike, spring water level, alkalinity, shoreline development factor (SDF), and lake hydrology (drainage or seepage lake). Rationale for selection of these variables is treated more fully in the Results section. Complete data sets for these five variables were available for 89 lakes.

The posterior probability (probability of a lake being placed within the given reproductive catagory) for each of the four reproductive categories (i) is given by the formula:

$$\text{Probability (that lake is in category i)} = \frac{\exp[-1/2(X-\overline{X}_i)'S^{-1}(X-\overline{X}_i)]}{\sum_{j=1}^{4} \exp[-1/2(X-\overline{X}_j)'S^{-1}(X-\overline{X}_j)]}$$

The symbols X, $\overline{X}_i$, and S represent the vector of measurements of the five ecological measurements for the lake, the vector of means for each variable for all lakes in each of the four reproductive categories, and the pooled variance-covariance matrix from the discriminant analysis respectively.

RESULTS

Variables Related to Muskellunge Reproduction

ANOVA identified nine significant relationships between ecological variables and muskellunge reproductive level after interdrainage variance was removed (Table 2). These included at least one variable from each of the five ecological data categories. Significant correlations (five positive and three negative) occurred between muskellunge reproductive level and eight of these variables, only conductivity yielding a nonsignificant correlation (Table 3). Additionally, there were significant intercorrelations among the nine parameters for about two-thirds of the possible paired comparisons.

Stepwise multiple linear regression indicated that the nine variables accounted for 57% of the variability of natural reproduction (Table 4). Northern pike abundance alone accounted for 42% of the variance. The best two-variable model included northern pike abundance and drainage lake system, which accounted for 50% of the variance.

Table 2. ANOVA results for ecological variables significantly related to level of muskellunge reproduction.

Parameter	No. of Lakes	F
Northern pike abundance	117	13.5***
Seepage lake	117	6.0***
Rising spring water	117	3.3***
Drainage lake	117	3.0**
Outlet volume	82	2.7**
Conductivity	89	2.6**
Shoreline development factor	112	2.4**
Cultural development index	117	2.2**
Alkalinity	117	1.9*

Probability levels: ***=$P<0.01$, **=$P<0.05$, *=$P<0.1$

Table 3. Correlation coefficients (r) among 9 ecological variables and level of muskellunge reproduction for the study lakes. Underscored values are significant at the 0.1 level.

Parameter	Drainage lake	Seepage lake	Rising spring water	Alkalinity	Abundance of pike	Cultural development index	Conductivity	Discharge volume	Shoreline development factor
Level of reproduction	.33	-.29	.26	.33	-.21	-.18	-.09[a]	.36	.17
Drainage lake		-.74	.21	.32	.18	-.01	.33	.18	.10
Seepage lake			-.16	-.32	-.11	.02	-.46	-.14	-.09
Rising spring water				.24	.30	.02	-.08	.76	.28
Alkalinity					.50	-.30	.91	.45	-.17
Northern pike abundance						-.04	.19	.23	-.11
Cultural Development index							.05	-.15	.05
Conductivity								-.12	-.15
Discharge volume									.45

[a] No data for the upper Mississippi River drainage

Table 4. Results of stepwise multiple linear regression on nine ecological variables significantly related to level of natural muskellunge reproduction. Variables are ordered vertically according to placement in the regression. N = 68.

Parameter	Cumulative R^2
Northern pike abundance	.417
Drainage lake	.501
Shoreline development factor	.527
Seepage lake	.531
Conductivity	.538
Rising spring water	.544
Outlet volume	.571
Alkalinity	.572
Cultural development index	.574

Association Between Muskellunge Reproduction and Northern Pike Abundance

Drainage-by-drainage correlation analysis of muskellunge reproductive level versus northern pike abundance produced significant negative relationships among lakes of the Chippewa River ($r = -0.64$), St. Croix River ($r = -0.73$), Lake Michigan ($r = -0.62$), and Lake Superior drainages ($r = -0.96$). A nonsignificant negative relationship occurred for the Wisconsin River drainage ($r = -0.13$) and a virtually random association was indicated for the Mississippi River drainage ($r = -0.03$).

ANOVA implicated 12 ecological variables as significant confounding factors in explaining the association (statistical interaction) of pike with muskellunge reproduction (Table 5). Of these, six have been previously implicated with the biology of the two species (Tables 2, 3): alkalinity, rising spring water, drainage lake, discharge volume,

Table 5. ANOVA results listing ecological variables significantly related to the association of muskellunge reproductive level versus northern pike abundance.

Parameter	F
Alkalinity	24.8***
Rising spring water	10.5***
Drainage lake	10.0***
Outlet volume	7.6***
pH	7.5***
Maximum depth	4.9***
Area	3.6**
Seepage lake	3.4**
Impoundment	3.2**
Cultural development index	2.5*
% agriculture	2.4*
Stabilized water level	2.3*

seepage lake, and cultural development index (number of commercial facilities plus number of dwellings divided by lake area). Two sets of variables are clearly redundant because they are inverse characteristics; viz, rising spring water and stabilized water level, drainage lake and seepage lake.

Table 6 presents group means for these variables, partitioned according to the four muskellunge-pike association categories. Lakes with vigorous natural populations of both species are best characterized as large impounded lakes (natural lakes with water level raised by a dam), with high alkalinity, rising spring water, and limited cultural influences (agriculture and other development); all are located in the Upper Mississippi River drainage (Fig. 1). Conversely, lakes with poor populations of both species were characterized as being of small size, low alkalinity, and low discharge volume. This category included the largest number of seepage lakes and a relatively high cultural development factor. Lakes in the remaining two categories were generally similar, being comprised predominantly of unmodified drainage lakes of intermediate size and alkalinity levels, limited rising spring water, and moderate to high cultural influences. In such lakes, limited muskellunge reproduction associated with strong northern pike populations may be promoted by agricultural influences and stabilized water levels.

Discriminant Analysis

One purpose of the discriminant analysis was to estimate potential muskellunge reproductive success in individual lakes as based on a simple array of readily measureable variables. From among the nine variables listed in Table 2, five were selected for the analysis: northern pike abundance, rising spring water level, alkalinity, SDF, and drainage lake system. The remaining variables identified in Table 2 were subjectively excluded because of considerations of redundancy (conductivity, seepage lake), limitations of available data (discharge volume), and complexity of generating data (cultural development index). Predicted levels of muskellunge reproduction were calculated for the 89 individual lakes. Fifty-eight percent of the lakes were classified identically with manager estimates, and 91% were classified within ±1 manager estimated reproductive level.

In general, there was stronger agreement at the higher reproductive levels than at the lower (Table 7). No lake having poor or low muskellunge reproduction by manager estimates (N=41) was

Table 6. Groups means of ecological variables associated with the association of muskellunge reproductive level versus northern pike abundance. Standard errors given in parentheses; sample sizes given in brackets.

Parameter	Muskellunge population			
	Self-sustaining		Non-self-sustaining	
	Northern pike population			
	Abundant[12]	Rare-absent[22]	Abundant [8]	Rare-absent[18]
Area	6876 (3885)	512 (247)	1005 (316)	288 (60)
Maximum depth	22 (3)	13 (2)	16 (3)	13 (1)
Alkalinity	152 (4)	53 (9)	80 (21)	40 (5)
pH	8.3 (0)	7.4 (0.2)	7.3 (0.3)	7.2 (0.1)
Impoundment	0.75 (0.13)	0.36 (0.10)	0.50 (0)	0.22 (0.10)
Seepage lake	0	0.05 (0.05)	0	0.28 (0.11)
Drainage lake	1 (0)	0.91 (0.06)	1 (0)	0.44 (0.12)
Rising spring water	0.66 (0.14)	0.14 (0.07)	0.12 (0.12)	0
Stabilized water level	0.08 (0.08)	0.14 (0.07)	0.50 (0.19)	0.17 (0.09)
Discharge volume	7 (2)	1.8 (1.0)	0.5 (0.2)	0.1 (0.1)
%agriculture	0	1.6 (0.6)	4.1 (1.7)	3.1 (1.4)
Cultural development index	0.003 (0.002)	0.011 (0.002)	0.015 (0.005)	0.014 (0.003)

classified by the analysis as having high reproductive potential. Similarly, among lakes estimated by managers as having moderate to high muskellunge (N=48) only four lakes were predicted to be of poor or low reproduction.

DISCUSSION

Muskellunge-Northern Pike Associations

Demonstrated statistical correlations between muskellunge reproductive success and several ecological variables suggest that the historical decline of the species in many midwestern lakes may be attributable to a complex of geographically-variable factors. Among these, the detrimental influence of the northern pike emerged as a critical element, particularly in the Chippewa River, St. Croix River, Lake Michigan, and Lake Superior drainage basins where the demise of self-sustaining populations has been most serious. On the other hand, there is little evidence here for negative influences of northern pike upon muskellunge reproduction in the Wisconsin River and Upper Mississippi River basins.

The northern pike is considered to have an ecological advantage over the muskellunge where the two species occur syntopically (Hess and Heartwell 1978). The invasion of muskellunge waters by northern pike has been associated with subsequent decline of the muskellunge (Inskip and Magnuson 1983) and has long been a concern in Wisconsin (Threinen and Oehmcke 1950). A notable example is one of Wisconsin's premier muskellunge lakes, Lac Courte Oreilles, which produced a world record

Table 7. Percentage agreement of 89 lakes classified by the five-factor discriminant function analysis with original manager estimates of muskellunge natural reproductive level in individual lakes. Number of lakes is given in parentheses.

Manager estimates	Predicted level of natural reproduction				
	Poor	Low	Moderate	High	Total
Poor	58 (14)	21 (5)	21 (5)	0	100 (24)
Low	24 (4)	47 (8)	29 (5)	0	100 (17)
Moderate	6 (1)	6 (1)	69 (11)	18 (3)	100 (16)
High	6 (1)	0	34 (11)	59 (19)	100 (32)

fish in 1949. The invasion of northern pike in the mid-1930s was followed by the decline of muskellunge reproduction. Concurrent with this invasion was the recreational development of shoreline adjacent to the muskellunge spawning area. Subsequently, highly eutrophic conditions developed, including decline of wild rice (*Zizania aquatica*) and other emergents, proliferation of dense submergent vegetation, and deposition of organic silts (Leon D. Johnson, Wisconsin Department of Natural Resources, personal communication). Stocking hatchery-reared fish became necessary to maintain Lac Court Oreille's muskellunge fishery. This was verified by an experimental 5-year period in which none were stocked. Population studies showed a gap in muskellunge year classes corresponding to years of no stocking (Johnson 1981).

Several mechanisms have been implicated in the competitive interaction between muskellunge and northern pike. Earlier spawning of northern pike on common spawning grounds and its subsequent predation on muskellunge larvae is thought to greatly reduce muskellunge populations in Wisconsin lakes (Threinen and Oehmcke 1950). In a laboratory study, Caplin (1982) demonstrated the predation of young-of-the-year (YOY) muskellunge by YOY northern pike, whereas the converse did not occur. Caplin extrapolated that the same type of predation may occur in nature and severely limit muskellunge recruitment. In addition, YOY northern pike gained efficiency as predators at a smaller size than muskellunge and, at equal size (3.5-5.5g), had higher feeding efficiencies. In another study, YOY northern pike exhibited less size selectivity for food, more aggressive feeding habits, and greater food conversion (Sorenson et al. 1966).

Morphoedaphic Influences.

The ecological mechanisms that favor the coexistence of the muskellunge and northern pike in some habitats, but not in others, are poorly understood. Considerations of ecological factors influencing lake fish communities implicate two major groups of factors, lake morphometry-hydrology and edaphic factors (Ryder et al. 1974). Several authors have implicated water body size as a determinant of fish community complexity (Barbour and Brown 1974; Emery 1978) when other ecological factors and geographical location are roughly similar. Literature on esocids suggests that a key factor in the coexistence of muskellunge and northern pike is habitat partitioning in sufficiently large and environmentally-heterogeneous water bodies. In the Niagara River spatial segregation of the species occurs during reproduction, muskellunge selecting lotic spawning habitat and northern pike more lentic areas (Harrison and Hadly 1978).

It is notable that, in the upper Midwest, the native distribution of muskellunge includes not only the northern lakes region of Wisconsin, Michigan, and Minnesota, but also extends southward in the large rivers draining these basins; namely, the Wisconsin, Chippewa, St. Croix, and Mississippi. In these large-river environments, the two species coexisted until recent times when cultural perturbations have seemingly shifted the ecological balance in favor of the northern pike. Within the study area, vigorous syntopic populations occur chiefly in lakes of the upper Mississippi basin. Leech Lake, Minnesota, serves as an example of a large lake affording spatial isolation of the two species during their early life histories. Whereas northern pike spawn in marshy near-shore areas, muskellunge spawn offshore in shallow bays in water 1 to 2-m deep over dense mats of stonewort (*Chara*) (Strand 1983).

It is of interest to pose the question of minimum water body size needed to support natural populations of the two species. In general, northern pike seem to be better adapted to smaller lotic and lentic systems, particularly those associated with marsh habitat (Trautman 1981). Tonn and Magnuson (1982) reported that small (mean, ca. 11 ha) winterkill-prone seepage lakes of northern Wisconsin support neither species, but rather are characterized by an umbrid-cyprinid community. In larger lakes (mean, ca. 31 ha) however, an esocid-centrarchid assemblage exists. These lakes have either high winter dissolved-oxygen levels, or access to high winter oxygen refuges through inlet or outlet streams. None of their study lakes ranging in size to 90 ha contained muskellunge.

Although our data set was biased toward larger lakes, several observations may be made about esocid populations in the smaller lakes. Only three lakes in this study contained neither species; they were among the smallest lakes, averaging 137 ha in area, and were either seepage or spring lakes. Marginal populations of both species of muskellunge alone occurred in 22 lakes as small as 38 ha and averaging approximately 220 ha; over two-thirds were drainage lakes. Marginal northern pike populations alone were present in five lakes of larger average size (mean, 521 ha), two being seepage lakes, two drainage lakes, and one a spring-fed lake. Lakes supporting both species, but with populations of one or the other tending to predominate, were even larger (means, 512-1005 ha), and nearly all were drainage lakes (Table 6). In summary, these data suggest that coexistence of the species is favored in environmentally-diverse drainage lakes of large area. Populations of both species are typically marginal in seepage lakes of small size. The status of populations in lakes of intermediate size may depend on local microhabitat factors such as edaphic conditions and cultural influences.

Consideration of the original distributional range of the muskellunge implies that it is primarily adapted to oligotrophic waters. Or, the muskellunge may have substituted the cooler habitat of oligotrophic lakes as a substitute for the rivers to which it had adapted. On the other hand, the broad circumpolar distribution of the northern pike, emcompassing both oligotrophic and cooler eutrophic waters, implies generally greater environmental tolerance on its part. Early life and reproductive adaptations further support this view of ecological distinctions between the species (Dombeck et al. 1984). For example, the muskellunge broadcasts its demersal nonadhesive eggs over the bottom, and newly hatched young remain quiescent at the bottom, becoming active only after yolk-sac consumption. These features have been interpreted as being adaptations for well-oxygenated environments (Balon 1975). In contrast, adaptations of the northern pike include deposition of highly-adhesive eggs onto aquatic vegetation above the bottom. Immediately after hatching, larvae are temporarily active before attaching to vegetation by means of a cephalic cement gland, behavior that would tend to promote survival under conditions of oxygen depletion.

Historical evidence indicates that with increasing cultural eutrophication, northern pike have tended to displace the muskellunge (Becker 1983). There is evidence that neither species is adapted to dystrophic waters or with edaphic factors associated with such waters. Neither occur in small bog lakes (Tonn and Magnuson 1982), and in the present study, three non-esocid lakes were characterized by low alkalinity (mean, 26 mg/liter), low pH (mean, 6.8), and low conductivity (mean 48 μmhos/cm). The most abundant coexisting populations of the two species occurred chiefly in the Upper Mississippi drainage lakes of Minnesota; these are hard-water (mean alkalinity, 152 mg/ liter) eutrophic lakes (mean pH, 8.3), owing to the basin's limestone bedrock features (Eddy and Underhill 1974). In contrast to eutrophic softwater lakes, which tend to accumulate fine organic sediments, these lakes are notable for extensive marl deposits and limited organic sediments, a factor that may favor the survival of early life stages of both species.

Various authors have associated declines in northern pike populations with cultural eutrophication along with other cultural habitat perturbations (Trautman 1981; Forney 1977). Although there is little previous evidence implicating eutrophication with the decline of muskellunge populations, the biology of the species suggests that it may be particularly sensitive to environmental changes commonly associated with eutrophication. From the present data, several observations may be made in this regard. Highest levels of muskellunge reproduction were correlated with little or no agricultural or other human developments within the direct drainage area (Table 6). On the other hand, lakes lacking muskellunge or supporting only limited natural populations were characterized by high cultural development factors.

Water-level Influence

In addition to morphoedaphic factors, there is evidence that water-level modifications can influence the well-being of these species and their ecological associations. Trautman (1981) suggested that wetland drainage has contributed to the demise of Ohio populations of both species. Donald M. Osterberg, State University College, Potsdam, NY (personal communication) believes that construction of dams on the St. Lawrence River and subsequent creation of lentic habitats has favored northern pike and contributed to a historical decline of muskellunge. Similarly, the decline of muskellunge in large midwestern rivers may be partly attributable to dam construction. The ecological effect of more uniform lentic conditions may have been the breakdown of the early life history spatial separation of the two species.

The literature implicates seasonal spring flooding of terrestrial and wetland habitats with reproductive success of northern pike (Carbine 1941; Johnson 1957). Similarly, Dombeck (1979) and Dombeck et al. (1984) have reported muskellunge spawning in flooded near-shore areas in several Wisconsin and Michigan lakes. This suggests that, in lakes, spatial isolation of early life history stages of the two species is associated with slightly different spawning habitat created by rising waters. Abnormally low water levels of the mid-1960's have been associated with poor muskellunge recruitment in the St. Lawrence River (John Casselman, Ontario Ministry of Natural Resources, Maple, Ontario, unpublished data). Conversely, stronger recruitment was noted during subsequent years of higher water levels as reflected by increased angler returns. Casselman believes that low water levels eliminate spawning and early life separation between muskellunge and northern pike. Results of this study identified rising spring water as a common feature in lakes maintaining naturally reproducing muskellunge (Table 2,3) as well as in lakes maintaining strong northern pike populations (Tables 5,6).

To the possible detriment of both species however, among many esocid lakes in the upper Midwest, there has been recent manipulation of water levels. Natural lakes have been modified into impoundments with lowhead spillway dams for general recreational purposes, and for flow augmentation, larger dams have been constructed. The effect of the low-head dams has been to increase lake area and to stabilize seasonal water level fluctuation by dampening spring flooding. Conversely, in flow-augmentation impoundments, a seasonal water-level pattern is maintained that is similar to natural water-level fluctuation but seasonally more extreme; i.e., water levels are lowered during low flow periods (late summer-winter) and filled during peak flow (spring). From evidence presented here, it may be inferred that such water-level manipulation can affect muskellunge reproductive success and associations of esocid populations. In general, both species seem to benefit from rising springtime water level (tables 2, 3, 4, 5, 6). Stabilized water levels, however, seem to be detrimental to muskellunge reproduction ($r=-0.22$, $P=0.03$, $N=97$) and of little influence upon northern pike ($r=0.06$, $P=0.56$, $N=114$).

MANAGEMENT IMPLICATIONS

The discriminant function developed here provides a technique for estimating the potential of individual lakes to support natural muskellunge reproduction. It is based upon five ecological variables that have been shown to be significant correlates of reproductive success and are available in many ecological survey reports or that may be generated with relatively little effort. Superficially, the strength and utility of the model might be questioned because of the relatively low agreement between manager-estimated and predicted reproductive levels (Table 7). However, the strength of the posterior probability within each reproductive level provides valuable information for the manager.

From information in Table 8, several examples of management application are shown. Lake Chippewa (Sawyer Co., WI) is a large (6,192 ha) soft-

Table 8. Example of management application of the posterior probabilities generated by the discriminant analysis model. Original manager-estimated muskellunge reproductive levels are underscored.

Reproductive level	Lake			
	Chippewa	Connors	Ghost	Butternut
		Posterior probabilities		
Poor (1)	0	__71__	8	__29__
Low (2)	1	18	19	25
Moderate (3)	13	8	__50__	33
High (4)	__88__	3	22	14

water flow-augmentation impoundment, the water level of which is lowered more than 3 m each winter and raised in early spring. It has traditionally had a good naturally-reproducing muskellunge fishery, with northern pike rare or absent until the past few years when a substantial increase in the northern pike population has been observed. There is good agreement between manager-estimated and predicted level of muskellunge reproduction (Table 8). The generated posterior probabilities indicate that Lake Chippewa probably has suitable elements for good muskellunge reproduction. Management options might be to monitor the populations's reproduction, identify, protect, and enhance the spawning and nursery areas and take steps to prevent the increase in northern pike numbers. Although Lake Chippewa has received token stockings of hatchery-reared muskellunge (possibly for social or political reasons), it is likely that natural reproduction is sufficient to maintain the fishery.

The second example, Connors Lake (Sawyer Co., WI) is a 174-ha soft-water lake, the water level of which is stabilized by a rock dam. Muskellunge were native to Connors Lake, northern pike are now common, and there is only limited muskellunge spawning habitat. Both the model and manager estimate agree that this lake has poor potential for successful muskellunge reproduction. Present management includes the stocking of hatchery-reared fingerling muskellunge. Other management options might include northern pike population control or water-level management. Lacking the ability to control these factors, the most feasible management solution might be to continue stocking hatchery-reared muskellunge at sizes when they are less vulnerable to esocid predation. These two examples illustrate how stocking determinations can be made relative to benefit/cost considerations. It seems that fish stocking would be beneficial in Connors Lake.

Ghost Lake (Sawyer County, WI), is a 151-ha soft-water stabilized impoundment with a natural muskellunge population but lacking northern pike. Natural muskellunge reproduction does occur; however, hatchery-reared fingerlings are stocked to maintain the desired population level. In Table 8, note that both the manager-estimated and predicted reproductive levels are in agreement that natural muskellunge reproduction is moderate. Management efforts to restore Ghost Lake's muskellunge fishery to a self-sustaining status might include water-level management or other reproductive habitat enhancement measures.

Butternut Lake (Prince County, WI), is a 467-ha soft-water natural lake with northern pike common. Muskellunge were native to this lake, but the population is maintained by fingerling stocking. Although the manager estimated low muskellunge reproduction, results of the discriminant function model were ambiguous (Table 8); no reproductive level had a posterior probability over 40%. We have empirically determined that the model is strongest for lakes with an individual probability above 60%, which corresponds to high agreement with the manager estimate. Conversely, for lakes with all probabilities below 40%, the discriminant model has little utility. In such cases, we recommend that the input data be examined for accuracy or that other factors not addressed by the model be considered. Relative to Butternut Lake, eutrophication from sewage may be the proximate factor limiting muskellunge reproductive potential. The model will likely not be useful where cultural eutrophication is overwhelming.

This model is not proposed as the sole basis for muskellunge management decisions in midwestern lakes, but rather, as another tool to be used in combination with other decision variables. The greatest utility of the model perhaps is to assist in determining stocking or habitat-improvement priorities. Caution in applying it beyond the geographic range for which it was developed is recommended.

Discriminant analysis of this sort would seem to have general utility to lake management objectives and strategies. Although other forms of multivariate analysis might also be applied (e.g., factor analysis), discriminant analysis is particularly useful when only qualitative or semiquantitative data is available (Kachigan 1982), as is often the case for management data bases. Additionally, categorizing lakes provides insights regarding ecological relationships over a broad geographical range and provides information concerning specific variables.

ACKNOWLEDGEMENTS

Funding was provided by United States Department of Agriculture Forest Service. We thank the many employees of the Michigan, Minnesota, and Wisconsin departments of Natural Resources who provided data used in study. The BASIC program for calculating posterior probabilities was written by M. Hanson. We thank R. W. Bachmann, K. D. Carlander, E. J. Crossman, R. A. Ryder, R. C. Schultz, and anonymous reviewers for their comments.

REFERENCES

Balon, E. K. 1975. Reproductive guilds of fishes: a proposal and definition. Journal of the Fisheries Research Board of Canada 32:821-864.

Barbour, C. D., and J. Brown. 1974. Fish species diversity in lakes. American Naturalist 108(962):473-489.

Becker, G. C. 1983. Fishes of Wisconsin. University of Wisconsin Press, Madison, Wisconsin, USA.

Caplin, D. L. 1982. An experimental study of interactions between young of the year northern pike (*Esox lucius*) and muskellunge *(Esox masquinongy)*. M.S. Thesis University of Wisconsin, Madison, Wisconsin, USA.

Carbine, W. F. 1941. Observations on the life history of the northern pike, *Esox lucius*, in Houghton Lake, Michigan. Transactions of the American Fisheries Society 71:149-164.

Dombeck, M. P. 1979. Movement and behavior of the muskellunge determined by radio telemetry. Wisconsin Department of Natural Resources Technical Bulletin 113.

Dombeck, M. P. 1984. Ecological factors affecting muskellunge (*Esox masquinongy*) reproduction in midwestern lakes. Ph.D. Dissertation. Iowa State University, Ames, Iowa, USA.

Dombeck, M. P., B. W. Menzel, and P. N. Hinz. 1984. Muskellunge spawning habitat and reproductive success. Transactions of the American Fisheries Society 113:205-216.

Eddy, S., and J.C. Underhill. 1974. Northern Fishes. University of Minnesota Press, Minneapolis, Minnesota, USA.

Emery, A. R. 1978. The basis of fish community structure: Marine and freshwater comparisons. Environmental Biology of Fishes 3:33-47.

Forney, J. L. 1977. Evidence of inter and intraspecific competition as factors regulating walleye (*Stizostedion vitreum vitreum*) biomass in Oneida Lake, New York. Journal of the Fisheries Research Board of Canada 34:1812-1820.

Harrison, E. J., and W. F. Hadley. 1978. Ecologic separation of sympatric muskellunge and northern pike. American Fisheries Society Special Publication 11:129-134.

Helwig, J. T., and K. A. Council, editors. 1979. SAS Users Guide. SAS Institute Inc., Cary, North Carolina, USA.

Hess, L., and C. Heartwell. 1978. Literature review of large esocids (muskellunge, northern pike, hybrid tiger muskellunge). Pages 139-175 in J. Dubé and Y. Gravel, editors. Proceedings of the 10th Warm Water Workshop. Northeastern Division, American Fisheries Society, Ministére du Tourism, de la Chasse et de la Pêche, Montréal, Québec, Canada.

Inskip, P. D., and J. J. Mangunson. 1983. Changes in fish populations over an 80-year period: Big Pine Lake, Wisconsin. Transactions of the American Fisheries Society 112:378-389.

Johnson, F. H. 1957. Northern pike year class strength and spring water levels. Transactions of the American Fisheries Society 86:285-293.

Johnson, L. D. 1981. Comparison of muskellunge *Esox masquinongy* populations in a stocked and unstocked lake in Wisconsin with notes on the occurrence of northern pike *Esox lucius*. Wisconsin Department of Natural Resources Research Report 110.

Kachigan, S. K. 1982. Multivariate statistical analysis. Radius Press, New York, New York, USA.

Miller, M. L. 1983. The status of muskellunge management in North America. Division of Wildlife, Ohio Department of Natural Resources, Columbus, Ohio, USA.

Oehmcke, A. A., L. Johson, J. Klingbiel, and C. Winstrom. 1974. The Wisconsin muskellunge. Its life history, ecology and management. Wisconsin Department of Natural Resources Publication 8-3600 (74).

Ryder, R. A., S. R. Kerr, K. H. Loftus, and H. A. Regier. 1974. The morphoedaphic index, a fish yield estimator-review and evaluation. Journal of the Fisheries Research Board of Canada 31:663-688.

Sorenson, L., K. Buss, and A. D. Bradford. 1966. The artificial propagation of esocid fishes in Pennsylvania. Progressive Fish-Culturist 28:133-141.

Strand, Robert F. 1983. Identification of principle muskellunge spawning areas in Leech Lake by radio telemetry. Minnesota Department of Natural Resources, Division of Fish and Wildlife, Fisheries Section Investigational Report 377.

Threinen, C. W., and A. A. Oehmcke. 1950. The northern invades the musky's domain. Wisconsin Conservation Bulletin 15:10-12.

Tonn, W. M., and J. J. Magnuson. 1982. Patterns in the species composition and richness of assemblages in northern Wisconsin lakes. Ecology 63:1149-1166.

Trautman, M. B. 1981. The fishes of Ohio. 2nd Edition. Ohio State University Press, Columbus, Ohio, USA.

Wisconsin Department of Natural Resources. 1982. Fish Management Handbook 3605.9. Madison, Wisconsin, USA.

APPENDIX I

The following program can be used to calculate the probability for natural muskellunge reproduction for a given lake by using the discriminant function model derived from this study. It is designed for use by resource managers on most personal computers and is written in a simple style with all input data prompted and output displayed on the screen.

Information needed to perform the calculation includes data from the lake to be classified as follows: rising spring water level and drainage lake as no = 0, yes = 1; alkalinity and SDF, actual values; and northern pike abundance as absent = 0, rare = 1, common = 2, and abundant = 3. The results are the percentage probability for a lake having a muskellunge reproductive level (1=poor, 2=low, 3=moderate, 4=high) calculated as follows:

Probability (%) that the lake is in reproductive category (i)

$$= \frac{\exp[-1/2(X-\overline{X}_i)'S^{-1}(X-\overline{X}_i)]}{\sum_{j=1}^{4}\exp[-1/2(X-\overline{X}_j)'S^{-1}(X-\overline{X}_j)]}$$

where X is the value for each of the five ecological parameters for the given lake, $\overline{X}_i$ is the mean value for each of the parameters at the ith reproductive level, and S^{-1} is the inverse of the pooled variance-covariance matrix for the five parameters. Both $\overline{X}_i$ and S^{-1} are included in the following program as constants derived from this study. Comment statements are preceded by "!".

```
100   !                              BASIC PROGRAM
110   !            FOR ESTIMATING MUSKELLUNGE REPRODUCTIVE POTENTIAL
120   !                           IN MIDWESTERN LAKES
130   !
140   ! BASED ON PAPER BY: DOMBECK, MENZEL, & HINZ
150   ! PROGRAMMED BY: MARTIN J. HANSON, P. E., 1984
160   !
170   ! ********************************INITIALIZE VARIABLES AND READ CONSTANTS  ********************************
180   !
```

```
190     DISPLAY ERASE ALL                                                    ! CLEAR SCREEN
200     !
210     GO SUB 1280                                                          ! WRITE SCREEN TEMPLATE
220     !
230     INTEGER I, J, K                                                      ! DEFINE VARIABLES
240     REAL X(4), RLM (3,4), DELTA (3,4), IPVCM (4,4), DOT 1(3,4) PP(3), DOT2(4), D
250     !
260     !
270     !
280     DATA 0.13, 0.48, 0.79, 0.80, 0.47
290     DATA 0.29, 0.43, 0.76, 0.50, 0.55        ! REPRODUCTIVE LEVEL
300     DATA 0.13, 0.54, 1.00, 0.30, 0.60        ! MEANS MATRIX (RLM)
310     DATA 0.31, 0.85, 0.94, 0.55, 0.65
320     !
330     !
340     !
350     DATA 8.21, -0.26, -0.97, -2.74, -2.79
360     DATA -0.26, 10.23, -0.68, -4.79, 2.66    ! INVERSE OF POOLED
370     DATA -0.97, -0.68, 10.25, -0.60, -0.83   ! VARIANCE-COVARIANCE MATRIX
380     DATA -2.74, -4.79, -0.60, 6.82, 0.14     !(IPVCM)
390     DATA -2.79, 2.66, -0.83, 0.14, 9.68
400     !
410     !
420     !
430     FOR I = 0 TO 3
440     FOR J = 0 to 4
450     READ RLM (I, J)                                                      ! LOAD REPRODUCTIVE LEVEL MEANS
460     NEXT J
470     NEXT I
480     !
490     !
500     !
510     FOR I = 0 to 4
520     FOR J = 0 to 4
530     READ IPVCM (I, J)                                                    ! LOAD INVERSE OF POOLED
540     NEXT J                                                               ! VARIANCE-COVARIANCE
540     NEXT J
550     NEXT I
560     !
570     !*****************************************DATA INPUT SECTION *****************************************
580     !
590     INPUT AT (7,31) SIZE (-40)" ":A$                    ! LAKE NAME
600     !
610     INPUT AT (9,61) SIZE (-5)" ":X(0)                   ! RISING SPRING WATER LEVEL (N=0, Y=1)
620     !
630     INPUT AT (10,61) SIZE (-5)" ":X(1)                  ! ALKALINITY (METHYL ORANGE)
640     X(1)=X(1)/100
650     !
660     INPUT AT (11,61) SIZE (-5)" ":X(2)                  ! DRAINAGE LAKE (N=0, Y=1)
670     !
680     INPUT AT (13,61) SIZE (-5)" ":X(3)                  ! NORTHERN PIKE ABUNDANCE
690     X(3)=X(3)/2                                         ! ABSENT=0, RARE=1, COMMON=2
700                                                         ! ABUNDANT=3
710     !
720     INPUT AT (14, 61) SIZE (-5)" ":X(4)                 ! SHORELINE DEVELOPMENT FACTOR
730     X(4)=X(4)/4
740     !
750     !****************************************COMPUTE REPRODUCTIVE LEVELS ****************************************
760     !
770     FOR I = 0 TO 3
780     FOR J = 0 TO 4
790     DELTA (I, J) = X(J) - RLM (I,J)                     ! COMPUTE DELTA MATRIX
800     NEXT J
810     NEXT I
820     !
830     !
840     !
850     FOR I = 0 TO 3
```

```
860    FOR J = 0 TO 4
870    DOT1(I,J)=0                                          ! COMPUTE DOT1 MATRIX
880    FOR K = 0 TO 4
890    DOT1(I,J) = DOT1(I,J) + DELTA(I,K) * IPVCM(J,K)
900    NEXT K
910    NEXT J
920    NEXT I
930    !
940    !
950    !
960    FOR I = 0 TO 3
970    DOT2(I)=0                                            ! COMPUTE DOT2 MATRIX
980    FOR J = 0 TO 4
990    DOT2(I) = DOT2(I) + DELTA (I,J) * DOT1(I,J)
1000   NEXT J
1010   NEXT I
1020   !
1030   !
1040   !
1050   D = 0
1060   FOR I = 0 TO 3
1070   PP(I) = EXP(-0.5 * DOT2(I))                          ! COMPUTE PP MATRIX
1080   !
1090   D = D + PP(I)                                        ! COMPUTE CONSTANT D
1100   NEXT I
1110   !
1120   !*******************************************OUTPUT SECTION *******************************************
1130   !
1140   FOR I = 1 TO 4
1150   IMAGE ##.#
1160   DISPLAY AT (22,17*I-4)USING 1150:(PP(I-1)/D)*100
1170   NEXT I
1180   !
1190   !
1200   !
1210   DISPLAY AT (24,44)"Y"
1220   INPUT AT (24,1) SIZE (-1) "Do you wish to make another analysis? (Y/N) ":C$
1230   IF C$ = "Y" THEN 570
1240   STOP
1250   !
1260   ! *****************************************DISPLAY CONTROL SUBROUTINE *****************************************
1270   !
1280   A$ = "*****************************************"
1290   B$="**********"
1300   DISPLAY AT (1,69)"Version 1.0"
1310   DISPLAY AT (2,1)A$;A$
1320   DISPLAY AT (3,1)B$
1330   DISPLAY AT (3,20)"PREDICTED NATURAL MUSKELLUNGE REPRODUCTION"
1340   DISPLAY AT (3,71)B$
1350   DISPLAY AT (4,1)A$;A$
1360   DISPLAY AT (7,21)"LAKE NAME:"
1370   DISPLAY AT (9,10)"Rising Spring Water Level (No=0, Yes=1)......"
1380   DISPLAY AT (10,10)"Alkalinity (Methyl Orange)............."
1390   DISPLAY AT (11,10)"Drainage Lake (No=0, Yes=1).............."
1400   DISPLAY AT (12,10)"Abundance of Northern Pike (Absent=0, Rare=1,"
1410   DISPLAY AT (13,10)"Common=2, and Abundant =3)............."
1420   DISPLAY AT (14,10)"Shoreline Development Factor..........."
1430   DISPLAY AT (16,15)"PROBABILITY OF NATURAL MUSKELLUNGE REPRODUCTION (%)"
1440   DISPLAY AT (17,29)"Posterior Probability"
1450   FOR I = 1 TO 4
1460   DISPLAY AT (19,I*17-5)"LEVEL";I
1470   DISPLAY AT (21,I*17-5)"_____"
1480   NEXT I
1490   DISPLAY AT (20,12)"(poor)"
1500   DISPLAY AT (20,30)"(low)"
1510   DISPLAY AT (20,45)"(moderate)"
1520   DISPLAY AT (20,63)"(high)"
1530   RETURN
1540   END
```

Negative Associations Between Abundances of Muskellunge and Northern Pike: Evidence and Possible Explanations

PETER D. INSKIP[1]

Center for Limnology
University of Wisconsin — Madison
Madison, Wisconsin 53706

ABSTRACT

Inverse trends in relative abundance of muskellunge and northern pike have been reported for numerous lakes and one large river. In each case, muskellunge appeared to decrease in abundance while northern pike appeared to increase. Most instances involved colonization of native muskellunge lakes by northern pike, but shifts in relative abundance of populations with a history of co-occurrence also have been noted. These trends have been interpreted as evidence of a negative interaction between the two species. Predation, competition, and hybridization are possible mechanisms of interference. Earlier spawning in the spring, shorter generation time, a more aggressive nature, and greater food conversion efficiency have been suggested as possible advantages for northern pike.

Predation by young-of-the-year (YOY) northern pike on YOY muskellunge was proposed as a likely mechanism of interaction nearly 35 years ago, and it remains the most attractive hypothesis, more on the basis of its intuitive appeal and explanatory power than on actual field evidence. Under this hypothesis, the apparent compatibility of muskellunge and northern pike in some waters can be explained by the availability of sufficient spawning habitat to permit spatial separation of the species, either because of differences in preferred spawning habitat, or because eggs are spread over a large enough area that encounters between YOY muskellunge and northern pike are minimized. The existence of one mechanism of interaction does not preclude the existence of others.

Circumstantial evidence of a negative interaction between the two species is persuasive and not to be discounted. However, negative associations in relative abundance might arise even in the absence of any direct species interaction. For example, changing environmental conditions could affect the two species quite differently if they have different tolerances or optima for certain environmental variables. Based on zoogeographic evidence, it appears that northern pike *might* be favored by cooler temperatures and more lentic conditions. Cultural development more often results in the conversion of lotic habitat to lentic habitat than vice versa. Sequelae of human settlement, such as increased turbidity, siltation, and accumulation of organic sediments might be expected to have a more severe impact on a species adapted for life in flowing water than on one which evolved for life in still water habitats. Differences in the environmental requirements of the two species have been postulated but not proven.

Effective muskellunge management in waters containing northern pike will require deeper understanding of how the two species interact, what factors modify the intensity and outcome of the interaction, and how environmental conditions and fishing pressure can otherwise favor one species over the other.

There is a general perception that northern pike abundance is a limiting factor for many muskellunge populations and that effective muskellunge management calls for the control of northern pike where the two species co-occur (Oehmcke et al. 1965; Becker 1983). This belief is grounded in the large number of reports, some documented, of declines in muskellunge populations associated with the expansion of northern pike populations (Threinen and Oehmcke 1950; Oehmcke 1951; MacCrimmon and Skobe 1970; Johnson 1981; Inskip and Magnuson 1983; Osterberg, D.M., personal communication). Phylogenetic affinities between muskellunge and northern pike, coupled with similarities in appearance and life history (Scott and Crossman 1973), make a strong species interaction plausible, or even expected. The objectives for this paper are to review the evidence that northern pike have displaced muskellunge, discuss possible mechanisms of interaction, and consider how environmental factors might modify the intensity and outcome of interaction or otherwise favor one species over the other and create the illusion of a strong species interaction.

The most compelling evidence that northern pike supplant muskellunge comes from observations made on lakes and lake systems that originally (as of the 1900s) contained muskellunge and to which

---

[1]Current address: Department of Epidemiology, Harvard School of Public Health, 677 Huntington Avenue, Boston, MA 02115

northern pike were intentionally or accidentally introduced, or eventually arrived naturally. Before discussing these observations, it is appropriate to review briefly the North American zoogeography of muskellunge and northern pike. This establishes a context in which to consider recent shifts in distribution and relative abundance and serves as a basis for hypotheses about differences in environmental requirements and optima.

ZOOGEOGRAPHY

Fossil evidence indicates that both muskellunge and northern pike were present in North America by the late Pleistocene (Smith 1954; Crossman and Harington 1970). Eastern populations of both species appear to have survived the Wisconsin glaciation in the Mississippi refugium, which included parts of the Mississippi, Missouri, and Ohio River systems (Weed 1927; McPhail 1963). With the retreat of the icesheets and associated northward movement of the north temperate climate, the ranges of both species also expanded northward. Present distributions show northern pike as the more successful, or at least more persistent, colonizer at high latitudes (Fig. 1). Muskellunge occur only slightly north of 50 N, while northern pike occur well north of 60 N (Crossman 1978). The Beringian refugium, which included parts of Alaska, the Yukon Territory and Siberia, may have served as a second origin of postglacial dispersal for northern pike (McPhail 1963); however, all populations east of the MacKenzie River in the Northwest Territories appear to have originated from the Mississippi stock (McPhail 1963; Morrow 1964), so eastward dispersal from the Beringian refugium does not account for their greater northern range. Native ranges extend southward to northern Alabama for muskellunge and to central Missouri for northern pike (Crossman 1978).

The native range of muskellunge included much of the Ohio River drainage and a few streams in the Tennessee River drainage, and muskellunge still occur in streams in West Virginia, Kentucky, Tennessee, North Carolina, and Ohio (Crossman

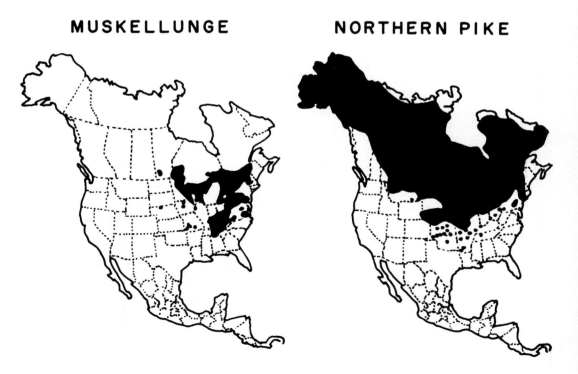

Figure 1. North American distributions of muskellunge and northern pike. Isolated occurrences outside of the contiguous ranges are the results of recent introductions. Redrawn from Becker (1983) with permission of the publisher.

1978). Northern pike are absent from most of these drainage systems (Crossman 1978).

As of the early 1900s, areas of co-occurrence between muskellunge and northern pike were few. Sympatry of populations in the Great Lakes and connecting waters and in parts of the upper Mississippi River drainage predated Caucasian settlement, but native distributions otherwise were largely disjoint. For example, Greene's (1935) survey of fish distribution in Wisconsin indicated that muskellunge occurred in many lakes in northcentral and northwestern Wisconsin, in the rivers draining these areas, and along the Lake Superior and Lake Michigan coasts but not in most of the remainder of the State (Fig. 2). Northern pike were generally absent from the two lake districts where muskellunge were common, but were widely distributed throughout the rest of the State (Fig. 2). In Minnesota, northern pike were broadly distributed, but muskellunge were "found in few numbers in several widely separated areas" (Eddy 1944:40). Muskellunge occur in the absence of northern pike in the Kawartha Lakes region of Ontario (Crossman, E.J., personal communication).

Differences between the native North American distributions of muskellunge and northern pike may have arisen due to different colonization opportunities, different environmental requirements, species interactions, or some combination of these influences. Greene (1935) interpreted the occurrence of muskellunge upstream of waterfalls in the Chippewa and Wisconsin River drainage systems as evidence of early postglacial ascent of these rivers, before current drainage patterns were established. Northern pike may have arrived later and been blocked by rapids or waterfalls. Whatever the reason for the near complementarity of native distributions of muskellunge and northern pike in Wisconsin, and for other differences between their respective North American ranges, it appears that relatively few populations, other than those in the lower Great Lakes, have a long history of co-occurrence.

RECENT CHANGES IN DISTRIBUTION AND RELATIVE ABUNDANCE

Human activity during the late nineteenth and early twentieth centuries has caused considerable

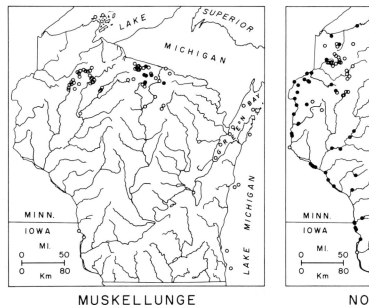
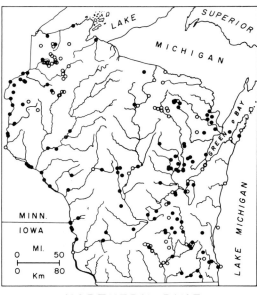

Figure 2. Reported occurrences of muskellunge and northern pike in Wisconsin as of 1935, including specimens actually examined (●), and those mentioned in the literature or other reports (○). Redrawn from Greene (1935) with permission of the publisher.

change in the distributions of muskellunge and northern pike (Carlander 1978; Crossman 1978). Stocking, in particular, has extended the area of overlap by bypassing geographic barriers to range extension. In addition, habitat modification has eliminated each species from certain areas (Trautman 1981) and has changed the conditions under which they co-occur. Fishing pressure almost certainly has increased in comparison to presettlement years. The two species appear to have responded differently to the aggregate of these cultural influences.

During the 1930s, before a negative interaction with muskellunge was suspected, northern pike were introduced to many Wisconsin lakes that had native muskellunge populations. Concern emerged shortly thereafter, with the observation that muskellunge populations appeared to decline in many of these lakes, and the practice of stocking northern pike in lakes containing muskellunge was discontinued in 1941 (Oehmcke 1951). However, northern pike continued to spread through drainage systems, and reports of declines in native muskellunge populations and of hybridization with northern pike accompanied this spread (Threinen and Oehmcke 1950; Oehmcke 1951). Northern pike now are ubiquitous in northern Wisconsin (Fig. 3).

The pattern reported for a chain of lakes near the headwaters of the South Fork of the Flambeau River in northcentral Wisconsin is illustrative. Northern pike were not known from these lakes as of 1936 (Elkins 1937) but were present in many of the lakes by the early 1940s and abundant by the middle 1940s (Oehmcke 1951). The colonizations are believed to have resulted from an unauthorized 1939 stocking (Oehmcke, A.A., personal communication). Angler complaints about a decline in the quality of the muskellunge fishing in these lakes were common by the end of the decade (Oehmcke 1951). A dam blocked upstream movement of northern pike into Big Pine Lake (Iron County), and catch records suggest that muskellunge remained abundant through the 1940s and early 1950s, well after declines had been reported for populations in neighboring lakes (Inskip and Magnuson 1983). The dam washed out in the middle 1940s, and northern pike were present in Big Pine Lake by 1948. By 1955, northern pike were relatively abundant and muskellunge relatively scarce, based on

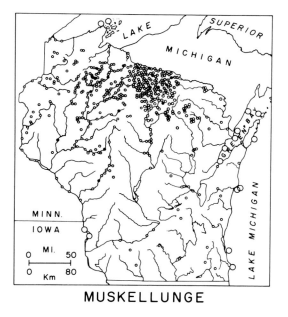
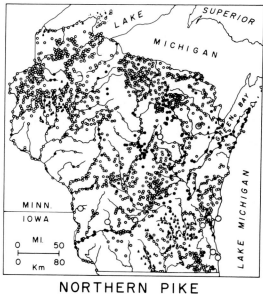

Figure 3. Reported occurrences of muskellunge and northern pike in Wisconsin as of 1983, including specimens actually examined (●), those mentioned in the literature and other reports (○), those found by the Wisconsin Fish Distribution Study (1974-75) (▲), and those reported by Greene (1935) (○). Redrawn from Becker (1983) with permission of the publisher.

catch data and angler recollections. A similar chronology has been reported for other lakes in Wisconsin (Threinen and Oehmcke 1950; Johnson 1982; Becker 1983). The sequence of events occurred in different calendar years in different lakes.

Perhaps the best-documented example of a negative association in abundance is that reported by Johnson (1981) for Lac Court Oreilles in Sawyer County, Wisconsin. Northern pike were not native to Lac Court Oreilles but were present by 1945. Too few were caught during the 1950s to permit reliable population estimates, but annual estimates were made beginning in 1961 (Fig. 4). They indicated a further increase in abundance of northern pike around 1970, implying one or more strong year classes during the late 1960s. Northern pike continued at this elevated abundance through 1977. Population estimates for muskellunge declined from the latter 1950s to the middle 1960s and again during the early 1970s, just after northern pike had become more abundant, but also after a five-year lull (1966-1970) in muskellunge stocking; thus, although population shifts are well-documented for Lac Court Oreilles, changes in northern pike abundance are confounded with changes in stocking practices, and we cannot be sure of the relative importance of the two factors.

A change in the relative abundance of muskellunge and northern pike also has been reported for the St. Lawrence River (Osterberg, D.M., personal communication), to which both species are believed to be native (Crossman 1978). Again, there has been a perception, based on "limited data", that the previous twenty years have seen a decrease in muskellunge abundance and an increase in the size of the northern pike population in the international section of the river (Osterberg, D.M., pers. comm.). This example differs from the others reported thus far in being for a river rather than a lake and for populations with a long history of co-occurrence. Also, it followed the construction of a hydroelectric dam that resulted in appreciable change in the hydrodynamic characteristics of this stretch of the river; thus, a changing environment accompanied this *reported* change in relative abundance of muskellunge and northern pike.

Lake Simcoe, Ontario supported a commercial fishery for muskellunge during the latter 1800s, but northern pike appear to have been absent or rare prior to 1920 (MacCrimmon and Skobe 1970). Subsequent trends in population sizes are not well-documented, but a sharp decline in the muskellunge population(s) is said to have occurred after 1930, contemporaneous with a pronounced increase in abundance of northern pike (MacCrimmon and Skobe 1970). Cultural disturbances occurring at this time included the introduction of carp and associated loss of rooted aquatic plants, and the dyking and draining of marshlands. MacCrimmon and Skobe (1970) noted that inverse relationships between muskellunge and northern pike had been reported for other southern Ontario lakes.

In summary, evidence that northern pike have displaced muskellunge is largely anecdotal and almost entirely circumstantial. The idea has a priori credibility because the two species are so similar. In some cases, we have observer accounts and, in others, there are data implying inverse trends in *relative* abundance. I am aware of only one lake (Lac Court Oreilles) for which estimates of absolute abundance are inversely related and, in this case, changes in stocking practices cloud the interpretation. The sheer number of anecdotal accounts of

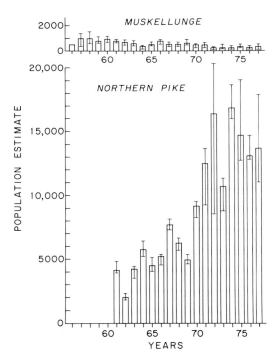

Figure 4. Population estimates for muskellunge and northern pike in Lac Court Oreilles, Wisconsin from 1956 to 1977. Ninety-five percent confidence limits also are shown. Redrawn from Johnson (1981) with permission of the publisher.

declines in muskellunge populations subsequent to colonization by northern pike tends to reinforce belief in a negative species interaction, but these observations should not all be considered independent. Knowledge of previously reported associations could influence perceptions and interpretations. This is not to say that there is insufficient reason for believing that northern pike have displaced muskellunge. The chronology recounted above for lakes near the headwaters of the Flambeau River provides reasonably strong evidence, because of the relative timing of apparent changes in upstream versus downstream lakes.

It should be noted that an inverse association in abundance of muskellunge and northern pike is not an universally observed pattern. For example, in Escanaba Lake (Vilas County, Wisconsin), substantial changes in abundance of northern pike did not appear to be accompanied or followed by opposite changes in abundance of muskellunge (Inskip and Magnuson, 1986). Comparison of lakes where a negative association is not apparent with those where a strong interaction is suggested might be informative about important mechanisms of interaction.

POSSIBLE MECHANISMS OF INTERACTION

Opportunities for interaction between muskellunge and northern pike exist at virtually all life history stages. Both species spawn in the spring, often in the same type of habitat, and early life histories are similar (Scott and Crossman 1973). Young-of-the-year (YOY) undergo parallel ontogenetic shifts in diet. Cannibalism has been reported among YOY (Bean 1908; Hunt and Carbine 1951) and among older fish (Schloemer 1938; Lawler 1965) of each species. Mechanisms of interaction based on interspecific predation or competition are plausible.

At this time, the consensus of opinion seems to be that the critical species interaction occurs early in life and is associated with concentrated spawning of muskellunge and northern pike in limited areas. This presumption is based, in part, on a fundamental similarity of spawning habits but also on the observation that waters where the two species have co-occurred for long periods of time typically are larger and have more diverse or extensive spawning habitat that those waters where displacement of muskellunge or low muskellunge populations have been reported (Johnson 1971; Harrison and Hadley 1978; Ryder, R.A., personal communication). The inference is that plentiful, heterogenous spawning habitat is conducive to more diffuse spawning and that muskellunge benefit when spawning is dispersed. Northern pike abundance and spring water levels were identified as strong predictors of muskellunge reproductive success from a statistical analysis of 117 lakes in Wisconsin, Minnesota and Michigan (Dombeck et al. 1986). Lakes in which muskellunge reproduction was rated as "good" characteristically had low northern pike populations and elevated spring water levels, conditions that might be expected to lessen the shared utilization of spawning habitat.

The advantage for northern pike is thought to derive from its earlier spawning time and consequent headstart in growth (Threinen and Oehmcke 1950). Northern pike spawn immediately after ice-out at temperatures of 4 to 11 C (Scott and Crossman 1973), and muskellunge shortly thereafter when the water has warmed to 9 to 15 C (Scott and Crossman 1973; Minor and Crossman 1978). In northern Wisconsin, this typically translates into a time difference of approximately two weeks, although spawning periods often partially overlap (Williamson 1942; Oehmcke 1951). The initial size differential could enable YOY northern pike to prey on muskellunge fry or, perhaps, to be more effective in feeding or gaining shelter from predators.

A pond experiment conducted by the Wisconsin Department of Natural Resources in 1947 (Oehmcke 1951) substantiated the idea that YOY northern pike possess some sort of advantage over YOY muskellunge. Twenty-five thousand northern pike fry were stocked in a hatchery pond with a plentiful supply of zooplankton (cladocerans). Eleven days later, an equal number of later-hatching muskellunge fry were stocked in the same pond. For the next two weeks, young muskellunge and northern pike were commonly seen, but, over the ensuing three weeks, muskellunge became relatively rare. When the pond was drained, roughly five weeks after the muskellunge had been stocked, 404 northern pike and four muskellunge were recovered, along with about 35,000 shiners. The northern pike were decidedly larger than the muskellunge.

Predation

Oehmcke (1951) inferred that predation by northern pike on muskellunge, and on smaller northern pike, was the most important cause of mortality in the experiment just described. Observations of YOY esocids in confined, artificial surroundings confirm that predation can result in dramatic declines in numbers within a very short time (Caplan 1982). Factors that govern the occurrence of predation among YOY esocids in the wild have not been elucidated. One would expect that the degree of overlap and concentration of spawning (see following section), relative sizes of fry, timing of dispersal of YOY from spawning grounds, the presence of other small fishes, and the availability of cover might be important variables.

In an experiment designed to examine the effect of relative size on the occurrence of interspecific predation, Caplan (1982) paired five 75 mm (fork length) muskellunge with five 55 mm northern pike in each of two circular laboratory tanks (180 cm diameter x 25 to 35 cm deep), and three 75 mm northern pike with five 55 mm muskellunge in each of two other tanks. Each tank also contained 200 to 300 cyprinids between 15 and 45 mm long. Forty-eight hours later, when the tanks were drained, Caplan (1982:37) observed that "with large muskies and small pike in the same tank, neither pike nor muskies were eaten. When large pike and small muskies were together, none of the pike but nearly all of the muskies were eaten." Caplan (1982) suggested that this finding might be attributable to differences in the length-weight relationships for the two species. For fish of a given length, northern pike were heavier than muskellunge. The 75 mm and 55 mm pike weighed approximately 2.8 and 1.1 g, respectively, as compared to 1.8 and 0.6 g for the 75 and 55 mm muskellunge. Matching the two species by length mismatched according to weight.

The relative timing of spawning might influence the size disparity between fry of the two species; environmental factors that control the timing of spawning of muskellunge and northern pike thus could influence the occurrence of the hypothesized predator-prey interaction.

Muskellunge eventually surpass northern pike in length, usually within the first two years of life (Carlander 1969). The predator-prey interaction might be reversed among older fish; however, larger esocids usually are much more dispersed than are fry on spawning grounds. Other potential prey fishes, such as yellow perch or walleye often far outnumber northern pike and could serve as a buffer against predation by muskellunge.

Availability of cover is another possible modifying factor for a predator-prey interaction, either because of interspecific differences in preferred cover type or because of varying likelihoods of predation by northern pike on muskellunge in different cover types. Distributions of 14 to 17 mm muskellunge and 25 to 30 mm northern pike were compared with respect to habitat type, defined by substrate (rock, sand, mud or vegetation) and cover density (two levels) (Caplan 1982). Distributions were examined for each species alone and for the two species together. Distributions of fry of each species in the absence of the other species were approximately uniform with respect to substrate and cover. When the two species were present together, muskellunge were disproportionately abundant in areas with rocky substrate, and nothern pike appeared to show a preference for the denser cover. Caplan (1982:22-24) remarked that "predation on the muskies by pike may have greatly influenced the distribution of muskies. Only 225 out of 900 muskies survived for 24 hours when placed with the pike. Predatory acts were frequently observed. The muskies were subject to predation in all of the quadrants and all of the substrate combinations, but the pike ate muskies at various rates depending on the pike's success in locating and capturing muskies over different substrates and cover densities. Since the muskies were small enough to lie in the cracks and holes among the rocks, the rate of predation was low on the rock substrate."

Interspecific Competition

The act of predation has a direct and immediate effect on the prey population. It causes a decline. One need only multiply this unit effect by a large number of acts of predation to see how the cumulative impact on the prey population could be considerable. Interspecific competition affects population size in a less direct way and implicitly involves increases in age-specific mortality rates or decreases in age-specific fecundity rates, or both, in comparison to what they would have been in the absence of interspecific competition. Here, the term "competition" is used in the sense recommended by Larkin (1956:330), namely, as "...the demand of more than one organism for the same resources of the environment in excess of immediate supply. In this sense it is a readily-defined phenomenon, one of the mechanisms involved in the resolution of *interactions* between populations. If one species eliminates another from an environment..., it may do so by competition, predation, environmental conditioning, etc. The whole relationship should not be called interspecific competition."

Muskellunge and northern pike might compete for food, cover, or spawning sites. Evidence for each of these is considered below. In each case, there is evidence of overlap in resource utilization based on data for allopatric populations, but little comparative information about sympatric populations. Where such overlap has been shown, it is difficult to establish that the availability of shared resources is a limiting factor, at the individual or population level. Furthermore, we must ask not only whether muskellunge and northern pike compete for a resource, but also, what is the advantage for northern pike that would enable them to displace muskellunge?

Competition for food. Food habits of muskellunge and northern pike of comparable size appear to be similar and competition for food is a possible mechanism of interference at all sizes. Insofar as interspecific competition for food is relevant to population control, however, the first several years of life would appear to be most important. Muskellunge must grow through a broad size range represented by multiple age classes of northern pike if they are to reach adult size. A given cohort of muskellunge may face competition first from northern pike of the same year-class, and later from older age-groups.

Food habits of both species vary as a function of size (Scott and Crossman 1973). Qualitative changes in diet occur most rapidly among fry, when the relative rate of growth in length is greatest. Fry switch from zooplankton (cladocerans and copepods) to aquatic insect larvae to fish within a matter of weeks (Elson 1940; Hunt and Carbine 1951; Fago 1977). After the first year of life, both species eat mostly fish, and larger esocids eat larger fish.

The type of fish that is eaten depends largely on what is available. Yellow perch, trout perch, gizzard shad, alewife, white sucker, and bluegill have been reported as constituting large fractions of the diet of northern pike in different lakes of the Upper Midwest of the United States and Canada (Seaburg and Moyle 1964; Lawler 1965; Johnson 1969; Wagner 1972; Wolfert and Miller 1978; Koshinsky 1979). Muskellunge food habits have not been documented so extensively. Yellow perch and white sucker have been shown to be locally important. (Hourston 1952).

Given the apparent flexibility in feeding habits of both species, it is difficult to make inferences about similarities or differences in food preferences based on data for allopatric populations. Food habits of co-occurring muskellunge and northern pike have not been described in detail. Stomach contents of angler-caught muskellunge and northern pike from Escanaba Lake (Vilas County, Wisconsin) were similar, but sample sizes were very small (Inskip and Magnuson, 1986. Fish predominated in the stomachs of both species. Yellow perch, walleye, and several species of cyprinids were most common among the remains that could be identified. A comparative study of food selection by yearling muskellunge and northern pike in tanks indicated a basic similarity in food preference among the types of prey offered (Weithman and Anderson 1977). Gizzard shad, goldfish, and golden shiner were selectively eaten relative to bluegill, largemouth bass, and channel catfish of comparable size. Although little in the way of comparative information is available for sympatric populations, it seems likely that there is considerable overlap in the types of prey consumed.

It is easier to demonstrate shared utilization of food resources by two species than to show that food supply is inadequate to meet their combined demand. An appropriate outcome, or performance, measure must be chosen. Growth rate seems to be an appropriate choice as a performance measure to use in evaluating for competition for food, assuming a correlation between growth and fitness (Maclean and Magnuson 1977).

Yearly indices of growth for muskellunge and northern pike from Escanaba Lake were positively associated (Inskip and Magnuson, 1986). This coordinate variation over a 25 year period (1957 to 1982), coupled with evidence of dietary overlap, inverse associations between growth indices and indicators of esocid abundance, and positive associations with an indicator of abundance of yellow perch suggest that muskellunge and northern pike shared a common food supply and that food shortages periodically depressed growth rates of both species. However, it is not evident that this apparent competition among individual fish was an important factor in determining population size.

Assuming for the moment that competition for food sometimes occurs between muskellunge and northern pike, how might periods of food scarcity work to the relative advantage of northern pike? Northern pike might be more aggressive or efficient in feeding or have a higher food conversion efficiency. Graff (1972) reported that muskellunge fry were less likely than northern pike fry to attack solitary prey organisms. Weithman and Anderson (1977:427) noted that "at relatively low levels of consumption near the maintenance feeding level, northern pike in tanks were more efficient at converting prey biomass into body tissue." Nevertheless, it is not possible to make any unconditional generalizations about relative food-conversion efficiencies in the absence of good comparative data over a range of temperatures and feeding regimens.

Populations of the two species might be differentially affected by reductions in growth rate and increases in total mortality rate. To the extent that northern pike reduce prey availability for muskellunge, they could prolong the time required for muskellunge to become sexually mature. If natural mortality rates also depend on food supply, the combined effects of an extended prereproductive phase and high juvenile mortality rates could result in low survival to reproductive age. The effect could be more pronounced for muskellunge than for northern pike because of the difference in size at maturity (Table 1); however, it is *not* clear that rates of growth, natural mortality, and fecundity for the two species have the same dependencies on food supply. Growth data for Escanaba Lake esocids, described above, suggest that growth rates of muskellunge and northern pike in certain size classes have varied in a similar fashion and that the variation was associated with food supply (Inskip and Magnuson, 1986). Data for northern pike also show the predicted changes in age at sexual maturity and mortality rate (Kempinger and Carline 1978). Analogous data for muskellunge are not available. This is a recurring problem: comparative inferences concerning muskellunge and northern pike are complicated by the lack of information for co-occurring populations, and of experimental data for individuals of the two species held under comparable conditions.

Competition for cover. Young-of-the-year esocids are susceptible to predation by esocid and nonesocid fishes and by other vertebrate and invertebrate predators (Scott and Crossman 1973). Hunt and Carbine (1951) documented predation by yellow perch on YOY northern pike. Predation by largemouth bass was an important factor limiting survival of muskellunge-northern pike hybrids stocked in Ohio reservoirs (Stein et al. 1981). Other studies have implicated predation as an important cause of mortality for stocked fingerling muskellunge (Johnson 1982). It is possible that protective cover is a limiting resource governing the survival of esocid fry, fingerlings, or both, and that northern pike, by virtue of their larger size, have an advantage in securing shelter.

Cover also may be important to esocids as a place from which to seek and pursue food. Not only do muskellunge and northern pike eat similar items, they appear to do so in the same areas, at least for fish of a given size. Both species are littoral zone predators. Competition for feeding cover could occur at any or all sizes, but northern pike would have a size advantage only during the first few weeks or months of life.

Telemetric studies have documented habitat use patterns of certain size classes of muskellunge (Minor and Crossman 1978; Dombeck 1979) and northern pike (Diana et al. 1977) in specific bodies of water, but not of both species in the same body of water at the same time. One would expect that habitat utilization would depend on site-specific characteristics, so it is difficult to generalize on the basis of one or two studies. Also, the presence of one species might influence habitat use by the other. Caplan's (1982) laboratory experiments with fry, described above, suggested one means (predation) by which this might occur.

Competition for spawning sites. Spatial overlap of spawning seems to be the rule among sympatric populations of muskellunge and northern pike in most lakes (Williamson 1942; Scott and Crossman 1973), but patterns of spawning-habitat utilization for co-occurring populations have not been extensively documented. Both species have been observed spawning in shallow (usually < 1 m) protected areas over bottoms covered with vegetation or plant debris, such as occur where marsh or terrestrial vegetation is inundated by spring floods, or in bays of lakes (Elson 1940; Williamson 1942; Clark 1950; Franklin and Smith 1963; Minor and Crossman 1978; Dombeck 1979; Trautman 1981). The two species have been observed spawning together (Eddy 1944; Black and Williamson 1947).

Spatial segregation of spawning of muskellunge and northern pike has been reported for several bodies of water, mostly in the Great Lakes drainage.

Table 1. Length and age at first spawning of muskellunge and northern pike in Northern Wisconsin lakes.

	Total length (cm)	Age	Reference
muskellunge			Johnson (1971)
male	55-65	3-5	
female	60-70	4-6	
northern pike			Priegel and Krohn (1975)
male	30-40	2	
female	35-45	2-3	

Muskellunge in the Niagara River spawned on shoals in the main river, where the current averaged 0.2 m/s, and YOY were found mostly where the current exceeded 0.1 m/s; northern pike migrated up slow-moving turbid streams and spawned in stiller water (Harrison and Hadley 1978). Osterberg, D.M. (pers. comm.) reported similar observations for populations in the St. Lawrence River; that is, muskellunge appeared to spawn where there was a distinct current, and northern pike utilized more stagnant water. Muskellunge in Presque Isle Bay, Lake Erie spawned in the open bay, while cohabiting northern pike utilized adjoining sloughs (Buss and Larsen 1961, cited in Buss et al. 1978). Muskellunge in Lake St. Clair (Haas 1978) and Leech Lake, Minnesota (Strand, 1986) also were thought to spawn in offshore areas. Apparent spawning depths were > 3 m in Lake St. Clair and 1 to 2 m in Leech Lake. In reference to populations in Minnesota, Eddy and Surber (1943:171) commented that muskellunge ordinarily "spawn in tributary streams and shallow lake channels rather than in the weeds as the northern pike does." This distinction is far from universal, but the preceding observations collectively suggest that there are interspecific differences in spawning habitat preference, at least for certain populations. Muskellunge do not appear to be so dependent on shallow, stagnant vegetated areas and make greater use of flowing water and deeper offshore areas. Manifestations of any differences in spawning habitat preference that do exist probably would depend on the types and quantity of that available when the fish are ready to spawn. Johnson (1971) observed that the degree of concentration of spawning of muskellunge and northern pike appeared to be related to the way in which a lake warmed in the spring. Spawning tended to be more concentrated when areas of a lake, such as a small bay, warmed more rapidly than the rest of the lake.

Northern pike might interfere with muskellunge reproduction by preventing muskellunge from spawning at all or by causing muskellunge to use inferior spawning habitat. Given the habits and relative sizes of spawning muskellunge and northern pike, it seems unlikely that this has been an important factor mediating changes in relative abundance. Spawning occurs in multiple acts, and the eggs of a single female may be spread over a large area (Scott and Crossman 1973). No parental care is provided. The temporal segregation of spawning of muskellunge and northern pike should minimize the frequency of interspecific encounters unless muskellunge congregate on spawning grounds well before spawning begins or northern pike remain on the grounds after their spawning ends. In such a case, agonistic encounters could occur, and muskellunge could be displaced from preferred spawning grounds; however, there is little reason to believe that this is the case. The average size of muskellunge spawners exceeds that of northern pike spawners, and it is difficult to imagine even an aggressive small northern pike deterring a larger muskellunge. Furthermore, muskellunge could return to a spawning area from which they had been excluded, once the northern pike had left. It is conceivable that muskellunge are somehow inhibited from spawning where eggs of northern pike have been deposited (for example, due to a pheromone), but, again, there is no evidence of such an effect.

Hybridization

Natural hybridization occurs between late spawning northern pike and early spawning muskellunge (Eddy 1944; Crossman and Buss 1965). Oehmcke (1969) reported that an estimated 40 to 50 percent of the "muskellunge" caught annually from Lac Vieux Desert, on the Wisconsin - Michigan border, were natural hybrids. Muskellunge and northern pike have been observed spawning together (Eddy 1944; Black and Williamson 1947), indicating that behavioral barriers to interbreeding are not well-developed. Crossman and Buss (1965: 1287) inferred that most reported instances of natural hybridization "resulted from the actual mating of the male of one species with the female of another (active hybridization)" rather than from "the haphazard mixing of sperm and egg (passive hybridization)." Concentrated spawning of muskellunge and northern pike within a small area would be expected to increase the likelihood of both types of hybridization. Because the F-1 hybrids are sterile or of greatly reduced fertility (Eddy 1944; Black and Williamson 1947), active hybridization, in particular, would signify wasted reproductive effort for the parental fish. Spatial and temporal overlap of spawning between a large population of northern pike and a small population of muskellunge could, conceivably, result in failed reproduction for a large percentage of the muskellunge population. This leaves unexplained, however, how northern pike would have become more abundant in the first place.

Disease

It is conceivable that the introduction of northern

pike to lakes containing muskellunge resulted in the exposure of the muskellunge to a new set of pathogens and that it had lesser resistance to one or more of these than northern pike. Sonstegard (1975, 1976, 1978) described a malignancy (lymphosarcoma) in northern pike that also occurs in muskellunge and is thought to be caused by a virus that is transmitted during spawning. Prevalence of the disease can be high in both species—muskellunge (21%); and northern pike (16%)—but the course of the disease is different for the two species (Sonstegard 1976). Spontaneous regressions appear to occur frequently in northern pike whereas progression to death appears to be almost inevitable for muskellunge. There is no evidence that any infectious disease process, let alone a specific oncogenic virus, has mediated the observed changes in relative abundance of muskellunge and northern pike, but the possibility deserves mention in the interests of completeness.

ENVIRONMENTAL CHANGE AS A FACTOR AFFECTING RELATIVE ABUNDANCE

Inverse associations in relative abundance of muskellunge and northern pike might arise even in the absence of any direct species interaction. For example, changing environmental conditions could affect two species quite differently if they have different tolerances or optima for certain environmental variables (Svardson 1976). Insight into how environmental requirements of muskellunge and northern pike might differ can be gained by comparing patterns of distribution in the recent and distant past. Such observations, recounted earlier, suggest possible differences in thermal requirements and in adaptations to riverine versus lacustrine environments. Specifically, it appears that northern pike might be favored by cooler temperatures and more lentic conditions.

The close correspondence between the 18 C isotherm for mean July daily air temperature (Thomas 1953) and the northern boundary of the range of muskellunge (Crossman 1978) suggests that climatic influences, acting on growth, survival or reproduction have limited their northward spread. The success with which northern pike recolonized far northern latitudes indicates that they are better able to cope with cold climates. Conversely, the native range of muskellunge extends farther south than does that of northern pike. Differences in spawning temperatures are consistent with the hypothesis that muskellunge are adapted to a warmer thermal regime; however, thermal requirements of the two species have not been subjected to thorough comparative study at a level of detail comparable to investigations of the thermal requirements of northern pike alone (Swift 1965; Johnson 1966; Lillelund 1966; Hassler 1970; Hokanson et al. 1973; Casselman 1978). Muskellunge eggs develop more slowly at low temperatures (10.6 - 12.2 C) than do those of northern pike (Walker 1968). Muskellunge fry will not "swim up" if temperatures remain under 13 C for an extended period (Johnson 1958). Northern pike larvae apparently are most tolerant of low temperatures, based on data presented by Hassler (1970).

There is no strong evidence that muskellunge are less stressed by high temperatures. Muskellunge (> 70 cm, fork length) in two Ontario lakes remained in shallow (< 1 m) water during very warm periods when water temperatures exceeded 25 C, even though cooler water was available (Minor and Crossman 1978). The preferred temperature of muskellunge in the lab was 24 C (Ferguson 1958), higher than would be expected for northern pike on the basis of growth data. Growth rate of laboratory-reared northern pike peaked at 19 to 21 C but declined rapidly at temperatures exceeding 23 C and ceased at 28 C for fish fed ad libitum (Casselman 1978). Krohn (1969) reported a mid-August die-off of northern pike in a Wisconsin lake following several very warm days; temperatures varied from 32 C at the surface to 24 C at a depth of 3.7 m, the lower limit of oxygenated water. Temperatures greater than 32 C were lethal within two days for fingerling northern pike and muskellunge (Scott 1964).

Short term climatic fluctuations or even specific weather patterns could mediate a switch in relative abundance of muskellunge and northern pike if, at some life-history stage, the two species are dissimilarly affected by high or low temperature extremes, by prolonged periods of warm or cold weather, or by temperature fluctuations.

A second difference in the native North American ranges of muskellunge and northern pike, in addition to the latitudinal difference, has already been alluded to. Muskellunge were, and still are, present in many southern rivers from which northern pike were, and are, absent. Crossman (1978:20) suggested that this is indicative of a more general pattern. "Throughout its range, the muskellunge occurs more frequently (at times other than spawning) in slow meandering streams and in the quieter portions of large moderately swift rivers. It may in fact be more of a river species than a lake species. Its presence in the south is limited to rivers, and it doubtless waited out at least the Wisconsin glaciation in these southern riverine environments."[2] As reported earlier, spawning of muskellunge and northern pike in several parts of the Great Lakes drainage is segregated in a way that is consistent with the idea that muskellunge evolved in lotic habitats and northern pike in lentic habitats. The perspective generates testable predictions about possible differences between muskellunge and northern pike, for example, with respect to their relative abilities to maneuver in a current or tolerances of siltation and low dissolved oxygen concentrations.

Early life history stages again seem to be most relevant. Perhaps the lesser surface area of YOY muskellunge of a given length, as compared to northern pike, is an advantage in moving water. On the other hand, northern pike might be better able to tolerate high rates of siltation or low dissolved oxygen concentrations. Northern pike fry have an adhesive structure on their heads, with which they can attach to aquatic vegetation or debris and remain off of the bottom (Scott and Crossman 1973). Muskellunge fry do not appear to have such an organ (Dombeck et al. 1984). The concentration of dissolved oxygen near the substrate-water interface appeared to be an important determinant of muskellunge reproductive success in eight Upper Midwestern lakes (Dombeck et al., 1986).

Because muskellunge and northern pike often spawn in shallow water, their eggs are very sensitive to water level changes, especially to declines in water level. A difference in spawning depth would influence their relative vulnerabilities. The two species also could be affected differentially by water level fluctuations because of the difference in spawning time (MacCrimmon and Skobe 1970).

Given the interspecific differences in size (and age) at sexual maturity, any factor that caused equal increases in total mortality rates might be expected to have different population level consequences for the two species. Fishing mortality, including mortality among released fish, is one component of total mortality. In Escanaba Lake, with no minimum size[3] or daily bag limit, the average annual exploitation rate for muskellunge between 1958 and 1969 was lower than that for northern pike[4] (25% versus 46%) (Kempinger et al. 1975). It is likely that the difference would be even greater elsewhere. Almost without exception, fishing regulations restrict the harvest of muskellunge more than that of northern pike. Yet, the time interval between when northern pike begin to recruit to the sport fishery and when they spawn for the first time is much shorter than the corresponding interval for muskellunge. Even in a catch and release fishery, juvenile muskellunge are vulnerable to fishing mortality longer than are juvenile northern pike. For a population undergoing rapid turnover, the life history of northern pike appears to be superior to that of muskellunge. Initiation of a sport fishery on a lake containing muskellunge and northern pike, or an increase in fishing pressure on a previously lightly-fished lake, should work to the *relative* advantage of northern pike.

DISCUSSION

That northern pike can have a detrimental effect on muskellunge populations has been presumed for approximately 45 years. The initial accounts of declines in native muskellunge populations coincident with, or subsequent to, colonizations by northern pike (Threinen and Oehmcke 1950; Oehmcke 1951) prompted the inference, and affinities between the two species gave it credibility. The fact that native Wisconsin distributions of muskellunge and northern pike were nearly complementary (Greene 1935) is consistent with the idea that they form an unstable species association. With each additional report of an apparent decline in a muskellunge stock associated with an increase in a northern pike population (Johnson 1981; Inskip and Magnuson 1983; Osterberg, D.M., personal communication), the presumption becomes more ingrained. Our understanding of the "machinery of interaction" (Larkin 1956:329) and of circumstances that modify the intensity and outcome of interaction have not advanced to the same degree as has the belief that the two species are somehow incompatible.

The preceding discussion addressed the possible importance of predation and interspecific competition for food, cover, and spawning sites in resolving interactions between populations of muskellunge and northern pike. This is a somewhat artificial compartmentalization, as the relevant biological

---

[2] Stocked muskellunge have grown well and provided good angling in southern reservoirs such as Norris Lake, Tennessee (Hall, G.E., personal communication).

[3] A 56 cm minimum length limit was in effect for northern pike during 1964 to 1972.

[4] Estimates of exploitation were based on muskellunge longer than 45 cm and on northern pike longer than 30 cm (1958 to 1963) and 56 cm (1964 to 1969).

processes often are interdependent. For example, consider a situation where spawning groups of both species use the same spawning grounds, in apparent disregard for each other. While there may be no identifiable interaction among the adult fish, their concentrated spawning activities could result in a density of fry in excess of what the immediate environment can support. Cannibalism (predation) or competition for food or cover might be the proximate mechanism of population loss, but conditions for these interactions would have been created by the shared use of the habitat by the spawning adults.

The hypothesis that predation by YOY northern pike on YOY muskellunge is the critical mechanism of interaction has gained wide acceptance, more because of its plausibility than direct field evidence. Muskellunge and northern pike often spawn in the same areas; northern pike spawn and hatch earlier; young northern pike become piscivorous very early in life, and cannibalism among YOY northern pike has been documented. Laboratory studies have shown that predation by YOY northern pike *can* be an important cause of mortality among YOY muskellunge (Caplan 1982), and results of a pond experiment were consistent with this idea (Oehmcke 1951). That interspecific predation is of controlling importance for wild muskellunge populations has not been demonstrated.

When considering any given mechanism of interaction as a possible basis for explaining changes in relative abundance, it is helpful to consider the following questions: (1) What is the evidence that the mechanism operates among individual muskellunge and northern pike? (2) Through what pathway might this interaction at the level of individual fish lower population size? (3) What is the evidence that it does? (4) What is the advantage for northern pike? (5) How might characteristics of the environment and the composition of the fish community modify the intensity and outcome of this type of interaction? What conditions promote more frequent or intense interaction? What conditions are relatively more favorable to muskellunge? (6) Is there evidence of racial differences between populations of muskellunge, or northern pike, on which the occurrence of this type of interaction would depend? These questions are posed not with the expectation that answers will soon, or ever, be available for an exhaustive list of possible mechanisms of interaction; however, precise management of muskellunge-northern pike fisheries, if it is possible at all, would appear to require answers to these questions. They define a framework in which we might seek to understand the species interaction.

To repeat, specific environmental variables could mediate changes in relative abundance of muskellunge and northern pike in two different ways. Environmental conditions could modify the occurrence and outcome of a particular type of species interaction (for example, see Svardson 1976). An environmental factor also could differentially, and independently, affect the abundance of each species. An underlying fabric of species interaction is not required.

On the basis of fossil evidence and twentieth century native distributions, Crossman hypothesized that muskellunge evolved in North America from northern pike (Crossman and Harington 1970). Whether the progenitor was northern pike or some third species, one wonders what sort of environmental conditions prevailed while speciation was taking place. An answer to this question might provide an understanding of conditions under which the life history of muskellunge is superior to that of northern pike. The fossil record cannot support any strong inferences in this regard. It is nonetheless helpful to adopt a historical (or, pre-historical) perspective when attempting to understand the species association. Such an outlook is potentially rich in insights and, at the very least, provides a framework in which to organize many of our observations. Zoogeographic issues are deserving of a more comprehensive treatment than they have received in this paper.

An evolutionary perspective explicitly alerts us to the possibility of intraspecies variation. Although we are nominally concerned with an interaction between two species, in any given body of water the concern is with an interaction between populations. It is virtually certain that racial differences exist for both species (for example, see Morrow 1964) and reasonable to presume that some of the differences are relevant to the species interaction. Indeed, such differences could *result* from coevolution of populations of muskellunge and northern pike. Extensive stocking probably has reduced much of the genetic variability among muskellunge stocks, but local or regional differences may still exist. Genetic variability presents the fishery manager not only with challenges but also with opportunities.

Zoogeographic analysis is only one of several possible nonexperimental approaches to understanding how environment affects a species' performance or modulates a species interaction. Dombeck et al. (1986) attempted to identify environmental characteristics that distinguished habitats where muskellunge did well in the presence of northern pike from those where they did poorly by means of a statistical analysis of a large number of lakes. Empirical approaches such as these can be effective in detecting pattern, and the results can provide useful management tools, even if they are only suggestive about mechanisms underlying the observed associations.

Experimental studies have the potential of yielding explanatory rather than descriptive understanding, but the muskellunge-northern pike association is refractory to experimental analysis. Only the early life history stages are amenable to study in the laboratory or in small artificial ponds, because of the food and space requirements of older fish and the relatively long generation time of muskellunge. Whole lake manipulations involving muskellunge and northern pike are routinely performed in the name of management (stocking, removal programs, harvest regulations), but seldom as part of designed experimental studies. Political considerations make this unfeasible in most settings. Furthermore, field studies of muskellunge populations are routinely plagued by problems associated with small sample sizes.

Notwithstanding these practical difficulties, there is a need for comparative study of selected aspects of the biology of muskellunge and northern pike. In particular, it seems both important and feasible to clarify the relative tolerances of early life history stages to extremes of physical and chemical variables, including temperature and dissolved oxygen concentration. It also would be helpful to have growth and life table data and information on food and space utilization for co-occurring populations. Such information will be hard to come by, but, if and when opportunities present themselves, they should be exploited.

It is doubtful that we will soon have a comprehensive mechanistic understanding of how muskellunge and northern pike interact and how these interactions are modified by physical, chemical and biological aspects of the environment. Such understanding remains a worthy goal for research, but, for the foreseeable future, it appears that management for this species pair will be based on perceptions, traditions, empirical tools and public pressure.

ACKNOWLEDGEMENTS

I wish to thank G. C. Becker (author), The Board of Regents of the University of Wisconsin System (copyright holder), and the University of Wisconsin Press (publisher) for the use of illustrations from "Fishes of Wisconsin". I also am grateful to the Wisconsin Department of Natural Resources for permission to reproduce illustrations from DNR publications by C. W. Greene and L. D. Johnson. Figures were redrawn from the originals by C. Hughes. E. J. Crossman and R. A. Ryder reviewed a draft of this manuscript and offered many constructive suggestions.

REFERENCES

Bean, T. H. 1908. The muskellunge of the Ohio basin. Transactions of the American Fisheries Society 37:145-151.

Becker, G. C. 1983. Fishes of Wisconsin. The University of Wisconsin Press. Madison, Wisconsin, USA.

Black, J. D., and L. O. Williamson. 1942. Artificial hybrids between muskellunge and northern pike. Transactions of the Wisconsin Academy of Sciences, Arts and Letters. 38:299-314.

Buss, K., and A. Larsen. 1961. The northern pike of Presque Isle Bay, Lake Erie. Pennsylvania Angler 30(9):4-6. (cited in Buss et al. (1978))

Buss, K, J. Meade, and D. R. Graff. 1978. Reviewing the esocid hybrids. American Fisheries Society Special Publication 11:210-216.

Caplan, D. L. 1982. An experimental study of interactions between young of the year pike (*Esox lucius*) and muskellunge (*Esox masquinongy*). M.S. thesis. University of Wisconsin--Madison, Madison, Wisconsin, USA.

Carlander, K.D. 1969. Handbook of freshwater fishery biology. Volume 1. Iowa State University Press, Ames, Iowa, USA.

Carlander, K.D., J.S. Campbell, and R.J. Muncy. 1978. Inventory of percid and esocid habitat in North America. American Fisheries Society Special Publication 11:27-38.

Casselman, J.M. 1978. Effects of environmental factors on growth, survival, and exploitation of northern pike. American Fisheries Society Special Publication 11:114-128.

Clark, C.F. 1950. Observations on the spawning habits of northern pike, *Esox lucius*, in northwestern Ohio. Copeia 1950:285-288.

Crossman, E.J. 1978. Taxonomy and distribution of North American esocids. American Fisheries Society Special Publication 11:13-26.

Crossman, E.J., and K. Buss. 1965. Hybridization in the family Esocidae. Journal of the Fisheries Research Board of Canada 22:1261-1292.

Crossman, E.J., and C.R. Harington. 1970. Pleistocene pike, *Esox lucius*, and *Esox* sp. from the Yukon Territory and Ontario. Canadian Journal of Earth Sciences 7:1130-1138.

Diana, J.S., W.C. McKay, and M. Ehrman. 1977. Movements and habitat preference of northern pike (*Esox lucius*) in Lac Ste. Anne, Alberta. Transactions of the American Fisheries Society 106:560-565.

Dombeck, M.P. 1979. Movement and behavior of the muskellunge determined by radiotelemetry. Wisconsin Department of Natural Resources Technical Bulletin 113, Madison, Wisconsin, USA.

Dombeck, M.P., B.W. Menzel, and P.N. Hinz. 1984. Muskellunge spawning habitat and reproductive success. Transactions of the American Fisheries Society 113:205-216.

Dombeck, M.P., B.W. Menzel, and P.N. Hinz. 1986. Ecological factors influencing muskellunge-northern pike interactions in midwestern lakes. American Fisheries Society Special Publication 15:122-134.

Eddy, S. 1944. Hybridization between northern pike (*Esox lucius*) and muskellunge (*Esox masquinongy*). Proceedings of the Minnesota Academy of Sciences 12:38-43.

Eddy, S., and T. Surber. 1943. Northern fishes. University of Minnesota Press, Minneapolis, Minnesota, USA.

Elkins, W.A. 1937. A fish yield study for certain lakes in the Chequamegon National Forest. Transactions of the American Fisheries Society 66:306-312.

Elson, P.F. 1940. Rearing maskinonge in a protected area. Transactions of the American Fisheries Society 70:421-429.

Fabricius, E., and K.J. Gustafson. 1958. Some new observations on the spawning behavior of the pike, *Esox lucius* L. Fisheries Research Board of Sweden, Institute for Freshwater Research, Drottningholm 39:23-54.

Fago, D.M. 1977. Northern pike production in managed spawning and rearing marshes. Wisconsin Department of Natural Resources Technical Bulletin 96, Madison, Wisconsin, USA.

Ferguson, R.G. 1958. The preferred temperature of fish and their midsummer distribution in temperate lakes and streams. Journal of the Fisheries Research Board of Canada 15:607-624.

Franklin, D.R., and L.L. Smith. 1963. Early life history of the northern pike, *Esox Lucius* L., with special reference to the factors influencing the numerical strength of year classes. Transactions of the American Fisheries Society 92:21-110.

Graff, D.R. 1972. Observations on interactions between juveniles of *Esox lucius* and *Esox masquinongy*. M.Ed. thesis. Pennsylvania State University, Pennsylvania, USA.

Greene, C.W. 1935. The distribution of Wisconsin fishes. Wisconsin Conservation Commission, Madison, Wisconsin, USA.

Haas, R.C. 1978. The muskellunge in Lake St. Clair. American Fisheries Society Special Publication 11:334-339.

Harrison, E.J., and W.F. Hadley. 1978. Ecologic separation of sympatric muskellunge and northern pike. American Fisheries Society Special Publication 11:129-134.

Hassler, T.J. 1970. Environmental influences on early development and year class strength of northern pike, *Esox lucius* (Linnaeus). Transactions of the American Fisheries Society 102:89-100.

Hokanson, K.E.F., J.H. McCormick, and B.R. Jones. 1973. Temperature requirements for embryos and larvae of the northern pike, *Esox lucius* (Linnaeus). Transactions of the American Fisheries Society 102:89-100.

Hourston, A.S. 1952. The food and growth of the maskinonge (*Esox masquinongy* Mitchill) in Canadian waters. Journal of the Fisheries Research Board of Canada 8:347-368.

Hunt, B.P., and W.F. Carbine. 1951. Food of the young pike, *Esox lucius* L., and associated fishes in Peterson's ditches, Houghton Lake, Michigan. Transactions of the American Fisheries Society 80:67-83.

Inskip, P.D., and J.J. Magnuson. 1983. Changes in fish populations over an 80 year period: Big Pine Lake, Wisconsin. Transactions of the American Fisheries Society 112:378-389.

Inskip, P.D., and J.J. Magnuson. 1986. Fluctuations in growth rate and condition of muskellunge and northern pike in Escanaba Lake, Wisconsin. American Fisheries Society Special Publication 15:176-188.

Johnson, L. 1966. Experimental determination of food consumption of pike, *Esox lucius*, for growth and maintenance. Journal of the Fisheries Research Board of Canada 23:1495-1505.

Johnson, L.D. 1958. Pond culture of muskellunge in Wisconsin. Wisconsin Department of Natural Resources Technical Bulletin 17, Madison, Wisconsin, USA.

Johnson, L.D. 1969. Food of angler-caught northern pike in Murphy Flowage. Wisconsin Department of Natural Resources Technical Bulletin 42, Madison, Wisconsin, USA.

Johnson, L.D. 1971. Growth of known-age muskellunge and validation of age and growth determination methods. Wisconsin Department of Natural Resources Technical Bulletin 49, Madison, Wisconsin, USA.

Johnson, L.D. 1981. Comparison of muskellunge (*Esox masquinongy*) populations in a stocked lake and unstocked lake in Wisconsin with notes on the occurrence of northern pike (*Esox lucius*). Wisconsin Department of Natural Resources Research Report 110, Madison, Wisconsin, USA.

Johnson, L.D. 1982. Factors affecting short-term survival of stocked muskellunge fingerlings in Wisconsin. Wisconsin Department of Natural Resources Research Report 117, Madison, Wisconsin, USA.

Kempinger, J.J., W.S. Churchill, G.R. Priegel, and L.M. Christenson. 1975. Estimate of abundance, harvest, and exploitation of the fish population of Escanaba Lake, Wisconsin. Wisconsin Department of Natural Resources Technical Bulletin 84. Madison, Wisconsin, USA.

Kempinger, J.J., and R.F. Carline. 1978. Dynamics of the northern pike population and changes that occurred with a minimum size limit in Escanaba Lake, Wisconsin. American Fisheries Society Special Publication 11:382-389.

Koshinsky, G.D. 1979. Northern pike at Lac La Ronge. Part 1. Biology of northern pike. Part 2. Dynamics and exploitation of the northern pike population. Saskatchewan Fisheries Laboratory Technical Report 79-80. Department of Tourism and Renewable Resources, Saskatoon, Saskatchewan, Canada.

Krohn, D.C. 1969. Summary of northern pike stocking investigations in Wisconsin. Wisconsin Department of Natural Resources Technical Bulletin 44, Madison, Wisconsin, USA.

Larkin, P.A. 1956. Interspecific competition and population control in freshwater fish. Journal of the Fisheries Research Board of Canada 13:327-342.

Lawler, G.H. 1965. The food of the pike, *Esox lucius*, in Heming Lake, Manitoba. Journal of the Fisheries Research Board of Canada 22:1357-1377.

Lillelund, V.K. 1966. Versuche zur erbrutung der Eier vom Hecht, *Esox lucius* L., in Abhangikeit von Temperatur und Licht. Archiv fur Fischereiwissenschaft 17:95-113.

MacCrimmon, H.R., and E. Skobe. 1970. The fisheries of Lake Simcoe. Ontario Department of Lands and Forests, Fish and Wildlife Branch, Toronto, Ontario, Canada.

Maclean, J., and J.J. Magnuson. 1977. Species interactions in percid communities. Journal of the Fisheries Research Board of Canada 34:1941-1951.

McPhail, J.D. 1963. The postglacial dispersal of freshwater fishes in northern North America. Doctoral dissertation, McGill University, Montreal, Quebec, Canada.

Minor, J.D., and E.J. Crossman, 1978. Home range and seasonal movements of muskellunge as determined by radiotelemetry. American Fisheries Society Special Publication 11:146-153.

Morrow, J.E. 1964. Populations of pike, *Esox lucius,* in Alaska and northeastern North America. Copeia 1964:235-236.

Oehmcke, A.A. 1951. Muskellunge yearling culture and its application to lake management. Progressive Fish Culturist 13:63-70.

Oehmcke, A.A. 1969. Muskellunge management in Wisconsin. Wisconsin Department of Natural Resources Bureau of Fish Management Report 19, Madison, Wisconsin, USA.

Oehmcke, A., L. Johnson, J. Klingbiel, and C. Wistrom. 1965. The Wisconsin muskellunge, its life history, ecology, and management. Wisconsin Conservation Department Publication 225, Madison, Wisconsin, USA.

Priegel, G.R., and D.C. Krohn. 1975. Characteristics of a northern pike spawning population. Wisconsin Department of Natural Resources Bulletin 86. Madison, Wisconsin, USA.

Schloemer, C.L. 1938. A second report on the growth of the muskellunge, *Esox masquinongy immaculatus* (Garrard), in Wisconsin waters. Transactions of the Wisconsin Academy of Sciences, Arts, and Letters 31:507-512.

Scott, D.P. 1964. Thermal resistance of pike *(Esox lucius* L.). muskellunge *(Esox masquinongy* Mitchill), and their F-1 hybrid. Journal of the Fisheries Research Board of Canada 21:1043-1049.

Scott, W.B., and E.J. Crossman. 1973. Freshwater fishes of Canada. Bulletin 184. Fisheries Research Board of Canada, Ottawa. 966 pp.

Seaburg, K.G., and J.B. Moyle. 1964. Feeding habits, digestive rates and growth of some Minnesota warmwater fishes. Transactions of the American Fisheries Society 93:269-285.

Smith, C.L. 1954. Pleistocene fishes of the Berends fauna of Beaver County, Oklahoma. Copeia 1954:282-289.

Sonstegard, R.A. 1975. Lymphosarcoma in muskellunge *(Esox masquinongy).* Pages 902-924 in W.E. Ribelin and G. Migaki, eds. Pathology of fishes. University of Wisconsin Press, Madison, Wisconsin, USA.

Sonstegard, R.A. 1976. Studies of the etiology and epizootiology of lymphosarcoma in *Esox (Esox lucius* L. and *Esox masquinongy).* Progress in Experimental Tumor Research 20:141-155.

Sonstegard, R.A. 1978. Lymphosarcoma in muskellunge and northern pike: guidelines for disease control. American Fisheries Society Special Publication 11:235-237.

Stein, R.A., R.F. Carline, and R.S. Haywood. 1981. Largemouth bass predation on stocked tiger muskellunge. Transactions of the American Fisheries Society 11:604-612.

Strand, R.F. 1986. Identification of principal spawning areas and seasonal distributions and movements of muskellunge in Leech Lake, Minnesota. American Fisheries Society Special Publication 15:62-73.

Svardson, G. 1976. Interspecific population dominance in fish communities in Scandinavian Lakes. Institute of Freshwater Research, Drottningholm, Sweden, Report 55:144-171.

Swift, D.R. 1965. Effect of temperature on mortality and rate of development of the eggs of the pike *(Esox lucius* L.) and the perch *(Perca fluviatilis* L.). Nature 206:528.

Thomas, M.K. 1953. Climatological atlas of Canada. Canada Department of Transport, Ottawa, Canada.

Threinen, C.W., and A.A. Oehmcke. 1950. The northern pike invades the musky's domain. Wisconsin Conservation Bulletin 15:10-12.

Trautman, M.B. 1981. The fishes of Ohio. Ohio State University Press, Columbus, Ohio, USA.

Wagner, W.C. 1972. Utilization of alewives by inshore piscivorous fishes in Lake Michigan. Transactions of the American Fisheries Society 101:55-63.

Walker, K.W. 1968. Temperature control in northern pike and muskellunge egg hatching. Proceedings of the North Central Fish Culture Workshop, Ames, Iowa. February 1968.

Weed, A.C. 1927. Pike, pickerel, and muskalonge. Chicago Field Museum of Natural History Zoology Leaflet 9:158-204.

Weithman, A.S., and R.O. Anderson. 1977. Survival, growth, and prey of Esocidae in experimental systems. Transactions of the American Fisheries Society 106:424-430.

Williamson, L.O. 1942. Spawning habits of muskellunge, northern pike. Wisconsin Conservation Bulletin 7:10-11.

Wolfert, D.R., and T.J. Miller. 1978. Age. growth, and food of northern pike in eastern Lake Ontario. Transactions of the American Fisheries Society 107:696-702.

Effects of Size at Stocking, Season, Largemouth Bass Predation, and Forage Abundance on Survival of Tiger Muskellunge

ROBERT F. CARLINE[1]

Ohio Cooperative Fishery Research Unit[2]
The Ohio State University
Columbus, Ohio 43210

ROY A. STEIN

Department of Zoology
The Ohio State University
Columbus, Ohio 43210

LAWRENCE M. RILEY[3]

Ohio Cooperative Fishery Research Unit
The Ohio State University
Columbus, Ohio 43210

ABSTRACT

We conducted studies in the laboratory, experimental ponds (0.8 and 2.0 hectares), and small impoundments (69 and 253 hectares) to define temporal patterns of tiger muskellunge mortality and to identify factors controlling their mortality and growth. Total mortality 50 days after stocking in ponds or impoundments ranged from 5 to 100% and averaged about 70% in 16 trials. Mortality of hybrids due to stresses associated with transport, handling, and temperature changes at stocking were reduced when fish were stocked in early autumn at water temperatures below 20 C. Mortality attributable to predation by largemouth bass ranged from 0-100% (mean 30%) and was related directly to largemouth bass densities and inversely to indices of abundance of largemouth bass prey. Predation on hybrids was most intense when small fish (168-184 mm long) were stocked in summer and was inconsequential when large hybrids (196-255 mm) were stocked in autumn. The presence of aquatic macrophytes did not influence intensity of predation by moderate to high densities of largemouth bass, even though hybrids preferred vegetated areas.

Overwinter survival of hybrids was directly related to their length. Largest hybrids (mean length 255 mm) had highest survival to spring (at least 40%) and to the following autumn (13%). Hybrid diet was related to prey type, density, and size. When gizzard shad were present, they were the most common prey, and growth of hybrids was faster in lakes with a gizzard shad-cyprinid forage base than in lakes with only centrarchids. Because hybrid growth was positively related to prey density and hybrid size was directly related to survival, an assessment of available forage shoudl provide guidance on number and size of hybrids to stock in a given lake. By stocking relatively large hybrids in autumn, short-term survival can be enhanced. However, factors affecting overwinter survival need to be identified before appropriate measures can be developed to increase survival to age 1.

Low survival of stocked tiger muskellunge in most Ohio reservoirs poses a major management problem. Low or variable survival of tiger muskellunge has been reported in other states (Johnson 1978; Hess 1981; Andrews and Laurion 1982) and, in general, has been a chronic problem with many esocid stocking programs (Flickinger and Clark 1978; Johnson 1982). Poor survival of stocked esocids has been ascribed to stress at stocking (Belusz 1978), rearing diet (Johnson 1978), and predation (Hess 1981; Stein et al. 1981; Johnson 1982). In general, causes of mortality have not been quantified.

In this paper we identify factors contributing to high mortality of stocked hybrids. We conducted studies of behavior, physiology, and population dynamics to quantify temporal patterns of mortality. Large-scale field studies combined with work in the laboratory and ponds provided a better understanding of factors controlling mortality and growth. An underlying premise was that growth and mortality are integrally linked. Recent work on size

---
[1] Present address: Pennsylvania Cooperative Fish and Wildlife Research Unit, The Pennsylvania State University, University Park, Pennsylvania 16802.
[2] The Unit is sponsored jointly by the United States Fish and Wildlife Service, Ohio Department of Natural Resources, and The Ohio State University.
[3] Present address: Arizona Game and Fish Department, Phoenix, Arizona 85023

and overwinter survival of smallmouth bass (Shuter et al. 1980), lipid reserves and overwinter survival of largemouth bass (Adams et al. 1982), and size-related predatory mortality of hybrid muskellunge (Stein et al. 1981) support the concept that growth rate and survival are related.

We first describe mortality of hybrids from time of stocking in summer or autumn until early winter and relate mortality to time of stocking, size of hybrids, and intensity of predation by largemouth bass. We then examine overwinter mortality of hybrids and these effects of prey type, size, and density on growth and long-term mortality.

METHODS

Over a five-year period we stocked 16 groups of hybrid muskellunge into two ponds and two impoundments (Table 1). All waters stratified thermally during summer and oxygenated epilimnions extended to depths of 2 to 3 m. Bird's Pond had no outflow and submersed macrophytes were sparse. Worthington Pond had intermittent outflow and supported extensive stands of cattails (*Typha* spp.) and submersed vegetation. Stonelick Lake had one inlet and an overflow dam; submersed macrophytes were limited to two small bays. Acton Lake had two inlets and an overflow dam; inshore cover for fish was limited to fallen trees. The largemouth bass was the only common piscivore in all four waters.

In the summers of 1981 and 1982 we tested the effects of aquatic macrophytes on habitat selection by hybrids and on predation intensity of largemouth bass. We divided the shoreline of Worthington Pond into six transects. Cattails were removed from three transects and the other macrophytes were mechanically removed or chemically treated to maintain vegetation-free sections. In 1981 we used electrofishing gear to determine the distribution of hybrids in vegetated and open transects. In 1982 SCUBA divers, maintaining depths of 1 and 3 m, swam around the entire perimeter of the pond, counted hybrids, and noted habitat characteristics.

Table 1. Locations and characteristics of Ohio study sites stocked with hybrid muskellunge.

Site	County	Surface area (hectares)	Maximum depth (m)	Range of Secchi disc readings in summer (m)
Bird's Pond	Delaware	0.8	4.0	1.0-2.5
Worthington Pond	Franklin	2.0	7.0	1.0-2.5
Stonelick Lake	Clermont	69	6.5	0.25-1.0
Acton Lake	Butler and Preble	253	10.0	0.25-1.0

Study lakes were routinely sampled from the time hybrids were stocked in summer or autumn until early December to obtain measures of prey composition and density. At Acton and Stonelick lakes, five shoreline sites (each about 20 m long) were sampled with a bag seine (11 m x 1.8 m x 5-mm bar mesh). We estimated the maximum size of forage fish that were vulnerable to predation by hybrids by multiplying mean hybrid length on each sampling date by 0.36, a value determined from an analysis of hybrid stomach contents. Numbers of fish captured with seines were divided by the area sampled to estimate inshore densities of available prey. Because Worthington Pond could not be easily seined, we used catch-per-unit-of-effort (CPUE) with electrofishing gear as an index of prey abundance there.

All hybrid muskellunge were raised on a dry pellet diet in fish culture facilities of the Ohio Department of Natural Resources. In most experiments, hybrids were graded to desired sizes the day before stocking and a sample was measured (total length to nearest mm) and weighed (nearest g). In those experiments where two sizes of hybrids were stocked (Table 2, stocks 10-15), we separated fish into two groups when fish were about 100 mm long; growth of one group was retarded by maintaining it at reduced temperatures. Three groups (stocks 1, 2, 8, Table 2) were given fine-wire jaw tags and two groups (stocks 14 and 15) were fin-clipped. None of the other stocks were marked. As per guidelines for transport in Ohio, hybrids were moved to lakes in aerated solutions of 0.5% salt and 10 mg/liter oxytetracycline. If lake temperatures exceeded transport tank temperatures by more than 5 C, lake water was added to the tank until temperatures were tempered to within 3 C of that in the lake. Tempering usually required 20 minutes. We estimated immediate post-stocking mortality by inspecting the stocking site for dead fish and maintaining a sample of fish in a nylon-mesh holding net (2 x 2 x 4 m) or floating wire-mesh cages (2.6 x 0.9 x 0.9 m) for up to 10 days.

After stocking, hybrids were routinely samples at night with a 300-volt, pulsed direct-current boom shocker. We estimated mortality from electrofishing CPUE data and population estimates. Daily instantaneous mortality rate was estimated from the slope of the relation between the natural logarithm of CPUE and number of days after stocking. These mortality rates were reported (Table 2) only if the relation was significant ($P<0.05$). About 50 days after stocking we estimated the number of surviving

Table 2. Summary of all hybrid muskellunge stockings, 1979-83, including total mortality and predatory mortality attributable to largemouth bass.

Site, stock number, and stocking date	Number stocked	Mean length in mm±1 SE	Method of estimation[a]	Total mortality 50 days post stocking (%)	Predatory mortality	Largemouth bass predators Minimum length (mm)	Largemouth bass predators Density (number per hectare)	Effective Predator Density (number per hectare)
Bird's Pond								
1. 15 Jul 1980	200	125±0.7	CPUE	100	100	150	75.0	69.4
2. 22 Jul 1980	200	137±0.6	CPUE	100	100	150	75.0	65.8
Stonelick Lake								
3. 14 Aug 1979	4,452	94±0.4	CPUE	99	-	-	-	-
4. 26 Aug 1980	4,300	179±1.1	CPUE	98	45	260	2.4	2.17
Action Lake								
5. 6 Aug 1979	14,725	107±0.8	CPUE	99	-	-	-	-
6. 3 Sep 1980	15,625	171±1.0	CPUE	97	26	200	4.0	2.60
Worthington Pond								
7. 22 Jul 1981	450	138±0.8	S	5	0	-	0	0
8. 25 Aug 1981	220	155±0.7	S	47	31[b]	250	12.2	11.57
9. 18 Aug 1982	225	154±0.6	CPUE	94	72[b]	210	74.7	29.6
Stonelick Lake								
10. 6 Oct 1981	1,610	184±0.6	S	70	19	200	4.7	3.69
11. 6 Oct 1981	700	243±3.4	CPUE	78	0	345	1.4	1.13
			S	81				
Acton Lake								
12. 13 Oct 1981	5,882	183±1.0	S	77	12	225	4.1	1.63
			P[c]	45				
13. 13 Oct 1981	2,500	255±1.6	S	62	0	345	1.2	0.68
			P[c]	19				
14. 12 Oct 1982	4,990	168±0.7	S[d]	80	9	225	4.7	3.65
			CPUE	61				
15. 12 Oct 1982	2,161	196±1.3	S[d]	80	0	275	2.0	1.15
			CPUE	61				
16. 3 Oct 1983	2,500	230±0.1	CPUE	53	9	315	2.1	1.5

[a] P=Petersen; S=Schnabel
[b] 15 days post-stocking
[c] Fish marked in late fall and recaptured following spring
[d] Both size groups of hybrids combined to derive a single mortality estimate

hybrids by the Schnabel multiple mark-recapture method (Ricker 1975). Population estimates were computed only if recaptures exceeded 10. To estimate hybrid survival from October to December 1981, we used the Petersen method (Ricker 1975). Hybrids were fin-clipped in December 1981; those captured in April 1982 were examined for fin clips. This estimate was computed from the number marked in December and the ratio of marked to unmarked hybrids in April. Such an estimate should be unbiased, if no differential mortality of marked and unmarked fish occurred (Ricker 1975).

When sampled, hybrids were weighed and measured; a sample was preserved for stomach content analysis. Instantaneous growth rates were computed as the difference between the natural logarithms of mean weights at the beginning and end of a sampling period.

We used electrofishing gear to capture age-1 and older largemouth bass, which were measured and finclipped; population sizes were estimated by the Schnabel method. Densities of largemouth bass that could prey on hybrids from a given stock were computed in two ways. In the first method, the smallest potential largemouth bass predator was estimated by dividing the length of the smallest stocked hybrid by the largest prey: predator length ratio computed from lengths of hybrids found in largemouth bass stomachs. Largemouth bass that were longer than this minimum predator length were used to estimate total predator density. In the second method we computed an Effective Predator Density that was based on both hybrid and largemouth bass length distributions. Hybrids were divided into 20-mm length groups and largemouth bass into 30-mm length groups (Table 3). The smallest largemouth bass in each length interval could prey on the smallest hybrid in the corresponding interval based upon

Table 3. An example demonstrating how Effective Predator Density was calculated for a group of small (mean length 183 mm) hybrid muskellunge stocked into Acton Lake, Ohio, 1981.

Hybrids			Smallest largemouth bass predators		Effective Predator Density
Length interval (mm)	Proportion of stock	Cumulative proportion	Length interval (mm)	Density (No. per hectare)	Density (No. per hectare)
(1)	(2)	(3)	(4)	(5)	(3)·(5)
140-159	0.02	0.02	225-254	1.93	0.04
160-179	0.29	0.31	255-284	0.69	0.21
180-199	0.63	0.94	285-314	0.26	0.24
≥200	0.06	1.00	>315	1.19	1.19
Totals				4.07	1.68

a prey: predator length ratio of 0.60-0.65. Densities of largemouth bass were assigned to each length interval on the basis of their length-frequency distribution and the estimated density of the entire largemouth bass population. To weigh the density of largemouth bass in each length interval, density was multiplied by the proportion of hybrids vulnerable to predation. The sum of these weighted densities was the Effective Predator Density.

Stomach contents of largemouth bass were removed by gastric lavage (Foster 1977). Food items, not identifiable in the field, were preserved and reexamined later. By assuming that hybrids in largemouth bass stomachs could still be identified after 24 hours, we could estimate daily predation rate as the number of hybrids per stomach. To estimate number of hybrids eaten, mean daily predation rate was multiplied by the length of the sampling period (up to 50 days) and population size of largemouth bass.

RESULTS

Short-Term Mortality

Total mortality. Losses of hybrids within the first 2 months after stocking were categorized as short-term mortality. Total mortality after 50 days ranged from 5 to 100% and exceeded 50% for most stocks. When both change in CPUE and population estimates were available to assess mortality, these estimates sometimes differed widely. For example, mortality estimates were 78 and 81% for stock 11 in contrast to 62 and 19% for stock 13 (Table 2). For the latter pair of estimates, we were unable to assess which estimate was most accurate. When two mortality estimates were available and the higher estimate was used, mean mortality for all stocks was 78%. If the lower estimate was used, the mean was 70%. These high mortality rates were due in part to predation by largemouth bass, which was a function of hybrid size at stocking and largemouth bass density.

Transport stress. Dead fish at the stocking site and in the holding nets were attributed to transport stress (included handling at the hatchery, confining fish in the transport truck at high densities, and increasing water temperature during transport and during transfer from truck to lake). Mortality was highest in a holding net at Stonelick Lake in August 1980; 29% of the hybrids died after 3 days. Lake surface temperature was 30 C, the highest temperature at which we stocked hybrids. However, some mortality in the holding net resulted from entanglement of fish in the nylon netting.

High temperature alone did not necessarily cause mortality. Several times we stocked fish when lake temperatures were 25-28 C and observed no mortalities. However, even in autumn, delayed mortality of up to 15% occurred in floating cages when fish were held for 10 days and water temperatures were below 20 C. Apparently, stresses other than temperature were responsible for these mortalities. In one instance 21% of the large hybrids (mean length 243 mm) died in floating cages in Stonelick Lake. A defective aerator in the transport truck injured a large number of fish, which probably contributed to this high mortality.

Predatory mortality. To define the density of potential largemouth bass predators for a given stock, we first examined length-frequency distributions of largemouth bass and ingested hybrids. Hybrids consumed by largemouth bass were 30-70% (mean 47%) of the largemouth bass length (Fig. 1). We used the 95th percentile of these data (0.62) to estimate length of smallest largemouth bass that could eat the smallest hybrid in a given stock.

Predation on hybrids was directly related to largemouth bass density (Fig. 2A). This strong correlation was due in part to the clumped distribution of data from ponds where largemouth bass densities were high and from impoundments where their densities were low. We speculated that some of the variability in this relation might be due to differences in the size structures of the largemouth bass populations. If, for example, most of the largemouth bass only narrowly exceeded the smallest predator length, the use of total largemouth bass density would overestimate the potential risk of predation to intermediate and large hybrids. To assess this possiblity, we correlated predatory mortality with Effective Predator Density (Fig. 2B). The relation

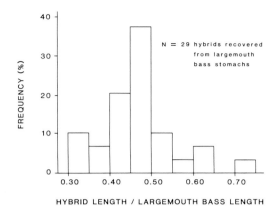

Figure 1. Frequency distribution of the total length of hybrid muskellunge recovered from stomachs divided by the total length of largemouth bass that had consumed them.

was only slightly better than when unadjusted largemouth bass density was the independent variable. Only one data point in this set was markedly altered by recomputing predator densities. Worthington Pond supported a large number of largemouth bass that were concentrated in the smallest predator length group, and the Effective Predator Density was about 40% of the unadjusted largemouth bass density. For the other largemouth bass populations, densities and Effective Predator Density were highly correlated ($r^2=0.99$); understandably, this adjustment did not measurably improve the relation between largemouth bass density and predatory mortality. However, this procedure may be useful for assessing size structures of predator and prey populations, particularly when relative risk of predation for groups with different mean lengths are to be compared.

Predation intensity and alternate prey. Abundance of alternate prey for largemouth bass could influence predation intensity on hybrids. Because largemouth bass ate prey as long as 150 mm (prey not sampled by our seining), we could not use these data to describe their available prey. Instead, we used the proportion with empty stomachs as an indirect measure of prey availability; this was estimated from 2,826 largemouth bass sampled during 10 stocking experiments in which predation on hybrids was estimated.

Predatory mortality of hybrids and percentage of largemouth bass with empty stomachs were directly related (Fig. 3), suggesting that largemouth bass

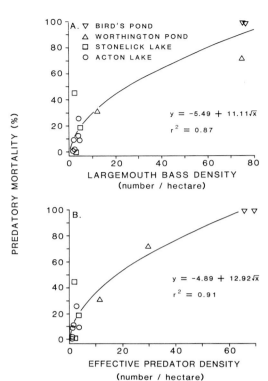

Figure 2. Panel A. Percentage of hybrid muskellunge mortality (50 days post-stocking) attributable to largemouth bass predation in relation to largemouth bass density in four Ohio lakes. Panel B. Percentage of hybrid muskellunge mortality attributable to largemouth bass predation in relation to Effective Predator Density.

were most likely to eat recently-stocked hybrids when other prey were not abundant. However, this relation was confounded, because largemouth bass density and percentage with empty stomachs also were correlated ($r^2=0.81$). Such a relation seems plausible, if high densities of largemouth bass heavily exploit their prey.

Hybrid size and mortality. Stein et al. (1981), who first demonstrated the importance of largemouth bass predation on survival of hybrid muskellunge, estimated that hybrids longer than 250 mm would be invulnerable to most largemouth bass in typical Ohio impoundments. To determine how hybrid size influenced predatory mortality, we stocked Acton and Stonelick lakes in October 1981 with two groups of hybrids that averaged about 180 and 250 mm (Table 2). We found no evidence of preda-

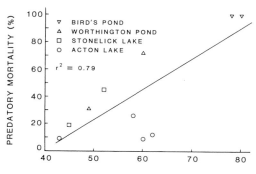

Figure 3. Percentage of hybrid muskellunge mortality (50 days post-stocking) attributable to largemouth bass predation in relation to the percentage of largemouth bass with empty stomachs in four Ohio lakes. Mortality data and stocking dates are given in Table 2.

tion on the large hybrids in either lake and only modest predation (12-19%) on the small fish. Even though stocking of large hybrids eliminated predatory mortality, total mortality was still high (Table 2).

In Stonelick Lake total mortality of the large hybrids exceeded that of the small ones. Part of this mortality was due to injury from defective aerators during transport; of a sample held in floating cages, 21% died. In Acton Lake total mortality of large hybrids was less than that of the small fish. We are uncertain of the actual difference, because population estimates from the two mark-recapture experiments differed considerably and estimated total mortality was 19 or 62% for the large group versus 45 or 77% for the small group. However, a significant (chi-square test; $P<0.01$) change in the ratio of small to large hybrids from time of stocking (1:0.42) to late autumn (1:0.73) suggested that large hybrids survived substantially better than did the small ones.

We attempted to repeat this experiment at Acton Lake in 1982, but had to stock smaller size groups because growth in the hatchery was slower than the previous years; mean lengths were 168 and 196 mm (Table 2). Even though some of the large hybrids were vulnerable to largemouth bass predation, none was found in predator stomachs. Predatory mortality of the small fish was only 9%, less than in previous stockings. Ratios of the two groups did not change from time of stocking to December, suggesting no differential mortality. However, total mortality was still substantial, either 61 or 80%, depending upon the method of estimation.

Stocking time and mortality. In 1979 and 1980, when Acton and Stonelick lakes were stocked during late summer, total mortalities approached 100% within 50 days. In 1979 we did not measure predation, but in 1980 predatory mortality was 26% in Acton Lake and 45% in Stonelick Lake (Table 2). In contrast, when lakes were stocked after autumn turnover at water temperatures near 18 C, total mortality after 50 days ranged from 45 to 80%, depending on our method of estimation, and predatory mortality ranged from 9 to 19% (mean 13%)--about one-third of the mean for fish stocked in summer.

Hiding cover and predation. Predatory efficiency of largemouth bass declines as density of submersed macrophytes increase (Savino and Stein 1982). High rates of predation on summer-stocked hybrids in Acton and Stonelick lakes and Bird's Pond could have been caused by a lack of adequate hiding cover for hybrids. None of these waters supported submersed vegetation; hiding cover was limited to small areas of cattails and fallen trees. To assess this possibility, we conducted experiments in Worthington Pond to determine (1) use of vegetative cover by hybrids in the absence and presence of largemouth bass, and (2) the influence of vegetation on susceptibility of hybrids to predatory mortality. The first objective was included to determine if high predatory mortality resulted because hybrids did not use vegetative cover or because largemouth bass could successfully prey on hybrids despite their association with cover.

During the first month of the experiment, submersed macrophytes were sparse and hybrids did not select vegetated transects (Fig. 4A). In fact on two occasions, catches were significantly (chi square test; $P<0.05$) greater in the open transects. In mid-August, bushy pondweed (*Najas minor*) became abundant in the vegetated transects and hybrid distribution began to shift. From mid-August to early September we captured significantly more hybrids in vegetated than in open transects on 5 of 9 days (Fig. 4A).

We stocked 90 largemouth bass (longer than 250 mm) to test their effect on hybrid distribution. Because hybrids present had grown beyond a vulnerable size, we added 220 smaller hybrids (mean length 168 mm). The addition of largemouth bass did not significantly (chi-square test; $P>0.05$) influence hybrid distribution. On four dates before largemouth bass were stocked, an average of 72% (of those sampled) of the small

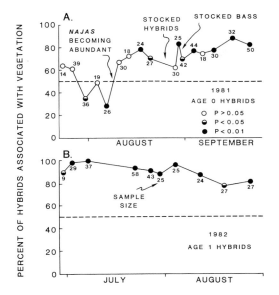

Figure 4. Panel A. Percentage of age 0-hybrid muskellunge captured with electrofishing gear in vegetated transects in Worthington Pond, Ohio, 1981. Panel B. Percentage of age-1 hybrid muskellunge observed by SCUBA divers in vegetated transects in Worthington Pond, 1982. Probabilities (P) refer to chi-square tests for equal distributions (dashed line) of hybrids in open and vegetated transects.

hybrids were captured in vegetated transects (Fig. 4A).

Despite extensive use of vegetation by these hybrids, predatory mortality was substantial, suggesting that vegetation did not reduce their vulnerability. Our estimate of 31% predatory mortality for the first 15 days was conservative because we were not able to sample the first night after stocking, when some predation probably occurred. Therefore, we assumed no predation on day 1. Of 90 largemouth bass stocked, only 22 survived, based on a mark-recapture estimate about a month after they were stocked. To compute predation, we assumed that only 22 largemouth bass were present during the experiment. Thus, predatory mortality of 31% was a minimum estimate.

In 1982 we conducted a similar experiment, but relied on observations by SCUBA divers to determine habitat selection by hybrids. Nearly all observations were of age-1 hybrids (total length 265-532 mm), which strongly selected vegetation (Fig. 4B). These fish were observed at depths of 0.5 - 3 m and were usually within 0.5 m of the bottom.

In mid-August 1982 we stocked 225 age-0 hybrids to retest the influence of vegetative cover on predation by largemouth bass. Several conditions differed from the 1981 experiment: (1) largemouth bass densities were about six times greater, (2) largemouth bass had been in the pond for 1 year or more, and (3) density of submersed macrophytes was greater. In 1982 predation rates on hybrids were high on the first day of stocking and we estimated that 72% of the hybrids had been eaten after 15 days. After 50 days, total mortality was estimated at 94%.

These experiments clearly demonstrated that both age-0 and age-1 hybrids were strongly associated with submersed macrophytes. However, the presence of macrophytes did not reduce predation by moderate to high densities of largemouth bass. Thus, high predation rates in Bird's Pond and Acton and Stonelick lakes probably were not due to the absence of vegetative cover for hybrids.

Rearing diet and mortality. Johnson's (1978) findings, that pellet-reared hybrids did not survive as well as fish-reared hybrids, led to an assessment by Gillen et al. (1981) of the ability of pellet-reared hybrids to eat live prey. They suggested that reduced predatory efficiency of pellet-reared hybrids could indirectly lead to mortality because of slow growth and increased vulnerability to other mortality agents. In laboratory experiments, pellet-reared hybrids readily ate live prey; in ponds hybrids successfully preyed on fathead minnows, but had little success in capturing bluegills. Tomcko et al. (1984), who examined how vegetation influenced predatory efficiency of pellet-reared hybrids, found that vegetation had little effect on the ability of hybrids to captures bluegills. Rather, hybrids successfully preyed on bluegills when bluegill densities were unusually high ($>5/m^2$), regardless of the presence of vegetation. Thus, pellet-reared hybrids can readily switch to live prey, although their predatory success will be a function of prey type and density. Our field data tend to corroborate these conclusions.

After stocking, we regularly examined stomach contents of hybrids to determine the proportion that was feeding. Proportions varied widely among lakes, but generally increased with time after stocking (Table 4). Highest proportions occurred in Worthington Pond in 1981 where high densities of juvenile gizzard shad, golden shiners, and bluegills were found. These high forage densities probably contributed both to feeding success and to rapid

Table 4. Percentage of hybrid muskellunge with food in their stomachs (sample size in parentheses) in relation to number of days after stocking. Samples were grouped by week from collections at Worthington Pond and Acton and Stonelick lakes, Ohio.

Site and year	Days after stocking					
	1-6	7-13	14-20	21-27	28-34	>35
Worthington Pond						
1981	96	85	100	100	100	
	(29)	(20)	(8)	(7)	(10)	
Acton Lake						
1980	0	7	--	--	--	
	(20)	(27)	--	--	--	
1981	--	13	--	43	29	25
	--	(32)	--	(14)	(24)	(10)
1982	26	33	45	60	70	67
	(40)	(21)	(20)	(5)	(20)	(30)
1983	0	40	90	--	90	60
	(9)	(15)	(10)	--	(10)	(15)
Stonelick Lake						
1980	--	0	0	14	--	--
	--	(15)	(20)	(7)	--	--
1981	--	0	--	0	28	31
	--	(38)	--	(14)	(18)	(35)
Means	30	25	58	43	63	46

growth. Hybrids increased in length from 140 to 280 mm in 8 weeks, the fastest growth we have observed. In contrast, at Stonelick Lake where gizzard shad were absent, feeding success of hybrids was lowest (Table 4) and growth was consistently slow. Thus, both experimental and field studies showed that pellet-reared hybrids readily ate live prey and that hybrid growth was a function of prey type and density.

Given these results, the influence of rearing diet on survival is not clear. Johnson (1978) noted that in four stocking trials with fish- and pellet-reared hybrids, the latter were smaller at stocking and had lower survival. He suggested that size rather than diet was the important factor influencing mortality. However, Andrews (1983) recaptured forage-reared hybrids at 20 times the rate of pellet-reared hybrids in two lakes even though the pellet-reared fish were slightly larger that those raised on fish (203 vs. 217 mm). All hybrids in our study were pellet-reared, and in several stocks survival was high. When short-term mortality was high, predation was usually important. Gillen et al. (1981) hypothesized that pellet-rearing hindered the ability of hybrids to capture live prey, which in turn led to poor survival. Experimental and field data strongly suggest that the hypothesis is incorrect. If pellet-rearing negatively affects survival, it must do so in some other manner. Perhaps pellet-reared hybrids behave differently than forage-reared hybrids because of the rearing environment, and these behavioral differences may increase their vulnerability to predation.

Long-Term Mortality

Our best data on long-term mortality, i.e., more than 2 months post-stocking, were derived from the 1981 and 1982 experiments at Acton Lake, in which we compared survival of large and small hybrids. Though Stonelick Lake also had been stocked with two size groups in 1981, we discontinued studies there because the lake was drawn down during the 1981-82 winter.

Large (mean length 255 mm) hybrids survived better than small (mean length 183 mm) ones during the first 2 months after stocking in 1981 (Table 5). Population estimates differed considerably between the Schnabel method conducted in late autumn and the Petersen method in which hybrids were marked in autumn and recaptured the following spring. The ratio of small to large hybrids declined (chi-square test; $P<0.001$) after stocking to late autumn, suggesting better survival of the large group.

Large hybrids survived markedly better over winter than small ones. By April only 6-8% of the small hybrids remained, whereas estimated survival of the larger ones was 52% by the Schnabel method and 41% by the change in CPUE method (Table 5). The ratio of small to large hybrids continued to shift toward large hybrids, which were more than twice as abundant as the small ones in April 1982.

From April to December 1982, survival of the large hybrids was higher than that of the small ones, but large numbers of both sizes were lost. By late autumn only 1% of the small fish and 13% of the large ones survived (Table 5). During 1982 we found no evidence of predation by largemouth bass on the small hybrids, all of which had grown beyond vulnerable size by December 1982. None of the large hybrids could be eaten by largemouth bass at any time during the year. Some hybrids emigrated downstream and some were caught by anglers, but we have no measure of these losses.

About 50% of the hybrids that survived to December 1982 were lost overwinter. Thus, 18 months after stocking, 7% of all fish remained (Table 5). Some of the clipped fins regenerated and we had difficulty in distinguishing the two groups; therefore, separate survival estimates for each group were not computed. On the basis of recogniz-

Table 5. Summary of survival estimates for the small (mean length 183 mm) and large (mean length 255 mm) hybrids stocked in Acton Lake, Ohio, on October 13, 1981. Chi square test was used to test for differences in the ratio of small to large hybrids in electrofishing samples and at stocking.

Time of estimate	Method of estimating numbers	Both groups m	Both groups N	Small fish m	Small fish N	Large fish m	Large fish N	Survival from time of stocking Small fish	Survival from time of stocking Large fish	Ratio of sm. to lg. hybrids in catches
At stocking					5,882		2,500			1 : 0.42
Dec. 1981	Schnabel estimate			331	1,351 (1,024-1,764)	220	944 (642-1,395)	0.23	0.38	1 : 0.73[a]
Dec. 1981	Petersen estimate for each group based on total marked in Dec. 1981 and recaptured in spring 1982			349	3,256 (1,714-7,329)	247	2,035 (1,375-2,993)	0.55	0.81	
April 1982	Single Schnabel estimate for both groups; number in each group estimated from proportions in catch	206	1,642 (1,000-3,650)		344		1,297	0.06	0.52	1 : 2.45[a]
April 1982	Change in CPUE for each group, based on data for autumn 1981 and spring 1982							0.08[b]	0.41[c]	
Dec. 1982	Single Schnabel estimate for both groups; number in each group estimated from proportions in catch	95	392 (233-616)		55		337	0.01	0.13	1 : 5.90[a]
April 1982	Schnabel estimate	49	179 (80-447)							

[a] Chi-square test; $P < 0.001$
[b] $r^2 = 0.86$; $P < 0.01$; for regression of $\log_e$ catch-per-unit-of-effort electrofishing with days after stocking
[c] $r^2 = 0.44$; $P < 0.05$; for regression of $\log_e$ catch-per-unit-of-effort electrofishing with days after stocking

able fin clips and backcalculation of length at annulus formation on scales, we concluded that most survivors were of the large group.

Large (mean length 196 mm) and small (mean length 168 mm) hybrids stocked in autumn 1982 had similar survival rates for the first 2 months after stocking. The ratio of small to large fish changed little through December; therefore, we computed a single estimate of survival for both groups (Table 6). Survival was 20% using the Schnabel method and 39% using the change in CPUE method. However, large fish survived better overwinter than small ones, as indicated by the significant decline in the ratio of small hybrids to large hybrids (chi-square test; $P < 0.05$). On the basis of the change in CPUE for autumn and spring samples, we estimated that 9% of the small hybrids and 16% of the large ones survived to spring. The trend of higher survival of the large group continued through 1983, although mortality of both groups was high. By December only 1% of the small ones and 5% of the large ones remained.

Several generalities emerged from our stocking experiments in Acton Lake. Long-term mortality of hybrids was always high and only a small proportion of this mortality could be contributed to largemouth bass predation. Survival of large hybrids was higher than that of small ones and the difference in survival became more apparent through time, because the ratios of these two groups continued to change in favor of the large fish. Survival after 14 months and mean length at stocking were related. Mean lengths at stocking of the four groups were 168, 183, 196, and 255 mm; their respective survival rates were 1,

Table 6. Summary of survival estimates for the small (mean length 168 mm) and large (mean length 196 mm) hybrids stocked in Acton Lake, Ohio, on October 12, 1982. Chi square test was used to test for differences in the ratio of small to large hybrids in electrofishing samples and at stocking.

Time of estimate	Method of estimating numbers	Number marked (m) population estimate (N) 95% confidence interval						Survival from time of stocking		Ratio of sm. to lge. hybrids in catches
		Both groups		Small fish		Large fish		Small fish	Large fish	
		m	N	m	N	m	N			
At stocking					4,990		2,161			1 : 0.43
Dec. 1982	Single Schnabel estimate for both groups; number in each group estimated from proportions in catch	170	1,429 (766-2,925)	110	998	60	432	0.20	0.20	1 : 0.49
Dec. 1982	Change in CPUE for both groups combined							0.39[a]	0.39	
April 1983	Change in CPUE for both groups combined, autumn and spring data used; different survival computed from proportions in spring catches							0.09[b]	0.16	1 : 1.23[c]
Dec. 1983	Single Schnabel estimate for both groups; number in each group estimated from proportions in catch	50	175 (107-389)	20	60	38	115	0.01	0.05	1 : 1.90[d]

[a] $r^2=0.42$; $P<0.05$; for the regression of $\log_e$ catch-per-unit-of-effort electrofishing and days after stocking
[b] $r^2=0.71$; $P<0.01$; for the regression of $\log_e$ catch-per-unit-of-effort electrofishing and days after stocking
[c] Chi-square; $P<0.05$
[d] Chi-squre; $P<0.001$

1, 5, and 13%.

Prey Availability and Hybrid Growth

Factors that influence growth of hybrids should be linked to survival because hybrid size is related to their vulnerability to predation and to overwinter survival. Consequently, we have placed considerable emphasis on defining the effects of prey size, type, and density on growth of hybrids.

Hybrids demonstrate a strong selection for prey size. In laboratory experiments hybrids most frequently eat fathead minnows that are about 40% of their length and bluegills that are about 28% of their length (Gillen et al. 1981). Size selectivity is also apparent in ponds, although hybrids tend to choose prey that are smaller than those eaten in laboratory experiments. When hybrids in ponds are offered appropriate sizes of either minnows or bluegills, instantaneous growth rates of hybrids are 2 to 15 times higher in ponds with minnows than in ponds with bluegills (Gillen et al. 1981; Tomcko et al. 1984).

Results from the Worthington Pond experiments illustrate the interactive effects of prey type, density, and size on hybrid growth. In summer 1981, there were high densities of prey fishes, when the first group of hybrids was stocked. These hybrids quickly began feeding on fish (Fig. 5, upper panel), primarily gizzard shad. Hybrids grew rapidly, increasing in length from 138 to 275 mm in 60 days. Age-0 centrarchids and golden shiners composed 90% of all available prey (Table 7); yet they were unimportant in hybrid diets (Fig. 5). Hence, hybrids were selecting gizzard shad. The low frequency of centrarchids in hybrid stomachs was not surprising, because hybrids had been unable to exploit moderately high densities of bluegills in ponds (Tomcko et al. 1984). Despite their high densities, golden shiners were not frequently eaten by hybrids.

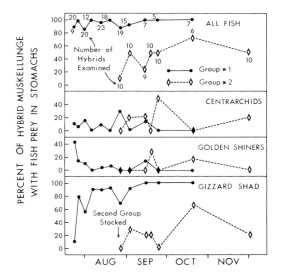

Figure 5. Percentage of hybrid muskellunge with fish in their stomachs. The first group of hybrids was stocked into Worthington Pond, Ohio, on July 22, 1981, and the second group was stocked on August 25, 1981.

These rapidly-growing hybrids in Worthington Pond maintained a sufficiently large size advantage over gizzard shad to enable them to feed on gizzard shad for the entire growing season (Fig. 6). On the basis of sizes of gizzard shad found in hybrid stomachs from all study sites, maximum length of ingested gizzard shad was 45% of hybrid length and the 95th percentile was 36% of hybrid length. If 36% of hybrid length is designated as maximum size of available prey, the proportion of gizzard shad vulnerable to hybrid predation can be approximated. On July 25 about 30% of the gizzard shad were vulnerable to hybrid predation (Fig. 6, upper panel). In 2 weeks hybrids had grown about 30 mm and 80% of the gizzard shad were small enough to be eaten. Thereafter, all gizzard shad were vulnerable to predation by this stock of hybrids.

The second group of hybrids, which was stocked in late August, grew slowly. Their length increase was only 23 mm from stocking until late October, a rate of 0.5 mm/day compared to 2.3 mm/day for the first group. Densities of available gizzard shad and golden shiners were 60-90% less than they had been during the first stocking (Table 7), because of this difference in hybrid sizes. This second group was not large enough to eat all gizzard shad; about 50% of the gizzard shad in late August and 20% in late September were vulnerable to these similar hybrids (Fig. 6). Reduced availability of gizzard shad was reflected in their diet. They ate fewer centrarchids

Table 7. Relative abundance of prey fishes from electrofishing samples in Worthington Pond, Ohio, 1981. Mean numbers of prey were computed from samples collected during the first 30 days after stocking, which was July 22 for Group 1 and August 25 for Group 2. Prey fishes that were 36% or less of mean hybrid length were considered vulnerable to predation.

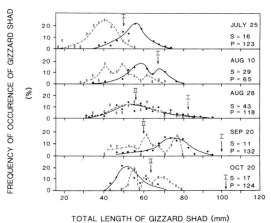

Figure 6. Length frequency distributions of gizzard shad collected by electrofishing (solid line) and from hybrid stomachs (dotted line) in Worthington Pond, Ohio. Only gizzard shad that were eaten by the July 1981 stock of hybrids are shown. Numbers of gizzard shad measured from electrofishing samples and hybrid stomachs are denoted by S and P, respectively. Maximum lengths of gizzard shad that could be eaten by July (I) and August (II) stocks of hybrids are indicated by arrows.

Hybrids	Number of samples	Mean number of prey fish captured per electrofishing transect (percentages in parentheses)			
		Gizzard shad	Golden shiners	Centrarchids	All prey
Group 1	5	19(10)	110(57)	63(33)	192
Group 2	4	6(4)	10(6)	147(90)	163

and golden shiners than did the first group (Fig 5). Thus, growth of hybrids in Worthington Pond was related to prey type, density, and size.

Prey type and density also influenced hybrid growth in our other study lakes. Instantaneous growth rates of hybrids in Acton and Stonelick lakes were positively related to mean density of available prey (Fig. 7). However, the data were widely scattered and no relation would be evident without the 1983 data from Acton Lake. Prey densities and hybrid growth rates were lowest at Stonelick Lake, which had no gizzard shad and few prey species other than centrarchids. Acton Lake supported a diverse forage base with intermediate to high densities and hybrid growth was highest. We also found a significant inverse relation ($r^2=0.77$; $P<0.01$) between instantaneous growth rates and percentage of hybrids with empty stomachs (Table 4); the first stock of hybrids in Worthington Pond was included in this analysis. Thus, both direct and indirect measures of prey density were positively related to hybrid growth rates.

In Acton Lake, food items in descending order of importance were gizzard shad, centrarchids, cyprinids, and other fishes (Table 8). However, gizzard shad ranked third and fourth in density of available prey; centrarchids were usually the most numerous. If we assume that the seine samples adequately reflected inshore densities, hybrids strongly selected gizzard shad. Although seine samples may

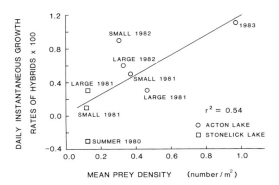

Figure 7. Daily instantaneous growth rates (60 days post-stocking) of hybrid muskellunge in relation to mean density of available forage fish in Acton and Stonelick lakes, Ohio. Hybrids were stocked in early autumn except for one stocking in Stonelick Lake in summer 1980.

reflect inshore densities, we do not believe that they accurately represented relative densities throughout the lake. Gizzard shad were widely distributed whereas the other species such as bluegills, spotfin shiners, bluntnose minnows, and logperch were restricted to the littoral area. Because of the apparent gear selectivity, we cannot conclude that gizzard

Table 8. Percent frequency of occurence of prey in stomachs of hybrid muskellunge collected from Acton and Stonelick lakes, Ohio. Percent composition and density of available inshore prey fishes were calculated from seine collections as the number that were 36% or less of mean length of the hybrids when samples were collected.

Lake, year and hybrid length (mm)	Number of stomachs	Percent of prey types in stomachs (st) and seine catches (se)								Total density of available prey (Number per m²)
		Gizzard shad		Centrarchids		Cyprinids		Other fish[a]		
		st	se	st	se	st	se	st	se	
Acton										
1981										
183	45	7	0	13	76	4	24	0	6[b]	0.37
255	45	22	6	7	65	2	28	0	1	0.46
1982										
168	86	22	2	13	59	1	38	2	1	0.30
196	81	25	4	5	55	2	40	9	1	0.33
1983										
230	80	52	6	10	38	1	59	0	3	0.97
Stonelick										
1981										
184	28	--	--	11	100	0	0	0	0	0.12
243	28	--	--	11	100	0	0	0	0	0.13

[a] Includes logperch, darters, suckers, redhorses, and bullheads.
[b] Less than 0.5

shad were preferred. We may have simply underestimated gizzard shad abundance.

In addition to low prey densities in Stonelick Lake, the preponderance of centrarchid prey contributed to slow hybrid growth. Only 11% of the small and large hybrids in 1981 had food in their stomachs (Table 8) and these fish together with the 1980 stock had the slowest growth of all stocks (Fig. 7). These results are consistent with the findings of Gillen et al. (1981) and Tomcko et al. (1984), who showed that hybrids were not efficient predators on centrarchids and grow slowly in the absence of alternate prey types.

DISCUSSION

Time of Stocking

Mortality of esocids due to stress from stocking procedures has been reported in several studies. High mortality occurs when esocids are stocked at water temperatures above 30 C (Hess 1981). Muskellunge stocked in summer do not survive as well as those stocked later, when water temperatures have declined (Johnson 1982). Similarly, we found that hybrids stocked in summer suffered higher mortality than did those stocked in autumn. Both Hess (1981) and Johnson (1982) recommended stocking when water temperatures are 15-20 C, a temperature range at which we found the least mortality. However, high water temperatures or rapid temperature increases do not necessarily cause high mortality. Although hybrid muskellunge can tolerate a temperature increase from 15 to 27 C within 8 minutes, the small additional stress of being in a net out of water for 30 seconds causes substantial mortality (M.E. Mather, The Ohio State University, pers. commun.). The interactive effects of high temperature, handling, and crowding increase the likelihood of death for recently-stocked esocids. Therfore, delaying stocking until after maximum summer temperatures should reduce deaths due to transport stress.

Losses to Predation

Considerable circumstantial evidence in the literature indicates that predation contributes significantly to mortality of stocked esocids. In the absence of predators, survival of northern pike (Beyerly 1978), muskellunge (Johnson 1982), and hybrids (Weithman 1975; Beyerly 1981) is usually high; yet in ponds or lakes with predators, esocid survival is substantially reduced. By examining stomachs of predators, we have quantified the importance of largemouth bass predation under different conditions. Though no other predators occurred in our lakes at densities high enough to influence hybrid mortality, other studies (Beyerle 1978, 1981; Hess 1981) have found that age-1 and older esocids suppress survival of juveniles. Also, walleyes may have been important predators in Johnson's (1982) study lakes. Predation on stocked esocids is probably not limited to any single predator species.

In this study losses to predation were higher for hybrids stocked in summer than for those stocked in autumn. Several factors may have contributed to this difference: (1) hybrids subjected to transport stress and high temperatures may be more vulnerable to predation, (2) at high temperatures largemouth bass consume more food (and perhaps more hybrids), and (3) thermal stratification of lakes and anoxic hypolimnions force largemouth bass and hybrids to concentrate in inshore areas.

Transport stress appeared most severe during summer, and predatory mortality was highest at Stonelick Lake when surface temperatures were >30 C. Mortality of hybrids in the holding net was also highest for this stock. Largemouth bass may have eaten hybrids that would have otherwise died because of transport and thermal stress. In addition, stressed fish may have been more vulnerable to largemouth bass predation, because they were less able to escape or their aberrant behavior attracted largemouth bass. Numerous examples in the literature document increased vulnerability of stressed fish to predation (for review see Coutant et al. 1979).

High water temperatures can increase predation owing to the effect of temperature on food consumption rates of largemouth bass, which are highest at 28 C (Niimi and Beamish 1974). All summer stockings were made when temperatures were 25-30 C. In contrast, water temperatures were less than 20 C during autumn stockings. Water temperatures declined slowly for the first 50 days after stocking in both summer and early autumn experiments; therefore, largemouth bass consumption rates also should have been declining. To better define the effect of a changing temperature regime on largemouth bass consumption rates, we used a bioenergetic model (Rice et al. 1983) and surface temperatures from Acton Lake to estimate maximum food consumption of a 300-mm largemouth bass. Simulations began on September 1 (summer stocking) or October 10 (autumn stocking) and were run for 50 days. Estimated food consumption rates of largemouth bass during late summer (3.4% of body weight per day) were double those of largemouth bass during autumn. If motivation to feed by largemouth bass is directly related to maximum food consumption rates, the risk of predation for hybrids stocked in summer should be double that of fish stocked in early autumn.

Because late summer limnological conditions restrict largemouth bass to the shallow epilimnion, high predation rates on hybrids may be increased further (Stein et al. 1981). During summer stratification, dissolved oxygen concentrations typically fell below 5 mg/liter between 2 and 3 m in Acton and Stonelicks lakes. If largemouth bass remained close to the bottom rather than suspended in open water, they would be concentrated in a relatively narrow area around the periphery of the lake, where hybrids were most frequently found. Predator concentration and habitat overlap with hybrids would probably promote high rates of predation. However, after autumn turnover, largemouth bass are probably more widely dispersed and risk of predation should be reduced.

Though we cannot separate the interactive effects of the above factors on largemouth bass predation, high total and predatory mortality observed for hybrids stocked in summer provide sufficient evidence to support autumn stocking after temperatures have fallen below 20 C.

We found a direct relation between predatory mortality and predator density. Though Andrews (1983) found a similar relation for hybrids stocked in five Indiana lakes, Johnson (1982) could not relate his index of predator abundance or mean length of predators to muskellunge survival. He suggested that this lack of relation may have been due to imprecise estimates of predator abundance. A second possible reason is an inappropriate choice of minimum predator length. Johnson (1982) considered only fish longer that 380 mm as predators. On the basis of lengths of largemouth bass and hybrids in their stomachs, we designated 62% of a largemouth bass' length as the largest hybrid that it could eat. If this length ratio is applied to Johnson's (1982) data, minimum predator length would be 242 mm when the smallest stocked muskellunge was 150 mm. Precise designation of minimum predator length is important because predator densities generally decline rapidly with increasing length. Thus, a small decrease in predator length can result in large adjustments in the computed predator density.

We found that intensity of predation was partly a function of hybrid size at stocking. If predation controlled survival of esocids in other studies, then size at stocking ought to influence survival. Indeed, mean length of stocked fish and short-term survival were consistently related in other esocid studies. Though demonstrated for northern pike (Vasey 1974; Flickinger and Clark 1978), muskellunge (Hess 1981; Johnson 1982; Serns and Andrews 1983), and hybrids (Hess 1981, Andrews and Laurion 1982; present study), two exceptions to this general relation exist. Beyerle (1981) stocked a reclaimed lake with muskellunge hybrids 89 mm long in June or July and with hybrids 178 mm long in August in each of three consecutive years. For all years, survival of the smaller fish exceeded that of the larger ones and survival of both groups in the second and third years declined as hybrid densities increased. Competition among age-0 fish and predation by older hybrids may have influenced survival of both size groups. In a second example (Andrews 1983), hybrids reared on fish survived better than pellet-reared hybrids, even though the pellet-reared fish were larger, 203 vs. 217 mm. When we stocked two different sizes of hybrids at the same time, predatory and total mortality were always highest for the small groups--a finding consistent with most other studies.

Several researchers have attempted to define the relation between size at stocking and survival (Flickinger and Clark 1978; Beyerle 1981; Johnson 1982; Serns and Andrews 1983), because rearing costs increase disproportionately with fish size. Although survival generally increases with size of stocked fish, actual survival rates vary greatly among studies. Variations in population structures and densities of predators may account for some of this variability.

If predation is the primary cause of mortality, we can estimate reduction in mortality as larger hybrids are stocked by computing the decrease in predator density. However, one must assume that mortality and predator density are directly related. We computed Effective Predator Densities (see Table 3) for four groups of hybrids (mean lengths 140-260 mm) exposed to largemouth bass populations with different size structures, but with the same density (7/hectare>135 mm). The pattern of decline in Effective Predator Density varied among the four largemouth bass populations (Fig. 8). In a heavily-exploited population of largemouth bass, such as that in Acton Lake in 1983, size structure was skewed toward small fish; hence, a small increase in

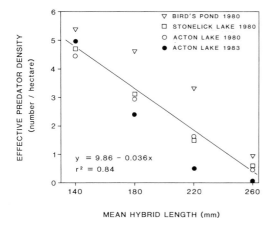

Figure 8. Effective Predator Density in relation to mean length of stocked hybrid muskellunge.

hybrid length resulted in a relatively rapid decline in Effective Predator Density. However, in a lightly-exploited population, such as that in Bird's Pond, Effective Predator Density remains relatively high for intermediate size hybrids. To approximate this decline in density, we fitted a straight line to all data points. Effective Predator Density decreased an average of about 20% with an increase in hybrid length of 25 mm. These data provide a basis for evaluating the increase in cost of rearing fish to a larger size with the expected increase in survival, providing that predation is the major component of mortality and it is directly related to predator density.

As in other work with esocids (Chapman and McKay 1984), we found tiger muskellunge to be closely associated with submersed vegetation. However, on the basis of our work and that of others, we find no evidence that aquatic macrophytes influence predatory mortality of stocked esocids. When we examined the influence of macrophytes on predation by largemouth bass (at moderate to high densities, 12 to 75/ha), we found that predation rates were high even though hybrids were strongly associated with vegetative cover. In related work, Johnson (1982) found no correlation between the area of a lake supporting macrophytes and muskellunge survival. He also stocked groups of muskellunge in vegetated and open areas, but found no difference in survival. Studies on largemouth bass and other fishes suggest that as macrophyte density increases, predatory efficiency declines sharply (Savino and Stein 1982). Perhaps further studies in lakes with moderate to large areas of macrophytes will demonstrate the extent of cover needed to reduce predatory mortality.

Size and density of available prey can affect survival of stocked esocids in several ways. Weithman and Anderson (1977) suggested that esocids should be stocked when other prey fishes are abundant so that chances of predation on stocked esocids would be reduced. Our data support this notion and suggest that intensity of largemouth bass predation was inversely related to prey availability (as indicated by percentage of empty largemouth bass stomachs). However, Johnson (1982) found no relation between density of large (100–150 mm long) forage and survival of muskellunge. Additional studies are needed to evaluate the relative importance of availability of alternative prey, esocid size, and predator density on predatory mortality of stocked esocids.

Prey availablility

Characteristics of the prey populations may be much more important to recently-stocked esocids than to their predators. Northern pike (Mauck and Coble 1971; Beyerle and Wiliams 1968), muskellunge (Krska and Applegate 1982), and hybrids (Gillen et al. 1981) strongly select for prey size and type. When given a choice both parent species and hybrids choose soft-rayed fishes over spiny-rayed ones (Beyerle and Williams 1968; Weithman and Anderson 1977); in reservoirs, soft-rayed fishes, particularly gizzard shad, are usually the most common food item (Vasey 1974; Casselman 1978; Axon 1981; present study).

Prey type and density also affect esocid growth. In experimental ponds with minnows or bluegills as prey, northern pike (Beyerle 1978) and hybrids (Gillen et al. 1981; Tomcko et al. 1984) grow faster in ponds with minnows. Growth rates of northern pike (Beyerle 1978), muskellunge (Johnson 1982), and hybrids (present study) are positively related to prey density. As esocids grow their vulnerability to predation continuously declines; hence, prey density should indirectly influence survival.

Most investigators who have examined prey density and survival of esocids indicate a positive relation. Beyerle (1978) concludes that survival of northern pike in ponds is related to density of soft-rayed fishes. In larger lakes, high survival of northern pike (Flickinger and Clark 1978) and muskellunge (Axon 1981) is linked to high prey densities. Andrews (1983) also concludes that survival of hybrids is related to prey availability;

among five lakes, hybrid survival was highest when proportions of centrarchid prey were lowest. Johnson (1982) found that density of small forage (50-100 mm) and survival of muskellunge were not correlated. In our study, the effects of hybrid size and largemouth bass density on survival could not be easily separated from possible direct effects of prey density on hybrid survival.

High prey densities and consequent rapid growth of esocids may influence survival in several ways. Among them are decreased vulnerability to predation and enhanced ability to persist with limited food supplies during winter. Overwinter survival of age-0 brook trout is positively related to their length in late fall (Hunt 1969). In laboratory experiments that simulated winter temperatures, survival of longer age-0 smallmouth bass exceeded that of shorter ones (Oliver et al. 1979). Upward shifts in length-frequency distribution from autumn to the following spring provide evidence for size-related mortality of age-0 smallmouth bass (Shuter et al. 1980) and largemouth bass (Adams et al. 1982) in lakes. Larger fish have greater energy stores that allow them to subsist longer than smaller fish, when food is limiting (Shuter et al. 1980; Adams et al. 1982).

Our data support the hypothesis that overwinter survival is directly related to length. The two smallest autumn stocks in Acton Lake in 1981 and 1982 were 168 and 183 mm long and survival was similar, 6 and 9%; the intermediate-size fish averaged 196 mm long and survival was 16%. The largest hybrids averaged 255 mm long and survival was 41 and 52% for the two methods of estimation. Largemouth bass accounted for only a minor part (9-12%) of the mortality of the smallest groups. None of the hybrids from the two larger groups was eaten by largemouth bass. Clearly, predatory mortality was not important and overwinter survival was directly related to esocid size. Laboratory and field experiments are needed to test the hypothesis that overwinter survival is a function of energy stores. If correct, then more attention should be given to condition of the fish at stocking. Beyond a certain size, perhaps 200-250 mm, length may be much less important than condition. Rearing practices that maximize condition factor may be crucial to promoting high overwinter survival.

Perhaps equally as important as condition at stocking is density of appropriate size and prey type. If vulnerable prey are scarce, growth of hybrids may be negligible and they may even lose weight. Although managers can do little to improve available forage, assessment of prey densities before stocking can provide guidance on number and size of fish to stock (providing a choice exists). In lakes with relatively low prey abundance, we believe it would be prudent to stock a few large fish. Larger hybrids can eat a wider size range of prey than small ones (Gillen et al. 1981) and should be better able to cope with a limited food supply.

In addition to paucity of available prey, emigration may have been responsible for some of our unaccountable losses. Some hybrids emigrated downstream from Acton and Stonelick lakes and were observed several kilometers downstream. Hess (1981) noted that emigration of parent species and hybrids from an impoundment was of sufficient magnitude to support a downstream fishery. Andrews and Laurion (1982) also found that hybrids emigrated from one natural lake to another via a connecting channel. We suggest that factors affecting emigration merit further study.

In summary, we found that hybrid muskellunge stocked in autumn survived better than those stocked in summer. Largemouth bass preyed most intensely on hybrids stocked in summer. Predatory mortality was positively related to largemouth bass density and inversely related to hybrid length. Overwinter survival of hybrids was directly related to their length. Hybrid growth was a function of prey type, size, and density. Hybrids grew fastest at high densities of gizzard shad and slowest when centrarchids were the primary forage. Conditions that promote rapid growth of hybrids should increase their long-term survival.

ACKNOWLEDGEMENTS

We thank the personnel of the Ohio Department of Natural Resources, particularly T. Nagel, P.J. Pfister, and D.B. Apgear, who provided fish and logistical support. We also thank the staff and students of the Aquatic Ecology Laboratory, The Ohio State University, for assisting in our many field trips. The efforts of R.S. Hayward and E.T. Rankin are especially appreciated. P.H. Eschmeyer, F.J. Margraf, F.J. Rahel, and D.L. Smith kindly reviewed the manuscript. This research was supported in part by funds from the Federal Aid in Fish Restoration Act under Dingell-Johnson Project F-57-R and from the Ohio Department of Natural Resources.

REFERENCES

Adams, S.M., R.B. McLean, and M.M. Huffman. 1982. Structuring of a predator population through temperature-mediated effects on prey availability. Canadian Journal of

Fisheries and Aquatic Sciences 39:1175-1184.

Andrews, S.J. 1983. Survival and growth of tiger muskellunge reared on live versus artificial diets. Mimeographed report. Fisheries Section, Indiana Department of Natural Resources, Indianapolis, Indiana, USA.

Andrews, S.J., and G. Laurion. 1982. Tiger muskellunge survival and habitat utilization in three northeast Indiana lakes. Mimeographed report. Fisheries Section, Indiana Department of Natural Resources, Indianapolis, Indiana, USA.

Axon, J.R. 1981. Development of a muskellunge fishery at Cave Run Lake, Kentucky, 1974-1979. North American Journal of Fisheries Management 1:134-143.

Belusz, L.C. 1978. An evaluation of the muskellunge fishery of Lake Pomme de Terre and efforts to improve stocking success. American Fisheries Society Special Publication 11: 292-297.

Beyerle, G.B. 1978. Survival, growth, and vulnerability to angling of northern pike and walleyes stocked as fingerlings in small lakes with bluegills or minnows. American Fisheries Society Special Publication 11:135-139.

Beyerle, G.B. 1981. Comparative survival and growth of a 8.9- and 17.8-cm (3.5- and 7.0-inch) tiger muskellunge planted in a small lake with forage fishes. Fisheries Research Report 1894, Michigan Department of Natural Resources, Ann Arbor, Michigan, USA.

Beyerle, G.B., and J.E. Williams. 1968. Some observations of food selectivity by northern pike in aquaria. Transactions of the American Fisheries Society 97:28-31.

Casselman, J.M. 1978. Effects of environmental factors on growth, survival, activity, and exploitation of northern pike. American Fisheries Society Special Publication 11:114-128.

Chapman, C.A., and W.C. MacKay. 1984. Direct observation of habitat utilization by northern pike. Copeia 1984:255-258.

Coutant, C.C., R.B. McLean, and D.L. DeAngelis. 1979. Influences of physical and chemical alternations on predator-prey interactions. Pages 57-68 in H. Clepper and R.H. Stroud, editors. Predator-prey systems in fisheries management. Sport Fishing Institute, Washington, District of Columbia, USA.

Flickinger, S.A., and J.H. Clark. 1978. Management evaluation of stocked northern pike in Colorado's small irrigation reservoirs. American Fisheries Society Special Publication 11:284-291.

Foster, J.R. 1977. Pulsed gastric lavage: an efficient method of removing the stomach contents of live fish. The Progressive Fish-Culturist 39:166-169.

Gillen, A.L., R.A. Stein, and R.F. Carline. 1981. Predation by pellet-reared tiger muskellunge on minnows and bluegills in experimental systems. Transactions of the American Fisheries Society 110:197-209.

Hess, L.J. 1981. Evaluation of esocid stockings in West Virginia. West Virginia Department of Natural Resources, Final Report, D-J Federal Aid Project F-23-R, Elkins, West Virginia, USA.

Hunt, R.L. 1969. Overwinter survival of wild fingerling brook trout in Lawrence Creek, Wisconsin. Journal of the Fisheries Research Board of Canada 26:1473-1483.

Johnson, L.D. 1978. Evaluation of esocid stocking program in Wisconsin. American Fisheries Society Special Publication 11:298-301.

Johnson, L.D. 1982. Factors affecting short-term survival of stocked muskellunge fingerlings in Wisconsin. Wisconsin Department of Natural Resources Research Report 117, Madison, Wisconsin, USA.

Krska, R.J. Jr., and R.L. Applegate. 1982. Food of young muskellunge in a power plant cooling reservoir. The Progressive Fish-Culturist 44:172-173.

Mauck, W.L., and D.W. Coble. 1971. Vulnerability of some fishes to northern pike (*Esox lucius*) predation. Journal of the Fisheries Research Board of Canada 28:957-969.

Niimi, A.J., and F.W.H. Beamish 1974. Bioenergetics and growth of largemouth bass (*Micropterus salmoides*) in relation to body weight and temperature. Canadian Journal of Zoology 52:447-456.

Oliver, J.D., G.F. Holeton, and K.E. Chua. 1979. Overwinter mortality of fingerling smallmouth bass in relation to size, relative energy stores, and environmental temperature. Transactions of the American Fisheries Society 108:130-136.

Rice, J.A., J.E. Breck, S.M. Bartell, and J.F. Kitchell. 1983. Evaluating the constraints of temperature, activity, and consumption on growth of largemouth bass. Environmental Biology of Fishes 9:263-276.

Ricker, W.E. 1975. Computation and interpretation of biological statistics of fish populations. Fisheries Research Board of Canada Bulletin 191, Ottawa, Canada.

Savino, J.F., and R.A. Stein. 1982. Predator-prey interaction between largemouth bass and bluegills as influenced by simulated, submersed vegetation. Transactions of the American Fisheries Society 111:255-266.

Serns, S.L., and L.M. Andrews. 1983. Survival and growth of muskellunge fingerlings stocked in four Vilas County, Wisconsin lakes. Wisconsin Department of Natural Resources Fish Management Report 116, Madison, Wisconsin, USA.

Shuter, B.J., J.A. MacLean, F.E.J. Fry, and H.A. Regier. 1980. Stochastic simulation of temperature effects on first-year survival of smallmouth bass. Transactions of the American Fisheries Society 109:1-34.

Stein, R.A., R.F. Carline, and R.S. Hayward. 1981. Largemouth bass predation on stocked tiger muskellunge. Transactions of the American Fisheries Society 110:604-612.

Tomcko, C.M., R.A. Stein, and R.F. Carline. 1984. Use of bluegill forage by tiger muskellunge: effects of predator experience, vegetation, and prey density. Transactions of the American Fisheries Society 113:588-594.

Vasey, F.W. 1974. Life history of introduced northern pike in Thomas Hill Reservoir. Missouri Department of Conservation, Final Report, D-J Federal Aid Project F-1-R-23, Jefferson City, Missouri, USA.

Weithman, A.S. 1975. Survival, growth, efficiency, preference, and vulnerability to angling of Esocidae. Master's thesis. University of Missouri, Columbia, Missouri, USA.

Weithman, A.S., and R.O. Anderson. 1977. Survival, growth, and prey of Esocidae in experimental systems. Transactions of the American Fisheries Society 106:424-430.

Response of Muskellunge to Establishment of Walleye in Chautauqua Lake, New York

STEPHEN R. MOORADIAN

New York State Department of Environmental Conservation
Olean, New York 14760

JOHN L. FORNEY AND MICHAEL D. STAGGS[1]

Department of Natural Resources, Cornell University
Ithaca, New York 14853

ABSTRACT

Muskellunge in Chautauqua Lake have supported a large and economically valuable sport fishery for over a century. Walleye were seldom observed prior to 1962 when the first successful year-class was produced. By the following decade the population had increased to about 28 age-1 and older walleye/ha. Adult stocks of muskellunge initially increased but by the late 1970's the catch-per-unit-effort in poundnets had fallen to about 60% of the mean catch in pre-walleye years. Lower muskellunge abundance was attributed to decreased survival of stocked fingerlings. As abundance declined mean weight of adult muskellunge increased from 2.8 to 3.9 kg. The increase in mean weight largely compensated for the decline in numbers and biomass of the stock remained near levels which prevailed before establishment of the walleye.

This paper describes changes in the population of muskellunge in Chantauqua Lake, New York between 1937 and 1983. During this period a remnant population of walleye began to reproduce successfuly and by the 1970's rivaled the muskellunge in abundance. We examined trends through time in abundance and growth of muskellunge and changes in spatial distribution for evidence of interaction with the burgeoning walleye population. Inferences drawn from postmortem analysis of natural experiments are inherently speculative, but may provide the insight needed to design more robust manipulative experiments.

Historically the muskellunge was the most abundant piscivorous species and the population supported a large and valuable fishery for over a century. Efforts to propagate muskellunge began in 1887 and by the early 1900's the practice of rearing and stocking fry was well established. Poundnets were set during the spawning period to obtain hatchery broodstock and the recorded catch provided a continuous index of adult stock abundance. A decline in the poundnet catch was the stimulus for an intensive biological study of the muskellunge in the early 1940's (Heacox 1946). A special muskellunge license was established and license reports provided annual catch statistics in subsequent years (Mooradian and Shepherd 1973). Data on muskellunge collected by the New York Department of Environmental Conservation (NYDEC) during the past two decades was reviewed by Bimber and Nicholson (1981) and information on walleye was analyzed by Staggs (1982). For this paper we draw on data in earlier publications and incorporate unpublished material from NYDEC files and results of ongoing studies.

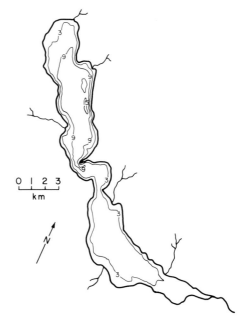

Figure 1. Bathometric map of Chautauqua Lake.

[1]Present address: Wisconsin Department of Natural Resources, Madison, Wisconsin 53711

Chantauqua Lake, located in southwestern New York State, has a surface area of 5,324 ha, a mean width of about 2 km and a length of 28 km. The lake is divided into two basins of nearly equal surface area by a central constriction which narrows the lake to about 0.5 km (Fig. 1). The north basin is mesotrophic, has a maximum depth of 23 m and temperature stratification may develop in summer (Mayer et al. 1978). The south basin is more eutrophic, maximum depth is 6 m and the basin is usually homothermal.

MATERIAL AND METHODS

Trends in abundance of adult muskellunge were determined from the mean annual catch in poundnets set by the Chatauqua Lake Hatchery for broodstock. Between 10 an 14 poundnets were set in the years 1910 - 66 and 5 -7 nets in later years. Total catch and number of nets fished were recorded, and after 1939 catch was reported by net. Nets were set in mid- to late-April and fished for 12 to 24 days. Examination of daily catch rates indicated that the netting period consistently encompassed the peak of the spawning. Most muskellunge captured in the spring were over 55 cm and nearly all were sexually mature.

Trends in growth of muskellunge between 1941 and 1982 were assessed from length increments of tagged muskellunge between release and recapture and observed lengths of fin-clipped muskellunge stocked as fingerlings. Muskellunge netted in 1941-45, 1961-65 and 1976-82 were tagged with a metal strap clamped near the base of the anterior edge on the dorsal fin. Total length and sex of tagged fish were recorded at the time of release in the spring and those recaptured in poundnets one year later were remeasured.

Fingerling muskellunge (10-25 cm) stocked in 1961 and later years were identified by clipping a single paired fin or both ventrals. The sequence of marks was repeated at 5-year intervals except in 1965 when fingerlings were not stocked and in 1976-77 when consecutive plantings were marked by removal of both ventral fins. Most of the poundnet catch was examined for fin-clips and total lengths of marked fish recorded.

Fin regeneration introduced some errors in assessment of muskellunge growth and survival from recoveries of marked fish. Comparison of ages assigned from fin-clips and scale markings suggested about 25% of the fish marked by removal of both ventral fins in 1976-77 had regenerated one of the paired fins and were incorrectly classified as right or left ventral clips (Table 1). Because scales were seldom taken from fin-clipped muskellunge prior to 1978, some recoveries from plantings marked by removal of both ventrals in earlier years were misidentified and assigned wrong age. These fish probably represented less than five percent of the total recoveries and their inclusion in samples is unlikely to obscure trends in growth and recruitment.

Walleye spawning in several tributaries to Chautauqua Lake were collected by electrofishing or caught in trapnets each year from 1966-80, except 1975. Lengths and scale samples were obtained from subsamples of the catch. In addition, walleye taken in 1966-69 and in 1976-80 were marked with sequentially numbered metal tags which loosely encircled the mandible. After 1977 abundance of walleye and other species were monitored from catch in variable-mesh gillnets set at four sites and fished at monthly intervals from June through October. Scales and measurements were taken and walleye examined for tags.

Walleye survival and rates of exploitation were estimated from tag recoveries in successive spawning runs and from numbers reported by anglers (Staggs 1982). Abundance of walleye in 1979-80 was estimated from the ratio of tagged to unmarked walleye taken in variable-mesh gillnets fished during the summer and from numbers recorded by fishermen participating in an angler diary program. Ob-

Table 1. Number of right ventral, left ventral and double ventral fin clips among muskellunge assigned ages from scale markings of 3-4 in 1980 and 4-5 in 1981. Fingerlings released in 1976-77 were marked by removal of both ventral fins.

Year Recaptured	Year-Class					
	1976 Fin clip			1977 Fin clip		
	LVRV	LV	RV	LVRV	LV	RV
1980	13	2	1	68	14	8
1981	37	5	4	60	15	7
Total	50	7	5	120	29	15
Probability of[a] failing to identify a LV clip		0.090			0.110	
Probability of[a] failing to identify a RV clip			0.122			0.194

[a]Calculated from formula presented by Russell, H.J. (1980), in "Analysis of double-tagging experiments: an update," Canadian Journal of Fisheries and Aquatic Sciences 37:114-116.

served and back-calculated lengths at aged based on the spring spawning run samples were used to assess walleye growth before 1980 (Staggs 1982) and back-calculated lengths from the summer gillnet samples were used for later years.

Information from fisheries surveys conducted before and after expansion of the walleye population was examined for evidence of changes in species composition of the fish community. Trends in age and growth of potential prey species were assessed from examination of scale samples (Staggs 1982).

RESULTS

Population Trends

Estimates of muskellunge abundance from the total catch and from the catch in a subset of five nets set at approximately the same locations in 1940-78 showed similar trends (Fig. 2). Catches in the five nets were higher in the 1940's but differences in the two indices narrowed in the 1960's and were indistinguishable after 1966 when only 5-7 nets were fished. Examination of hatchery records showed that about one-third of the sites fished in the 1940's had been replaced by the 1960's, and nets eliminated were generally those set at less productive sites. If progressive culling of less productive set occurred prior to 1940, reported catches may have underestimated the relative abundance of muskellunge in the early 1900's.

Early records suggested a progressive decline in abundance of adult muskellunge and the lowest poundnet catches ocurred in the late 1930's (Fig. 2). Heacox (1946) attributed the decline to increased exploitation by anglers, and the imposition of more restrictive angling regulations in 1941 was followed by an increase in adult stock. Catches averaged about 150 per net in the late 1940's and 1950's, then rose to over 400 per net in the 1960's. The irruption was brief, and by the late 1970's catches had dropped to levels only slightly higher than those which prevailed in the late 1930's. Most muskellunge caught in poundnets were ages 4-10, which suggests the abrupt increase in stock abundance in the late 1960's was attributable to successful natural reproduction or high survival of fingerling muskellunge stocked in the late 1950's and early 1960's.

Rearing and stocking of fingerling muskellunge began in 1928, and starting in 1961 all fingerlings were marked to evaluate their contribution to the fishery (Mooradian and Shepherd 1973). By the mid-1970's, over 70% of the muskellunge taken in poundnets were fin-clipped. Fingerling plantings in the early 1960's were more successful than later plantings judging from their contribution to the poundnet catch at age 5 (Table 2). The proportion of stocked fingerlings recaptured per 100 poundnet days of effort decreased from an average of 0.029 in 1961-69 to 0.006 in 1970-78. Analysis of stocking records failed to reveal any strong correlations between relative survival of plantings and mean length of fingerlings or date stocked (Forney 1982). A switch from extensive pond to intensive trough culture in 1973-75 may have contributed to the low

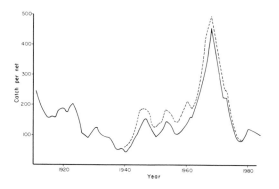

Figure 2. Three-year running average of mean annual catch of muskellunge in poundnets. Solid line total catch; dashed line subset of 5 nets.

Table 2. Relative survival of fin-clipped fingerling muskellunge stocked in 1961-1978 estimated from their contribution to the catch in poundnets at age 5.

Year stocked	Number stocked	Catch/100 poundnet days	Relative survival
1961	10,300	141	0.014
1962	9,220	317	0.034
1963	17,530	309	0.018
1964	7,000	226	0.032
1966	1,120	72	0.064
1967	5,818	236	0.041
1968	15,785	96	0.006
1969	9,018	208	0.023
1970	1,740	11	0.006
1971	17,700	86	0.005
1972	15,100	204	0.014
1973	14,770	146	0.010
1974	14,770	13	0.001
1975	16,650	19	0.001
1976	21,790	269	0.012
1977	15,189	64	0.004
1978	26,095	50	0.002

survival of fingerlings stocked in recent years. Although fingerlings confined to troughs and fed fish attained about the same size as pond-reared muskellunge stocked in earlier years, more subtle physiological and behavioral differences may have existed.

Expansion of the walleye population began in the 1960's about 50 years after attempts to establish walleye by stocking were abandoned. Over five million walleye fry were stocked between 1903 and 1908, but only 4 adults were taken in poundnets in later years (Bean 1912). No walleye were observed during a biological survey and creel census in 1937 (Moore et al. 1937). Later unpublished records show that individual walleyes were caught in 1943 and 1946, and an additional three walleye were taken in the 1950's. Whether sporadic reproduction by survivors of the early fry stockings or unauthorized releases maintained the small population is unknown.

The first successful year-classes of walleye were produced in 1962 and 1964 (Fig. 3). Members of these two cohorts comprised 93% of the spawning run catch in 1966-68 and only 2% of the fish caught were hatched prior to 1962. No quantitative estimates of abundance were made in the 1960's but the increase in adult stock was probably rapid. First-year tag returns from anglers averaged 1.4%, which suggested low rates of exploitation, and estimates of annual survival from tag recoveries approached 100% (Staggs 1982). As stock abundance increased recruitment stabilized in the 1970's, as shown by the relative contribution of year-classes to the spawning run (Fig. 3). The increase in numbers of walleye was accompanied by a decline in annual survival to 67% and a corresponding increase in minimum rates of exploitation to 9.9%.

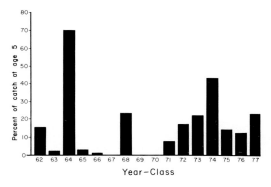

Figure 3. Relative strength of walleye year-classes based on percentage contribution to the catch at age 5 in spawning runs.

Abundance of walleye was estimated from the numbers of tagged fish released in 1979 and 1980 and the ratio of marked to unmarked walleye in the angling catch and gillnets in both years. Confidence intervals on individual walleye population estimates were broad but the mean of the four estimates indicated a population of about 150,000 age-1 and older walleye (Staggs 1982). Numbers were equivalent to a density of 28 fish per ha, or a biomass of about 20 kg/ha based on the mean weight of walleye in the 1980 gillnet catch. The standing stock in Chautauqua Lake compared favorably with the mean of 16 kg/ha for age-3 and older walleye in 20 north temperate lakes (Carlander 1977).

Information on the response of prey species to the increase in walleye abundance was fragmentary. High densities of muskellunge coupled with the expansion of the walleye may have temporarily disrupted the fish community in the late 1960's. Anglers reported an abrupt decline in catches of yellow perch and both white and/or black crappie. However, changes in species abundance were probably transitory. Black crappie, yellow perch, pumpkinseed, in order of decreasing abundance, were the species most frequently caught in variable-mesh gillnets in 1937 (Greeley 1938). In 1978-83 the order of abundance was white crappie, black crappie, yellow perch, pumkinseed and bluegill. Changes in species ranking were evident but their relation to changes in predator abundance was speculative. Age compostion of the gillnet catch showed the increase in white crappie was attributable to the appearance of a single dominant year-class in 1977; by 1983 a shift to black crappie was evident.

Growth

Recoveries of tagged muskellunge from successive releases indicated a progressive increase in growth (Table 3). All size-classes of muskellunge tagged in 1961-65 grew more rapidly than in 1941-45, and a further increase in growth was evident from annual length increments of muskellunge tagged in 1976-81. Only males were recaptured in adequate numbers to evaluate trends in growth.

No explanation can be offered for the severely depressed growth of males in the early 1940's. Adult stocks as measured by the catch in poundnets were at their lowest recorded level in the late 1930's (Fig 2). Although the population was expanding in the early 1940's abundance of muskellunge was substantially lower than in the 1960's. Accelerated growth during a period of population expansion

Table 3. Mean annual length increments (mm) of male muskellunge tagged in April-May and recaptured one year later classified by length at release. Number of recaptures in parentheses.

Period	\multicolumn{8}{c}{Length at release}							
	600-629	630-659	660-689	690-719	720-749	750-779	780-809	810-829	830-859
1941-45	47	41	28	22	19	12	12	6	2
	(40)	(35)	(47)	(56)	(50)	(24)	(22)	(21)	(22)
1961-65	57	55	41	33	20	19	22	24	17
	(21)	(49)	(126)	(116)	(89)	(57)	(21)	(4)	(6)
1976-81	70	58	58	40	32	26	28	21	17
	(1)	(9)	(12)	(18)	(16)	(18)	(11)	(8)	(7)

Table 4. Catch of muskellunge in five poundnets set at the same locations each year expressed in kg per net.

Year	Catch	Year	Catch	Year	Catch
1945	463	1960	786	1976	450
1946	530	1961	727	1977	308
1947	464	1962	480	1978	217
1948	391	1963	486	1979	409
1949	251	1964	827	1980	410
1950	385	1965	740	1981	558
1951	335				
1952	445				
1953	414				
1954	706				
1955	373				
Mean	432		674		392

implies an increase in abundance or availability of prey. If a change in community structure occurred, it probably predated expansion of the walleye population. Examination of annual increments of muskellunge tagged in 1961-65 did not reveal any trend in growth during the early 1960's when the first strong year-classes of walleye appeared.

After 1960, annual plantings of fingerling muskellunge were marked and survivors of these cohorts were recovered in poundnets. Mean observed length of males at age 5 increased from about 70 cm to 77 cm between 1967 and 1983 (Fig. 4). Comparable gains in length-at-age were evident among other age-groups of males and for adult females taken during the same period (Bimber 1982).

More rapid growth contributed to an increase in mean weights of muskellunge taken in poundnets from about 2.8 kg in the 1940's to 3.9 kg in the late 1970's. Length-weight regressions (Bimber 1978) were used to estimate weights from lengths in years when most fish were measured. Mean annual weight of the catch in the 5 poundnets used as an index of abundance was 432 kg in 1945-55 and 392 kg in 1976-81 (Table 4). Assuming vulnerability of muskellunge to capture in poundnets was independent of body size, biomass of the adult stock was about the same during the pre- and post-walleye periods.

Growth of walleye followed a pattern characteristic of an invader. Members of the dominant 1962, 1964, and 1968 year-classes grew rapidly, and observed lengths of males in spawning runs approached or exceeded 47.5 cm at age 4 (Fig. 4). Growth of subsequent year-classes declined and mean lengths at age 4 were generally less than 42 cm. Back-calculated lengths of walleyes taken in gillnets indicated that the slower regime, which had characterized the late 1970's, persisted through 1983; however, growth was variable and dependent on prey abundance (Staggs 1982). The transition from rapid to relatively slow growth was abrupt, which suggested early year-classes might have depleted some segment of the prey resource. Over 90% of the prey in stomachs of walleye examined in 1978-79 were age-0 yellow perch (Einhouse 1981); stomachs of walleye taken in earlier years were not examined.

Muskellunge Distribution

Tag returns reported by anglers from muskellunge released in 1961-65 and 1976-81 were examined for evidence of changes in spatial distribution of the population. Muskellunge were tagged from nets set at the same 5 locations during both periods and recaptures made in 1961-68 and 1976-83 were pooled (Table 5). The presence of two or more subgroups of muskellunge with different spawning areas and summer ranges was evident from the distribution of tag recoveries. Muskellunge tagged in the spring at the site located in the south

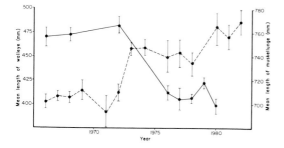

Figure 4. Observed length at age 5 of male muskellunge (---) and age 4 male walleye (___). Data points are mean ± 2 std errors.

Table 5. Number of muskellunge tagged from poundnets in 1961-65 and 1976-81 and recoveries by anglers from June through October in three areas of Chautauqua Lake. Between period differences in distribution tested for heterogeneity by Chi-square.

Basin and period	Release Number	Recaptures by Basin			Percent recaptured	Chi-square
		North	Inter-	South		
North						
1961-65	1393	104	14	20	9.9	12.13[a]
1976-81	931	29	7	20	6.0	
Inter-						
1961-65	1707	84	38	72	11.4	9.87[a]
1976-81	963	18	19	41	8.1	
South						
1961-65	789	8	6	63	9.8	3.29[b]
1976-81	755	2	1	43	6.9	

[a] $X^2.01 = 9.2$, 2 D.F.
[b] North and Interbasin pooled

basin were generally recaptured in the south basin while muskellunge tagged at more northern sites occupied both basins in summer. Haas (1978) found spatially segregated subpopulations of muskellunge which migrated to widely separated spawning grounds.

A change in distribution of tag returns between the first and second tagging periods was evident from recoveries of muskellunge tagged in north and interbasin area. About 75% of the adults tagged in the north basin in April May, 1961-65, were caught in the north basin during the post-spawning period. Among fish tagged at the same two sites in 1976-81 only 52% were recovered in the north basin. Recoveries of muskellunge tagged from nets set in the interbasin area in 1961-65 were about equally divided between the north and south basins, but later releases were recovered mostly in the south basin.

The pattern of tag returns suggests a shift in the distribution of the population during the post-spawning period. Muskellunge move extensively during the spring but occupy a restricted home range in summer (Minor and Crossman 1978; Dombeck 1979). In recent years a large proportion of the muskellunge spawning in the northern portion of the lake have migrated to summer ranges in the south basin. Although a switch in fishing effort from the north to the south basin would have increased the number of north basin tags caught in south basin, this interpretation was incompatible with evidence that anglers returned about the same percentage of tags from fish released in the north and south basins during both tagging periods (Table 5). A shift in fishing effort to the south basin would have caused a disproportionate increase in exploitation of muskellunge tagged in the south basin because these fish remained concentrated in south basin during the post-spawning period.

DISCUSSION

The rapid expansion of the walleye population and the synchronous decline in stocks of adult muskellunge during the 1970's led to speculation that these events were casually related (Bimber and Nicholson 1981). In retrospect, the first large year-class of walleye appeared about five years before the catch of adult muskellunge reached a record peak in the late 1960's. The muskellunge which contributed to the peak adult abundance were those produced in the late 1950's and early 1960's. Thus a period in which muskellunge recruitment was abnormally successful generally coincided with the appearance of the first successful walleye cohort in 50 years. Although the timing suggests that a temporary disruption of the fish community may have enhanced recruitment of both predators, variation in recruitment of muskellunge was evident in earlier years.

Trajectory of the muskellunge population was probably determined more by rates of recruitment than rates of exploitation. Although angling regulations imposed in 1941 to reduce exploitation were followed by an increase in stock abundance, recovery of the population may have been more strongly influenced by increased recruitment than reduced exploitation. The increased catch in poundnets was accompanied by an abrupt drop in average length of muskellunge (Heacox 1946), which suggests one or more strong year-classes were recruited in the early 1940's. The dramatic increase in stock abundance in the 1960's, at the time when fishing effort measured by sales of muskellunge licenses was stable (Mooradian and Shepherd 1973), further illustrated the strong influence of recruitment on population trends.

Fingerling stocking supported the muskellunge fishery in the 1970's, and most recruits in earlier years were probably of hatchery origin. Plantings in the 1960's contributed more fish to the poundnet catch than later releases of the marked fingerlings. Walleye biomass was increasing in the late 1960's and the reduction in growth of walleye suggested the prey supply was limited. Possibly the diet of walleye broadened to include an occasional muskellunge, but the relation between walleye abundance and muskellunge fingerling survival was confounded by changes in fish cultural practices which could have influenced post-stocking survival of muskellunge. Stomachs of 270 walleye collected in September-December, 1978-80, did not contain any fingerling muskellunge (Einhouse 1981). However, encounter rates would be low and occurrence of predation difficult to detect.

Addition of northern pike to Escanaba Lake where walleye were predominant led to near extinction of several prey species (Kempinger and Carline 1977). In contrast, establishment of walleye in Chautauqua Lake where muskellunge were dominant had little detectable effect on the fish community. Walleye fed mostly on yellow perch, but no change in the size or age structure of the yellow perch population was evident through 1980 (Staggs 1982). Year-classes of yellow perch sampled in the 1930's and 1940's generally attained a length of 18 cm at age 3, and length-at-age of cohorts collected in the late 1970's was virtually the same. Growth rates of several species of centrarchids seemed equally unresponsive to the addition of a predator (Staggs 1982). Stable growth of prey species implies community stability, but variation in predator growth suggests that subtle changes in community structure did occur.

Growth of muskellunge increased as the walleye population expanded. If competition between these species occurred it was not manifested in the growth of age-5 muskellunge, which showed a progressive increase in length at a time when biomass of walleye was increasing. Muskellunge consume larger prey than walleye (Einhouse 1981), and size partitioning of the prey resource would minimize competition. Overlap in food-resource utilization was probably reduced by divergence in growth and mean body size of walleye and muskellunge during the 1970's. However, there was no evidence that divergence in growth was a response to interspecific competition.

Growth increments of tagged muskellunge showed a progressive increase in successive tagging periods (Table 3). Thus the trend toward more rapid growth of muskellunge may have been established before the first strong year-classes of walleye appeared and before muskellunge density declined in the 1970's. The argument that accelerated growth of muskellunge was an intraspecific, density-dependent response cannot be reconciled with the apparent insensitivity of growth in the 1930's to a similar decline in muskellunge abundance. Perhaps growth was responding to long-term trends in lake productivity or aquatic flora (Mayer et al. 1978) but no link between growth and environmental factors has been identified.

The shallow south basin with extensive beds of rooted aquatic macrophytes more closely matched the preferred habitat of muskellunge (Scott and Crossman 1973) than the deeper, less eutrophic north basin, which might be characterized as more typical walleye habitat (Ryder and Kerr 1978). Despite rather marked differences in habitat, ecological segregation of species by basin was not pronounced. Variable-mesh gillnets set in June-October, 1978-80, caught about the same number of walleye (540) in the north basin as in the south basin (664). During the same period gillnets caught 17 muskellunge in the north basin and 15 in the south basin. Although more tagged muskellunge were caught by anglers in the south basin after walleye became established there was no evidence the shift in distribution of tagged fish was a response to interspecific competition.

Twenty years after the establishment of walleye, average weight of muskellunge had increased but survival of stocked fingerlings had declined. Higher stocking rates and more rapid growth largely compensated for lower recruitment, and biomass of muskellunge remained near levels which prevailed in the 1940's and 1950's. Whether expansion of the walleye contributed to lower recruitment and accelerated growth of muskellunge can only be determined from a longer time series.

REFERENCES

Bean, T.H. 1912. Annual report of the fish culturist. Pages 227-280 *in* New York State Conservation Commission 2nd Annual Report.

Bimber, D.L. 1978. An analysis and historical overview of the muskellunge population and sport fishery of Chautauqua Lake, New York. Master's thesis, State University College, Fredonia, New York, USA.

Bimber, D.L. 1982. Longevity, growth and mortality of muskellunge in Chautauqua Lake, New York. New York Fish and Game Journal 29:134-141.

Bimber, D.L. and S.A. Nicholson. 1981. Fluctuations in the muskellunge (*Esox masquinongy* Mitchill) population of Chautauqua Lake, New York. Environmental Biology of Fishes 6:207-211.

Carlander, K.D. 1977. Biomass, production and yields of walleye (*Stizostedion vitreum vitreum*) and yellow perch (*Perca flavescens*) in North American lakes. Journal of the Fisheries Research Board of Canada 34:1602-1612.

Dombeck, M.P. 1979. Movement and behavior of the muskellunge determined by radio-telemetry. Wisconsin Department of Natural Resources, Technical Bulletin 113.

Einhouse, D.W. 1981. Summer-fall movements, habitat utilization, diel activity and feeding behavior of walleye in Chautauqua Lake, New York. Master's Thesis, State University College, Fredonia, New York, USA.

Forney, J.L. 1982. Muskellunge-walleye interactions in Chautauqua Lake. Final Report. New York Department of Environmental Conservation, Federal Aid in Fish Restoration Project FA-5-F. Albany, New York, USA.

Greeley, J.R. 1938. Fishes of the area with annotated list. *In* A biological survey of the Allegheny and Chemung watersheds. New York State Conservation Department, 27th Annual Report (1937). Supplement 12.

Haas, R.C. 1978. The muskellunge in Lake St. Clair. American Fisheries Society Special Publication 11:334-339.

Heacox, C. 1946. The Chautauqua Lake muskellunge: research and management applied to a sport fishery. Transactions North American Wildlife Conference 11:419-425.

Kempinger, J.J. and R.F. Carline. 1977. Dynamics of the walleye (*Stizostedion vitreum vitreum*) population in Escanaba Lake, Wisconsin, 1955-72. Journal of the Fisheries Research Board of Canada 34:1800-1811.

Lee, R.M. 1920. A review of the methods of age and growth determination by means of scales. Fishery Investigations, London, Series II. 4:32 p.

Mayer, J.R., W.M. Barnard, W.J. Metzger, T.A. Storch, T.A. Erlandson, J.R. Luensman, S.A. Nicholson, and R.T. Smith. 1978. Chautauqua Lake - watershed and lake basin. Pages 1-120 *in* J.A. Bloomfield, editor. Lakes of New York State, volume II. Academic Press, New York, New York, USA.

Minor, J.D. and E.J. Crossman. 1978. Home range and seasonal movements of muskellunge as determined by radiotelemetry. American Fisheries Society Special Publication 11:146-153.

Moore, E. and staff members. 1937. A creel census at Chautauqua Lake, New York. Transactions of the American Fisheries Society 67:130-138.

Mooradian, S.R. and W.F. Shepherd. 1973. Management of muskellunge in Chautauqua Lake. New York Fish and Game Journal 20:153-157.

Ryder, R.A. and S.R. Kerr. 1978. The adult walleye in the percid community - a niche definition based on feeding behavior and food specificity. American Fisheries Society Special Publication 11:39-51.

Scott, W.B. and E.J. Crossman. 1973. Freshwater fishes of Canada. Bulletin 184. Fisheries Research Board of Canada, Ottawa. 966 pp.

Staggs, M.D. 1982. Changes in growth, recruitment, and survival in the walleye (*Stizostedion vitreum vitreum*) population of Chautauqua Lake, New York, 1960-80. Master's Thesis, Cornell University, Ithaca, New York, USA.

Fluctuations in Growth Rate and Condition of Muskellunge and Northern Pike in Escanaba Lake, Wisconsin

PETER D. INSKIP[1] AND JOHN J. MAGNUSON

Center for Limnology
University of Wisconsin — Madison
Madison, Wisconsin 56706

ABSTRACT

Indices of growth rate and condition of muskellunge from Escanaba Lake, Wisconsin varied in parallel with the corresponding indices for northern pike during the period 1957 to 1982. Abundances of yellow perch, northern pike and muskellunge, and July temperature appear to have been among the controlling environmental variables. Growth rate of yearling esocids was negatively associated ($r = -0.47$ and -0.37 for muskellunge and northern pike respectively; $N = 16$ years) with a harvest/effort indicator of relative abundance of northern pike and positively associated ($r = +0.49, +0.37$) with average July air temperature. Condition of larger (40-70 cm) esocids was negatively associated ($r = -0.56, -0.36$; $N = 26$ years) with harvest/effort of muskellunge and positively associated ($r = +0.56, +0.76$) with harvest/effort of yellow perch. Yellow perch were common relative to other food items in stomachs of muskellunge and northern pike harvested from Escanaba Lake during 1977 to 1979. It is likely that the availability of yellow perch was the most important biotic factor mediating the synchronous fluctuations in growth rate and condition of the two esocids. The abundance of muskellunge in Escanaba Lake between 1957 and 1982 was lower and more stable than that of northern pike, based on harvest/effort data, and abundances of the two species did not appear to be inversely related. Competition for food appears to have occurred among individual muskellunge and northern pike, but it is not apparent that this interaction had a controlling effect on muskellunge population size.

Declines in muskellunge populations have been reported for many lakes to which northern pike have been introduced or gained access (Threinen and Oehmcke 1950; Oehmcke 1951; Johnson 1981; Becker 1983; Inskip and Magnuson 1983; Inskip, 1986). The recurrence of the pattern argues against a merely coincidental association. A negative interaction between these congeners seems likely.

Competition for food is one mechanism by which these closely related predators might negatively interact. Both species become piscivorous at an early age (Elson 1940; Hunt and Carbine 1951), and food habit studies for populations in the northern midwest (Hourston 1952; Seaburg and Moyle 1964; Johnson 1969) and for individuals in experimental systems (Weithman and Anderson 1977) indicate that food preferences are similar. Detailed studies of the food habits of co-occurring muskellunge and northern pike have not been conducted due largely to the muskellunge's limited range, low abundance, and low catchability.

Decreases in abundance due to competition for food most likely would be mediated through decreases in growth rate and condition of individual fish, with associated decreases in population age-specific survival and fecundity rates. Survival rate and age-specific fecundity of northern pike are positively related to growth rate and condition (Carbine 1943; Kempinger and Carline 1978). Comparable decreases in growth rate and condition of muskellunge and northern pike might have different population level consequences because of differences in the size and age at which sexual maturity is attained. Muskellunge mature at a larger size (Hourston 1952; Johnson 1971; Priegel and Krohn 1975) and, where the two species co-occur, at an older age. The shorter generation time of northern pike might prove advantageous under conditions where growth of both species is severely limited. High mortality rates among juvenile muskellunge, coupled with a prolonged pre-reproductive phase, would result in low survival to spawning age.

Data and samples collected by the Wisconsin Department of Natural Resources (DNR) through regular monitoring of the Escanaba Lake (Vilas County, Wisconsin) sport fishery provided the opportunity to compare temporal fluctuations in growth measures for co-occurring muskellunge and northern pike. Additional long-term information on

---

[1]Current address: Department of Epidemiology, Harvard School of Public Health, 677 Huntington Avenue, Boston, MA 02115

the abundances of esocids and other fishes, and on local air temperature, provided a possible basis for explaining patterns of variation in growth indices. Kempinger and Carline (1978) documented fluctuations in growth and natural mortality rates of northern pike in Escanaba Lake between 1958 and 1977. Densities of northern pike and of likely prey species also oscillated, and the pattern of covariability of growth, mortality, and abundances indicated that the northern pike population experienced occasional food shortages (Kempinger and Carline 1978). If muskellunge and northern pike relied on the same food resources, the growth rate of muskellunge would be expected to have declined during these same years. The objective for this study was to evaluate the extent to which growth rate (and condition) of muskellunge and northern pike in Escanaba Lake were under common environmental control.

The muskellunge is believed to be native to Escanaba Lake, but the northern pike was introduced (Kempinger et al. 1975). Northern pike and muskellunge fry were stocked each year from 1937 to 1941, and muskellunge were stocked in 1961, 1965, and 1966 (Table 1) (Kempinger et al. 1975). Northern pike were rare until 1956, when a strong year class was produced. They outnumbered muskellunge in the angler harvest every year from 1957 to 1979 (Kempinger and Carline 1977; S. Serns, Wisconsin DNR, personal communication). Natural hybridization has occurred between the two species, and hybrids were relatively common in some years (Hoff and Serns, 1986). Known hybrids were excluded from the growth and condition analyses presented in this paper.

STUDY SITE

Escanaba Lake is located in the Northern Highland State Forest, in north-central Wisconsin. Surrounding property is undeveloped, and the shoreline is forested. Inlets and outlets are intermittent. Escanaba Lake is of moderate surface area (119 ha) and depth (mean =4.3m; maximum = 8 m) (Kempinger and Carline 1977) in comparison to neighboring lakes. Indicators of fertility are low, but typical for the region. Black et al. (1963) listed a conductivity of 50 μmhos/cm, a total alkalinity of 25 mg/1, and a pH of 7.3. These variables had not changed appreciably by the latter 1970s, based on values given by Kempinger and Carline (1977). Temperature can vary several degrees from surface to bottom in deeper parts of the lake, but a sharp, stable thermocline does not develop. Typical maximum summer temperatures are between 24 and 28 C at the surface (S. Serns, Wisconsin DNR, personal communication). Dissolved oxygen concen-

Table 1. Summary of esocid stockings in Escanaba Lake since 1933 (adapted from Kempinger (1972) and Kempinger et al. (1975)).

Year Class	Date stocked	Number	Length at stocking (cm)	Number harvested[1]
Muskellunge:				
1937	1937	114,000	"fry"	?
1938	1938	39,000	"	?
1939	1939	86,000	"	?
1940	1940	63,000	"	?
1941	1941	7,000	"	?
1960	6 June 1961	197	25-35 yearling	57
1961	23 Aug 1961	291	15-25 fingerling	13
1965	30 Aug 1965	151	20-30 fingerling	0
1965	13 Sep 1965	150	20-30 fingerling	0
1965	9 May 1966	155	25-30 yearling	2
Northern pike:				
1937	1937	307,000	"fry"	?
1938	1938	50,000	"	?
1939	1939	90,000	"	?
1940	1940	50,000	"	?
1941	1941	50,000	"	?

[1] Based on a mandatory creel census (permit fishing)

trations are usually greater than 5 mg/l throughout the lake (Kempinger and Carline 1977). The fish community includes many of the coolwater and warmwater species that occur in the region, including muskellunge, northern pike, walleye, yellow perch, white sucker, several species of centrarchids, and several species of cyprinids (Kempinger et al. 1975). Smallmouth bass were common during the 1940's and 1950's but now are scarce.

Escanaba Lake has been the focus of a long term experiment concerning the effects of liberalized regulations on a sport fishery (Kempinger et al. 1975). The Wisconsin DNR has waived daily bag and minimum-size limits, and there is no closed season. Anglers are required to obtain a free permit before beginning a day's fishing and to register their catch at the end of the day. This policy has been in effect since 1946, except for a 56 cm minimum-length limit on northern pike during the period 1964 to 1972.

METHODS

Indices of growth rate and condition were calculated for muskellunge and northern pike. The correlation between yearly indices for the two species gave a measure of the synchrony of fluctuations in growth rate and condition. Correlation and multiple regression were used to quantify associations between environmental variables and growth indices. Stomach contents of northern pike and muskellunge captured in Escanaba Lake during 1977, 1978 and 1979 were examined in order to compare and contrast their diets and to guide formulation of regression models.

Sources of Data

Scale samples and length-weight data from Escanaba Lake muskellunge and northern pike were obtained from collections made by the Wisconsin DNR. Collections include scales from fish caught by anglers and those captured in 1.2 cm bar-mesh fyke nets. Scales were removed from an area above the lateral line, midway along the length of the body. Total lengths were measured to the nearest 0.25 cm, and fish were weighed to the nearest 4.5 g. Weights were measured only for angler-caught fish.

Relative abundances of fishes were estimated for each year using harvest and effort data compiled by the DNR. The "fishing year" was considered to begin and end with ice-out in spring, which typically occurs during the last half of April (Kempinger et al. 1975). The DNR estimated absolute abundances of sport fishes for a subset of the years included in this study (Kempinger et al. 1975; Kempinger and Carline 1977, 1978), and these data are available for comparison. Absolute abundance was estimated for muskellunge (including muskellunge x northern pike hybrid) longer than 46 cm (total length), northern pike > 30 cm, walleye > 28 cm, and yellow perch > 15 cm. From 1964 to 1972, a 56 cm minimum-length limit restricted the harvest of northern pike. The harvest rate in the absence of a size limit was projected from a regression equation relating harvest per 1000 angler hours to spring population estimates, based on six years when the size limit was not in effect (H/E = 12.31 + 0.0148 (P); $R^2 = 0.56$).

Average air temperatures for June, July, and August were used as indicators of yearly differences in summer water temperatures. Air temperatures were recorded at the U.S. Weather Station at Rest Lake, Wisconsin, approximately 30 km away. (NOAA 1954-1982). A comparison of monthly mean water temperature in Escanaba Lake and monthly mean air temperature at Rest Lake for 1958, 1962, and 1977 indicated that major differences among years in atmospheric thermal regime were reproduced in the Lake (Inskip 1980).

Wisconsin DNR personnel at the Escanaba Lake Contact Station removed stomachs from angler-caught muskellunge and northern pike when fishermen returned to register their catch. Labelled stomachs were wrapped in cheese cloth and preserved in 10% formalin for later inspection. Samples were obtained from June through November in 1977, June through August in 1978, and May through June in 1979.

Age Determination and Scale Measurements

Scales were examined for all 908 muskellunge that were collected between 1954 and 1978. The median number of fish for each calendar year was 32, with a range of 9 to 90. For northern pike, 30 fish were randomly selected from the total collection for each calendar year from 1956 to 1978.

Scale images were impressed on plastic slides and examined at 39X magnification. The annual occurrence of growth checks on scales has been confirmed for muskellunge in Wisconsin (Johnson 1971) and for northern pike in Michigan (Williams 1955). Principle criteria used to assign annuli were similar to those used for northern pike in other studies (Williams 1955; Frost and Kipling 1958).

The relationship between total length and anterior scale radius appeared to be linear for both species. Backcalculations were performed using a modification of the direct proportion method incorporating a correction factor (Lagler 1956). Correction factors of 7.5 cm and 4.5 cm were used for muskellunge (Johnson 1971) and northern pike (Franklin and Smith 1960) respectively.

Calculation Of Relative Growth And Condition

Two indices were calculated -- one for yearling fish, based on scale readings; the other for older, angler-caught fish based on length-weight data.

Relative growth index for yearlings. A growth index was derived to compare second-year growth increments of muskellunge and northern pike. For each fish, the calculated length (L) at annulus I was subtracted from the calculated length at annulus II, and the difference ($\triangle L$) was plotted versus the calculated length at annulus I (L_I). Regression equations were fitted to estimate the expected $\triangle L$, or $\hat{\triangle} L$, as a function of L_I for each fish. The relative growth index was defined as the ratio of observed to expected increments ($\triangle L / \hat{\triangle} L$). Average values of $\triangle L / \hat{\triangle} L$ were determined for each year-class. Using only second-year growth of two-year old fish avoided problems of small sample size of older fish, sexual dimorphism of growth of older fish, inaccurate scale readings for older fish, and differences among ages owing to Lee's Phenomenon. The equations for calculating $\hat{\triangle} L$ were $22.1 + 0.000 (L_I)$ for muskellunge and $25.2 - 0.207 (L_I)$ for northern pike.

Relative condition index. Length-weight data were treated in an analogous manner for estimates of relative condition of angler-caught fish between 40 and 70 cm total length. The natural logarithm of weight was regressed on the natural log of total length for 1049 northern pike captured between 1957 and 1977 and, separately, for 312 muskellunge captured from 1947 to 1977, inclusive. An index of relative condition was defined as the ratio of ln w (observed) to $\hat{\ln} w$ (expected).

Esocids between 40 and 70 cm were used for the index to provide suitable yearly samples over a reasonably narrow length interval. All muskellunge falling within this range were included. Northern pike, which were considerably more abundant in the angler catch, were stratified by 2.54 cm intervals from 40 to 70 cm. For each month from May to September, of each year from 1957 to 1977, up to three fish were randomly selected from each stratum for determination of relative condition. From 1964 to 1972, the 56 cm length limit restricted the harvest of northern pike to fish larger than the minimum.

Expected values of ln w for muskellunge were calculated using the equation $\hat{\ln} w = -14.663 + 3.4049 (\ln TL)$, with an R^2 of 0.954. The equation for northern pike was $\hat{\ln} w = -11.812 + 2.9659 (\ln TL)$; $R^2 = 0.934$. Length and weight data for muskellunge (N=82) and northern pike (N=290) captured during 1978 to 1982 were not used in the derivation of the regression equations, but the same equations were used to calculate the expected value of ln w for these years.

Multiple Regressions Relating Indices Of Growth And Condition To Environmental Variables

Years in which sample sizes were small were not identical for muskellunge and northern pike. Regressions were computed for years in which sample sizes for both species exceeded two fish and, additionally, for muskellunge for all years for which samples of muskellunge exceeded two fish. Regressions were computed using the relative abundance and air temperature variables in different combinations. Only the equation giving the highest multiple correlation coefficient R^2 after adjusting for degrees of freedom is reported.

Extraneous Sources of Variation in Growth Indices

Differences in size-selectivity of sampling gear (hook and line versus fyke net), seasonal variation in condition, differences in condition between large and small esocids within the 40 to 70 cm length interval, and dissimilar growth rates for stocked versus naturally-reproduced fish were considered as possible sources of variation in average growth indices among years. After examination, none of these potential confounding factors appeared to be important.

RESULTS

Associations Between Indices of Growth

Relative growth ($\triangle L / \hat{\triangle} L$) indices for yearling muskellunge and northern pike largely covaried between 1957 and 1977 (Fig. 1a). Relative growth

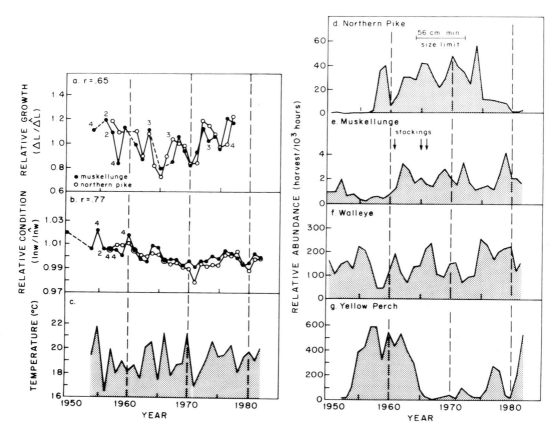

Figure 1. Long term changes in (a) relative growth and (b) relative condition of Escanaba Lake muskellunge and northern pike, (c) average July air temperature at Rest Lake, and the relative abundances of (d) northern pike, (e) muskellunge, (f) walleye, and (g) yellow perch. Numbers next to points in (a) and (b) are sample sizes for years in which the index was based on fewer than five fish (northern pike or muskellunge). Dashed lines indicate years with missing data. Note differences among scales of relative abundance.

was particularly high in 1957, 1963, 1972, and 1977 and low in 1962, 1965, 1970, and 1971. Indices for the two species diverged for the 1958, 1967, and 1976 growing seasons.

Relative condition (ln w/lnŵ) profiles are also similar between species, but the general pattern differs from that shown by relative growth indices (Fig. 1b). Relative condition of muskellunge and northern pike was high in 1960, 1961, 1964, 1965, 1976 to 1978, 1981, and 1982 and low during the latter 1960s and early 1970s and in 1979 to 1980. Minimum values occurred in 1971 for both species, but this depression was especially pronounced for northern pike. It marked the culmination of a nearly continuous decline in condition of northern pike from the 1960 maximum. Relative condition of muskellunge was nearly as low in 1962 and 1963 as in 1971, but condition of northern pike remained generally high through 1965. Correlation coefficients for the association between relative growth and relative condition were low for muskellunge ($r = 0.00$) and northern pike ($r = 0.30$).

Stomach Contents

Most of the food items in stomachs of muskellunge and northern pike collected during 1977, 1978, and 1979 were fish (Fig. 2). Many carcasses were digested beyond recognition. Among those

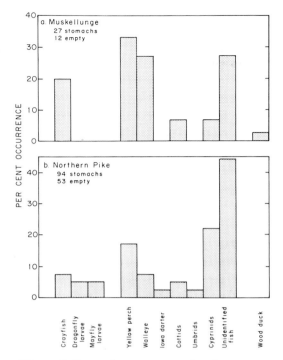

Figure 2. Percent occurrence of various organisms in stomachs of angler-caught (a) muskellunge and (b) northern pike that contained food.

that could be identified, yellow perch and walleye were most common in muskellunge, while yellow perch and several cyprinid species predominated in northern pike.

Associations Between Growth Indices and Selected Environmental Variables

Summer temperature and the abundances of northern pike, muskellunge, yellow perch and walleye were postulated as possible modifiers of esocid growth in Escanaba Lake. Historical data were used to estimate trends in these variables (Fig. 1 c-g). Correlation coefficients for the association between relative (harvest/effort) and absolute estimators of abundance were 0.77, 0.75, and 0.89 for muskellunge, northern pike, and yellow perch respectively, but only 0.06 for walleye. None of the relative abundance or temperature variables alone could account for a major fraction of the variability in relative growth indices (Table 2; Fig. 3). Relative growth indices for both species were negatively associated with the harvest/effort indicator of abun-

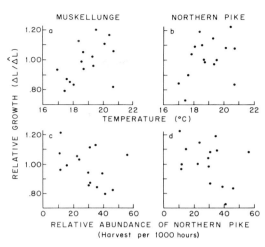

Figure 3. Relationship between relative growth ($\triangle L/\widehat{\triangle L}$) of muskellunge and northern pike and (a,b) average July air temperature, and (c,d) the relative abundance (harvest/1000 hours) of northern pike.

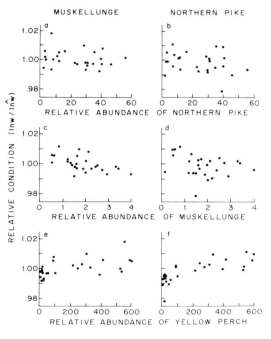

Figure 4. Relationship between relative condition ($\ln w/\widehat{\ln w}$) of muskellunge and northern pike and the relative abundances (harvest/1000 hours) of (a,b) northern pike, (c,d) muskellunge, and (e,f) yellow perch.

Table 2. Correlation matrix for relative growth of muskellunge and northern pike, relative abundances (harvest per 1000 angler hours) of muskellunge, northern pike, walleye and yellow perch, and average summer air temperatures at Rest Lake. Multiple regression equations relating relative growth indices to environmental variables also are shown. Partial t-statistics are given below the corresponding regression coefficients.

	Relative Growth		Relative Abundance (harvest/10^3 hours)				Average Air Temp. (C)		
	Musk.	N. Pike	Musk.	N. Pike	Walleye	Y. Perch	June	July	August
Relative Growth									
Muskellunge	--								
Northern pike	.65*	--							
Relative Abundance									
Muskellunge	.05	-.16	--						
Northern pike	-.46*	-.37	.13	--					
Walleye	-.01	-.21	-.37	-.49*	--				
Yellow perch	-.09	.23	-.05	-.13	-.34	--			
Average Air Temp.									
June	.10	-.23	-.02	-.13	.40	-.15	--		
July	.49*	.37	-.14	-.03	-.01	-.13	.02	--	
August	-.04	-.06	.30	-.09	-.18	-.10	-.16	.07	--
June - August	.30	-.01							

*coefficient significantly different from zero (one-sided p<0.05)

Multiple Regression Equations	Number of Years	R^2
Northern pike:		
Relative Growth = .5877 - .006347(NP) - .0010708(W) + .04017(July Temp.)	16	.46
(-2.49) (2.07) (1.63)		
Muskellunge:		
Relative Growth = .4507 - .006907(NP) + .04708(July Temp.) - .0008869(W) - .0001767(YP)	16	.55
(-2.77) (2.09) (-1.66) (-1.19)		
Relative Growth = .2542 - .004979(NP) + .05175(July Temp.) - .0006905(W)	18	.47
(-2.54) (2.27) (-1.57)		

dance of northern pike but seemingly independent of the abundance of muskellunge. Correlations with harvest/effort of walleye and yellow perch were low. The relative growth of both esocids was positively associated with average July air temperature but not with June or August average air temperatures.

Higher (absolute value) correlation coefficients were obtained for the relative condition indices (Table 3) than for the relative growth indices. Significant (P<0.05) negative associations existed between relative condition of muskellunge and northern pike and the relative abundance (harvest/effort) of muskellunge but not between relative condition and harvest/effort of northern pike (Fig. 4 a-d; Table 3). Condition of both esocids was positively associated with the indicator of relative abundance of yellow perch at low rates of yellow perch harvest (0 to 150 fish per 1000 hours), but at higher levels the relationship appeared to flatten (Fig. 4 e,f). A curvilinear model would give a better fit than the straight line model that underlies the correlation coefficient. Associations between relative condition of muskellunge and northern pike and average summer air temperatures were not stable (one sided P>0.05) (Table 3).

Several of the explanatory variables were correlated with each other. Prominent negative associations existed between harvest/effort indicators of abundance for walleye and northern pike ($r = -0.49$), walleye and muskellunge ($r = -0.37$), and walleye and yellow perch ($r = -0.34$) for the 16 years included in the relative growth analysis (Table 2). The harvest/effort of walleye was positively associated with average June air temperature ($r = +0.40$). Negative associations also existed between the harvest/effort of walleye and of northern pike ($r = -0.43$), walleye yellow perch ($r = -0.38$) and muskellunge and yellow perch ($r = -0.35$) for the 26 years included in the relative condition analysis (Table 3).

Multiple regression was used in order to consider the joint effects of these temperature and relative abundance variables. Harvest/effort of northern pike, average July air temperature, and harvest/effort of walleye were useful explanatory variables for relative growth of both muskellunge and northern pike (Table 2). Relative condition regressions

Table 3. Correlation matrix for relative condition of muskellunge and northern pike, relative abundances (harvest per 1000 angler hours) of muskellunge, northern pike, walleye and yellow perch, and average summer air temperatures at Rest Lake. Multiple regression equations relating relative condition indices to environmental variables also are shown. Partial t-statistics are given below the corresponding regression coefficients.

	Relative Condition		Relative Abundance (harvest/10^3 hours)				Average Air Temperature (C)		
	Musk.	N. Pike	Musk.	N. Pike	Walleye	Y. Perch	June	July	August
Relative Condition									
Muskellunge	--								
Northern pike	.77*	--							
Relative Abundance									
Muskellunge	-.56*	-.36*	--						
Northern pike	-.19	-.24	.00	--					
Walleye	-.10	-.27	.03	-.43*	--				
Yellow perch	.56*	.76*	-.35*	-.27	-.38*	--			
Average Air Temp.									
June	-.13	-.22	-.05	.20	.22	-.26	--		
July	-.02	-.02	-.15	-.01	.12	-.08	.06	--	
August	.03	.19	.01	-.05	-.29	.00	-.05	-.18	--

*coefficient significantly different from zero (one-sided p<0.05)

Multiple Regression Equations	Number of Years	R^2
Northern pike:		
Relative Condition = .992929 + .000026003(YP) (5.65)	26	.57
Muskellunge:		
Relative Condition = 1.00311 + .000011827(YP) - .002727(M) (2.59) (-2.50)	26	.46
Relative Condition = 1.00680 - .003480(M) + .000010379(YP) - .00006851(NP) (-3.14) (2.00) (-1.11)	29	.52

DISCUSSION

for both species included the harvest/effort of yellow perch (Table 3). Harvest/effort of muskellunge was effective in explaining some of the residual variability in relative condition indices for muskellunge, but not for northern pike. Inclusion of the relative abundance of northern pike marginally increased the adjusted R^2 for muskellunge relative condition. Signs of all regression coefficients were in the same direction as the corresponding bivariate correlation coefficients.

Chance is an unlikely explanation for the positive associations between growth indices for muskellunge and northern pike, and no sampling biases that can account for the parallel fluctuations could be identified. Apparently, these measures were indicative of real trends in growth rate of yearling esocids and condition of larger fish. If true, this implies either that growth of muskellunge depended on one or more of the same factors that governed growth of northern pike, or that growth of muskellunge and northern pike covaried due to biologically-independent associations with separate, but covarying, environmental factors. Neither possibility can be excluded, but the following observations bear on the interpretation: (1) Stomach contents of muskellunge and northern pike captured during 1977, 1978, and 1979 were similar, but sample sizes were small. Yellow perch were a common food item in stomachs of both species; (2) yearly indices of relative condition of muskellunge and of northern pike were positively associated with abundance of yellow perch, and (3) negatively associated with abundance of muskellunge; (4) relative growth of yearling muskellunge and northern pike were negatively associated with abundance of northern pike; (5) relative condition of muskellunge and northern pike were high in the late 1950s when northern pike and likely prey species both were plentiful; (6) relative growth of yearling muskellunge and northern pike were positively associated with average July air temperature; (7) correlation coefficients for these associations were not particularly high (0.35< $|r|$ <0.60), except for the association between condition of northern pike and abundance of yellow perch ($r=0.76$).

A credible explanation for these observations is that muskellunge and northern pike in Escanaba Lake utilized a shared food base (including yellow perch) and that food shortages periodically depressed growth rate and condition of individuals of both species. July air temperature also appears to have been a controlling factor. If competition is defined as "the demand of more than one organism for the same resources of the environment in excess of immediate supply" (Larkin 1956), it appears that competition for food occurred among muskellunge and northern pike in Escanaba Lake during the period covered by this study. Effects of crowding among esocids, apart from competition for food, cannot be excluded; however, it should be noted that relative condition of esocids was high during the late 1950s, when northern pike and likely prey species both were plentiful. This argues against a crowding effect independent of food availability as the primary explanation for the variation in growth indices.

No information is available concerning habitat use by muskellunge and northern pike in Escanaba Lake; for example, whether individuals of the two species are spatially segregated or occur together, or are sedentary versus wide-ranging. However, if yellow perch, or any other shared prey species, range throughout habitats frequented by muskellunge and those used by northern pike, then predation by one esocid would lower resource availability for others. Exploitation (scramble) competition (Emlen 1973) thus could occur however different the habitat use patterns of the two species.

It is unclear to what extent muskellunge and northern pike were responsible for creating food shortages. Perhaps the two species utilized similar food resources, but prey availability was largely determined by other factors, such as predation by other species or weather-induced variation in reproductive success. The synchrony of changes in growth indices might reflect a common response to a (seemingly) independently varying food supply. For example, walleye in Escanaba Lake also prey heavily on yellow perch (Thuemler 1969, cited in Kempinger et al. 1975; Morsell 1970). In most years, walleye have been more abundant, by numbers and biomass, than muskellunge and northern pike combined (Kempinger et al. 1975; Kempinger and Carline 1977). They may have been more influential in cropping the yellow perch population. This cannot be decided on the basis of available information. The lack of an association between relative condition and the relative abundance (harvest/effort) of walleye could be due to poor representation of walleye abundance. In any case, evidence that walleye consumed more yellow perch than did muskellunge and northern pike would not be evidence that competition for yellow perch did not occur between muskellunge and northern pike. It *would* establish the importance of a third, perhaps more influential, competing species. Even if predation by esocids was of minor importance in comparison to other factors affecting prey abundance, it still would have a negative effect on availability.

Yellow perch appear to have been an object of competition between muskellunge and northern pike, but other fishes may have been as well. Long-term data on abundance of cyprinids are not available. Changes in abundances of pumpkinseed, bluegill, and black crappie from 1957 to 1977 occurred parallel with fluctuations in abundance of yellow perch, and regression analysis cannot dissociate their individual associations with esocid growth indices. Centrarchids were not found in stomachs of muskellunge and northern pike captured from 1977 to 1979, but centrarchids were rare in Escanaba Lake during those years (Kempinger and Carline 1977). Pumpkinseed were much more abundant in Escanaba Lake during the late 1950's than in the 1970's and may have been more frequently eaten by esocids at that earlier time; however, Seaburg and Moyle (1964) and Johnson (1969) found that northern pike selectively preyed on yellow perch when perch and sunfish both were present. Muskellunge growth rate was positively associated with the abundance of yellow perch in two Wisconsin lakes (Gammon and Hasler 1965). The availability of yellow perch appears to have played a central role in mediating changes in growth rate and condition of esocids in Escanaba Lake.

Relative condition of large esocids was more positively associated with the harvest/effort measure of yellow perch abundance than was the relative growth of yearlings. Yellow perch of the sizes harvested by sport fishermen probably are too large to be eaten by yearling esocids and may be competitors rather than prey. Large yellow perch are cannibalistic (Eschmeyer 1938); they may compete with young esocids for small yellow perch. Large esocids and yellow perch feed on quite different particle sizes, so direct competition is unlikely and the relationship is more purely that of predator and prey. Differences between the ecological requirements of yearling esocids, which typically began their second growing seasons at total lengths of 20 to

32 cm, and the 40 to 70-cm fish might explain the low correlations between relative growth and relative condition for each species.

Harvest/effort data were used to estimate changes in relative abundance of sport fishes. As noted previously, correlations with estimates of absolute abundance were reasonably high ($0.75 < r < 0.90$) for yellow perch, muskellunge and northern pike, but low ($r = 0.06$) for walleye. The harvest/effort indicator clearly is unreliable for walleye. Even the higher correlation coefficients for the other three species, however, are not so high as to preclude bias in the harvest/effort measure. Constancy of exploitation rates from year to year would support the validity of the harvest/effort indicator of abundance; however, exploitation rates were variable for all four species in Escanaba Lake during the period 1946 to 1969 (Kempinger et al. 1975). Kempinger and Carline (1977) observed that the rate of exploitation of walleye in Escanaba Lake appears to have been higher in years in which food supply (yellow perch) and growth rate were low. Esocids also might have been more catchable, or more targeted by fishermen, when yellow perch were scarce. This would confound the relationships between indices of relative growth and condition and the harvest/effort of muskellunge and northern pike. However, the multiple regression analysis showed that harvest/effort of muskellunge was a useful explanatory variable for muskellunge relative condition, even controlling for harvest/effort of yellow perch.

Note that the harvest/effort measure is based on cumulative harvest and cumulative effort, each summed over a recording year. It is not an indicator of instantaneous relative abundance (as, for example, at the start of the growing season), because the effort is not instantaneous. It is more properly viewed as a weighted average of harvest rates (h_i/e_i) for successive short intervals of time ($\triangle t$) within the year, with weights equal or proportional to angler effort (e_i) expended during the interval; that is, harvest/effort for the year equals

$$\frac{\sum_{i=1}^{n} e_i \left(\frac{h_i}{e_i}\right)}{\sum_{i=1}^{n} e_i}$$

where n equals 365 days/$\triangle t$.

The number of fish at risk of being harvested in any given interval depends on the number present at the start of the recording year and any intervening recruitment and mortality, including mortality due to predation. A more precise analysis than is possible here would consider the seasonality of growth in relation to seasonal abundance.

The effect of esocid predation in reducing the supply of yellow perch is implicitly accounted for by the indicator of perch abundance. If the sole effect of esocid abundance on esocid growth was via the effect of esocid predation on abundance of yellow perch, then one would not expect to see a negative association between growth and esocid relative abundance after controlling for the relative abundance of yellow perch. That this was not the case may be due to undetected confounding; that is, harvest/effort of esocids may have been statistically associated with the actual controlling variable(s). Alternatively, muskellunge and northern pike might behave in ways that limit each other's growth apart from the consumption of yellow perch. The stomach analysis showed that the diets of muskellunge and northern pike included prey besides yellow perch. Competition might have occurred for one or more of these alternative prey. It is also possible that some direct behavioral interaction was involved.

Esocid density in Escanaba lake has been determined largely by the abundance of northern pike since the late 1950s. The abundance of muskellunge would seem of secondary importance in determining the intensity of demand for shared food resources. The relative growth of yearlings of each species did correlate more highly with the harvest/effort of northern pike than with that of muskellunge; however, relative condition of larger esocids was more negatively associated with the harvest/effort of muskellunge than with that of northern pike, and indicators of abundances of muskellunge and northern pike were only weakly correlated. The harvest/effort indicator of relative abundance combines all sizes large enough to be caught by sport fishermen into one category. The average weight of individual muskellunge in the angler harvest was roughly twice that of northern pike in most years (Inskip 1980). Relative abundance estimates based on all sizes combined may underrepresent the abundance of muskellunge relative to that of northern pike in the size range used for the relative condition analysis.

Ambient thermal regime is a possible modifying factor for a competitive interaction between muskellunge and northern pike. Latitudinal differences in native ranges (Crossman 1978), together with observations on thermal optima, preferenda, or distribution of northern pike (Casselman 1978) and muskellunge (Ferguson 1958; Minor and Crossman 1978; Dombeck 1979) support the hypothesis that muskellunge have a higher thermal optimum for growth. Northern pike grow well at temperatures up to 23 C (Casselman 1978), which is just below the typical maximum summer water temperature in Escanaba Lake. If muskellunge have a still higher thermal optimum for growth, one would predict that growth indices for muskellunge would show a more positive association with an appropriate index of summer water temperature.

Results here do not substantiate the hypothesis. July typically is the month of warmest air and water temperatures. If one presumes a sinusoidal pattern of warming and cooling, then average July air temperature might be a useful gross indicator of the thermal experience of fishes in Escanaba Lake. Relative growth was positively associated with average July air temperature for muskellunge ($r = +0.49$) and for northern pike ($r = +0.37$), and the two coefficients were not significantly different from each other ($P > 0.05$, $N = 16$ years). There were years in which July air temperatures were high and relative growth was low (most notably, 1970), but no years in which July air temperatures were low and relative growth was high. Inferences must be guarded because of the small sample sizes; nonetheless, results suggest that warm July air temperatures (mean > 17.5 to 18.0 C) are necessary for favorable growth of esocids in Escanaba Lake. Strong correlations between summer temperatures and relative condition would not be expected because each yearly index is an average of $\ln w / \ln \hat{w}$ for individual fish captured in different months, including some captured in late spring or early summer.

Parallels exist between the results of this study and those reported by Kipling (1983) for a long-term study of growth of northern pike (year-classes 1944 to 1978) in Windermere, England. The average weight of four-year-old female northern pike was positively associated with an index of the fishes' thermal experience (cumulative number of degree days over 14 C). However, the association was much stronger for some year classes than for others. Kipling suggested that the availability of Eurasian perch, the primary prey of northern pike in Windermere (Frost 1954), modified the effect of temperature. When perch of appropriate size (< 15 cm) were plentiful, and there was a mix of warm and cool years, conditions were permissive for the effect of temperature to be expressed, and temperature then appeared to be the most important variable controlling the growth of northern pike.

Data from Escanaba Lake suggest that competition for food occurred periodically among muskellunge and northern pike but do not indicate what the population consequences have been. Based on the harvest/effort indicator of relative abundance, however, muskellunge did not become rarer in Escanaba Lake following the establishment of northern pike. If anything, muskellunge appear to have *increased* in abundance during the 1960's relative to the 1950's.

Three factors--stocking, hybridization, and small sample sizes--complicate use of the muskellunge harvest data as a basis for inferring population consequences of competition with northern pike. The muskellunge stockings might have elevated the population to greater levels than would have occurred otherwise. The 1961 stockings contributed appreciably to the angler harvest (18 to 35%) over the ensuing five years, but the 1965 and 1966 stockings did not (Kempinger 1972). The relative abundance (harvest/effort) of muskellunge still had not dropped to pre-1960 levels by 1982, long after the last stocking. Apparently, natural reproduction sustained the muskellunge population during this period. At the same time, however, natural hybridization was occurring between muskellunge and northern pike, and muskellunge harvest records include the hybrid fish. Kempinger (1972) reported that 30 out of 97 of the "muskellunge" handled during spring nettings between 1965 and 1970 were hybrids. The percentage of hybrids does not appear to be increasing. Between 1978 and 1982, only 18 out of 168 of the "muskellunge" harvested by anglers were noted as being hybrids.

These muskellunge data for Escanaba Lake are too meager to permit an evaluation of temporal changes in mortality rate or age at first reproduction as was done by Kempinger and Carline (1978) for northern pike. Hoff and Serns (1986) reported an average annual mortality rate of .345 for age 2 to age 13 muskellunge in Escanaba Lake based on the aggregated muskellunge harvest for the years 1946 to 1981; this is less than the lowest annual mortality estimate (0.59) for ages 2 to 7 northern pike given by Kempinger and Carline (1978) for a subset of seven of these years.

Whatever the history of the muskellunge population might have been in the absence of northern pike and with no stocking, the observed pattern for Escanaba Lake does not conform to the sequence of events reported for other northern Wisconsin lakes, where sharp increases in abundance of northern pike appear to have been followed by declines in native muskellunge populations. Muskellunge were not particularly common in Escanaba Lake even before northern pike became established. Characteristics of Escanaba Lake apart from its northern pike population appear to be more important in limiting muskellunge population size.

ACKNOWLEDGMENTS

We are grateful to the Wisconsin Department of Natural Resources and, in particular, to DNR personnel involved with the Northern Highland Fishery Research Project for their support and cooperation during this study. S. Serns, J. Kempinger, O. Baccus and M. Hoff were extremely helpful in providing data, samples and information on Escanaba Lake. We thank S. Serns, J. Kempinger and B. Reynolds for reviewing drafts of this paper. C. Hughes drew the illustrations. This research was supported, in part, from funds supplied by the Federal Aid in Fish Restoration Act, under Dingell-Johnson project F-83-R and National Science Foundation Long Term Ecological Research Program Grant Number DEB 8012313.

REFERENCES

Becker, G.C. 1983. Fishes of Wisconsin. The University of Wisconsin Press, Madison, Wisconsin, USA.

Black, J.J., L.M. Andrews, and C.W. Threinen. 1963. Surface water resources of Vilas County. Wisconsin Conservation Department, Madison, Wisconsin, USA.

Carbine, W.F. 1943. Egg production of the northern pike, *Esox lucius* L., and the percentage survival of eggs and young on the spawning grounds. Papers of the Michigan Academy of Sciences, Arts and Letters 29: 123-137.

Casselman, J.M. 1978. Effects of environmental factors on growth, survival, activity and exploitation of northern pike, *Esox lucius*. American Fisheries Society Special Publication 11:114-128.

Crossman, E.J. 1978. Taxonomy and distribution of North American esocids. American Fisheries Society Special Publication 11:13-26.

Dombeck, M.P. 1979. Movement and behavior of the muskellunge determined by radio-telemetry. Wisconsin Department of Natural Resources Technical Bulletin 113, Madison, Wisconsin, USA.

Elson, P.F. 1940. Rearing maskinonge in a protected area. Transactions of the American Fisheries Society 70:421-429.

Emlen, J.M. 1973. Ecology: An evolutionary approach. Addison-Wesley Publishing Company, Reading, Massachusetts, USA.

Eschmeyer, W. 1938. Further studies of perch populations. Papers of the Michigan Academy of Sciences, Arts and Letters 23:611-631.

Ferguson, R.G. 1958. The preferred temperature of fish and their midsummer distribution in temperate lakes and streams. Journal of the Fisheries Research Board of Canada 15:607-624.

Franklin, D.R., and L.L. Smith. 1960. Notes on development of scale patterns in the northern pike, *Esox lucius*, L. Transactions of the American Fisheries Society 89:83.

Frost, W.E. 1954. The food of pike, *Esox lucius* L., in Windermere. Journal of Animal Ecology 23:339-360.

Frost, W.E., and C. Kipling. 1958. The determination of the age and growth of pike (*Esox lucius* L.) from scales and opercular bones. Journal du Conseil 24:314-341.

Gammon, J.R., and A.D. Hasler. 1965. Predation by introduced muskellunge on perch and bass, 1: Years 1-5. Transactions of the Wisconsin Academy of Sciences, Arts and Letters 54:249-272.

Hoff, M.H., and S.L. Serns. 1986. The muskellunge fishery of Escanaba Lake, Wisconsin under liberalized fishing regulations, 1946-1981. American Fisheries Society Special Publication 15:249-256.

Hourston, A.S. 1952. The food and growth of the masquinonge (*Esox masquinongy* Mitchill) in Canadian waters. Journal of the Fisheries Research Board of Canada 8:347-368.

Hunt, B.P., and W.F. Carbine. 1951. Food of the young pike, *Esox lucius* L., and associated fishes in Peterson's ditches, Houghton Lake, Michigan. Transactions of the American Fisheries Society 80:67-83.

Inskip, P.D. 1980. Annual and seasonal variability in growth of sympatric muskellunge (*Esox masquinongy* Mitchill) and northern pike (*Esox lucius* Linnaeus). M.S. thesis, University of Wisconsin–Madison, Madison, Wisconsin, USA.

Inskip, P.D. 1986. Negative associations between abundances of muskellunge and northern pike: Evidence and possible explanations. American Fisheries Society Special Publication 15:135-150.

Inskip, P.D., and J.J. Magnuson. 1983. Changes in fish populations over an 80 year period: Big Pine Lake, Wisconsin. Transactions of the American Fisheries Society 112:378-389.

Johnson, L.D. 1969. Food of angler-caught northern pike in Murphy Flowage. Wisconsin Department of Natural Resources Technical Bulletin 42, Madison, Wisconsin, USA.

Johnson, L.D. 1971. Growth of known-age muskellunge in Wisconsin and validation of age and growth determination methods. Wisconsin Department of Natural Resources Technical Bulletin 49, Madison, Wisconsin, USA.

Johnson, L.D. 1981. Comparison of muskellunge (*Esox masquinongy*) populations in a stocked lake and unstocked lake in Wisconsin with notes on the occurrence of northern pike (*Esox lucius*). Wisconsin Department of Natural Resources Research Report 110, Madison, Wisconsin, USA.

Kempinger, J.J. 1972. Muskellunge fishing on Escanaba Lake with liberalized fishing regulations, 1946-1971. Unpublished report. Wisconsin Department of Natural Resources, Madison, Wisconsin, USA.

Kempinger, J.J., and R.F. Carline. 1977. Dynamics of the walleye (*Stizostedion vitreum vitreum*) population in Escanaba Lake, Wisconsin, 1955-72. Journal of the Fisheries Research Board of Canada 34:1800-1811.

Kempinger, J.J., and R.F. Carline 1978. Dynamics of the northern pike population and changes that occurred with a minimum size limit in Escanaba Lake, Wisconsin.

American Fisheries Society Special Publication 11:382-389.

Kempinger, J.J., W.S. Churchill, G.R. Priegel, and L.M. Christenson. 1975. Estimate of abundance, harvest, and exploitation of the fish population of Escanaba Lake, Wisconsin 1946-69. Wisconsin Department of Natural Resources Technical Bulletin 84, Madison, Wisconsin, USA.

Kipling, C. 1983. Changes in the growth of pike (*Esox lucius*) in Windermere. Journal of Animal Ecology 52:647-657.

Lagler, K.F. 1956. Freshwater fishery biology. William C. Brown Company, Dubuque, Iowa, USA.

Larkin, P.A. 1956. Interspecific competition and population control in freshwater fish. Journal of the Fisheries Research Board of Canada 13:327-342.

Minor, J.D., and E.J. Crossman. 1978. Home range and seasonal movements of muskellunge as determined by radiotelemetry. American Fisheries Society Special Publication 11:146-153.

Morsell, J.W. 1970. Food habits and growth of young-of-the-year walleye from Escanaba Lake. Wisconsin Department of Natural Resources Research Report 56, Madison, Wisconsin, USA.

NOAA (National Oceanographic and Atmosphere Administration). 1954-1982. Climatological Data: Wisconsin. Volumes 59-87. National Climatic Center, Asheville, North Carolina, USA.

Oehmcke, A.A. 1951. Muskellunge yearling culture and its application to lake management. Progressive Fish Culturist 13:63-70.

Priegel, G.R., and D.C. Krohn. 1975. Characteristics of a northern pike spawning population. Wisconsin Department of Natural Resources Technical Bulletin 86, Madison, Wisconsin, USA.

Seaburg, K.G., and J.B. Moyle. 1964. Feeding habits, digestive rates, and growth of some Minnesota warmwater fishes. Transactions of the American Fisheries Society 93:269-285.

Threinen, C.W., and A.A. Oehmcke. 1950. The northern pike invades the musky's domain. Wisconsin Conservation Bulletin 15:10-12.

Thuemler, T.F. 1969. The food habits of walleye *Stizostedion vitreum vitreum* (Mitchill), in Escanaba Lake, Vilas County, Wisconsin. M.S. thesis, University of Wisconsin--Milwaukee, Milwaukee, Wisconsin, USA.

Weithman, A.S., and R.O. Anderson. 1977. Survival growth, and prey of Esocidae in experimental systems. Transactions of the American Fisheries Society 106:424-430.

Williams, J.E. 1955. Determination of age from scales of northern pike (*Esox lucius* L.). Doctoral dissertation, University of Michigan, Ann Arbor, Michigan, USA.

Prey Selection by Three Esocid Species and a Hybrid Esocid

ROBERT ENGSTROM-HEG

New York Department of Environmental Conservation
Stamford, New York 12167

RICHARD T. COLESANTE

New York Department of Environmental Conservation
Constantia, New York

GERALD A. STILLINGS

Indiana Department of Natural Resources
Avoca, Indiana

ABSTRACT

Laboratory experiments indicated that muskellunge-northern pike hybrids tended to be bottom-oriented in their behavior, and to be highly selective feeders, eating white suckers in preference to golden shiners, and golden shiner in preference to yellow perch. Northern pike showed a similar pattern, but selection between suckers and shiners was less clear-cut. Muskellunge were not significantly selective, and appeared to be mid-water feeders. Chain pickerel tended to be surface feeders and to select against white suckers.

In describing the feeding habits of the larger North American Esocidae, Scott and Crossman (1973) emphasize the point that they are relatively unselective feeders; they accept as food practically anything that is alive and of the right size, including fish, amphibians, reptiles, birds and mammals. Nevertheless, esocids are primarily piscivorous, and not totally unselective. Beyerle and Williams (1968) found that northern pike would select cyprinid and catostomid prey before spiny-rayed fishes, and spiny-rayed fishes over bullheads. Fathead minnows and lake chubsuckers which "gathered near the bottom", were selected in preference to golden shiners that tended to school at mid-depth. Northern-pike selected small centrarchids over larger ones. Weithman and Anderson (1977), in studies of northern pike, muskellunge and their F1 hybrid, found that all three esocids would select clupeid or cyprinid prey in preference to centrarchids, and centrarchids over channel catfish. Beyerle (1978) reported that northern pike, tended to deplete stocked populations of minnows but not of bluegills in small Michigan Lakes.

The present studies were directed toward determining whether there were differences in feeding behavior and food selection of muskellunge, northern pike, their F1 hybrid, and chain pickerel. If such differences occurred among these esocids, we were interested in their implications for management of these fish and for their possible use in panfish control.

METHODS

Individual esocids used in these tests ranged from 25 to 40cm total length. Muskellunge and hybrids were entirely aquarium-reared fish; northern pike and chain pickerel were aquarium-acclimated fish with some natural foraging experience. All esocids had been fed at least a month on fish, predominantly golden shiners, and then fasted for 24 hours before each test. Four or five individuals of each species were placed in covered $1.94m^3$ fiberglass tanks maintained at 15-18 C. Equal numbers (10-16) of weighed and measured fish of three prey species were then added to the tanks. Prey species used in various tests included golden shiners, white suckers, yellow perch, pumpkinseed, and brown bullhead. Individuals ranged from 45 to 90 mm total length. Since the tanks provided essentially no cover or shelter, prey fish were accessable to predation at all times.

Prey fish remaining and those presumed eaten were counted, usually at 24-hour intervals for up to 7 days. Tests were terminated at the end of the 24-hour period during which the prey community had become reduced to a single species. Given the limitations on time and tank space, this design had an advantage over one in which prey fish were replaced upon being eaten, in that it would clearly show any transfer of predatory pressure from a preferred to a less preferred prey species as the former were reduced or eliminated.

Results were analyzed by 24 or 48-hour periods using Ivlev's (1961) electivity index, $(r_i - p_i)/(r_i + p_i)$, where r_i is the proportion of a food item in the diet of a predator, and p_i is its proportion in the potential prey community. Application of this index to a laboratory test involves a difficulty in that p_i changes during the course of the test as fish are eaten. We used the mean of the initial and final population of each species to estimate their relative abundance in the population during a given period. An arithmetic mean was appropriate inasmuch as the esocids tended to consume prey fish at a fairly constant daily rate. Reduction of prey numbers was more nearly linear than exponential.

It should be noted that E does not have a symmetrical distribution. Where there is no selection ($r_i = p_i$), E has a value of zero. Where a potential prey species is not eaten ($r_i=0$), E has a value of -1.00. A value approaching +1.00 would be possible only if the predator fed exclusively or nearly so, on a species that formed a miniscule proportion of the potential prey population ($r_i=1.0$, $p_i \to 0$). Where a prey species constitutes 1/3 of the community, as at the beginning of one of our tests, the maximum value of E is $(1.00 - 0.33)/(1.00 + 0.33) = 0.50$. Positive values near 0.50 are indicative of a high degree of selection.

Statistical significance of the results was determined by using the chi-square test in two ways. The first (Tables 1 thru 4) tested whether the cumulative numbers eaten at the end of each time period differed significantly from the original proportion. Where three prey species were used, chi-square values were determined for each pair of species (1df) and for the three species (2df). The second (Figs. 1 thru 5) tested the hypothesis that during a given period, the relative occurrence of prey species in the diet did not differ by more than chance from its proportion in the community.

RESULTS

Predator Behavior

All esocids exhibited the strike behavior described by Webb and Skadsen (1980) for the northern pike - muskellunge hybrid. Northern pike and hybrids tended to be bottom oriented, and to stalk individual isolated prey. They would often pursue after a missed strike. They showed little interest in schooled fish if dispersed fish were present. Muskellunge appeared to be more of a lurking than a stalking predator. They travelled less, spent more time hovering at mid-depth, made shorter strikes, and rarely pursued after a miss. The behavior of the chain pickerel was similar to that of the muskellunge except that they more often hovered near the surface and struck at surface-swimming fish.

Prey Selection

Hybrids (Table 1, Figs. 1 and 2) - When confronted with white suckers, golden shiners and yellow perch, hybrids consistently ate most of the suckers before eating significant numbers of shiners, and most of the shiners before starting on perch. In the first two tests, they experienced some difficulty swallowing perch, which averaged slightly larger than the suckers and appreciably larger than the golden shiners. In the third test, where the sizes of all prey fish were similar, however, the results were essentially the same. In an attempted test with white suckers, golden shiners and brown bullheads, the esocids attempted to feed on the bullheads, but could not swallow them because of their pectoral spines. After the first day, they ceased striking at them. Suckers were again selected over shiners. Two tests were run with white suckers, golden shiners, and pumpkinseeds. In the first test, the order of selection was (1) white suckers, (2) golden shiners (3) pumpkinseeds, but selection against pumpkinseeds was less marked than for yellow perch. The second test failed to show significant selectivity.

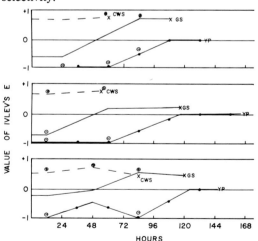

Figure 1. Electivity indices, $E=(r_i-p_i)/(r_i+p_i)$, for northern pike - muskellunge hybrids confronted with white suckers (CWS), golden shiners (GS) and yellow perch (YP). Plus signs + indicate r_i significantly greater than p_i; minus signs − indicate r_i significantly less than p_i (chi-square, 1df, .05).

Table 1. Summary of prey selection experiments with northern pike-muskellunge hybrids.

Date	Prey[a]	Number	Weight (g)	Length (mm)	Mean E		Chi-Square Significance Level[b]			Species Preferred
							48hr	96hr	Final	
10/13/74	CWS	10	2.9	66	+0.70	CWS-GS	NS	NS	NS	CWS
	GS	10	1.0	53	−0.35	GS-YP	NS	.05	NS	GS
	YP	10	4.5	81	−1.00	CWS-YP	.01	NS	NS	CWS
						Combined	.01	.01	.01	CWS
11/20/74	CWS	16	3.6	74	+0.56	CWS-GS	.01	.01	NS	CWS
	GS	16	1.5	58	+0.08	GS-YP	NS	NS	.02	GS
	YP	16	4.4	81	−0.81	CWS-YP	.01	.01	.01	CWS
						Combined	.01	.01	.01	CWS
12/9/74	CWS	15	5.9	86	+0.51	CWS-GS	.05	.02	NS	CWS
	GS	15	4.6	84	+0.09	GS-YP	NS	NS	.01	GS
	YP	15	4.4	84	−0.72	CWS-YP	.01	.01	.01	CWS
						Combined	.01	.01	.01	CWS
1/8/75[c]	CWS	15	4.7	84	+0.36	CWS-GS	.05	NS	NS	CWS
	GS	15	4.1	81	−0.40					
1/14/76	CWS	15	4.1	76	+0.28	CWS-GS	NS	NS	NS	CWS
	GS	15	1.7	61	−0.13	GS-PSS	NS	NS	NS	GS
	PSS	15	1.5	50	−0.25	CWS-PSS	NS	NS	.05	CWS
						Combined	NS	NS	NS	CWS
2/7/76	CWS	15	5.2	84	+0.05	CWS-GS	NS	NS	NS	Uncertain
	GS	15	2.0	69	+0.05	GS-PSS	NS	NS	NS	GS
	PSS	15	1.8	48	−0.13	CWS-PSS	NS	NS	NS	CWS
						Combined	NS	NS	NS	Uncertain

[a] CWS = white sucker
GS = golden shiner
PSS = pumpkinseed
YP = yellow perch
NS = not significant

[b] Chi-square is based on null hypothesis of cumulative selection by species at end of period.

[c] Brown bullheads were present but not eaten.

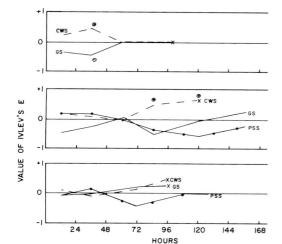

Figure 2. Electivity indices for northern pike-muskellunge hybrids. Top graph: prey were white suckers (CWS) and golden shiners (GS). Brown bullheads, also present, were not accepted as food. Lower two graphs: prey were white suckers, golden shiners, and pumpkinseeds (PSS).

Northern pike (Table 2, Fig. 3) - Northern pike selected prey species in the same order as the hybrids, but selectivity was not as strong, and was not statistically significant for white suckers and golden shiners.

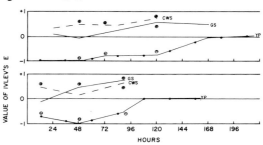

Figure 3. Electivity indices for northern pike confronted with white suckers (CWS), golden shiners (GS) and yellow perch (YP).

Muskellunge (Table 3, Fig. 4) - In the first two tests, muskellunge did not select significantly between white suckers and golden shiners, and did not

Table 2. Summary of prey selection experiments with northern pike.

Date	Prey	Number	Weight (g)	Length (mm)	Mean E		48 hr	96 hr	Final	Species Preferred
							Chi-Square Significance Level			
10/7/75	CWS	15	1.5	58	+0.42	CWS-GS	NS	NS	NS	CWS
	GS	15	1.6	58	+0.34	GS-YP	NS	.01	.0a	GS
	YP	15	1.8	59	−0.74	CWS-YP	.01	.01	.01	CWS
						Combined	.01	.01	.01	CWS
9/28/76	CWS	15	1.9	58	+0.41	CWS-GS	NS	NS	NS	CWS
	GS	15	1.9	61	+0.31	GS-YP	NS	.01	.01	GS
	YP	15	1.9	60	−0.94	CWS-YP	.05	.01	.01	CWS
						Combined	.05	.01	.01	CWS

Table 3. Summary of prey selection experiments with muskellunge.

Date	Prey	Number	Weight (g)	Length (mm)	Mean E		48hr	96hr	Final	Species Preferred
							Chi-square Significance Level			
11/13/74[a]	CWS	10	0.5	48	−0.04	CWS-GS	NS	NS	NS	Uncertain
	GS	10	1.1	41	+0.02					
12/9/74[a]	CWS	15	0.9	58	−.23	CWS-GS	NS	NS	NS	GS
	GS	15	1.5	50	−.02					
10/21/76	CWS	15	2.1	61	−0.06	CWS-GS	NS	NS	NS	Uncertain
	GS	15	1.7	61	−0.05	GS-YP	NS	NS	NS	GS
	YP	15	2.0	61	−0.24	CWS-YP	NS	NS	NS	CWS
						Combined	NS	NS	NS	Uncertain

[a] Yellow perch, mean weight 4.3 grams, were present but not eaten.

accept yellow perch as prey, probably because of their larger size. In a third test using smaller perch there was no significant selection, though perch had the lowest mean E value.

Chain pickerel (Table 4, Fig. 5) - Chain pickerel did not select significantly between golden shiners and pumpkinseeds, but did select significantly against white suckers. It is not clear from these tests whether this was because of the larger size of the suckers, or their bottom-dwelling habit.

DISCUSSION

The feeding behavior of northern pike-muskellunge hybrids in these tests corresponded closely with that of northern pike as described by Beyerle and Williams (1968). The hybrids here tended to be more selective than the northern pike, but this may simply reflect differences between individual fish. Observed behavior was quite similar. Muskellunge were clearly less selective and less bottom-oriented. Chain pickerel appeared to be surface-oriented and selected against bottom-dispersed fish. In a sympat-

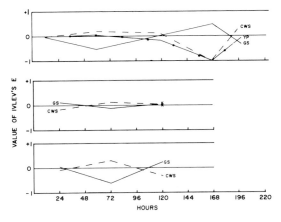

Figure 4. Electivity indices for muskellunge confronted with white suckers (CWS), golden shiners (GS) and yellow perch (YP). Two tests in which perch were not eaten are treated as two-species trials. No values of r_i differed significantly from p_i.

Table 4. Summary of prey selection experiments with chain pickerel.

Date	Prey	Number	Weight (g)	Length (mm)	Mean E		Chi-Square Significance Level			Species Preferred
							48hr	96hr	Final	
1/14/76	CWS	15	4.2	76	-0.54	CWS-GS	NS	NS	NS	GS
	GS	15	1.6	61	+0.32	GS-PSS	NS	NS	NS	GS
	PSS	15	1.8	51	+0.12	CWS-PSS	NS	NS	.05	PSS
						Combined	NS	NS	.05	GS
2/7/76	CWS	15	4.9	84	-0.57	CWS-GS	.01	.01	.01	GS
	GS	15	2.2	68	+0.46	GS-PSS	NS	NS	NS	GS
	PSS	15	1.8	50	+0.01	CWS-PSS	NS	NS	NS	PSS
						Combined	.01	.01	.01	GS

ric situation, the three esocids would probably occupy and feed in slightly different habitats.

Beyerle and Williams (1968) predicted that abundant soft-rayed forage would act as a buffer between northern pike and panfish, and that maximum growth of northern pike would occur in situations where there were large populations of both large and small soft-rayed food fishes (minnows and suckers, for example). Results of the current tests indicate that this prediction can be extended to the pike-muskellunge hybrid, but that it is less applicable to muskellunge or chain pickerel. Hybrids have been found to be less vulnerable to angling than northern pike, though slightly more so than muskellunge (Beyerle 1973, Weithman and Anderson, 1976). It appears likely that their relative vulnerability will tend to increase as anglers learn to use lures, baits and techniques that take advantage of their distinctive feeding behavior.

Prey selection in esocids depends on behavior of the prey as well as of the predator. In the laboratory situation, northern pike and hybrids selected bottom-dispersed fish over those that tended to school in mid-water, and soft-rayed over spiny-rayed fish. Bullheads were initially attractive, but were rejected because of their pectoral spines. All other prey species tested were eaten by all esocids, but usually not until almost all of any strongly preferred species had been consumed. The combined effect of size and the presence of spines on the prey fish probably influenced food preference, but did not prevent consumption except in the case of the bullheads and of the perch presented to muskellunge in the first two tests (Table 3). Hybrids selected white suckers over other prey species when the suckers were the largest prey fish present (Table 1). Feeding strategies of northern pike and hybrids differed markedly from those of muskellunge and chain pickerel. It appears

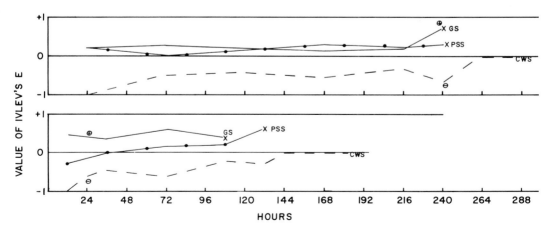

Figure 5. Electivity indices for chain pickerel confronted with white sucker (CWS), golden shiners (GS) and pumpkinseeds (PSS).

likely that these differences would persist in the wild, though they would be modified somewhat by the presence of aquatic vegetation and other cover, and by seasonal abundance of prey fish of suitable size. The differences should be considered when esocids are stocked wholly, or in part, to control panfish population density. Stocking of hybrids or northern pike to control perch or centrarchid populations might be less immediately successful if there were large numbers of suckers available. Muskellunge and chain pickerel might be more effective predators where there are dense weed beds, northern pike and hybrids where there is a higher proportion of open water that would allow pursuit of prey and a better view of the bottom.

Where stocking is intended to provide a sport fishery for esocids, better results will be attained by stocking them in waters with an abundance of their preferred prey. One criterion for stocking hybrids in New York is that receiving waters should support a population of soft-rayed fish, preferably suckers.

REFERENCES

Beyerle, G. B. 1973. Comparative growth, survival, and vulnerability to angling of northern pike, muskellunge, and their hybrid tiger muskellunge stocked in a small lake. Michigan Department of Natural Resources Fisheries Research Report.

Beyerle, G. B. 1978. Survival, growth and vulnerability to angling of northern pike and walleyes stocked as fingerlings in small lakes with bluegills or minnows. American Fisheries Society Special Publication No. 11:135-139, Washington, D.C. U.S.A.

Beyerle, G. B. and J. E. Williams, 1968. Some observations of food selectivity by northern pike in aquaria. Transactions of the American Fisheries Society 97:28-31.

Ivlev, V. S. 1961. Experimental Ecology of the Feeding of Fishes. Yale University Press, New Haven, Connecticut, U.S.A.

Scott, W.B. and E.J. Crossman. 1973. Freshwater Fishes of Canada. Bulletin 184. Fisheries Research Board of Canada, Ottawa, Canada. 966 pp.

Webb, P. W. and J. M. Skadsen 1980. Strike tactics of *Esox*. Canadian Journal of Zoology 58:1462-1469.

Weithman, A.S. and R. O. Anderson. 1976. Angling vulnerability of Esocidae. Southeastern Association of Game and Fish Commissioners Thirthieth Annual Conference. 99-102

Weithman, A. S. and R. O. Anderson. 1977. Survival, growth and prey of Esocidae in experimental systems. Transactions of the American Fisheries Society 106:424-430.

MANAGEMENT CONSIDERATIONS

Musky Management — A Changing Perspective From Past To Present

DELANO R. GRAFF, Chief

Division of Fisheries
Pennsylvania Fish Commission

The modern perspective on muskellunge is to regard them as a very desirable sport fish managed for their trophy value. Muskies are considered something special and held in high esteem by both anglers and fishery managers. There is indeed--to borrow a phrase from author Larry Ramsell--a "Muskie Mystique" (Ramsell 1982). This has not always been the case. The attitude of anglers and the perspective of fisheries scientists relative to muskies and their management have changed a great deal from past to present. Muskies, as with many other sport fish, were once far more abundant than they are now. For much of the 19th Century and into the early 20th Century, muskies were regarded, at best, as something to kill and eat and, at worst, voracious predators to be disposed of before they ate all the "good" fish.

Very early in the game of man versus nature in North America our forefathers came to the same conclusion as thousands of anglers have since: hook and line is one of the least effective ways to capture muskies. This, however, was not a deterrent to these inventive exploiters of natural resources. Muskellunge were shot, speared, seined, and taken in pound nets, fyke nets, and trap nets. Over most of their range they were harvested commercially until the early 1900s and, in some areas, into the 1930s. Techniques such as winter spearing and netting during spawning season were commonly used for commercial harvest.

The effect of virtually unrestricted exploitation during the 19th and 20th Centuries was compounded by the prevailing attitude that expansion of industry and agriculture was far more important than a few fish that moved into a marsh or frequented a local river. Actions such as ditching and draining of wetlands and swamps for agriculture or allowing industrial wastes to enter streams and rivers were considered natural consequences of progress. Draining, siltation, and pollution contributed to a substantial loss of habitat for muskies, but no one considered this of much, if any, economic significance. Trautman (1957) speculated that the destruction of habitat was so extensive in some areas that, even if there has been no commercial fishing or no period of unrestricted exploitation, muskellunge populations would still have declined dramatically.

Whatever the causes, it is a fact that muskellunge were reported to be abundant in the 1800s and they were among the early species to show a noticeable decline in abundance (Kirtland 1851). That decline led to the first attempts at restoration and set into motion the management trends of today.

The first efforts at muskellunge management consisted of stocking artificially propagated fry to bolster natural populations. Some states have been stocking muskellunge for a long time: New York started in 1887 in Chautauqua Lake; Wisconsin started in 1899; Pennsylvania made a few stockings in the 1890s but gave it up because propagation was too difficult; Ontario started in 1927; and Pennsylvania came back to muskellunge propagation in 1953, as did Ohio (Buss 1960). Since then, the practice has spread.

Fry stocking was practiced in the early stages of musky management due to difficulty in producing larger-size fish. The availability of large numbers of hatchery-produced fry led fishery managers to introduce muskellunge into new waters (new in the sense that they had not, historically, held muskies). This brought an interesting and challenging social element into muskellunge management. There were some popular misconceptions that muskies would eat all the bass and walleyes; that each muskie was a veritable eating machine that could destroy thousands of other fish.

Popular writers--and perhaps a few technical writers--had given muskies a colorful reputation as a ferocious killer, the "Tiger of the Lily Pads" sort of thing. Outdoor writers wrote things like, "I know of no better way of ruining good fish than the introduction of this vicious killer." David Starr Jordan called esocids "mere machines for the assimilation of other organisms" (Jordan and Evermann 1923). (The idea died hard.) The point is perhaps best made by the title of a 1952 publication, *The Pike of Quebec: The Ogres of Our Waters* (Legendre 1952). Fishery managers of the 1950s also contributed to the problem by stressing the positive fishery management role of large esocids as a means of controlling stunted fish populations. This had an unintended negative effect because public perception was, if it eats panfish, it'll eat bass!

At the same time that some managers were having problems with what muskellunge would do to other species, other fishery scientists were beginning to wonder what effect other species, particularly northern pike, were having on muskellunge. Managers had often noticed that as muskellunge populations were declining northern pike populations were increasing. There are sound reasons for an assumption of competition between northern pike and muskies, with the advantage going to northerns because of earlier spawning and more aggressive feeding habits. A 1950 article from Wisconsin succinctly expressed what was emerging as a major concern in musky management, *The Northern Invades The Muskie's Domain* (Threinen and Oehmcke 1952).

By the 1940's and 1950's there were also increasing concerns about the role of angling in the decline of muskies. There had been a rather generally accepted belief that muskies were so difficult to catch that sport fishing was not a factor in the decline of populations. To be sure, seasons, size, and creel limits were imposed as early management policies, but often the size limits were low (e.g., 24 inches), and the creel limits by today's standards were rather liberal.

At the onset of the 1960's--and what can be considered the modern era of musky management--fisheries professionals were dealing with:
1. A fish that had declined appreciably from historical levels of abundance.
2. A fish that presented considerable difficulty to fish culturists and was not always available for stocking in the sizes and numbers that fishery managers might prefer.
3. A fish that existed at relatively low densities as an adult and was difficult to sample reliably with nets, except during spawning.
4. A fish that was being managed primarily through stocking but with some question as to the contribution hatchery fish were making to total populations.
5. A fish that was considered an effective predator with the capability of influencing the composition of fish communities.
6. A fish that was perhaps at a competitive disadvantage when sharing habitat with northern pike.
7. A fishery which was increasing in popularity among better-informed and better-equipped anglers and whose impact on population structures was coming into question.

These perspectives set the stage for modern trends in musky management, and from this point in the early 1960's, what has changed? There is still concern about the decline of naturally produced muskellunge populations; but the real change is in the appreciation of the complexity of the problem, of the wide array of factors that interact to contribute to the decline. There has been a remarkable change in the relationship of fish culture and musky management and also in the manager's ability to assess the contribution of hatchery fish to musky populations. Musky culture has advanced to the point where the culturist can reliably provide the numbers and sizes of fish requested by the manager. Pond, or extensive culture, was established as a reliable source of hatchery muskellunge through the widespread adoption and development of techniques used in Wisconsin (Johnson 1958); intensive culture was pioneered in Pennsylvania a short time later in the early 1960's (Sorenson et al. 1966). Both techniques provide fish in the numbers and sizes needed.

A major change in both culture and management resulted from the advance of intensive culture to production of artificial diets (Graff and Sorenson 1970). The use of artificial diets has not made a substantial contribution to culture of purebred muskies, but it has significantly increased the role of tiger muskies (the hybrid northern pike x muskellunge) in musky management.

A new approach to musky management--widespread use of tiger musky fingerlings in stocking programs, particularly in waters that were not, historically, "pure musky" waters--developed as a result of advances in fish culture techniques; fishery managers also identified a more prominent role for the hybrid, moving it from an occasional part to a major constituent of some musky fisheries. The manager's perception of the contribution of hatchery fish to musky populations has changed dramatically in the past 20 or 25 years. There is little disagreement now that the single most effective management effort has been the maintenance of musky populations through stocking of 6-inch or larger fingerlings.

Today's question is not does stocking pay, but

how can stocking be made to pay better? The importance of stocking is acknowledged, but it is also recognized that, in many instances, a substantial percentage of stocked fingerlings are lost (unaccountable mortality) shortly after stocking (Johnson 1975). Different stocking techniques--cove stocking, night stocking, shoreline plantings, open water plantings--do not produce consistent increases in survival.

Fishery managers are aware of the role of stocking due to advances in the ability to sample musky populations. Development and improvement of electrofishing gear enable fishery managers to sample immature and adult fish; the techniques used by Leon Johnson in Wisconsin influenced ways to sample and assess populations. In addition to muskies, fishery managers can also better sample associated fish populations.

Attitudes about the impact of muskies on other fish populations have changed. Muskies do eat a variety of other fishes, but they do not eliminate other gamefish populations. With good estimates on population densities from food conversion and stomach samples, the muskellunge is no longer perceived as a veritable vacuum cleaner, moving through the water ingesting other fish by the thousands. It's generally accepted by anglers and professional workers alike that muskies select prey based on abundance, availability, and size. In large systems, muskies are only one component of a community, and they are not going to perform fishery management miracles by eating all the undesirable or slow-growing fish.

Substantial progress has been made on predator-prey questions. Managers are not as concerned now about what muskies eat as about what eats muskies! Northern pike are still very much a concern, but there are conflicting opinions: Are pike the problem always, sometimes, or only in certain habitat? Are they a symptom of habitat change? What ecological factors determine the results of interaction between northerns and muskies? No one is thrilled about pike in musky water, but the question is still open. Maybe pike are not the only predator or competitor. Some work in Ohio has implicated largemouth bass as the culprit in the loss of stocked tiger muskies (Stein et al. 1982).

Technology in other fields (in addition to electrofishing) has contributed to changing perspectives in musky management. Computers, able to handle difficult, multivariant analyses and multiple linear regressions, have enabled researchers to look at all the factors involved in muskellunge reproduction in a way not possible 20 years ago. Radio telemetry--skillfully used as a research tool by Dr. Crossman and his colleagues--has provided new insight into muskellunge movement and how and when muskies use various habitats (Minor and Crossman 1978).

This growing appreciation of the importance of critical aspects of musky habitat may enable us to do more than preserve such things as spawning areas (to prevent further habitat loss); it may provide managers with the ability to enhance or alter habitat to actually restore features crucial to muskellunge survival (M. Dombeck, personal communication).

In all the change of attitudes and perceptions, perhaps none has been more striking than that of the manager's and angler's perspective on angling. Angling (harvest) is now acknowledged to be an important factor in determining the size and age structure of musky populations. It is a major component of mortality in larger and older fish, shifting the population to younger fish. Relatively low creel limits and a size limit of at least 30 inches are the "order of the day" for modern musky managers. Perspective here is clouded by the hybrid (tiger) musky's vulnerability to angling being much higher than the purebred fish. What may be appropriate management of hybrids may not be the best approach for purebreds.

Anglers are also changing. No longer is every legal 'lunge killed and the head nailed on the barn door—grisly evidence of angling prowess. Catch-and-release, a philosophy which stresses the unharmed release of muskies and puts strict constraints on harvest, is growing in popularity. Muskies, Inc., is espousing this philosophy; it's refreshing and may have important implications for future management strategies. Trout managers have long incorporated catch-and-release as a tool for maintaining a high quality fishery; it's conceivable we will see the same approach for management of exceptional muskellunge fisheries.

Perception today is that the muskellunge is a very important animal; in many waters, the economically most important. It is so important that it justifies the bringing together of representatives of research, management, administration, and the angling public to meld their findings and observations into a synthesis of the best information available on muskies, a striking change in perception of the importance of an animal that was once shot, clubbed, speared, and netted! Muskies are like the lady in the cigarette ad, "You've come a long way, baby!"

REFERENCES

Buss, K. 1960. The muskellunge. Pennsylvania Fish Commission Special Purpose Report, Bellefonte, PA, USA.

Graff, D. R. and L. Sorenson. 1970. The successful feeding of a dry diet to esocids. Progressive Fish-Culturist 32:31-35.

Johnson, L. D. 1958. Pond culture of muskellunge in Wisconsin. Wisconsin Conservation Department Technical Bulletin 17 (54 pp), Madison, Wisconsin, USA.

Johnson, L. D. 1975. How many muskies aren't there anymore? Wisconsin Conservation Bulletin 40:20-21.

Jordan, D. S. and B. W. Evermann. 1923. *American Food and Game Fishes*. Doubleday, New York. 574 pp.

Kirtland, J. P. 1851. Esox estor Les. maskallonge. Cleveland Family Visitor. Cited by Trautman, M. B. 1957. *The Fishes of Ohio,* Ohio University Press, Columbus. 683 pp.

Legendre, V. MS. 1952. The pike of Quebec: The ogres of our waters. Quebec Dept. Fish Game Off. Biol. 42 pp. Cited by Crossman, E. J. and C. D. Goodchild. 1978. *An Annotated Bibliography of the Muskellunge Esox masquinongy (Osteichthyes: Salmoniformes)*. Royal Ontario Museum, Ontario. 131 pp.

Minor, J. D. and E. J. Crossman. 1978. Home range and seasonal movement of muskellunge as determined by radiotelemetry. American Fisheries Society Special Publication 11:146-158.

Ramsell, L. A. 1982. A compendium of muskie angling history, Echo Printing, Alexandria, Minnesota, USA.

Sorenson, L. K., K. Buss and A. D. Bradford. 1966. The artificial propagation of esocid fishes in Pennsylvania. Progressive Fish-Culturist 28:133-141.

Stein, R. A., R. F. Carline, and R. S. Hayward. 1982. Largemouth bass predation on stocked tiger muskellunge. Transactions of the American Fisheries Society 110:604-612.

Threinen, C. W. and A. Oehmcke. 1952. The northern invades the musky's domain. Wisconsin Conservation Bulletin 15, Madison, Wisconsin, USA.

Trautman, M.B. 1957. *The Fishes of Ohio.* Ohio University Press, Columbus. 683 pp.

Philosophy of Muskellunge Management

PAUL J. WINGATE

Minnesota Department of Natural Resources
500 Lafayette Road
St. Paul, Minnesota 55146

ABSTRACT

Major areas of past, present, and future muskellunge management philosophy are briefly described with suggested directions for future management. Most muskellunge management has been to maintain or produce a trophy fishery, to protect or restore an endemic fishery, to diversify angling opportunity and/or to control nongame fish populations. These goals have been attained through various techniques which include: regulations, special stamps, stocking, and promoting natural reproduction, various genetic strains and hybrid muskellunge. Future management needs include more reliable estimates of harvest, pressure and economic values of the muskellunge fishery, much of which could be voluntarily supplied by anglers. Muskellunge genetics and the role each available strain might have in management plans may well be the single most important area in future muskellunge management.

The rationale for muskellunge management, both present and past, has been to maintain or produce a trophy fishery, to protect or restore an endemic fishery, to diversify angling opportunity and/or to control nongame fish populations. All of these are valid concepts still in use today. Many management efforts have been successful and goals have been attained without completely knowing why something worked or the direction in which we are going. We need to be evaluating our objective(s), how to reach them and how effective we are in reaching them. A more concerted effort in this direction would be beneficial for all managers and anglers.

DISCUSSION

There is no precise definition of what constitutes a trophy muskellunge. A trophy means different things to different anglers, and his perception of the fish on his hook on a given day is often influenced by his experiences, successes and the size limits. For the novice, any muskellunge is a trophy, while for the veteran angler the minimum size for a trophy increases with experience and past catches. For many anglers, especially the novice, the minimum-size limit denotes the beginning of a trophy size, while other anglers believe size regulations have little to do with trophy size or biological needs. Only with higher minimum-size limits do the two concepts start to become synonymous.

One angler's trophy may be another angler's runt, which needs to be released to grow and be caught again another day. To many muskellunge anglers, a fish is not considered a trophy until it reaches 50 inches in length. In addition, these anglers also realize that a catch rate of one fish per 100 fishing hours of any size muskellunge is a good fishery. Veteran anglers can help the novice angler to develop a different trophy concept that involves catch and release.

Musky anglers can also aid in dispelling the belief among many nonmusky anglers that muskellunge eat everything in the lake and are thus responsible for poor fishing for other species. These attitudes need to be rectified to ease implementation of muskellunge management measures such as lake designation, restocking and new introductions.

Regulations

Harvest regulations have been the primary management technique used to reach and maintain the desired goal of a trophy fishery. Bag limits, closed seasons, gear restrictions, technique restrictions and minimum-length limits are among the regulations utilized. Bag limits are employed to better distribute the harvest, to limit the harvest of trophy fish and to provide a goal for fishermen (to attain the creel limit with large fish). Closed seasons protect spawning adults, coincide with fishing seasons for other species and reduce the time available for exploitation. Gear and technique restrictions include the number of lines allowed and whether an angler is permitted to troll. Length limits nominally protect immature fish and may allow mature adults to spawn at least once, restrict harvest until trophy size is reached (require fishermen to fish for "trophies"), prevent over-exploitation and provide for maximum growth. They can be made to coincide with northern pike size limits and traditional past management practices and to mollify pressure from user groups.

Some of the above regulations have been promulgated and used without any biological basis or follow up studies to determine whether they were effective. Some of the studies initiated to answer questions on regulations have not been definitive because of inadequate sample sizes. Minimum-size limits less than 30 inches have been used, which do not allow muskellunge to reach sexual maturity before being legally available for harvest; nor do they allow for differential growth rates between sexes. Larger minimum-size limits often protect male muskellunge throughout their life cycle when they could realistically be harvested by anglers. Minimum-size limits appear to promote the catch-and-release ethic among novice anglers and may allow the survival of more fish to this limited size.

Regulations designed to enhance a trophy fishery must take into consideration the productivity of the watershed, latitude (length of growing season), forage base, habitat and the population status of the muskellunge. Different regulations may be necessary on different lakes or on stocked versus natural muskellunge populations, to provide a trophy fishery.

A slot size limit for muskellunge might be an appropriate way to allow novice anglers to keep smaller fish, especially males, protect the prime spawning sizes and allow the harvest of trophy fish (mostly females). Another method to allow the harvest of males, while still protecting females to trophy size, would be to allow smaller minimum-size limits for males than for females. This now appears possible with an external sexing technique being developed by Crossman and Lebeau (personal communication).

Stamps

In recent years, there has been a proliferation of special fishing-license stamps to benefit a particular species, and pressure is developing in some areas to have a muskellunge stamp in addition to the regular fishing license. Though expensive to the angler, it might be worthwhile to promote such a stamp, even though there will be added management problems, e.g. should everyone be required to purchase a stamp, or only those fishing specifically for muskellunge, and how do you determine who they are? This would particularly be a problem with the large number of catch and release muskellunge anglers. With many other species in a lake, this kind of stamp would create an enforcement problem.

The major purpose of a stamp is to provide more revenue for management of a particular species. If there is a large demand for a species, such as walleye in Minnesota, this would create few additional problems, but if the number of anglers for a species is comparatively small, then the management agency could be spending more revenue than the stamp would generate. Other anglers will probably react negatively to a small group receiving proportionately large sums for their favored species, or to a large number of lakes being designated for that species.

As an example, Minnesota has approximately 5,000 musky fishermen out of 2 million licensed anglers; it spends $50,000 annually on muskellunge stocking and has designated 26 lakes totaling 35,000 acres as muskellunge waters. An additional 200,000+ acres contain muskellunge. In comparison with the number of fishermen, this is a large amount of water available for muskellunge fishing. A muskellunge stamp will certainly call attention to these conditions, and in the end, the musky angler will probably lose, if he persists in pushing for a special stamp for his sport. The benefit-cost ratio does not have to be positive for all management programs but the consequences with respect to adverse publicity and ensuing controversies should be carefully considered before their implementation.

Stocking

Stocking is a tool that has been extensively used in muskellunge management. The cost effectiveness of past and present muskellunge rearing and stocking has varied among management agencies and has depended on egg source, rearing facilities, necessity of stocking, type of waters stocked, return of stocked fish to the creel, survival of stocked fish, and angler interest (Miller 1983). The cost of rearing muskellunge to fry or advanced fingerling size varies from as much as $17.84 to as little as $0.55 per fish (Ragan 1983).

Reliable estimates of catch rates, total harvest, muskellunge mortality rates, angler usage, and angler expenditures are not readily available in most agencies, and 60 percent of the state agencies questioned about harvest and angler expenditures had no available data (Ragan 1983). The average musky angler harvests 0.01 to 1.2 fish per year with most estimates being toward the lower end of the scale (Ragan 1983). Until these figures are available, it will not be possible to precisely determine cost effectiveness of muskellunge stocking programs. Over the years, muskellunge management programs have seldom shown a positive benefit-cost ratio, but recent work indicates some changes in a positive direction (Belusz and Witter 1986; Menz and Wilton 1983).

The goal of many agencies in stocking muskellunge is to reintroduce the species where they have been extirpated due to changing environmental conditions or over-exploitation. Care should be exercised to stock the same strain which was originally present to increase the chances of recreating a naturally reproducing population. If the original strain is not available, a closely related one would be the next best option. This is the ideal situation since one muskellunge management goal is to cease stocking as soon as the population is able to sustain itself. With introduced populations outside their native range, this is often not possible.

Stocking to sustain a trophy population is often not cost effective, unless there is very high usage. Stocking on top of a natural population is seldom productive and usually a waste of resources; however, supplemental stocking seems to be worthwhile in situations where the natural population is declining and/or very heavily exploited. Survival of stocked fish is usually an unknown entity, which directly influences the cost and the success of the program (Axon 1978).

Stocking can give the manager better control over the numbers of fish present which enables him to control population size as well as fishing pressure. By and large, the angler does not perceive a difference between natural and stocked fish unless different strains are used (Ohio, Mississippi River and Wisconsin strains appear phenotypically different), or different maximum sizes are attained. In cases where stocking is necessary to sustain a muskellunge population, other factors, such as deteriorating habitat or northern pike competition, should be investigated to discover the need for this stocking. In many cases, there will not be a muskellunge fishery without stocking, but this tool should be used with caution. Naturally reproducing populations should remain the major goal.

Genetics

In recent years, there has been renewed emphasis on genetic research. It was once thought there were several species of muskellunge, then races, then all the same. We now know that there are vast differences in growth rates, intraspecific relationships, habitat preferences, spawning area preferences, distribution, and maximum sizes attained. Many of these differences may be genetic. These factors need to be considered when evaluating waters to be stocked and goals to be attained.

Much research needs to be done to locate and further delineate the available stocks and strains of muskellunge before they are further diluted or eliminated by continued indiscriminate stocking or by managing muskellunge waters for other species. Management programs must be changed to reflect this new knowledge or we are shirking our professional responsibilities. Armed with this information, managers may be able to select the most appropriate strain to create a trophy fishery or to add diversity. Caution, however, must be exerted by the manager to preserve the genetic integrity of known muskellunge strains (Koppelman and Philipp 1986).

At some stage in the genetic studies of muskellunge diversity, it might be necessary or advisable to set up a genetic clearing house to catalogue the available strains and their locations throughout North America. Management agencies must be aware of what strains they have and of any possible genetic contamination that they may be introducing into their populations. Fish stocking with different strains does not always produce the desired result.

Hybrid Muskellunge

Hybrid or tiger muskellunge have been used for many years by state agencies in their esocid management programs. Tiger muskellunge are easier and less expensive to culture, provide a better return to the angler's creel, have better growth rates and apparently do not reproduce. They can also control nongame fish, provide species diversity in lakes and provide both a muskellunge and northern pike fishery in urban fishing situations and in chemically rehabilitated lakes.

Some anglers have voiced concern over the use of tiger muskellunge because of genetic dilution of the pure-strain muskellunge. They are also concerned about financial dilution of the pure strain muskellunge program. There is an existing philosophy that a tiger muskellunge cannot be a trophy fish. While these concerns may have some validity, in some situations the tiger muskellunge appears to be more successful than the pure strain. Management of high-use urban lakes and rehabilitated lakes, where fish diversity may be more important than fishing for an occasional trophy, are two such examples. On the other hand, ice fishing and the extensive use of live bait work against the effectiveness of these hybrids because of the ease with which they succumb to angling. In some cases, the hybrid appears to better withstand competition from northern pike than do pure muskellunge, and it may be more readily accepted by the public where there is concern over the stocking of muskellunge, since it appears not to reproduce. Restraint should be used by all managers who may be considering the stocking of tiger muskellunge. In the right situation, they can be an important addition, but they should not replace a pure-strain muskellunge management program.

SUMMARY

Certain areas should be emphasized in future muskellunge management programs. Better methodology is needed to effectively sample the relatively few large fish in a large body of water; better estimates of fishing pressure, harvest and economic values of the program would be helpful, as would more voluntary angler information on catch and release mortality; regulations require evaluation from biological and social viewpoints and angler acceptance of regulations needs further study; hatchery and stocking techniques should be refined to decrease management costs; life history investigations on both pure-strain and tiger muskellunge would improve management; habitat protection must be accelerated; and delineation and evaluation of muskellunge strains need further work.

The question to be asked is whether present management philosophies are consistent with what is wanted and can reasonably be expected. Do all management programs have to have a positive benefit:cost ratio? Does the addition of an alternative trophy species (muskellunge) justify a negative benefit:cost ratio? Perhaps the objective in muskellunge management should be to set desirable and realistic goals without worrying about the benefit:cost ratio. Perhaps the creation of a new, diverse trophy fishery, or protecting an endemic population, is worth the additional cost. Each agency, with or without input from muskellunge anglers, must make this decision and manage accordingly.

While we know a great deal about the muskellunge, much work remains to be done. The priority for managing muskellunge in many resource agencies is not very high due to the paucity of information about muskellunge fishermen, muskellunge life history, the trophy nature of the fish and misconception by some of the public of the fish's voracity. Future work must be accomplished by cooperating anglers, academicians and resource agencies to find the needed solutions to better manage this trophy sport fish in a manner that is financially reasonable and publicly acceptable. I feel this symposium has shown how these three groups working together can make a significant contribution toward reducing the information gap that presently exists. With continued cooperation, we will be able to better clarify our goals and research direction for future muskellunge management.

ACKNOWLEDGEMENTS

I would like to express sincere appreciation to the Symposium Steering Committee for making available abstracts and papers of the Symposium participants and to R.F. Strand, D.H. Schupp and J.M. Hoenig for extensive reviews and editing. Appreciation is extended to J.R. Axon, D.R. Graff and C.E. Johnson who critically reviewed the manuscript before the Symposium.

REFERENCES

Axon, J.R. 1978. An evaluation of the muskellunge fishery in Cave Run Lake, Kentucky. American Fisheries Society Special Publication Number 11:328-333.

Belusz, L.C., and D.J. Witter. 1986. A survey of muskellunge angler characteristics, expenditures and benefits. *In* G.E. Hall and M.J. Van Den Avyle, editors. Reservoir Fisheries Management: Strategies for the 80's. Proceedings Reservoir Fisheries Management Symposium. Lexington, Kentucky, June 14-16, 1983.

Koppelman, J., and D.R. Philipp. 1986. Genetic applications in muskellunge management. American Fisheries Society Special Publication 15:111-121.

Menz, F.C., and D.P. Wilton. 1983. An economic study of the muskellunge fishery in New York. New York Fish and Game Journal. 30:12-29.

Miller, M.L. 1983. The status of muskellunge management in North America. Results of a mail survey conducted during fall 1981. Ohio Department of Natural Resources, Division of Wildlife. Mimeo.

Ragan, J. 1983. Future research and management needs for the muskellunge and hybrid muskellunge. North Dakota Game and Fish Department. Mimeo.

A Comparison of Muskellunge and Hybrid Muskellunge in a Southern Wisconsin Lake

DALE A. BREGE

Department of Natural Resources
Box 383
Montello, Wisconsin 53949

ABSTRACT

A comparison of age and growth, return to the creel, and harvest size characteristics was made for muskellunge and hybrid muskellunge in Little Green Lake, Wisconsin. Most hybrids reached 30 inches by their fourth summer while most muskies attained that length by their fifth summer. Since 1963, in a voluntary musky registration program, an average of 108 muskellunge and 125 hybrid muskellunge have been recorded annually. Anglers have registered 11.6% of the musky stocked and 14.6% of the hybrid musky stocked; their average size was 33.1 and 31.6 inches, respectively. Total harvest was 2.03 lb/acre for musky and 1.98 lb/acre for hybrids. Exploitation rates were 26.6% for musky and 33.3% for hybrid musky. Live bait was used to catch 55.1% of the hybrids while artificial baits accounted for 87.0% of the musky caught. Since 1972, 147 muskellunge and 128 hybrids have been caught and released. Due to the hybrid's greater vulnerability to angling, muskellunge may be the better choice for stocking intensively-fished waters.

In their efforts to expand the traditional range of muskellunge fishing opportunities, fish managers have based their decisions relative to esocid stocking mainly on survival of stocked fish and propagation costs.

Beyerle (1973) compared growth, survival, and angling vulnerability of northern pike, muskellunge, and hybrid muskellunge resulting from fingerlings stocked in a 14-acre Michigan lake; he found that northern pike had the highest survival but slowest growth rate, hybrid musky were intermediate in both survival and growth rate, and muskellunge had the lowest survival but the fastest growth rates. He concluded that the northern pike were 2.7 times more vulnerable to angling than hybrid muskellunge. Weithman and Anderson (1977) found that survival of young-of-the-year esocids was 24% for muskellunge, 58% for northern pike and 74% for hybrid musky.

In Wisconsin, Klingbiel (1981) determined that rearing of muskellunge fingerlings cost $11.96/lb as opposed to $7.61/lb for hybrid fingerlings, primarily because the muskies were reared on live minnows and the hybrids on dry pellets. Hesser (1978) stated that the sterile tiger musky (hybrids) are often utilized because they are more adaptable to hatchery techniques.

Other than these short term experiments on small bodies of water, there are no available data that provide a basis for comparison of growth, survival to adult, angler exploitation, and harvest size characteristics for musky versus hybrid musky. The management of Little Green Lake, by the Wisconsin Department of Natural Resources (DNR), has been conducive to comparisons between these two fish. This paper reports the results of data collection and analysis over a 20-year period.

STUDY SITE

Little Green Lake, located in southeastern Wisconsin (Figure 1), is a fertile, eutrophic lake with a surface area of 466 acres, a maximum depth of 24 feet and a mean depth of 10 feet. Average total alkalinity, conductivity, and pH are approximately

Figure 1. Map of Wisconsin showing the original distribution of muskellunge (dotted) and the location of Little Green Lake (arrow and dot).

124 ppm, 242 micromhos/cm and 8.1, respectively. Thermal stratification occurs from July through September; the hypolimnion has little dissolved oxygen by late summer. Secchi disk readings can be as high as 10.0 feet, but periodic summer algae blooms reduce readings to as low as 0.5 feet and raise pH readings to near 9.0. Dense areas of aquatic macrophytes, primarily *Potamogeton crispus*, are present. Excessive algae and macrophyte growths are chemically treated by the local lake association.

Little Green Lake was chemically treated with toxaphene in July 1956 to eradicate a carp and bullhead problem. The lake was restocked that year with walleye, largemouth bass, smallmouth bass, muskellunge, and panfish. Hybrid muskellunge were first introduced in 1970.

Early muskellunge stocking rates varied; however, quotas since 1978 have been fairly consistent with annual stockings of 400 musky and 800 hybrid musky fingerlings (Table 1). Maintenance stocking rates for the State of Wisconsin are generally two large fingerling per acre, biennially, but due to the scarcity of musky waters in southern Wisconsin and high angler use of the lake, stocking rates for Little Green Lake were elevated to the current quotas. Fish management surveys have not indicated any evidence of natural musky reproduction.

Table 1. Summary of musky and hybrid musky stocked in Little Green Lake, WI, 1957-1982.

Year	Musky Number	Musky Size	Hybrid musky Number	Hybrid musky Size
1957	300	8.7"	0	-
1958	1100	6-7"	0	-
1959	1000	4-6"	0	-
1960	1500	8-13"	0	-
1961	384	5-12"	0	-
1964	2000	9-13"	0	-
1966	3250	7-9"	0	-
1967	1040	9-11"	0	-
1968	1200	7-9"	0	-
1969	1200	13-15"	0	-
1970	1200	9-11"	2604	13-15"
1971	1200	10-13"	0	-
1972	0	-	1200	13-15"
1973	500	11-13"	800	12-15"
1974	400	10-12"	828	10-15"
1975	200	10-12"	800	12-15"
1976	400	12-15"	826	12-15"
1977	1040	10-12"	800	8-10"
1978	400	8-10"	800	8-10"
1979	436	8-10"	800	8-10"
1980	400	11-13"	800	10-12"
1981	400	8-10"	800	8-10"
1982	400	7-10"	800	7-10"
Subtotals:				
1957-1969	12,974	-	0	-
1970-1982	6,976	-	11,858	-
Totals	19,950		11,858	

Harvest regulations in effect at Little Green Lake during the study period included a 30-inch minimum size limit and a bag limit of one per day. In 1983, however, Wisconsin raised the minimum size limit to 32 inches.

METHODS

A voluntary muskellunge registration program for anglers was initiated in 1963. Resort owners on the lake kept charts on which anglers recorded such information as name and address; fish length, weight, and species; date, time, and location caught; bait used; fin clip; and if the fish was kept or released. Beginning in 1983, cooperators were asked to take scale samples.

The latest DNR fishery survey was conducted in 1978. Fyke nets were fished four days in April (72 net lifts total); four seine hauls with a 1600-feet, 3/4-inch mesh, were made in early May; and three complete shoreline trips were made with a 230-volt DC electrofishing unit on three nights in late April. During these operations, 159 musky and 82 hybrid musky (includes legal and sublegal fish) were tagged with numbered Floy T-anchor spaghetti tags.

Harvest information on muskellunge numbers and size was collected using the voluntary angler-registration program. Angler exploitation was calculated from tag returns. The legal-size musky population was estimated prior to the 1978 season opening by using Bailey's modified Petersen formula for mark and recapture studies as described by Ricker (1975). The mortality rate for true musky was determined by calculations from a catch curve (Ricker 1975). The population size and mortality rate for hybrids were not calculated due to insufficient numbers of captured fish. Age and growth data of angler-caught fish was calculated from marked known-age fish.

RESULTS

Data from all field operations in 1978 indicated a substantial difference in the ability to sample hybrids and muskellunge (Table 2). Fyke nets were most effective for sampling legal-size musky; of the total number of legal-size musky sampled (137), 102, or 74%, were taken in fyke-netting operations. Seine hauls were effective for collecting sublegal hybrids and legal musky; electrofishing was inefficient at sampling either hybrids or muskies, regardless of size. The large number of sublegal hybrids taken in seine hauls was composed of 40.0% Age I fish, 56.2% Age II, and 3.8% Age III and older.

Table 2. Hybrid and true muskellunge caught with different sampling gear during 1978 from Little Green Lake, WI.

Fish	Size	Fyke nets	Seine	Electrofishing
Muskellunge	Legal	102	26	9
Muskellunge	Sublegal	6	2	5
Hybrid	Legal	7	7	5
Hybrid	Sublegal	3	73	7
	Total	118	108	26

None of the three methods was efficient at sampling legal-size hybrids. Although 18 legal hybrids were tagged, none were recaptured during subsequent field sampling efforts. A large population of legal hybrids was considered to be present in the lake since anglers reported the harvest of 139 during the 1978 fishing season. Apparently, a lack of spawning tendencies and less movement in shallow water areas during April and early May makes the hybrid muskellunge a difficult fish to sample. In a Pennsylvania lake, Hesser (1978) also had difficulty in trapnetting tiger muskellunge even though a significant population was known to exist.

Growth of stocked muskies has been good with hybrid muskellunge reaching 30 inches at Ages III-V and true muskellunge at Ages III-VI. The growth rates shown in Table 3 would be somewhat biased for Age III and IV muskies and Age III hybrids because only the faster-growing fish would have been of legal size. Most of the hybrid muskies reached 30 inches by their fourth summer of growth and the majority of muskellunge by their fifth summer. The fastest growth noted for hybrids came from their first planting (1970) when two fish reached 37.0 and 38.2 inches, respectively, at Age IV. In a previous study at Little Green Lake, legal-sized hybrids were reported at Age II+ (Hacker 1973).

A catch curve was calculated for muskies Age V-VIII using data from the 1978 fyke net samples (Congdon 1979). The total annual mortality was estimated at 44.1% ($r = 0.88$). Insufficient numbers of hybrid musky were sampled to obtain a mortality estimate; however, based on harvest data and their more active behavior, the mortality rate of the hybrid was assumed to be somewhat higher than for muskellunge.

The population of legal-sized musky present prior to the opening of the 1978 fishing season was estimated to be 460 fish or approximately one per acre. Hybrid population size was not estimated due to insufficient sample numbers; however, based upon stocking rates and angler harvest in succeeding years, the hybrid musky population was probably at least equal to that of the muskellunge population.

During the field operations of 1978, 128 legal-sized muskies and 18 legal-sized hybrids were tagged. Tag returns by anglers in 1978 were 34 (26.6%) for musky and 6 (33.3%) for hybrids (Congdon 1979). Although resort owners were conscientious about the accuracy of their records, there were undoubtedly a few anglers who did not register their fish; therefore, the exploitation rates are probably minimal values. However, there is good agreement between the percent tag returns, the 1978 population estimates, and the reported angler harvest for 1978. Using the 26.6% exploitation rate for musky and the legal-sized population estimate of 460 muskies the number of legal fish expected to be harvested in 1978 was 122; the actual number registered was 114.

From 1963 through 1982, a total of 2,168 muskellunge, were registered in the voluntary creel census for an average of 108 per year and a range of 64 to 202 (Table 4). During this period, anglers registered 11.6% of the total number of musky stocked. Musky stocked after 1979 were not included in the total since the majority of fish would have not yet entered the legal size fishery. The highest exploitation of any one year-class was reported by Hacker (1973, updated) for the 1960 year-class when 27.3% were creeled.

From 1972 through 1982, a total of 1,378 angler-caught hybrid musky were registered for an average of 125 per year and a range of 20 to 339. The low 1975 figure was expected since the majority of the 1970 year class had already been caught and no hybrids were stocked in 1971. The only harvest of hybrid musky recorded by year-class at Little Green

Table 3. Average Length (in) of hybrid and true muskellunge from Little Green, WI; sample sizes are in parentheses.

Age Group	Hybrid	True Musky
I	*12.4 (15)	-
II	*19.6 (71)	-
III	29.7 (34)	31.0 (6)
IV	32.3 (14)	31.5 (7)
V	34.5 (5)	32.0 (26)
VI	36.7 (4)	33.5 (26)
VII	-	36.8 (11)
VIII	-	38.4 (9)
IX	-	39.6 (4)
X	44.0 (1)	42.8 (4)
XI	-	43.9 (4)
XII	-	43.9 (3)
XIII	-	45.8 (1)

*Data taken from scale samples of 1978 survey; all other data taken from angler-registered, known-age fish.

Table 4. Muskellunge catch at Little Green Lake, WI 1963-1982.

Year	Variety	Number caught	Length $\bar{x}$ (in)	Weight $\bar{x}$ (lb)	Largest (in)	Total weight	Lb per acre
1963-1971	Musky	979	33.2	9.4	46.0	9138	2.18
1972	Musky	202	32.7	8.5	45.8	1718	3.69
	Hybrid	167	30.9	6.4	33.0	1070	2.30
1973	Musky	116	34.0	10.1	45.0	1173	2.52
	Hybrid	339	31.6	7.4	36.0	2511	5.39
1974	Musky	70	32.6	7.8	47.0	546	1.17
	Hybrid	35	32.3	7.5	38.3	263	0.56
1975	Musky	95	32.4	8.2	47.5	780	1.67
	Hybrid	20	32.3	7.4	40.0	148	0.32
1976	Musky	94	32.7	7.9	47.5	743	1.59
	Hybrid	130	31.4	7.2	35.8	937	2.01
1977	Musky	81	32.4	8.2	39.8	664	1.42
	Hybrid	234	31.4	7.4	38.9	1733	3.72
1978	Musky	114	33.5	8.4	41.8	957	2.05
	Hybrid	139	32.6	7.9	39.0	1100	2.36
1979	Musky	52	33.9	9.2	42.0	479	1.03
	Hybrid	68	32.1	7.1	38.0	483	1.04
1980	Musky	89	34.4	9.8	43.0	873	1.87
	Hybrid	104	31.3	7.7	38.2	802	1.72
1981	Musky	121	33.7	8.9	41.8	1078	2.31
	Hybrid	49	32.5	8.0	39.5	392	0.84
1982	Musky	155	34.0	9.0	44.8	1398	3.00
	Hybrid	93	31.5	7.5	40.0	699	1.50
1972-82 Totals or Ave.	Musky	1189	33.3	8.8	-	10,409	2.03
	Hybrid	1378	31.6	7.4		10,138	1.98

Table 5. Number of fish in various size-classes of muskellunge recorded through voluntary angler registration, Little Green Lake, WI 1972-1982.

Year	Variety	30-33.9	34-37.9	38-41.9	≥42	Total
1972	Musky	161	33	4	4	202
	Hybrid	167	0	0	0	167
1973	Musky	68	30	15	3	116
	Hybrid	320	19	0	0	339
1974	Musky	59	6	3	2	70
	Hybrid	31	3	1	0	35
1975	Musky	75	14	0	6	95
	Hybrid	17	2	1	0	20
1976	Musky	74	17	2	1	94
	Hybrid	124	6	0	0	130
1977	Musky	61	19	1	0	81
	Hybrid	212	21	1	0	234
1978	Musky	72	34	8	0	114
	Hybrid	114	24	1	0	139
1979	Musky	26	22	3	1	52
	Hybrid	56	11	1	0	68
1980	Musky	52	26	9	2	89
	Hybrid	88	13	3	0	104
1981	Musky	72	38	11	0	121
	Hybrid	37	11	1	0	49
1982	Musky	107	39	7	2	155
	Hybrid	86	5	2	0	93
Total	Musky	827 (69.5%)	278 (23.4%)	63 (5.3%)	21 (1.8%)	1189
	Hybrid	1252 (90.9%)	115 (8.3%)	11 (0.8%)	0 (0%)	1378

Lake was for 1970 when 20.8% of the hybrids stocked were registered in the voluntary creel census. Anglers have harvested 14.6% of the total number of hybrids stocked since 1972. Harvest at Little Green Lake averaged 2.03 lb/acre for musky and 1.98 lb/acre for hybrids. Earlier, Hacker (1973) reported musky harvest from this lake averaged 2.6 lb/acre over a 10-year period; however, since the introduction of hybrids, the combined musky and hybrid harvest has averaged 4.01 lb/acre from 1972-1982 (Table 4).

The average size of muskellunge harvested from Little Green Lake was 33.3 inches and 8.8 pounds, whereas, the average hybrid taken was 31.6 inches and 7.4 pounds (Table 4). The largest muskellunge taken measured 47.5 inches and the largest hybrid musky was 40.0 inches. Most of the hybrid muskies were harvested soon after they reached legal size. Anglers recorded 90.9% of the hybrids in the 30-33.9 inch size-class with only 0.8% in the 38-41.9 inch-class; none were 42 inches or larger (Table 5). Muskellunge had more large fish registered; there were 69.5% in the 30-33.9 inch class, 5.3% in the 38-41.9 inch range and 1.8% were 42 inches or larger.

Hybrids appeared to be more susceptible to live bait fishing than muskies. From 1972-82, 55.1% of the hybrids were caught on live bait and 44.9% on artificial baits. Only 13.0% of the muskies have been taken on live bait, whereas, 87.0% were taken on artificial baits (1963-82).

The first muskellunge recorded as released was a 31-inch fish in 1967. Since 1972, there have been 147 musky and 128 hybrids released, which represents 12.4% and 9.3% of their catch, respectively. Release percentages have been steadily increasing over the last 10 years; since 1972, they have ranged from 4.2% (1975) to 23.9% (1982) for muskellunge and from 0% (1975) to 25.0% (1980) for hybrids. Survival of released fish is dependent on hooking severity, angling stress, and handling. Weithman and Anderson (1976) estimated that hooking mortality of released yearling northern pike, muskellunge, and hybrid muskellunge caught with artificial lures from experimental ponds was negligible (1.7%); however, fishing with live bait (primarily suckers) has long been blamed for observed mortalities on Little Green Lake.

MANAGEMENT IMPLICATIONS

Fishermen on Little Green Lake have accepted the hybrid muskellunge as a trophy sport fish. The hybrid appears to offer some advantages over the true musky in producing an acceptable sport fishery. The hybrid costs less to rear, has a faster initial growth rate, and appears to provide a better return to the creel. Since the hybrid musky is sterile, the population is entirely dependent on stocking; if the introduction of the hybrid musky creates an undesired effect, it can be eliminated by discontinuing the stocking. The faster growth rate seems to be a temporary advantage; the hybrid does reach legal size one year earlier, but this growth advantage does not continue much beyond Age V, at which age most of the hybrids are already harvested. If the management of a lake is aimed at consistently producing fish of a larger trophy size, then the muskellunge might be a wiser choice, especially in intensively fished waters. Harvest data by size class from Little Green Lake showed a definite advantage of the musky in producing larger fish when confronted with heavy angling pressure.

Exploitation and harvest data showed that hybrids are harvested soon after they reach legal size. The vulnerability of hybrids to live sucker baits is considered important. It may be desirable at some future time to establish an "artificials only" regulation which would likely result in decreasing sublegal hybrid-hooking mortality, a lowering of the hybrid's vulnerability, and increasing the average size of hybrids harvested.

ACKNOWLEDGEMENTS

The earlier work or Vern Hacker, Michael Primising, James Congdon, and all the assistants who worked on Little Green Lake is gratefully acknowledged. Sincere appreciation is extended to the resort owners and fishermen who cooperated in recording the muskellunge data. Critical review of the manuscript was provided by David Hanson, Gordon Priegel, and John Klingbiel. The author wishes to thank Lois Rehse and Margaret Fischbach for typing the manuscript.

REFERENCES

Beyerle, G.B. 1973. Comparative growth, survival, and vulnerability to angling of northern pike, muskellunge, and the hybrid tiger muskellunge stocked in a small lake. Fish Research Report 1799, Michigan Department of Natural Resources, Lansing, Michigan, USA.

Congdon, J.C. 1979. Fish Management Survey of Little Green Lake, Green Lake County. Wisconsin Department of Natural Resources, Madison, Wisconsin, USA.

Hacker, V.A. 1973. The results of a ten year voluntary muskellunge creel census at Little Green Lake, Green Lake County, Wisconsin, 1963-1972. Fish Management Report 58, Wisconsin Department of Natural Resources, Madison, Wisconsin, USA.

Hesser, R.B. 1978. Management implications of hybrid esocids in Pennsylvania. American Fisheries Society, Special Publication 11:302-307.

Klingbiel, J. 1981. The status of the muskellunge program: an informational report to the Natural Resources Board. Administrative Report 11, Wisconsin Department of Natural Resources, Madison, Wisconsin, USA.

Ricker, W.E. 1975. Computation and interpretation of biological statistics of fish populations. Bulletin 191, Fisheries Research Board of Canada, Ottawa, Canada.

Weithman, A.S., and R.O. Anderson. 1976. Angling vulnerability of Esocidae. Proceedings of the 30th Annual Conference Southeastern Association of Game and Fish Commissioners 30:99-102.

Weithman, A.S., and R.O. Anderson. 1977. Survival, growth, and prey of Esocidae in experimental systems. Transactions of the American Fisheries Society 106:424-430.

Muskellunge Habitat
With Guidelines For Habitat Management

MICHAEL P. DOMBECK[1]

Chequamegon National Forest
157 North Fifth Avenue
Park Falls, Wisconsin 54552

ABSTRACT

Loss of alteration of habitat, especially reproductive habitat, is considered to be a major cause of a decline of native muskellunge populations. Active habitat management programs specifically for muskellunge are virtually nonexistent. This paper summarizes muskellunge habitat needs and problems and suggests guidelines for its management.

Adult muskellunge are strongly associated with submergent vegetation while juveniles seem to prefer emergents; both are associated with woody debris such as submerged tree tops, stumps, and logs in rivers and littoral zones of lakes. Dissolved oxygen (DO) depletion from accumulation of flocculent organic mucks and dense aquatic vegetation in spawning areas has been associated with poor natural reproduction. Invasion of formerly-exclusive muskellunge waters by northern pike has also been associated with the decline of natural muskellunge populations.

Proposed muskellunge habitat management guidelines include: control of northern pike populations; maintenance of balanced fish populations; identification of spawning areas, and acquisition of adjacent lands; monitoring of DO at spawning sites and rehabilitation of unsuitable substrates; addition of spawning substrates; timber management of riparian zones adjacent to spawning areas; and management of watershed to maintain high water quality.

In many areas where the muskellunge once maintained healthy natural populations, it has been extirpated or is sustained only through artificial propagation and stocking. Loss or alteration of habitat, especially reproductive habitat, is considered to be a major cause in the decline of native muskellunge populations (Trautman 1981; Oehmcke et al. 1974). Habitat loss in Kentucky streams is currently the most important issue in managing muskellunge in that state (Axon and Kornman 1986). In 1969, Wisconsin identified five major approaches to muskellunge management of which habitat preservation and protection were considered the most important (Oehmcke 1969). More recently, a survey of muskellunge management in North America found that only two of 16 states and three provinces which responded reported habitat protection/improvement as a principle component of their programs (Miller 1983). Current muskellunge management strategies are largely confined to stocking hatchery-reared fingerlings and harvest regulations. The scarcity of active muskellunge habitat management programs is likely due to paucity of knowledge rather than lack of interest.

Habitat information across the species range is limited, and detailed knowledge of specific habitat requirements for all life stages remains incomplete. Recent telemetry studies have defined habitats utilized by adult muskellunge in Ontario (Minor and Crossman 1978), Wisconsin (Dombeck 1979), Minnesota (Strand 1986), and Iowa (Miller and Menzel 1986 b) and identified spawning areas (Dombeck 1979, Strand 1986). Studies detailing habitats of river populations are available only for Kentucky streams (Brewer 1980, Axon and Kornman 1986). Comprehensive habitat studies of lake populations focus on spawning and early life habitats in Michigan, Wisconsin, and Minnesota (Dombeck et al. 1984a; Dombeck et al. 1986) and nursery habitat in Georgian Bay (Craig and Black 1986).

The objective of this paper is to summarize information concerning muskellunge habitat, and provide the resource manager with ideas and guidelines for its management. The emphasis here is on spawning and early life habitat because this critical period influences survival or abundance and is the period when greatest mortalities occur (May 1974; Balon 1975). Although the specific stage at which muskellunge year-class strength is established is unknown, it is likely determined during the first year of life. For purposes of discussion, the life stages follow the definitions for northern pike given by Inskip (1982).

---

[1] Present address: USDA-Forest Service, Fisheries and Wildlife Management, 630 Sansome Street, San Francisco, CA 94111.

MUSKELLUNGE HABITAT

Adult

The muskellunge has the most limited range of North American esocids being restricted to waters of eastern North America. An inventory of coolwater fish habitat estimated that the area of muskellunge and hybrid inhabit was over 3.3 million acres, or about 1% of the total freshwater area of North America; this commmpares to 54% inhabited by the congeneric northern pike (Carlander et al. 1978). Recent introductions have greatly expanded the muskellunge's range (Crossman 1978); however, these have only rarely resulted in self-sustaining populations, even within the specie's native range (Hess and Heartwell 1978).

The muskellunge occurs chiefly in the Great Lakes, St. Lawrence River, upper Mississippi River, and southern Hudson Bay drainages extending south in larger rivers of the Appalachians to Alabama (Crossman 1978). Despite its comparatively limited distribution, the muskellunge occupies a wide variety of habitat types ranging from small midwestern lakes less than 50 acres to the Great Lakes and Kentucky streams less than 50 ft wide to the St. Lawrence River. Typical habitat descriptions given in species accounts state that in Canada the muskellunge inhabits warm, heavily vegetated lakes, stumpy weedy bays, and slow heavily vegetated rivers and is regularly found among dense growths of pondweeds (*Potamogeton* spp.) (Scott and Crossman 1973). Oehmcke et al. (1974) gave a similar description but add that muskellunge are sometimes found in virtually weedless, sterile lakes. Additionally, they stated that muskellunge usually remain in waters less than 15 ft deep but occasionally are found in depths of 50 ft. Michigan recognizes two varieties of muskellunge which occupy different habitats. The northern muskellunge inhabits large, fertile, shallow lakes and reservoirs while the Great Lakes muskellunge inhabits the Great Lakes and closely associated waters (Shrouder 1975). In Ohio, Trautman (1981) gave similar habitat descriptions for the Great Lakes muskellunge and noted that the Ohio muskellunge inhabits low gradient streams (<3 ft/mi) in unglaciated sections of the state.

Telemetry studies have demonstrated the species close association with submergent vegetation and woody structure, and depth and thermal preference. In two Ontario lakes, summer habitat (home area) utilized by adult muskellunge primarily ranged from 4-9 ft deep, with males generally occupying larger but shallower (4-5 ft) areas (Minor and Crossman 1978). The home area of a large male in Stony Lake, Ontario was characterized by less than 30% vegetation, pondweeds and milfoil (*Myriophyllum* spp.) and irregular shaped rock in shelves and outcroppings. In contrast, home areas in Nogies Creek, Ontario were covered with dense (75%) vegetation primarily pondweeds and coontail (*Ceratophyllum* spp.) with abundant stumps and logs. Similar summer muskellunge habitat in two Wisconsin lakes was reported by Dombeck (1979) where fish generally occupied areas less than 6 ft deep associated with submergent vegetation (primarily pondweeds), stumps, and logs. Transmitter-tagged muskellunge in West Okoboji Lake, Iowa, were located 78% of the time in water 6-25 ft deep; only 37% of the lake's area lay within those contours (Miller and Menzel, 1986 b); they noted further that in late summer the fish selected shallower water near the weedline where they remained suspended above or between distinct patches of vegetation. All three telemetry studies observed that in summer muskellunge tended to occupy littoral habitats where water temperatures (77-83 F) were higher than midlake temperatures.

As water temperatures cool and weedbeds deteriorate during autumn, the muskellunge appears to be somewhat nomadic and less strongly associated with specific habitat types (Dombeck 1979; Miller and Menzel in press a). Fish abandoned summer home areas and extensive movements were typical, perhaps influenced by movements and location of prey species during this period of increased activity. During winter months their movements are more localized with home areas smaller and deeper than in summer habitat (Minor and Crossman 1978; Dombeck 1979). Minor and Crossman (1978) found that muskellunge inhabited waters 7-14 ft deep with some logs and stumps and 30-50% submergent vegetation. In contrast, muskellunge in Leech Lake, Minnesota, occupied winter areas nearly twice as large and distinctly separate from summer areas (Strand 1986).

Studies of river populations in Tennessee (Parsons 1959), West Virginia (Miles 1978), and Kentucky (Brewer 1980) reported muskellunge occurring chiefly in pools having a mean maximum depth of 5 ft, mean width of 25 ft or more, and gradients of 3-8 ft/mi. A comprehensive review of muskellunge habitat in Kentucky streams is given by Axon and Kornman 1986), in this same volume.

Spawning/Embryo

Information concerning muskellunge spawning,

for the most part, consists of qualitative descriptions based on observations made during spawn-taking operations. Nevin (1901) reported spawning in areas of greatest log, stump, and brush density in Wisconsin flowages, while Oehmcke et al. (1974) reported that in Wisconsin lakes they spawn in shallow bays (<3 ft) over muck bottoms covered with detritus and dead vegetation. Similarly, fish in Nogies Creek, Ontario were observed spawning in waters less than 3 ft over matted vegetation and tree leaves that had fallen the previous autumn (Minor and Crossman 1978). Deeper (3-6 ft) open-water spawning areas were identified in Leech Lake, Minnesota, covered by dense beds of stonewort (*Chara* sp.), which were growing over flocculent marl substrate (Strand 1986). Stonewort is also present in spawning areas in several self-sustaining northern Michigan lakes (Dombeck et al. 1984a). Stonewort usually occurs in hard-waters and is associated with the deposition of calcarious materials (Fassett 1957). Spawning areas in Leech Lake most closely resemble those utilized by the Great Lakes muskellunge in Lake St. Clair, Michigan, where spawning occurs in openwater over 9 ft deep (Haas 1978). Shallow vegetated bays, similar to known spawning areas in Wisconsin and eastern Ontario, exist in Leech Lake; however, telemetry studies and intensive netting have shown that this habitat is not utilized by muskellunge for spawning (Strand 1986).

Comparative studies of spawning habitat in eight muskellunge lakes in Michigan, Minnesota, and Wisconsin associated dissolved oxygen concentration (DO) at the bottom-water interface with reproductive success (Dombeck et al. 1984a). Four lakes with self-sustaining populations had high DO (mean, 6.0-8.4 ppm) while the others with little or no natural reproduction had low DO (mean 0.4-2.4 ppm). Dissolved oxygen depletion was attributed to night-time respiration of dense submergent vegetation and/or decomposition of flocculent organic silts. Suitable egg incubation substrates were aerated by annual reservoir drawdown (late fall-early spring) or were composed or inorganic materials (sand/gravel or marl). A statistical analysis of environmental parameters of 117 lakes in the same three states associated muskellunge reproductive success with limited northern pike abundance, rising springtime water levels, high alkalinity, high shoreline development factor, high discharge volume, low cultural development of adjacent lands, and drainage lake systems (Dombeck et al. 1986). The large lakes of Minnesota's upper Mississippi River drainage were an exception since they support strong self-sustaining populations of both muskellunge and northern pike. In Leech Lake, for example, reproductive isolation may be the key; muskellunge spawn in offshore areas while northern pike utilize shallow areas nearshore areas (Strand 1986).

Similarly, separation of spawning habitat between sympatric populations of the two esocids occurs in the Niagara River, New York, where muskellunge spawn in lotic waters (currently velocity, 0.66 ft/second) and northern pike utilize lentic habitats (Harrison and Hadley 1978). Brewer (1980) observed muskellunge spawning in Kinniconick Creek, Kentucky, in a pool averaging about 2 ft deep over sandstone-based rubble and bedrock; he stated that spawning typically occurs at either the upper or lower ends of low-gradient pools. Most pools contained decaying leaves and/or rooted vegetation, with substrates consisting primarily of rubble and gravel. In contrast, three muskellunge spawning areas in the St. Lawrence River were located in shallow bays with little current and noted exposure to the prevailing westerly winds (Al Schiavone, New York State Department of Environmental Conservation, Watertown, New York, personal communication).

Fry/Juvenile

Information on habitats utilized by fry and juvenile muskellunge is very limited, compared to other life stages. Fry and juveniles in Kentucky streams occupied pool habitats similar to adults (Brewer 1980). In Niagara River, New York, currents exceeded 0.33 ft/second in habitats where young-of-the-year were most frequently found (Harrison and Hadley 1978). Juveniles in the St. Lawrence River were found in currents greater than 0.3 ft/second (pers. comm. Donald M. Osterberg). Summer nursery habitat of juveniles (young-of-the-year) in Georgian Bay consisted of a narrow band (6 ft) of dense emergent vegetation, adjacent to a wider band (60 ft) of less dense emergents (Craig and Black 1986). Similarily, transmitter-tagged, 12.6-inch muskellunge typically inhabited beds of bullrushes *(Scirpus spp.)* and submerged tree-tops (pers. comm. David A. Hanson).

HABITAT PROBLEMS

Habitat alterations impacting muskellunge populations are numerous, some being quite obvious while others may be very subtle. Most states and provinces have legislation protecting aquatic habitats from drastic modifications. Long-term subtle habitat changes can be equally serious since they occur largely unnoticed. In addition, gradually-declining natural populations are often supplemented by hatchery-reared fingerlings to maintain the desired population levels, further masking the effect of habitat deterioration. Habitat changes and perturbations most often cited as impacting muskellunge populations are summarized in Table 1.

Obviously, there are many such habitat modifications where the impacts on the muskellunge are not yet fully realized or understood. No data is available concerning the impacts of many point- and nonpoint-pollutants, such as acid rain, heavy metals, environmental contaminants, thermal effluents, etc. Therefore, we can only extrapolate from pertinent studies on other species about probable impacts on muskellunge. However, any environmental change, which results in a deviation from optimal muskellunge habitat, can only contribute to the decline of the species.

HABITAT MANAGEMENT GUIDELINES

The following guidelines are recommended for application to existing and potential muskellunge waters. They should be used until further experience reveals more efficient methods for improving and preserving the habitat. Habitat improvement measures should be applied on a case-by-case basis following careful evaluation by a professional fisheries biologist. No measure, or combination of measures, should be applied indiscriminately to any system. These guidelines are directed primarily toward enhancing natural reproduction, with the assumed management objective of restoring or maintaining self-sustaining muskellunge fisheries (no stocking required). In numerous waters many nonhabitat parameters exist which limit muskellunge populations. These include overharvest (Hanson 1986); competition with or predation by northern pike (Caplin 1982; Inskip 1986); or predation by stunted panfish populations (Scott and Crossman 1973). Therefore, species management

Table 1. A summary of habitat perturbations/modifications impacting muskellunge.

Perturbation/modification	life stage	references
Stream impoundment		
Blocks spawning migration	spawning-embryo	Trautman 1981
		Axon & Kornman 1986
Eliminates riverine habitat	adult	Brewer 1980
Draining wetlands, filling marshes	spawning-embryo	Bimber 1978
Eliminates spawning areas		Trautman 1981
Channelization	spawning-adult	
Eliminates habitat		Trautman 1981
Decreasing springtime water-level		
Strands spawners & young and exposes eggs	all	Scott & Crossman 1973
Eutrophication		
Excessive weed growth & buildup of organic substrates	spawning-embryo	Dombeck et al. 1984a
Low winter dissolved oxygen	adult	Dombeck 1979
Sediments		
Loss of spawning habitat	spawning-embryo	Muncy et al. 1979
		Axon & Kornman 1986
Suspended solids		
Impairs sight feeding	fry-adult	Farnworth et al. 1979
		Muncy et al. 1979
Elimination of vegetation		
Elimination of habitat	fry-adult	Bimber & Nicholson 1981
		Trautman 1981
	fry	Craig & Black 1986
Loss of woody cover	fry-adult	Axon & Kornman 1986
Acid mine drainage	all	Axon & Kornman 1986
Brine drainage from oil wells	all	Axon & Kornman 1986
Gravel dredging	all	Axon & Kornman 1986
Shoreline modification		
Clearing trees in riparian zone	all	Axon & Kornman 1986
For recreational use	spawning-fry	Craig & Black 1986

suggestions are included. An evaluation of the entire aquatic ecosystem is necessary if the efforts of habitat improvements are to ultimately result in increased numbers of catchable size fish for the angler.

These habitat management guidelines are based upon a synthesis of the literature, inference, and experience with other species; results have not been specifically documented for muskellunge. They are designed to provide the field-level resource manager with ideas for consideration in planning and implementing habitat improvement measures for muskellunge.

Control Northern Pike Populations

The invasion of muskellunge waters by the northern pike has been associated with the subsequent decline of the muskellunge (Inskip and Magnuson 1983). The northern pike is considered to have an ecological advantage over the muskellunge when they occur together (Hess and Heartwell 1978). Dombeck et al. (1984a) suggested that the northern pike is better adapted to reproduce in more eutrophic waters. Even though strong self-sustaining populations of both species occur in some waters such as the upper Mississippi River drainage (Dombeck et al. 1986), in many other waters the northern pike has invaded and become the dominant large predator (Threinen and Oehmcke 1950). The negative associations between these two esocids and potential competitive mechanisms are reviewed by Inskip (1986).

Due to the ecological similarity between the muskellunge and northern pike, it seems likely that spawning habitat improvement measures for muskellunge may also improve northern pike reproduction and ultimately increase numbers of northern pike. It's, therefore, important to know the characteristics of northern pike populations in waters where such improvement measures are being considered. Management options to reduce northern pike might include liberalized regulations, population manipulation and controlled water-level drawdown during and prior to spawning to reduce reproduction. In small lakes, additional management options might include northern pike removal or restricting their access to spawning areas.

Maintain Balanced Fish Populations

In addition to northern pike, many other fishes prey on the very young, nearly immobile muskellunge fry (Scott and Crossman 1973). Overabundant panfish populations should be reduced using conventional population manipulation techniques such as reclamation, netting or liberalized regulations.

Identify Location of Spawning Areas

If reproductive habitat is to be protected, its location must be known. Consider acquisition of lands adjacent to and influencing spawning habitat. Modification of shorelines on many waters has resulted in degradation or destruction of spawning habitat. It is imperative that remaining high-quality spawning habitat be preserved.

Monitor Dissolved Oxygen Concentrations

Low DO at the bottom-water interface in spawning areas has been associated with poor natural muskellunge reproduction (Dombeck et al. 1984a). Should daily DO go below the critical level of 3.2 ppm (Seifert et al. 1973) during spring or early summer, it is likely that high early-life mortality will result. A simple technique for measuring the microstratification of DO at the bottom-water interface has been described by Dombeck et al. (1984b).

Rehabilitate Spawning Substrate

Where bottom DO falls below 3.2 ppm at dawn during the time of egg and larval development, efforts should be made to determine cause(s) of DO depletion. Depending on logistics, some rehabilitative measures might include dredging high BOD substrates, or lake bottom sealing, and water level management. Dredging and sealing high BOD bottoms with artificial fabric has been utilized in lake rehabilitation (Dunst et al. 1974) and depending on local conditions may be a management option.

Water-level management (drawdown) to aerate high BOD substrates results in dessication and consolidation of organic material (Born et al. 1973). The subsequent chemical oxidation greatly reduces BOD and causes physical shrinkage of substrates, providing a more suitable environment for egg and larval development. Oxidation of substrates also "locks-up" nutrients and slows the natural eutrophication process (Dunst et al. 1974). Where winters are harsh, over-winter drawdown causes aquatic plants to freeze-out (Beard 1973; Nichols 1974) and may be used to control dense aquatic vegetation problems.

Placement of Spawning Substrates

Placement of spawning substrates, particularly in combination with bottom sealing, may be an option to improve early life habitat. Sand or gravel may be placed over fabric used in bottom sealing (Nichols 1974). Judicious placement of woody materials can improve both spawning and fry-adult habitat. Nevin (1901) reported muskellunge spawning in areas of greatest log, stump, and brush density. Muskellunge are strongly associated with fallen trees in streams (Axon and Kornman 1986; Kornman 1983) and logs and stumps in lakes (Minor and Crossman 1978; Dombeck 1979).

Design Dams to Facilitate Water-Level Management

The flooding of terrestrial areas during early spring can provide high-quality muskellunge spawning habitat. The ability to manage water-level at low cost is a valuable management tool. In addition to the above mentioned benefits, occasional drawdowns can increase the life of reservoirs and facilitates shoreline vegetation management and erosion control (Dunst et al. 1974; Born et al. 1973).

Manage Forested Riparian Zones Adjacent to Spawning Areas

The adjacent forests should be managed for coniferous species and timber adjacent to muskellunge spawning areas allowed to overmature and die naturally. The input of deciduous leaves into spawning sites, especially those with already high BOD substrates, adds to the organic loading of sediments. In laboratory experiments Dombeck et al. (1984a) found DO depletion and muskellunge egg mortality associated with leaves. Coniferous needles are less often blown into the water and have a much slower decomposition rate.

Due to its growth form and resistance to decay, cedar (*Thuja* spp.) appears to be a highly desirable tree for riparian areas, and management for cedar might be a suitable long-term objective. Other desirable species appear to be white pine (*Pinus strobus*) and eastern hemlock (*Tsuga canadensis*). Where the site index limits management of these species, a longterm objective might be the uneven-age management of northern hardwoods to increase the pine and hemlock component. Fisheries managers should be aware of the important relationship between riparian zone vegetation management and fish habitat.

Manage Watershed to Maintain High Water Quality

Organic loading from natural and human sources should not exceed levels that will result in depression of DO in spawning areas (Dombeck et al. 1984a) Sediment deposition can eliminate weedbeds utilized by juveniles and adults and modify spawning areas, while increases in suspended solids may impair visibility and feeding opportunities (Farnworth et al. 1979; Muncy et al. 1979; Axon and Kornman 1986). Axon and Kornman (1986) observed strongest muskellunge populations in the least developed watersheds of Kentucky streams while Dombeck et al. (1986) found natural reproduction negatively associated with cultural developments in midwestern lakes. Practices such as road-building, logging, mining, and agriculture should not be allowed to increase sediment and nutrient input by the surface runoff. Sediment traps should be utilized in tributary streams where necessary to reduce sediment loading caused by such activities. The area of direct drainage to the lake should be managed to protect shorelines from erosion and modification.

There presently exists only a superficial understanding of muskellunge-habitat relationships. Research concerning the specific physical and biological requirements, particularly of early-life stages of this species, should receive high priority.

REFERENCES

Axon, J.R., and L.E. Kornman. 1986. Characteristics and management of native muskellunge streams in Kentucky. American Fisheries Society Special Publication Number 15:263-272.

Balon, E. K. 1975. Reproductive guilds of fishes: a proposal and definition. Journal of the Fisheries Research Board of Canada 32:821-864.

Beard, T. D. 1973. Overwinter drawdown. Impact on aquatic vegetation in Murphy Flowage, Wisconsin. Wisconsin Department of Natural Resources Technical Bulletin 61.

Bimber, D. L. 1978. An analysis and historical overview of the muskellunge population and sport fishery of Chautauqua Lake, New York. M. S. thesis, State University College, Fredonia, New York, USA.

Bimber, D. L. and S. A. Nicholson. 1981. Fluctuations in muskellunge, *Esox masquinongy* Mitchill, populations of Chautauqua Lake, New York. Environmental Biology of Fishes 6:207-211.

Born, S. M., T. L. Wirth, E. M. Brick, and J. O. Peterson. 1973. Restoring the recreational potential of small impoundments. Wisconsin Department of Natural Resources Technical Bulletin 71.

Brewer, D. L. 1980. A study of native muskellunge populations in eastern Kentucky streams. Fisheries Bulletin of the Kentucky Department of Fish and Wildlife Resources, Number 64.

Caplin, D. L. 1982. An experimental study of interactions between young of the year pike *Esox lucius* and muskellunge *Esox masquinongy*. M. S. thesis, University of Wisconsin, Madison, Wisconsin, USA.

Carlander, K. D., J. C. Campbell, and R. J. Muncy. 1978. Inventory of percid and esocid habitat in North America. American Fisheries Special Publication 11:27-38.

Craig, R.E., and R.M. Black. 1986. Nursery habitat for muskellunge in southern Georgian Bay. American Fisheries Society Special Publication Number 15:79-86.

Crossman, E. J. 1978. Taxonomy and distribution of North American esocids. American Fisheries Society Special Publication 11:13-26.

Dombeck, M. P. 1979. Movement and behavior of the muskellunge determined by radio telemetry. Wisconsin Department of Natural Resources Technical Bulletin 113.

Dombeck, M. P., B. W. Menzel, and P. N. Hinz. 1984a. Muskellunge spawning habitat and reproductive success. Transactions of the American Fisheries Society 113:205-216.

Dombeck, M. P., R. W. Bachmann, and B. W. Menzel. 1984b. A method for measuring oxygen in aquatic microzones. Journal of Freshwater Ecology 2:535-540.

Dombeck, M.P., B.W. Menzel, and P.N. Hinz. 1986. Natural muskellunge reproduction in midwestern lakes. American Fisheries Society Special Publication Number 15:122-134.

Dunst, R. C., S. M. Born, P. D. Uttormark, S. A. Smith, S. A. Nichols, J. O. Peterson, D. R. Knaver, S. L. Serns, D. R. Winter, and T. L. Wirth. 1974. Survey of lake rehabilitation techniques and experiences. Wisconsin Department of Natural Resources Technical Bulletin 75.

Farnworth, E. G., M. C. Nicols, C. N. Vann, L. G. Wolfson, R. W. Bosserman, P. R. Hendrix, F. B. Golley, and J. L. Cooley. 1979. Impacts of sediment and nutrients on biota in surface waters of the United States. United States Environmental Protection Agency, EPA-600/3-79-105, Athens, Georgia, USA.

Fassett, N. C. 1957. A manual of aquatic plants. University of Wisconsin Press, Madison, Wisconsin, USA.

Haas, R. C. 1978. The muskellunge in Lake St. Clair. American Fisheries Society Publication 11:334-339.

Hanson, D.A. 1986. Population characteristics and angler use of muskellunge in nine northern Wisconsin lakes. American Fisheries Society Special Publication Number 15:238-248.

Harrison, E. J. and W. F. Hadley. 1978. Ecological separation of sympatric muskellunge and northern pike. American Fisheries Society Special Publication 11:129-134.

Hess, L. and C. Heartwell. 1978. Literature review of large esocids (muskellunge, northern pike, hybrid tiger muskellunge). Pages 139-175 in J. Dubé and Y. Gravel editors. Proceedings of the 10th Warm Water Workshop. Northeastern Division, American Fisheries Society, Québec Ministère du Loisir, de la Chasse et de la Pêche Direction de la recherche faunique, Montréal, Québec, Canada.

Inskip, P. D. 1982. Habitat suitability index models: northern pike. United States Fish and Wildlife Service Biological Services Program FWS/OBS-82/10.17.

Inskip, P.D. 1986. Negative associations between abundance of muskellunge and northern pike: Perspective on a species pair. American Fisheries Society Special Publication Number 15:135-150.

Inskip, P. D. and J. J. Magnuson, 1983. Changes in fish populations over an 80-year period: Big Pine Lake, Wisconsin. Transactions of the American Fisheries Society 112:378-389.

Kornman, L. E. 1983. Muskellunge streams investigations at Kinniconick and Tygarts creeks. Kentucky Department of Fish and Wildlife Resources Bulletin 68.

May, R. C. 1974. Larval mortality in marine fishes and the critical period concept. Pages 3-5 in J. H. S. Blaxter, Editor. The early life history of fish. Springer-Verlag Press, New York, New York, USA.

Miles, R. L. 1978. Life history of muskellunge in West Virginia. American Fisheries Society Special Publication 11:298-301.

Miller, M. L. 1983. The status of muskellunge management in North America. Division of Wildlife, Ohio Department of Natural Resources, Columbus, Ohio, USA.

Miller, M.L. and B.W. Menzel. 1986a. Summer range, spring movements, and homing of muskellunge (*Esox masquinongy*) in West Okoboji Lake, Iowa, as determined by ultrasonic telemetry. Environmental Biology of Fishes.

Miller, M.L. and B.W. Menzel. 1986b. Warm season movement, activity, and habitat use patterns of muskellunge in West Okoboji, Iowa. American Fisheries Society Special Publication Number 15:51-61.

Minor, J. D. and E. J. Crossman. 1978. Home range and seasonal movements of muskellunge as determined by radio-telemetry. American Fisheries Society Special Publication 11:146-153.

Muncy, R. J., G. J. Atchison, R. V. Bulkley, B. W. Menzel, L. G. Perry, and R. C. Summerfelt. 1979. Effects of suspended solids and sediment on reproduction and early life of warmwater fishes: A review. United States Environmental Protection Agency, EPA-600/3-79-042, Corvallis, Oregon, USA.

Nevin, J. 1901. The propagation of muskellunge in Wisconsin. Transactions of the American Fisheries Society 30:90-93.

Nichols, S. A. 1974. Mechanical and habitat manipulation for aquatic plant management: A review of techniques. Wisconsin Department of Natural Resources Technical Bulletin 77.

Oehmcke, A. A. 1969. Muskellunge management in Wisconsin. Wisconsin Department of Natural Resources Fish management Report 11.

Oehmcke, A. A., L. Johnson, J. Klingbiel, and C. Winstrom. 1974. The Wisconsin Muskellunge. Its life history, ecology, and management. Wisconsin Department of Natural Resources Publication 8-3600(74).

Parsons, J.W. 1959. Muskellunge in Tennessee streams. Transactions of the American Fisheries Society 88:136-140.

Scott, W. B. and E. J. Crossman. 1973. Freshwater fishes of Canada. Bulletin 184. Fisheries Research Board of Canada, Ottawa, 966 pp.

Seifert, R. E., W. A. Spoor, and R. F. Syrett. 1973. Effects of reduced oxygen concentration on northern pike, *Esox lucius*, embryos and larvae. Journal of the Fisheries Research Board of Canada 30:849-852.

Shrouder, J. D. 1975. The muskellunge in Michigan. Michigan Department of Natural Resources Fisheries Division Pamphlet 45.

Strand, R.F. 1986. Identification of principal spawning areas and seasonal distribution and movements of muskellunge. American Fisheries Society Special Publication Number 15:62-73.

Threinen, C. W. and A. A. Oehmcke. 1950. The northern invades the musky's domain. Wisconsin Conservation Bulletin 15:10-12.

Trautman, M. B. 1981. The fishes of Ohio, revised edition. Ohio State University Press, Columbus, Ohio, USA.

Survival of Stocked Muskellunge Eggs, Fry, and Fingerlings in Wisconsin Lakes

DAVID A. HANSON, MICHAEL D. STAGGS, STEVEN L. SERNS,
LEON D. JOHNSON, AND LLOYD M. ANDREWS

Wisconsin Department of Natural Resources
Box 309
Spooner, Wisconsin 54801

ABSTRACT

Angler harvests of muskellunge exceeds natural recruitment in most muskellunge waters in Wisconsin; consequently, populations are supplemented by hatchery propagation. This paper summarizes survival estimates from 5 egg, 6 fry, and 74 fingerling stockings. Factors influencing survival were examined by comparing estimates of fall fingerling density and two indices of survival to a set of independent factors using simple correlation and stepwise regression.

Stocking fertilized muskellunge eggs on gravel and sand substrate resulted in a measurable year-class of muskellunge in two of five trials, and fall fingerlings were observed from two of six fry stockings. Survival of stocked fingerlings to fall averaged 38.7% and ranged from 0.0% to 95.7%.

In multiple regression analyses of 59 fingerling stockings, a significant positive relationship was found between observed survival to fall and length at stocking. A negative relationship between fall survival and days at large was also identified but this was expected. The geometric mean daily survival was independent of the number of days in the lake and provided a better index for examining the effect of the independent factors on fingerling survival. Mean daily survival was positively correlated with length at stocking and days at large. The fall population density of fingerlings was highly correlated with the initial stocking rates. Tests of the above derived models with nine later stockings found the effect of length at stocking to be variable.

This study suggested that best survival to first fall was achieved by stocking the largest fingerlings, however, similar results may have been achieved by stocking proportionately greater numbers of smaller fingerlings. There was no evidence that high stocking rates adversely affected fingerling survival.

Fishable muskellunge populations exist in 703 lakes and 48 streams in Wisconsin (WDNR 1982). In most of these waters, angler harvest exceeds natural recruitment (WDNR 1979), so an average of 1,000,000 fry and 140,000 fingerlings are stocked annually to supplement natural reproduction (Klingbiel 1981).

The high cost of hatchery propagation and low survival of stocked muskellunge makes dependence on stocking an expensive management option. Hatchery cost of raising muskellunge fingerlings in Wisconsin was estimated to be nearly $12.00/pound in 1977-78 (Klingbiel 1981). Poor survival of stocked fingerlings in Wisconsin and other states has been well documented (Belusz 1975, 1978; Johnson 1976, 1978; Serns and Andrews 1986). Similar low survival of hybrid muskellunge has been reported by Beyerle (1981) and Stein et al. (1981), although hybrid muskellunge are reported to survive better than true muskellunge when stocked into similar waters (Beyerle 1973; Weithman and Anderson 1977). Stocking of either muskellunge or hybrid fry is initially less expensive than fingerling stocking but fry stocking has been successful in only limited situations (Hess and Heartwell 1979; Porter 1977). Recent successes in planting artificially-fertilized salmonid eggs (Federation of Fly Fishermen 1975 Swanson 1982; Harshbarger and Porter 1982) have suggested egg stocking as a potentially less expensive muskellunge management alternative, but little work on this has been done (Forney 1969; Johnson 1969).

With increasing angler demand for quality muskellunge fisheries and declining management funds, Wisconsin has begun to evaluate the success of its muskellunge stocking program from a cost-benefit perspective. The objective of this study was to combine existing data on fingerling muskellunge survival from Johnson (1976) and Serns and Andrews (1986) with recently collected data and identify which factors were important in determining stocking success. Future work will incorporate hatchery cost and survival information into a comprehensive model of the Wisconsin muskellunge hatchery program.

METHODS

Known numbers of muskellunge eggs, fry, and fingerlings were stocked 85 times in 31 different study lakes ranging in size from 17 to 407 ha (Table 1). To distinguish stocked fish from any natural reproduction, one of the paired fins was removed from fingerlings larger than 80 mm total length. The fin-clipping was assumed to have no effect on short-term survival (McNeil and Crossman 1979). Mean total length of fingerlings from 74 stockings was 222 mm and ranged from 58 to 318 mm; stocking rates averaged 13.5 fish/ha (range 2.4 to 75.5; Table 1).

Table 1. Stocking information and lake characteristics for egg, fry, and fingerling stockings.

Lake name (County)	Lake area(ha)	Date	Total length (mm)	Temp (C)	No. stocked	No. stocked/ha
		EGG				
Bass (Oneida)	29.9	05/11/83	-	13	210,000	7012
Bass (Washburn)	58.3	04/30/84	-	10	451,500	7748
Cisco (Bayfield)	38.4	05/05/84	-	09	290,100	7546
Deep (Washburn)	17.4	04/30/82	-	11	167,000	9596
Patterson (Washburn)	76.1	05/07/83	-	12	577,000	7584
		FRY				
Bass (Oneida)	29.9	05/21/84	20	16	71,000	2371
Bass (Washburn)	58.3	05/14/83	20	16	108,000	1853
Cisco (Bayfield)	38.4	05/25/83	20	15	71,250	1853
Cyclone (Washburn)	36.8	06/01/82	20	19	32,400	880
Harmon (Washburn)	38.9	05/24/82	20	17	100,000	2574
Patterson (Washburn)	76.1	05/23/84	20	17	188,000	2471
		FINGERLING				
Arrowhead (Vilas)	40.1	06/29/76	99	23	396	9.9
		08/10/76	203	21	396	9.9
		09/21/76	295	16	396	9.9
		06/29/77	140	20	396	9.9
		07/26/77	213	23	396	9.9
		09/14/77	307	16	396	9.9
Bass (Price)	34.0	08/22/73	221	22	415	12.2
Boot (Rusk)	35.2	09/21/71	292	16	600	17.0
		08/29/72	234	21	510	14.5
Brandy (Vilas)	44.5	06/29/76	99	21	440	9.9
		08/10/76	203	23	440	9.9
		09/21/76	295	14	440	9.9
		06/29/77	140	20	440	9.9
		07/26/77	213	22	440	9.9
		09/14/77	307	16	440	9.9
Clear (Sawyer)	31.2	09/12/72	282	18	437	14.0
		09/12/73	269	20	384	12.3
		09/11/74	284	17	400	12.8
		09/16/81	249	18	354	11.4
Crane Chase (Price)	34.8	08/25/71	241	20	490	14.1
		08/22/72	221	22	499	14.3
		08/29/73	226	21	417	12.0
Des Moines (Burnett)	92.7	06/02/77	58	22	7,000	75.5
		08/02/77	234	21	698	7.5
		09/01/82	249	19	916	9.9
Harmon (Washburn)	38.9	06/13/76	58	21	800	20.6
		08/09/76	234	26	200	5.1
Island (Sawyer)	27.5	09/19/72	287	17	407	14.8
Johnson (Vilas)	31.6	06/29/76	99	23	312	9.9
		08/10/76	203	22	312	9.9
		09/21/76	295	17	312	9.9
		06/29/77	140	21	312	9.9
		07/26/77	213	24	312	9.9
		09/14/77	307	17	312	9.9
L. Holly (Sawyer)	17.0	08/27/75	239	24	200	11.8
		06/14/76	58	21	1,000	58.8
		08/08/76	236	27	100	5.9

TABLE 1, *(continued)*

Lake name (County)	Lake area (ha)	Date	Total Length (mm)	Temp (C)	No. stocked	No. Stocked/ha
L. Sand (Barron)	40.9	09/05/73	259	18	421	10.3
		08/28/74	254	23	396	9.7
		08/20/75	231	22	400	9.8
		06/13/76	58	21	800	19.6
		08/09/76	236	26	200	4.9
		08/22/79	236	19	397	9.7
L. Sand (Sawyer)	31.6	09/05/73	257	19	377	11.9
Leisure (Washburn)	30.4	09/05/74	267	18	290	9.6
		08/20/75	234	24	300	9.9
		06/13/76	58	21	1,500	49.4
		08/08/76	236	26	150	4.9
		08/30/82	246	19	400	13.2
Mathews (Washburn)	106.4	06/14/77	58	20	8,000	75.2
		08/02/77	234	21	798	7.5
Mud/Callahan (Sawyer)	237.2	08/31/82	246	17	2,000	8.4
Pear (Washburn)	19.8	09/04/74	264	20	226	11.4
		08/27/75	241	23	202	10.2
Pulaski (Rusk)	51.0	09/15/71	277	19	703	13.8
Sparkling (Vilas)	51.4	06/29/76	99	21	508	9.9
		08/10/76	203	22	508	9.9
		09/21/76	295	17	508	9.9
		06/29/77	140	18	508	9.9
		07/26/77	213	23	508	9.9
		09/14/77	307	17	508	9.9
Trego (Washburn)	182.5	09/21/83	315	13	900	4.9
Twenty-Six (Burnett)	93.1	06/13/77	58	20	7,000	75.2
		08/03/77	206	21	700	7.5
Whitefish (Sawyer)	318.1	09/22/82	318	16	750	2.4
		FINGERLING - 1984*				
Butternut: LCO (Price)	407.1	09/12/84	295	14	993	2.4
MQ		09/12/84	246	14	1000	2.5
Clear: large (Sawyer)	31.2	09/06/84	287	17	150	4.8
small		09/06/84	236	17	150	4.8
Nancy: Leech (Washburn)	312.4	09/20/84	297	19	1,187	3.8
Towanda: LCO (Vilas)	59.1	09/20/84	290	UK	300	5.1
Leech		09/20/84	297	UK	300	5.1
Twenty-Six: large (Burnett)	93.1	09/06/84	287	16	450	4.8
small		09/06/84	236	16	450	4.8

*1984 fingerling stockings not included in stepwise regression analysis but used to test derived model. Broodstock or size classes given with lake name (LCO=Lac Court Oreilles, MQ=Minocqua).

Fertilized eggs, fry, and fingerlings less than 80 mm TL were not marked for future identification but were only stocked in lakes with few or no adult muskellunge and no known natural reproduction. Five egg stockings were made with eggs collected from wild populations of muskellunge, artificially fertilized and allowed to water harden prior to transportation to the lake to be stocked. Eggs were then scattered at densities of 7,000 to 9,600 eggs/surface ha on sand and gravel substrate in water depths less than 1.2 m. In 1982, swim-up fry were stocked en masse at public access sites; in other years they were scatter-planted near shoreline vegetation over sand and gravel substate. Fry stocking rates averaged 2000 fry/ha (range from 880 to 2574; Table 1) in six stockings.

Stocking success was evaluated by mark-recapture population estimates during late fall of the year of stocking. Fingerlings were sampled at night around the shoreline with an A.C. boat-mounted electroshocker. Before 1973, population estimates were made using the Bailey modification of the Peterson method (Ricker 1975); more recent estimates were made from multiple samples and based on the maximum-likelihood solution to the Schnabel model (Robson and Regier 1978).

In several stocking evaluations there were too few recaptures to make unbiased population estimates; most of these were small fingerling stockings made concurrently with stockings of large fingerlings. In these lakes only one mark-recapture study was carried out to estimate the success of both stockings, so if a population estimate of the larger fingerlings was made, the population of the smaller fingerlings was also estimated using the ratio of the number of individual fish handled:

$$N_{sf} = \frac{C_{sf} N_{lf}}{C_{lf}}$$

where C_{sf} was the number of individual small fingerlings handled during the mark-recapture period; C_{lf} was the corresponding number of large fingerlings handled; and N_{lf} was the Schnabel population estimate of the large fingerlings. Variance of this population estimate could not be calculated because the covariance between the ratio C_{sf}/C_{lf} and the large fingerling population estimate, N_{lf}, was unknown.

Trials with too few recaptures and no concurrent successful mark-recapture estimate have survival reported as a minimum known survival based on the number of unique individuals handled. These trials were not used in the correlation or regression analyses.

Estimates of fall fingerling density and two indices of survival were compared to a set of independent factors using simple correlation and stepwise regression. The independent factors were total length at stocking, water temperature at stocking (°C), Julian date of stocking (number of days after January 1), stocking rate (number/ha), number stocked, lake area (ha), and number of days at large (days between stocking and population estimation). Selection of these factors was based primarily on availability of data and it was recognized that input of other biotic and abiotic variables may have improved the analysis. The 65 fingerling stockings from 1971-83 were used to initially derive models and 9 stockings from 1984 were then used to test the models.

One index of survival was simply the percentage of stocked fish surviving at the time of the population estimate. However, this observed fall survival was expected to have a nonlinear relationship with the number of days at large. Since the observed survival after N days could be represented as the product of N nonmeasured daily survival rates, (S_i):

$$S = S_1 \times S_2 \times S_3 \times .... \times S_N$$

the geometric mean daily survival over the N-day period could be calculated as:

$$\overline{S} = \sqrt[N]{\prod_{i=1}^{N} S_i} = \sqrt[N]{S}$$

Mean daily survival was a better index for examining the effects of the independent factors on survival because it did not decline with passing time as did simple observed survival. Note that calculating the geometric mean daily survival gave exactly the same result as converting the observed survival to an N-day force of mortality (Z), partitioning the mortality equally between the N days, and calculating the daily survival (Ricker 1975). The geometric mean daily survival derivation, however, made no assumptions about the distribution of mortality between the individual days.

For each combination of number stocked and days at large, there is a minimum observable daily survival rate below which the observed fall survival would be zero. For example, after a stocking of 1000 fish, a constant daily survival rate of 89.1% would leave only a single survivor after 60 days. Since an observed survival of zero might represent a low, but constant, daily survival or some catastrophic fingerling loss, trials with no observed survivors were omitted from the analyses of daily survival.

A final concern was that both survival indices and the fall fingerling density were based on a single population estimate and the variances of the population estimates differed considerably among trials. A weighted regression using the inverse of the variance as the weight accounted for the survival variability and was used to verify results obtained from the standard regression analyses.

The variance of the observed fall survival was,

$$V(S) = \frac{V(N)}{N_0^2}$$

where N_0 was the original number released and $V(N)$ was the variance of the population estimate. The variance of the geometric mean daily survival was approximated using a first order Taylor series expansion as,

$$V(\overline{S}) = \frac{V(S) \, S^{\left(\frac{2-2D}{D}\right)}}{D^2}$$

where D was the number of days geometrically averaged and V(S) and S as calculated above. Note that the variance of the mean daily survival did not reflect the variances between the actual individual daily survivals which could not be estimated.

Some of the stocking evaluations were from experiments evaluating possible differences in survival due to genetic strain (Hanson 1983), anesthesia, sedation, conditioning, and stocking location (Johnson 1976). None of these factors or manipulations were systematically present throughout the entire data set, and no bias due to these experiments could be isolated.

Six stockings in Sparkling Lake met with unusually low success and no population estimates could be made (Table 2). Results from Sparkling Lake were excluded from correlation and stepwise regression analyses and are discussed separately.

RESULTS AND DISCUSSION
Survival of Stocked Eggs and Fry

Two of five plantings of fertilized eggs on sand and gravel substrate resulted in a measurable year class of muskellunge. The estimated densities of fall fingerlings were 2.9 fish/ha in Deep Lake and 0.6 fish/ha in Patterson Lake. Mean total lengths of the surviving fingerlings were 261 mm and 292 mm, respectively. Observed survival and mean daily survival from egg to fall fingerling was 0.029% and 95.0% in Deep Lake and 0.007% and 93.5% in Patterson Lake (Table 2). No fingerlings were observed in Bass Lake (Oneida County), Bass Lake (Washburn County) or Cisco Lake.

Table 2. Population and survival estimates for egg, fry, and fingerling stockings (Estsimated variances given in parentheses). Data from lakes with multiple stockings are presented in same order as Table 1.

Lake name	Days at large	Number handled	Population estimate		% Fall survival		% Mean daily survival	
			EGG					
Bass	150	0	-	-	0.000	-	0.0	-
Bass	149	0	-	-	0.000	-	0.0	-
Cisco	148	0	-	-	0.000	-	0.0	-
Deep	159	35	49	(47)	0.029	(0.00)	95.0	(0.01)
Patterson	143	30	42	(37)	0.007	(0.00)	93.5	(0.01)
			FRY					
Bass	142	4	-	-	0.006	-	93.3	-
Bass	139	0	-	-	0.000	-	0.0	-
Cisco	127	0	-	-	0.000	-	0.0	-
Cyclone	85	0	-	-	0.000	-	0.0	-
Harmon	109	0	-	-	0.000	-	0.0	-
Patterson	133	19	31	(55)	0.016	(0.00)	93.7	(0.03)
			FINGERLING					
Arrowhead	117	41	91	(756)	23.0	(48.2)	98.8	(0.07)
	75	73	124	(400)	31.3	(25.5)	98.5	(0.04)
	33	111	164	(289)	41.4	(18.4)	97.4	(0.09)
	120	31	60	(20)	15.2	(1.3)	98.4	(0.00)
	93	80	135	(64)	34.1	(4.1)	98.8	(0.00)
	43	227	254	(225)	64.1	(14.4)	99.0	(0.02)
Bass	28	95	179	(1645)	43.0	(95.5)	97.0	(0.62)
Boot	21	262	371	(902)	61.8	(25.1)	97.7	(0.14)
	29	150	254	(1610)	49.8	(61.9)	97.6	(0.28)
Brandy	116	1	3[a]	-	0.6	-	95.7	-
	74	14	32	(306)	7.3	(15.8)	96.5	(0.51)
	32	97	265	(4032)	60.2	(208.3)	98.4	(0.54)
	125	1	3[a]	-	0.7	-	96.1	-
	98	11	69	(4489)	15.7	(231.8)	98.1	(0.95)
	48	125	335	(1722)	76.1	(89.0)	99.4	(.07)
Clear	26	124	151	(173)	34.6	(9.0)	96.0	(0.10)
	19	140	258	(1803)	67.1	(122.3)	97.9	(0.72)
	19	149	225	(416)	56.2	(26.0)	97.0	(0.21)
	34	78	151	(578)	42.7	(46.1)	97.5	(0.21)
Crane Chase	34	195	338	(1825)	68.9	(76.0)	98.9	(0.14)
	33	86	96	(28)	19.2	(1.1)	95.1	(0.03)
	26	124	176	(457)	42.3	(26.3)	96.7	(0.20)
Des Moines	90	0	0[a]	-	0.0	-	0.0	-
	29	122	291	(1907)	41.7	(39.2)	97.0	(0.25)
	45	79	136	(342)	14.9	(4.1)	95.9	(0.08)
Harmon	86	0	0[a]	-	0.0	-	0.0	-
	30	30	45	(61)	22.5	(15.2)	95.1	(0.30)

TABLE 2, *(continued)*

Lake name	Days at large	Number handled	Population estimate		% Fall survival		% Mean daily survival	
			FINGERLING					
Island	28	160	389	(5269)	95.7	(378.5)	99.8	(0.53)
Johnson	121	12	21	(420)	6.7	(43.2)	97.8	(0.62)
	79	61	111	(1056)	35.6	(108.5)	98.7	(0.13)
	37	117	186	(676)	59.6	(69.4)	98.6	(0.14)
	119	4	6[a]	-	1.9		96.7	-
	92	21	23	(1)	7.4	(0.1)	97.2	(0.00)
	42	112	178	(281)	57.1	(28.8)	98.7	(0.05)
Lower Holly	33	65	81	(53)	40.5	(13.2)	97.3	(0.07)
	85	0	0[a]	-	0.0		0.0	-
	55	12	22	(80)	22.0	(79.7)	97.3	(0.52)
Little Sand	33	105	124	(202)	29.5	(11.4)	96.4	(0.11)
	19	195	328	(1123)	82.8	(71.6)	99.0	(0.28)
	24	156	220	(237)	55.0	(14.8)	97.5	(0.08)
	85	1	1[a]	-	0.1		92.5	-
	29	88	100	(34)	50.0	(8.6)	97.6	(0.04)
	28	136	246	(1641)	61.9	(104.1)	98.3	(0.33)
Little Sand	37	87	134	(542)	35.6	(38.1)	97.2	(0.21)
Leisure	14	94	144	(135)	49.7	(16.0)	95.1	(0.30)
	34	85	241	(2729)	80.3	(303.2)	99.4	(0.40)
	84	3	8[a]	-	0.5		94.0	-
	29	37	100	(689)	66.7	(306.2)	98.6	(0.80)
	40	77	147	(491)	36.7	(30.7)	97.5	(0.14)
Mathews	86	0	0[a]	-	0.0		0.0	-
	27	52	188	(3900)	23.6	(61.2)	94.8	(1.36)
Mud/Callahan	42	127	193	(1852)	14.6	(4.6)	95.5	(0.11)
Pear	14	98	148	(211)	65.5	(41.3)	97.0	(0.46)
	27	78	108	(133)	53.5	(32.6)	97.7	(0.15)
Pulaski	27	243	338	(762)	48.0	(15.4)	97.3	(0.09)
Sparkling	119	2	-	-	0.4	-	95.5	-
	77	3	-	-	0.6	-	93.6	-
	35	5	-	-	0.4	-	87.6	-
	124	2	-	-	1.0	-	95.6	-
	97	0	-	-	0.0	-	0.0	-
	47	13	-	-	2.6	-	92.5	-
Trego	38	470	691	(835)	76.8	(10.3)	99.3	(0.01)
Twenty-Six	95	1	2[a]	-	0.0		91.5	-
	34	79	119	(154)	17.0	(3.1)	94.9	(0.08)
Whitefish	39	269	669	(5857)	(89.2)	(104.1)	99.7	(0.09)
			FINGERLING - 1984[b]					
Butternut: LCO	42	283	601	(3163)	60.5	(32.1)	98.8	(0.05)
MQ	42	177	445	(3781)	44.5	(37.8)	98.1	(0.10)
Clear: large	43	37	74	(286)	49.5	(127.0)	98.4	(0.27)
small	43	44	86	(293)	57.0	(130.1)	98.7	(0.21)
Nancy: Leech	44	161	238	(299)	20.0	(2.1)	96.4	(0.03)
Towanda: LCO	41	77	179	(1168)	59.5	(129.7)	98.7	(0.10)
Leech	41	88	158	(452)	52.7	(50.2)	98.5	(0.10)
Twenty-Six: large	42	129	247	(795)	54.8	(39.3)	98.6	(0.07)
small	42	93	147	(227)	32.6	(11.2)	97.4	(0.06)

[a] Population estimates determined by the ratio method.
[b] 1984 fingerling estimates not used in stepwise regression but used to test derived model. Broodstock or size class given with lake name (LOC)=Lac Court Oreilles, MQ=Minocqua).

Two of six fry stockings resulted in a measurable year-class of fingerlings. Observed survival, mean daily survival, and fall density of the two successful fry stockings were lower than from the two successful egg stocks; these values from fry stocking were 0.006%, 93.3% and 0.1 fish/ha for Bass Lake (Oneida County) and 0.016%, 93.7% and 0.4 fish/ha for Patterson Lake, respectively (Table 2). Mean total lengths of the surviving fingerlings were 269 mm in Bass Lake and 287 mm in Patterson Lake.

Survival was observed in two of the four stockings when fry were scatter-planted but no survival was observed from two stockings when fry were stocked en masse.

Success of egg stocking is not unprecedented. Forney (1969) stocked eyed muskellunge eggs in a controlled marsh on a shaded screen, 15 cm from the bottom in water 30 cm deep. He reported that survival to July was 1.1%, which exceeded survival from fry stocking in the same marsh the following year.

The success of the two egg stockings may be due in part to bottom substrate. Johnson (1969) and Dombeck et al. (1984) demonstrated higher survival of muskellunge eggs on inorganic substrates versus organic substrates. Dombeck et al. (1984) found lakes with high B.O.D./low dissolved oxygen conditions at the bottom-water interface of muskellunge spawning sites had little natural recruitment of muskellunge. Typical spawning site conditions in Wisconsin are muck and detritus (Dombeck 1979; Oehmcke 1974), which could account for the poor natural reproduction observed in many waters. The stocking of fertilized eggs on sand and gravel substrates in Deep and Patterson lakes produced a greater fall fingerling density than typically observed from natural reproduction or fry stocking in many Wisconsin waters (Hanson, unpublished data). This success was achieved even though the number of eggs stocked in these lakes was less than the total egg deposition from four average-size females (Scott and Crossman 1973).

The failure of three egg-stocking experiments, however, indicated that factors other than bottom substrate influenced the survival of muskellunge from egg to fingerling. Interestingly, both egg and fry stockings in Patterson Lake were successful. Patterson Lake had an unusually high white sucker population (WDNR, unpublished data) and during the fry stocking, white suckers were actively spawning in the area of the fry release. A large white sucker population also inhabits Deep Lake where the other successful egg stocking occurred. Axon and Kornman (1986) reported golden redhorse were an indicator of muskellunge populations and habitat in Kentucky streams. Hanson (1986) observed that muskellunge growth was positively associated with adult catostomid abundance in nine Wisconsin lakes. While none of these results explicitly demonstrated an interaction between muskellunge and catostomids, future research may confirm the presence of an important predator-prey relationship.

Survival of muskellunge fry stocked into intensively-managed rearing ponds is highly variable in Wisconsin (Hanson, unpublished data). The timing of pond stocking with zooplankton blooms and adequate abundance of acceptable-sized forage, usually fry of some other species, are thought to be the limiting factors in hatchery pond survival. Given highly-variable survival of fry stocked under intensively-managed conditions, it is not unreasonable to expect poor survival in natural lakes, where forage is less abundant and predators are present.

Failure of some fry stockings may simply be due to stocking of insufficient numbers. Fry stocking rates in this study (mean of 2000 fry/ha) were such that if mean daily survival had been the mean from the two successful egg planting trials (94.2%), the expected fall density of fingerlings after 120 days would have averaged 1.5 fingerlings/ha. Since such fall densities were not found, it seems likely that the stocked fry suffered catastrophic post-stocking mortality in addition to the on-going daily mortality. Until further knowledge is gained on factors affecting fry survival, it is recommended that if fry stockings are used, they be at greater densities than occurred in this study to compensate for unknown sources of mortality.

Factors Affecting Fingerling Survival

Observed survival of stocked fingerlings ranged from 0.0% to 95.7% and averaged 38.7% in the 68 evaluations with population estimates. Mean daily survival ranged from 91.5% to 99.8% and averaged 97.4% in the 64 lakes with non-zero observed survival (Table 2).

The independent factor of primary interest was length at stocking. In most muskellunge propagation programs, selection of a stocking size dictates stocking date and temperature at stocking, while stocking rates are usually dependent on production capacities and demand for fingerlings. Based on simple correlations, observed survival was positively related to length at stocking (Table 3; Fig. 1), and negatively related to days at large. Theory suggests that, aside from an initial stocking mortality, observed survival should decline exponentially with time. Since there was a high negative correlation between length at stocking and days at large, analyses based solely on observed survival would not discriminate between effects of these two factors.

The use of the geometric mean daily survival appeared to remove the systematic effect of days at large, since the number of days at large was not correlated with mean daily survival (Table 3). Mean daily survival was also correlated with length at

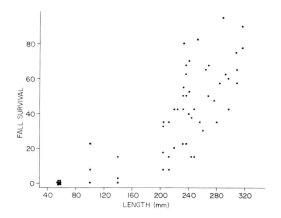

Figure 1. Plot of observed survival to fall (%) versus mean length at stocking (mm) for fingerling stockings.

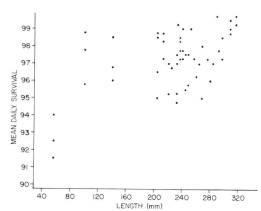

Figure 2. Plot of mean daily survival (%) versus length at stocking (mm) for fingerling stockings.

stocking (Fig. 2), but appreciably less than the correlation of observed survival and length.

A correlation matrix of observed survival, mean daily survival, and the independent factors indicated that stocking rate and number stocked had significant negative correlations with both survival indices (Table 3). Observed and mean daily survival had a significant positive correlation with Julian date of stocking, and observed survival had a significant negative correlation with temperature at stocking. These correlations, however, were probably the result of significant correlations between the following independent variables: length and date of stocking, length and number stocked, and length and temperature at stocking. The remaining correlations among variables are presented in Table 3 for consideration by researchers in the design of future studies but are not discussed.

Because many of the independent factors were correlated with each other, stepwise multiple re-

Table 3. Simple correlation coefficients (above) and significance levels (below) between observed survival, mean daily survival, fall density (No./ha) and independent factors; N=59 except for correlations involving mean daily survival (N=55) and fall density (N=27).

	Observed survival	Daily survival	Length	Number stocked	Stocking rate	Lake area	Temp.	Stocking date	Days at large
Mean daily survival	0.700 0.001	-	-	-	-	-	-	-	-
Total length	0.787 0.001	0.546 0.001	-	-	-	-	-	-	-
No. stocked	-0.370 0.004	-0.545 0.001	-0.509 0.001	-	-	-	-	-	-
Stocking rate	-0.435 0.001	-0.564 0.001	-0.648 0.001	0.878 0.001	-	-	-	-	-
Lake area	0.062 0.640	-0.016 0.907	0.102 0.444	0.312 0.016	0.039 0.768	-	-	-	-
Temperature	-0.331 0.010	-0.147 0.283	-0.391 0.002	-0.045 0.736	-0.006 0.965	-0.340 0.008	-	-	-
Stocking date	0.788 0.001	0.487 0.001	0.964 0.001	-0.472 0.001	-0.575 0.001	0.073 0.582	-0.470 0.001	-	-
Days at large	-0.715 0.001	-0.146 0.287	-0.768 0.001	-0.274 0.036	0.327 0.012	-0.031 0.813	0.224 0.088	-0.807 0.001	-
Fall density	0.661 0.001	0.542 0.004	0.134 0.504	-0.409 0.034	-0.558 0.002	-0.486 0.010	0.041 0.840	0.166 0.010	-0.518 0.006

Table 4. Variables selected to be significantly related in a stepwise multiple regression to observed fall survival (N=59), geometric mean daily survival (N=55), and fall density (N=27).

Dependent variable	Independent variable	Regression coefficient	Significance level
Observed fall survival (%)	Intercept	5.0855	-
	Length at stocking (mm)	0.2022	0.0001
	Days at large	-0.2183	0.0335
(Model R^2 = 64.9%)			
Mean daily survival (%)	Intercept	90.9585	-
	Length at stocking (mm)	0.0225	0.0001
	Days at large	0.0303	0.0002
	Number stocked	-0.0006	0.0020
(Model R^2 = 56.9%)			
Observed fall density (No./ha)	Intercept	-3.6605	-
	Stocking rate (No./ha)	0.4974	0.0038
	Days at large	-0.1155	0.0448
	Length at stocking	0.0281	0.1098
(Model R^2 = 50.2%)			

gression was used to choose models containing only the most significant variables (Table 4). Only two variables had a significant influence on observed fall survival, length at stocking (+) and days at large (-) accounted for 64.9% of the variation in observed survival. Three variables had a significant influence on mean daily survival, length at stocking (+), days at large (+) and number stocked (-) accounted for 56.9% of the variance in mean daily survival.

Length at stocking was the most important positive influence on both observed survival and mean daily survival. Most authors agree that larger fingerlings show increased survival (Johnson 1976, Serns and Andrews 1986, Stein et al. 1981), but Johnson (1976) has suggested that the gains in survival are not sufficient to offset the increased cost of raising fingerlings to larger sizes.

The effect of increased days at large (earlier stockings) was a reduction in observed survival but an increase in mean daily survival over the period. Of course, observed fall survival does decline with earlier stockings regardless of either length at stocking or an increased mean daily survival, so earlier stockings must be made at higher densities to achieve similar fall populations.

The number of fingerlings stocked was negatively correlated with mean daily survival. This effect was presumably the result of the positive correlation between fingerling size and number stocked and probably has no biological significance; in six of seven trials when 1,000 or more fingerlings were stocked the mean length was only 58 mm.

A comparison of unweighted and weighted stepwise regressions on a subset of 46 stocking evaluations for which population estimate variances could be calculated indicated that, if the variances of the population estimates producing the survival values could be reduced, the resulting models would have improved fits. Though this result is intuitive, it is not usually statistically verified. The unweighted stepwise regression selected length at stocking, days at large and number stocked as the most important variables in a model accounting for 49.8% of the variation in observed survival (Table 5). When a weighted regression using the inverse of the survival estimate variances as the weight was applied to the same trials, the same variables accounted for 77.5% of the weighted variance in observed survival and 80.4% of the variance, if lake area is added to the model. Similarly, the unweighted stepwise regression procedure selected length at stocking, days at large and number of stocked in a model accounting for 36.3% of the variation in mean daily survival (Table 5). However, a weighted model R^2 of 41.5% based on only length at stocking and days at large, was found in the weighted regression.

Since the relationship between length and observed survival appeared curvilinear (Fig. 1), analyses of factors affecting survival were also done using $\log_{(10)}$ transformations. Observed fall survival values had higher correlations with the independent factors using transformed data than occurred with non-transformed data, but there were no differences with log-transformed mean daily survival. Since stepwise regression selected the same factors (length at stocking and days at large) at approximately the same significance levels as analyses with untransformed data, and non-trans-

Table 5. Comparison of unweighted and weighted stepwise regression for 46 stocking evaluations. The inverse of the estimated variance of the dependent variable was used as the weighting factor.

Dependent variable	Weighting	Independent variable	Regression coefficient	Significance level
Observed fall survival	None	Intercept	-25.3347	-
		Length at stocking	0.3613	0.0001
		Number stocked	-0.0180	0.0357
		Days at large	-0.2808	0.0388
		(Model R^2 = 49.8%)		
Observed fall survival	1/Variance	Intercept	-41.5807	-
		Length at stocking	0.3865	0.0001
		Number stocked	-0.0401	0.0083
		Days at large	-0.2930	0.0001
		Area	0.2153	0.0163
		(Model R^2 = 80.4%)		
Mean daily survival	None	Intercept	91.2520	-
		Length at stocking	0.0231	0.0002
		Number stocked	-0.0014	0.0196
		Days at large	0.0284	0.0028
		(Model R^2 = 36.3%)		
Mean daily survival	1/Variance	Intercept	89.1226	-
		Length at stocking	0.0280	0.0001
		Days at large	0.0295	0.0001
		(Model R^2 = 41.5%)		

formed data lead to simpler interpretations, analyses of transformed data are not further discussed.

Data from six evaluations in Sparkling Lake were omitted from the above analyses. Minimum observed survival averaged 0.8% in Sparkling Lake and ranged from 0.0 to 2.6%. Mean daily survival for the five non-zero observed survivals averaged 93.0%. While all sizes of fingerlings stocked in Sparkling Lake survived poorly, larger fingerlings survived better than smaller fingerlings. These results suggested similar mechanisms affecting survival were present in Sparkling Lake but at a different intensity than in other lakes. Serns and Andrews (1986) further discuss results from Sparkling Lake.

Factors Influencing Fall Density

Analyses of factors affecting fall fingerling densities were not independent of the analyses of fingerling survival because survival and stocking rate uniquely determined fall density. A separate analysis of fall density, however, was expected to provide additional information on stocking success because survival was apparently unrelated to stocking rate in stepwise regression analyses.

Stocking evaluations in lakes which received multiple stockings of different-sized fish in the same year but on different dates were excluded. In simple correlations fall density was positively related to both stocking rate (Fig. 3) and fall survival, and negatively related to days at large and the actual number stocked (Table 3). The apparent contradiction between the effects of stocking rate and number stocked was probably caused by the influence of a few large lakes (i.e., Whitefish and Trego), which were stocked at low densities (but high numbers due to larger lake size) and had high survival rates (due to large fingerling size at stocking).

A stepwise multiple regression identified stocking rate, days at large, and length at stocking as the most important predictors of fall fingerling density (Table 4). The model R^2 was only 50.2%, however, indicating that fall density was not well determined by initial stocking information. The selection of the three variables was consistent with the results from the analysis of fingerling survival, except that length at stocking was the least important variable. The reduction in significance of length at stocking was attributed to the lower ranges of stocking lengths (221-318 mm) in the subset of trials used in the fall density analysis.

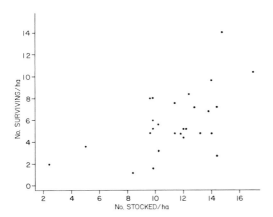

Figure 3. Plot of fall fingerling density (No./ha) versus stocking rate (No./ha) for fingerling stockings.

There has been speculation that increased stocking rates may have an adverse effect on the success of fingerling stockings (e.g., Porter 1977). Over the ranges of stocking rates observed in this study, there was no indication of maximum or optimum rates.

TESTING OF THE MODELS

Based on the fall survival and mean daily survival models derived above, two stockings were made in 1984 to specifically test the effect of length at stocking. Equal numbers of two size groups of fingerlings were stocked into both Clear and Twenty-Six lakes on the same date (Table 1). The mean length and standard deviation of the large-size group was 287 mm and 18.2 mm compared with 236 mm and 18.7 mm for the smaller-size group, respectively. The fingerlings were raised in separate ponds in the hatchery to eliminate any effect of grading a single hatchery lot into different size groups. Since days at large and number stocked were equal for both size-classes in each lake, any difference in survival would be attributable to length at stocking. According to the multiple regression models (Table 4), the larger fingerlings were expected to show a 10% higher observed fall survival and a 1% higher daily survival. In Twenty-Six Lake, the larger fingerlings had a 22% higher fall survival and a 1.2% higher daily survival as predicted; however, in Clear Lake the smaller fingerlings showed a 7% higher fall survival and 0.3% higher daily survival (Table 6).

Additional stockings were made in 1984 to assess survival differences between fingerlings from several different broodstocks (Table 1). Stockings were again made in equal numbers on the same date; however, in Butternut Lake the two stocks were different in mean length. The larger Lac Court Oreilles broodstock fingerlings showed a 25% higher fall survival and a 0.7% higher daily survival than the smaller Minocqua fingerlings (Table 6). If it can be assumed that broodstock source does not affect survival, then the results of this trial provided evidence for the positive effect of length at stocking. Stockings of Lac Court Oreilles and Leech Lake fingerlings in two other lakes did not vary greatly in length at stocking. Stockings in Towanda Lake showed survival very close to that predicted by the models; however, survival of the Leech Lake fingerlings in Nancy Lake was much lower than expected. Nancy Lake is known to have a large population of stunted northern pike and many of the recaptured muskellunge bore scars from predator attacks. Results from the 1984 trials provide some experimental evidence that length at stocking is an important factor in the success of fingerling stockings.

In contrast to the variable results on length of stocking, the 1984 trials provided good evidence that the fall density model was indeed correct. In

Table 6. Results of observed fall and geometric mean daily survival for 1984 fingerling stockings as predicted from multiple regression model and actually measured. Broodstock sources were Leech, Lac Court Oreilles (LCO), and Minocqua (MQ) Lakes.

Stocking		Length (mm)	Fall survival (%)		Daily survival (%)	
			Measured	Predicted	Measured	Predicted
Butternut:	LCO	295	60.5	55.6	98.8	98.3
	MQ	246	44.5	45.7	98.1	97.2
Clear:	LCO	287	49.5	53.7	98.4	98.6
	LCO	236	57.0	43.4	98.7	97.5
Nancy:	Leech	297	20.0	55.5	96.4	98.3
Towanda:	LCO	290	59.5	54.8	98.7	98.6
	Leech	297	52.7	56.2	98.5	98.7
Twenty-Six:	LCO	287	54.8	53.9	98.6	98.4
	LCO	236	32.6	43.6	97.4	97.3

Table 7. Fall density (No./ha) for 1984 fingerling stockings predicted from multiple regression model and actually measured.

Lake	Stocking rate	Days at large	Mean length	Fall Density Measured	Fall Density Predicted
Butternut	4.9	42	270	2.6	1.5
Clear	9.6	43	262	5.1	3.5
Nancy	3.8	44	297	0.8	1.5
Towanda	10.2	41	294	5.7	4.9
Twenty-Six	9.7	42	262	4.2	3.7

lakes receiving multiple stockings, stocking rate was the sum of the two stockings and the mean length at stocking was the weighted average of the two stockings. Although specific predictions from the multiple regression equation differed up to 1.5 fish/ha, the ranking of lakes by measured fall densities was preserved by the predicted densities (Table 7). Stocking rate appeared to be a consistent predictor of fall density.

MANAGEMENT IMPLICATIONS

Success of egg and fry stockings in this study were generally poor. Only two of five egg stockings and two of six fry stockings resulted in measurable survival. Average observed and daily survival from successful egg stockings was higher than from successful fry stockings, which suggests that egg stocking may be a better alternative than the more commonly-used fry stocking. However, neither technique provided consistent results compared with large fingerling stocking. Until further research can identify factors limiting survival of stocked eggs and fry, such stockings should be made only when surplus eggs or fry are available; in waters where previous egg or fry stockings have been successful; or when large fingerling stocking is not a feasible management alternative. If stocked fry suffer catastrophic post-stocking mortality in addition to ongoing daily mortality, stocking rates in this study may have been too low to produce a fall fingerling population. If fry stockings are to be made they should contain increased numbers of fry to allow for unknown sources of mortality.

Large fingerlings, in general, survive better than small fingerlings and such stockings seem the most effective way to increase muskellunge populations. Within the range of fingerling stocking rates analyzed in this study (2.4 to 17.0 fish/ha), there was no apparent effect of stocking rate on survival but fall density was highly correlated with stocking rate.

Results from testing of the models verified both the expected usefulness and limitations of the models. These models were poor predictors of outcomes of a given stocking likely due to exclusion of important variables affecting survival, such as predator abundance, forage availability, or habitat limitations; the inadequate measurement of included variables such as population estimate variances; or poor characterization of the stocked population size structure (e.g. by simple arithmetic means). Nevertheless, the models can be useful in the development of general cost-effective stocking strategy for an overall stocking program. Specific details of such a strategy will require additional information on potential hatchery production, demand for fingerlings, and costs of raising fingerlings to various sizes.

Results from this study highlighted the need to improve the experimental design in future studies of stocked fingerling survival. Such evaluations should attempt to minimize the high variance in population estimates through increased sampling effort or improved sampling methods. A problem identified with previous studies that were superimposed on regular stocking schedules is that many of the variables, including length at stocking, observed survival, temperature, date of stocking, days at large and daily survival, turn out to be highly intercorrelated. To adequately separate the effects of these variables, future studies must be designed either like the 1984 stocking of Clear and Twenty-Six lakes, where several variables were held constant in order to evaluate a single factor, or with factorial designs which have most combinations of levels of the independent variables, and thus little or no intercorrelation.

ACKNOWLEDGEMENTS

The authors wish to thank R. Masterjohn, D. Stafford, H. Carlson, J. Smith, R. Stueck, K. Arnold, L. Groth, R. Wendt, and R. Theis for their assistance in data collection. The support of the Spooner and Woodruff Fish Hatchery personnel, particularly L. Sather, J. Bush, M. Balcher, E. Mertz, C. Gahler and G. Lensert, is appreciated. J.

Fabert assisted in the data analysis and C. Johnson provided support which was critical to this study.

This study was supported in part by funds supplied by the Federal Aid in Fish Restoration Act under Dingell-Johnson Project F-83-R. Thanks are due to L. Christenson, J. Klingbiel, E. Lange, H. Snow and several anonymous reviewers for their comments on an earlier draft of this manuscript.

REFERENCES

Axon, J.R. and L.E. Kornman. 1986. Characteristics and management of native muskellunge streams in Kentucky. American Fisheries Society Special Publication 15:263-272.

Belusz, L.C. 1975. The use of isolation coves in assessing muskellunge stocking mortality. Proceedings Annual Conference of Southeastern Game & Fish Commission. 29:251-253.

Belusz, L.C. 1978. An evaluation of the muskellunge fishery of Lake Pomme de Terre and efforts to improve stocking success. American Fisheries Society Special Publication 11: 292-297.

Beyerle, G.B. 1973. Comparative growth, survival, and vulnerability to angling of northern pike, muskellunge, and the hybrid tiger muskellunge stocked in a small lake. Michigan Department of Natural Resources Fisheries Research Report 1799. Ann Arbor, Michigan USA.

Beyerle, G.B. 1981. Comparative survival and growth of 8.9- and 17.8 cm (3.5 and 7.0-inch) tiger muskellunge planted in a small lake with forage fishes. Michigan Department of Natural Resources Fisheries Research Report 1894. Ann Arbor, Michigan USA.

Dombeck, M.P. 1979. Movement and Behavior of the Muskellunge determined by radio-telemetry. Wisconsin Department of Natural Resources Technical Bulletin No. 113. Madison, Wisconsin USA.

Dombeck, M.P., B.W. Menzel and P.N. Hinz. 1984. The influence of substrate on muskellunge egg mortality. Transactions of the American Fisheries Society. 113:205-216.

Federation of Fly Fishermen. 1975. Promotional literature. Bartlesville, Oklahoma, USA.

Forney, J.L. 1969. Cornell University, Warm Water fisheries management in New York lake and ponds Project F17-R-13 job progress report. Albany, New York, USA.

Hanson, D.A. 1983. Wisconsin Department of Natural Resources. Statewide Fisheries Research Project F-83-R-17. Job progress report, study 228. Madison, Wisconsin, USA.

Hanson, D.A. 1986. Population characteristics and angler use of muskellunge in eight northern Wisconsin lakes. American Fisheries Society Special Publication 15:238-248.

Harshbarger, T.J. and P.E. Porter. 1982. Embryo survival and fry emergence from two methods of planting brown trout eggs. North American Journal of Fisheries Management 2:84-89.

Hess. L and C. Heartwell. 1979. Literature review of large esocids (Muskellunge, Northern Pike, hybrid Tiger Muskellunge). Pages 139-175 in J. Dubé and Y. Gravel, editors. Proceedings of the 10th Warm Water Workshop Northeastern Division, American Fisheries Society. Québec Ministère du Loisir, de la Chasse et de la Pêche, Direction de la recherche faunique, Montréal, Québec, Canada.

Johnson, L.D. 1969. Wisconsin Conservation Department. Statewide fisheries research project F-83-R. Job progress report, Study 201. Madison, Wisconsin, USA.

Johnson, L.D. 1976. Wisconsin Department of Natural Resources. Statewide fisheries research project F-83-R-12. Job Progress Report, Study 209. Madison, Wisconsin, USA.

Johnson, L.D. 1978. Evaluation of esocid stocking program in Wisconsin. American Fisheries Society Special Publication No. 11:298-301.

Klingbiel, J. 1981. The status of the muskellunge program. An informational report to the Natural Resources Board. Wisconsin Department of Natural Resources Administrative Report 11. Madison, Wisconsin, USA.

McNeil, F.I. and E.J. Crossman. 1979. Fin clips in the evaluation of stocking programs for muskellunge (*Esox masquinongy*). Transactions of the American Fisheries Society. 108:335-343.

Oehmcke, A.A., L. Johnson, J. Klingbiel and C. Winstrom. 1974. The Wisconsin Muskellunge. Its life history, ecology and management. Wisconsin Department of Natural Resources Publication 8-3600(74). Madison, Wisconsin USA.

Porter, L. 1977. Review of selected literature of muskellunge life history, ecology, and management. Minnesota Department of Natural Resources, Section of Fisheries Special Publication No. 119.

Ricker, W.E. 1975. Computation and interpretation of biological statistics of fish populations. Bulletin 191. Fisheries Research Board of Canada, Ottawa. 382 pp.

Robson, D.S. and H.A. Regier. 1978. Estimation of Population Number and Mortality Rates. Pages 124-158 in W.E. Ricker, editor. Methods for assessment of fish production in fresh waters. IBP Handbook No. 3, 3rd Edition.

Scott, W.B. and E.J. Crossman. 1973. Freshwater Fishes of Canada. Bulletin 184. Fisheries Research Board of Canada, Ottawa. 966 pp.

Serns, S.L. and L.M. Andrews. 1986. Survival and growth of 4-, 8-, and 12- inch stocked muskellunge fingerlings in four Vilas County, Wisconsin lakes. American Fisheries Society Special Publication 15:229-237.

Stein, R.A., R.F. Carline and R.S. Hayward. 1981. Largemouth bass predation on stocked tiger muskellunge. Transactions of the American Fisheries Society 110:604-612.

Swanson, B.L. 1982. Artificial turf as a substrate for incubating lake trout eggs on reefs in Lake Superior. Progressive Fish-Culturist 44:109-111.

WDNR (Wisconsin Department of Natural Resources). 1979. Muskellunge management plan (sport: all waters). pp7-1 through 7-7 in Fish and Wildlife Comprehensive Plan, Part I: Management strategies 1979-1985. Wisconsin Department of Natural Resources, Madison, Wisconsin, USA.

WDNR. 1982. Wisconsin muskellunge waters. Wisconsin Department of Natural Resources Publ. 1-3600(32):40pp.

Weithman, A.S. and R.O. Anderson. 1977. Survival, growth and prey of Esocidae in experimental systems. Transactions of the American Fisheries Society 106(5):424-430.

Youngs, W.D. and D.S. Robson. 1978. Estimation of population number and mortality rates. Methods for assessment of fish production in fresh waters. T. Baganel (ed). IBP Handbook No. 3, 3rd Edition.

Comparative Survival and Growth of Three Sizes of Muskellunge Fingerlings Stocked in Four Northern Wisconsin Lakes

STEVEN L. SERNS AND LLOYD M. ANDREWS

Wisconsin Department of Natural Resources
Box 440
Woodruff, WI 54568

ABSTRACT

The survival and growth of 4-, 8-, and 12-inch muskellunge fingerlings stocked at the rate of 4/acre in four northern Wisconsin lakes in 1976 and 1977 were investigated. Fish from the six separate stockings were given a distinctive fin-clip, and data on survival and growth were obtained in subsequent electrofishing and fyke-netting collections.

In Arrowhead Lake, in all three size groups, survival was similar for fish stocked in 1976, but for those stocked in 1977 survival was highest for 12-in fingerlings. Survival was also highest for the 12-in fingerlings stocked in Brandy and Johnson lakes in both years, but in Sparkling Lake survival was negligible both years for all stocked fingerlings. Growth in Arrowhead Lake was better than growth of fingerlings in hatchery rearing ponds. The good survival and growth of fingerlings stocked in Arrowhead Lake may be attributed to the low density of northern pike and other predators and an abundant supply of young-of-year yellow perch, a good potential food source.

Small muskellunge fingerlings (4 in) can contribute to a sport fishery when stocked in a lake with a low density of predators and a good supply of suitable forage. Where moderate-to-large northern pike populations exist it is recommended that only larger (12 in plus) fingerlings be stocked.

The increasing expense of hatchery production requires that propagation costs be reduced without sacrificing survival of stocked fish. Costs of propagating muskellunge could be reduced by harvesting and distributing hatchery-reared fingerlings at sizes smaller than 12 in. Much of the expense of the program is the result of obtaining live food to raise the fingerlings to larger lengths.

Until recently, little research was directed at determining the survival of muskellunge fingerlings in relation to their size at stocking. Johnson (1982) reported no significant difference between the survival of small (6.5 to 11.0 in) vs. large muskellunge fingerlings (8.5 to 13.0 in) in studies conducted between 1972 and 1975. Differences in total length between the maximum length of the "small group" and the minimum length of the "large group" were usually only 0.1 in in most trials and certainly confounded these results. In other work, Johnson (1982) reported better survival of 9.5-in fingerlings when compared with 2.5-in fish in eight lakes ranging in size from 42 to 5,000 acres.

Both fish managers and hatchery personnel suggested research be conducted to provide more information on survival and growth of various-sized stocked muskellunge. This study was designed to examine the survival and identify possible sources of mortality of 4-, 8- and 12-in muskellunge fingerlings stocked in four northern Wisconsin lakes. These size groups were chosen for the following reasons. First, the number of muskellunge in the hatchery ponds is usually thinned in late June, when the fish are about 4 in, to decrease losses due to cannibalism and to reduce total feed costs. Second, in recent years some fishery managers have requested 8-in fingerlings for stocking the lakes they manage while others wanted 12-in fish.

STUDY SITES

Four Vilas County lakes were selected for this study because of their similar size (<150 acres) and water chemistries (Table 1). Arrowhead, Brandy, and Johnson Lakes are connected by a stream which flows out of Arrowhead Lake, through Brandy Lake, and into Johnson Lake. Sparkling Lake is similar in size and water hardness (as indicated by alkalinity and conductivity values) to the other lakes but is a seepage (landlocked) lake having water of much higher transparency (Table 1).

Brandy Lake had the greatest density of rooted aquatic vegetation when compared to the other study lakes, followed by Johnson, Arrowhead, and Sparkling Lakes (Table 2). Only six macrophyte species were observed in Sparkling Lake, and each was classified as sparse. Determinations as to the relative abundance of rooted aquatic vegetation were made subjectively, based on visual observations.

The fish communities observed during spring and fall electrofishing collections (fall 1976 through fall 1979) varied considerably among the study lakes.

Table 1. Morphological, chemical, and physical data on the four study lakes in Vilas County, Wisconsin, from Black et al (1963).

Lake	Surface area (acre)	Maximum depth (ft)	Water source	Methyl orange alkalinity (ppm)	pH	Conductivity (umhos @ 77 F)	Secchi disc transparency (ft)
Arrowhead	99	43	spring	38	6.8	99	10.5
Brandy	110	44	drainage	36	6.6	87	9.5
Johnson	78	42	drainage	45	7.4	98	6.0
Sparkling	127	64	seepage	25	7.4	92	30.5

Table 2. The relative abundance* of rooted aquatic macrophytes in four Vilas Co., Wisconsin lakes on 20 July 1976.*

		Lake			
Scientific name	Common name	Arrowhead	Brandy	Johnson	Sparkling
Equisetum sp.	Horsetail	s	-	-	-
Isoetes sp.	Quillwort	-	-	c	s
Typha sp.	Cattail	c	c	c	-
Sparganium sp.	Burreed	s	-	c	-
Najas flexilis	Slender naiad	a	a	c	s
Potamogeton pectinatus	Comb pondweed	c	c	-	-
P. amplifolius	Largeleaf pondweed	a	a	a	-
P. gramineus	Grassleaf pondweed	c	c	s	s
P. epihydrus	Ribbonleaf pondweed	a	c	s	-
P. robbinsii	Fern pondweed	-	a	a	-
P. praelongus	Whitestem pondweed	-	a	s	-
P. zosteriformis	Flatstem pondweed	c	c	c	-
Sagittaria sp.	Arrowhead	c	c	s	-
Elodea canadensis	Waterweed	s	c	a	-
Vallisneria americana	Eel grass	s	-	-	-
Dulichium arundinaceum	Pond sedge	s	-	s	-
Eleocharis sp.	Spikerush	s	-	s	-
Scirpus sp.	Bulrush	s	s	s	-
Calla palustris	Water arum	-	-	-	s
Eriocaulon septangulare	Pipewort	-	-	s	-
Pontederia cordata	Pickerel weed	c	c	-	-
Juncus spp.	Rush	s	a	-	-
Polygonum amphibium	Water knotweed	-	-	s	s
Ceratophyllum demersum	Hornwort	-	c	c	-
Nymphaea odorata	Fragrant water lilly	a	-	a	-
Nuphar variegatum	Bullhead pond lily	c	c	-	-
Myriophyllum exalbescens	Spiked water milfoil	-	a	a	-
Utricularia vulgaris	Great bladderwort	s	-	-	-
Lobelia sp.	Lobelia	-	-	-	s
Relative abundance of all rooted aquatic macrophytes		c	a	c-a	s

*Determined by visual observation
**s = sparse, c = common, a = abundant

Table 3. Data on the muskellunge fingerlings stocked in four Vilas Co., Wisconsin Lakes in 1976 and 1977 and surface water temperatures at the time of stocking.

Year stocked	Date stocked	Size group (in)	Mean total length ($\pm$ 2 SD) at stocking (in)	Fin Clip[b]	No. stocked in each lake[a]				Surface temperature (°F) at time of stocking			
					A	B	J	S	A	B	J	S
1976	6/29	4	3.9 ± 0.6	LV	396	440	312	508	73.0	69.5	73.0	70.0
	8/10	8	8.0 ± 1.2	RV	396	440	312	508	70.0	73.0	71.5	72.0
	9/21	12	11.6 ± 1.6	BV	396	440	312	508	60.0	58.5	62.0	62.0
1977	6/29	4	5.5 ± 0.6	RP	396	440	312	508	68.0	68.0	70.0	63.5
	7/20	8	8.4 ± 0.6	LP	396	440	312	508	74.0	72.0	75.0	74.0
	9/14	12	12.1 ± 1.0	AN	396	440	312	508	61.0	60.0	62.0	62.0

[a] A = Arrowhead, B = Brandy, J = Johnson, and S = Sparkling.

[b] LV = left ventral, RV = right ventral, BV = both ventrals, RP = right pectoral, LP = left pectoral, and AN = anal.

The fish assemblages of Brandy and Johnson lakes could be characterized as primarily esocid-centrarchid-catostomid complexes (northern pike, largemouth bass, bluegill, and white sucker), whereas the fish community of Arrowhead Lake was dominated by esocid-percid species. The Sparkling Lake fishery was basically an esocid-percid-centrarchid community (muskellunge, walleye, yellow perch, and smallmouth bass). Electrofishing runs in the falls of 1976 and 1977 indicated high densities of young-of-year yellow perch in Arrowhead Lake in both years, while there appeared to be low numbers of small-size forage fish in the other three lakes. Fyke-net sampling in May of 1981 and 1982 indicated low densities of adult northern pike in Arrowhead and Sparkling Lakes (CPUE = 0.0-0.1) when compared with Johnson and Brandy Lakes (CPUE = 4.2-12.7).

METHODS

Each lake was stocked in 1976 and 1977 with 4-, 8- and 12-in muskellunge fingerlings at the rate of 4/surface acre (Table 3); one exception was that in 1977 fish in the smallest group averaged 5.5 in. Fingerlings belonging to each size group each year were uniquely marked by fin-clipping without anesthesia and were held overnight in a hatchery trough before stocking. At the time of planting, a sample of the fingerlings (thirty 4-in, twenty 8-in, and ten 12-in) were held for 48 hours in a 4x3x3 ft cage in each lake without food to assess immediate mortality resulting from stress from fin clipping, transportation, handling, and stocking. Surface water temperatures were recorded to the nearest 0.5° F in each lake on each stocking date.

During 1976 the entire shoreline of each lake was electrofished with a 230-volt, AC electrofishing boat about 1 month after each size-group was stocked to estimate mortality. In addition, the number of surviving muskellunge of each size-group stocked in 1976 was estimated by mark-recapture techniques (Schnabel method; Ricker 1975) in each lake every fall and spring, beginning with the fall of 1976 and continuing through the fall of 1979. The survival of each length group stocked in 1977 was also estimated by the same method in the fall of 1977 and the spring and fall of 1978 and 1979.

Annual survival rates and monthly instantaneous total mortality rates from the time of stocking to each subsequent sampling date were determined for each size-group in each lake by comparing the mark-recapture population estimates calculated at various time intervals after stocking with the total number stocked (Ricker 1975). Catch/effort (number caught/net-day) data were determined for adult muskellunge from the 1976 and 1977 stockings in spring fyke-net (1-in square mesh) surveys conducted from 1981-1983.

All stocked fish captured for the first time in any electrofishing survey from 1976-79, or fyke-netting surveys in 1981 (12-30 April), 1982 (4-10 May), and 1983 (30 April-6 May) were measured for total length to the nearest 0.1 in and given a temporary (top caudal) fin clip. For each sampling date, the mean length of each size-group stocked and the growth increment from stocking to subsequent recapture were determined.

RESULTS

Immediate mortality was negligible for all size-groups held in live cages for 48 hr. None of the caged fingerlings in any size-group held in Arrowhead or Johnson Lakes in either year died within 48 hr; of the 4- and 8-in caged fish in Brandy Lake, 7 and 5%,

Table 4. Population, survival, and mortality estimates calculated for muskellunge fingerlings stocked in four Vilas County, Wisconsin lakes in 1976 and 1977 and sampled spring and fall through 1979. Top to bottom figures for each fall and spring sampling period are population (Schnabel), percent annual survival in ( ), and monthly instantaneous total mortality (Z) estimates.

Lake	Year stocked	Size group at stocking (in)	Fall 1976	Spring 1977	Fall 1977	Spring 1978	Fall 1978	Spring 1979	Fall 1979
Arrowhead	1976	4	91 (23) 0.37	91 (23) 0.14	6 (2) 0.26	24 (6) 0.13	14 (4) 0.12	*	*
		8	124 (31) 0.46	131 (33) 0.12	33 (8) 0.22	17 (4) 0.15	11 (3) 0.15	*	*
		12	164 (41) 0.88	161 (41) 0.12	24 (6) 0.22	22 (6) 0.15	9 (2) 0.15	*	*
	1977	4			60 (15) 0.47	7 (2) 0.38	18 (5) 0.19	5 (1) 0.20	*
		8			135 (34) 0.36	26 (7) 0.30	30 (8) 0.17	10 (3) 0.18	*
		12			254 (64) 0.44	108 (27) 0.17	70 (18) 0.13	54 (14) 0.10	*
Johnson	1976	4	21 (7) 0.67	4 (1) 0.42	*	*	*	*	*
		8	111 (36) 0.41	80 (26) 0.15	12 (4) 0.22		12 (4) 0.12	*	*
		12	186 (59) 0.51	62 (20) 0.22	*	*	*	*	*
	1977	4			*	*	9 (3) 0.22	*	*
		8			23 (7) 0.87	9 (3) 0.39	8 (3) 0.24	*	*
		12			178 (57) 0.56	67 (21) 0.21	27 (9) 0.18	*	*
Brandy	1976	4	*	*	*	*	*	*	*
		8	32 (7) 1.05	4 (1) 0.45	*	*	*	*	*
		12	265 (60) 0.51	36 (8) 0.31	*	*	*	*	*
	1977	4			*	*	*	*	*
		8			69 (16) 0.62	46 (10) 0.25	5 (1) 0.30	*	*
		12			335 (76) 0.27	158 (36) 0.13	6 (1) 0.32	*	*
Sparkling	1976 and 1977	4 8 12	Inadequate numbers of fish captured in both years for population estimates.						

*Too few captured to allow population estimate using multiple mark-recapture technique.

Table 5. Total catch of adult muskellunge stocked in 1976 and 1977 in spring fyke-net surveys conducted from 1981-1983 in four Vilas Co., Wisconsin lakes.

Lake	Year stocked	Size group at stocking (in)	Spring 1981		Spring 1982		Spring 1983	
			Catch	Net-days effort	Catch	Net-days effort	Catch	Net-days effort
Arrowhead	1976	4	7	35	15	42	8	42
		8	6	35	16	42	10	42
		12	3	35	11	42	7	42
	1977	4	0	35	8	42	4	42
		8	3	35	11	42	9	42
		12	5	35	27	42	16	42
Johnson	1976	4	*		2	12	*	
		8	*		2	12	*	
		12	*		0	12	*	
	1977	4	*		0	12	*	
		8	*		0	12	*	
		12	*		2	12	*	
Brandy	1976	4	0	18	0	12	0	36
		8	0	18	0	12	0	36
		12	0	18	1	12	0	36
	1977	4	2	18	0	12	1	36
		8	0	18	0	12	1	36
		12	0	18	1	12	5	36
Sparkling	1976	4	0	14	1	15	*	
		8	0	14	0	15	*	
		12	0	14	0	15	*	
	1977	4	0	14	0	15	*	
		8	0	14	0	15	*	
		12	0	14	0	15	*	

*No netting done.

respectively, died within 48 hr in 1976; 3 and 20%, respectively, of the 4- and 12-in fingerlings held in Sparkling Lake in 1976 perished within 48 hr. There was no mortality of the other size-groups in 1976 or any of the size-groups in 1977 in Brandy and Sparkling Lakes.

An attempt was made to capture the stocked fingerlings by electrofishing one month after they were introduced. However, none of the 4-in fingerlings were captured in late July and not enough 4- or 8-in fingerlings were caught in early September in any of the lakes to allow for a population estimate. Several of the fingerlings stocked at 4 and 8 in were caught during the autumn shocking period (late October-early November).

Initial survival (to first fall) of 4- and 8-in fingerlings and long-term survival (one year+) of 12-in fingerlings was generally higher in Arrowhead Lake than in the other three bodies of water (Tables 4 and 5). While initial survival of 12-in fingerlings was good in Johnson and Brandy Lakes, few were caught as adults in the springs of 1981 and 1982. Both short- and long-term survival of muskellunge fingerlings in Sparkling Lake was poor. There appeared to be little difference in survival of the three size-groups stocked in Arrowhead Lake in 1976, however, the 12-in fingerlings stocked in 1977 exhibited better survival than either the 4-in or 8-in fingerlings. Monthly instantaneous mortality rates for fish sampled in fall 1976 were highest for 12-in fish (approximately 1.5 mo after stocking), compared to rates for 8-in fish (3 mo after stocking) and 4-in fish (5 mo after stocking). However, in subsequent monthly samples estimates of instantaneous mortality were similar for all three size-groups. Mortality estimates for the three size-groups stocked in 1977 were similar in fall samples that year but were lower for the 12-in fish during subsequent sampling periods (Table 4).

Water temperatures at the time of stocking (Table 3) did not appear to affect survival of the muskellunge fingerlings in this study (Tables 4 and 5). Short- and long-term survival was higher for 4-in fish stocked in Arrowhead Lake at warmer temperatures in 1976 than for the 1977 stocking; however, the 8-in fish from the 1976 plant also survived better than the 1977 stocking, but the water temperatures at the time they were stocked in 1976 were cooler than in 1977.

Mean total lengths of muskellunge collected in Arrowhead Lake at various times after stocking were higher for the fingerlings stocked at 4 in than those stocked at 8 and 12 in. In the other three lakes no one group had higher mean lengths than another when sampled at various time intervals after stocking (Table 6).

DISCUSSION

Survival

Factors which have been associated with the survival of stocked muskellunge fingerlings in previous studies are: density of potential predators, density of rooted aquatic vegetation, forage availability, and water temperature at the time of stocking (Belusz 1978, Johnson 1982). Increased density of adult northern pike has been implicated in the decline of muskellunge populations and also in the survival of stocked muskellunge fingerlings (Oehmcke 1951, Scott and Crossman 1973, Johnson 1981). In laboratory studies, Caplan (1982) found that young-of-year northern pike compete with young-of-year muskellunge for food and they prey on the muskellunge as well.

The better survival of muskellunge fingerlings stocked in Arrowhead Lake in 1976, compared with survival in the other study lakes, may be related to the low density of adult northern pike, moderate aquatic vegetation density, and presence of large

Table 6. Mean total lengths at recapture and growth increments since stocking for fingerling muskellunge stocked in 1976 and 1977 in four Vilas County, Wisconsin lakes. Top to bottom figures for each fall and spring sampling period are mean total length in inches, number in sample, and average growth increment (inches) since stocking.

Lake	Year stocked	Size group at stocking (in)	Fall 1976	Spring 1977	Fall 1977	Spring 1978	Fall 1978	Spring 1979	Fall 1979	Spring 1981	Spring 1982	Spring 1983
Arrowhead	1976	4	12.5	13.1	17.9	19.2	23.6	28.3	*	30.4	32.6	33.7
			(41)	(41)	(4)	(11)	(6)	(3)		(7)	(15)	(8)
			8.6	9.2	14.0	15.3	19.7	19.9		26.5	28.7	29.8
		8	11.7	12.4	17.5	17.6	22.2	23.2	27.0	28.3	31.6	33.0
			(73)	(57)	(15)	(11)	(6)	(5)	(1)	(6)	(16)	(10)
			3.7	4.4	9.5	9.6	14.2	15.2	19.0	20.3	23.6	25.0
		12	12.0	12.3	17.3	18.3	23.1	23.3	*	27.9	31.2	32.0
			(111)	(54)	(10)	(14)	(5)	(4)		(3)	(11)	(7)
			0.4	0.7	5.7	6.7	11.5	11.7		16.3	19.6	20.4
	1977	4			11.8	12.3	18.8	18.4	24.0	*	30.4	31.6
					(31)	(7)	(6)	(5)	(1)		(8)	(4)
					6.3	6.8	13.3	12.9	18.5		24.9	26.1
		8			11.8	12.2	18.4	18.3	*	26.6	29.4	30.5
					(80)	(18)	(14)	(6)		(3)	(11)	(9)
					3.4	3.8	10.0	9.9		18.2	21.0	22.1
		12			12.3	12.6	17.7	17.2	22.1	23.8	28.3	30.7
					(227)	(63)	(43)	(23)	(2)	(5)	(27)	(16)
					0.2	0.5	5.6	5.1	10.0	11.7	16.2	18.6
Johnson	1976	4	10.7	11.0	15.5	15.8	*	*	*		32.1	
			(12)	(3)	(1)	(2)					(2)	
			6.8	7.1	11.6	11.9					28.2	
		8	10.7	11.1	15.6	14.5	20.3		23.3		32.3	
			(61)	(29)	(6)	(4)	(6)	*	(1)		(2)	
			2.7	3.1	7.6	6.5	12.3		15.3		24.3	
		12	11.9	12.1	15.8	14.8	22.1	*	*		*	
			(117)	(34)	(2)	(1)	(1)					
			0.3	0.5	4.2	3.2	10.5					
	1977	4			10.4	10.7	16.6	*	*		*	
					(4)	(3)	(5)					
					4.9	5.2	11.1					
		8			10.3	10.6	16.1	*	*		*	
					(21)	(5)	(4)					
					1.9	2.2	7.7					
		12			12.2	12.4	16.3	*	21.1		30.2	
					(112)	(35)	(16)		(4)		(2)	
					0.1	0.3	4.2		9.0		18.1	

Table 6 (con't)

Lake	Year stocked	Size group at stocking (in)	Fall 1976	Spring 1977	Fall 1977	Spring 1978	Fall 1978	Spring 1979	Fall 1979	Spring 1981	Spring 1982	Spring 1983
Brandy	1976	4	10.7 (1) 6.8	*	*	*	*	*	*	*	*	*
		8	10.6 (14) 2.6	12.0 (4) 4.0	15.2 (2) 7.2	14.7 (2) 6.7	20.0 (1) 12.0	*	*	29.8 (2) 21.8	*	*
		12	11.8 (97) 0.2	12.2 (12) 0.6	14.4 (1) 2.8	14.8 (1) 3.2	20.0 (1) 8.4	*	*	*	32.7 (1) 20.9	*
	1977	4			9.7 (1) 4.2	11.0 (2) 5.5	17.2 (1) 11.7	*	*	*	*	33.0 (1) 27.5
		8			10.8 (11) 2.4	10.8 (7) 2.4	15.1 (3) 6.7	*	*	*	*	33.9 (1) 25.5
		12			12.4 (125) 0.3	12.5 (62) 0.4	16.3 (4) 4.2	*	*	*	28.2 (1) 16.1	33.8 (5) 21.7
Sparkling	1976	4	10.7 (2) 6.8	*	*	*	*	*	*	*	25.7 (1) 21.8	
		8	11.3 (3) 3.3	18.1 (4) 10.1	23.6 (1) 15.6	*	*	*	*	*	*	
		12	12.6 (5) 0.9	*	*	*	*	*	*	*		
	1977	4			12.4 (2) 6.9	12.6 (1) 7.1	17.6 (1) 12.4	*	*	*	*	
		8			*	*	*	*	*	*	*	
		12			12.6 (13) 0.5	12.8 (1) 0.7	19.1 (3) 0.7	*	*	*	*	

*None captured on sampling date.

numbers of young-of-year yellow perch there.

Flickinger and Clark (1978) reported that the stocking success of 2-in northern pike appeared to be directly related to availability of small forage fish, whereas these correlations were not evident with 15-in northern pike. Beyerle (1981) stated that it may be biologically and economically more advantageous to stock small (3.5 in) rather than large (7 in) tiger muskellunge in lakes with low predator densities. Gillen et al. (1981), in a study of tiger muskellunge predation on minnows and bluegills, stated that stocking programs using small predators must be coordinated with the size and abundance of prey available because of the narrow size range of prey vulnerable to small predators.

The presence of high numbers of adult northern pike in Brandy and Johnson Lakes and the shortage of small forage fishes may have contributed to the poor initial survival of the 4- and 8-in muskellunge fingerlings in Brandy Lake and poor long-term survival of all three size-groups in both lakes. Several muskellunge collected in the falls of 1976 and 1977 had open wounds on them, possibly from attacks by the abundant adult northern pike. Apparently, the large numbers of adult northern pike and low numbers of forage fish outweighed the benefit of dense rooted aquatic vegetation that should have helped protect muskellunge fingerlings in these two lakes.

The somewhat lower survival of the 4- and 8-in fingerlings stocked in Arrowhead Lake in 1977 versus those stocked in 1976 was not due to the use of pectoral fin clips instead of pelvic (ventral) clips used on fish stocked in 1976. In Brandy Lake, survival of 4- and 8-in fish stocked in 1977 (pectoral clipped) was greater than for fish of the same length stocked in 1976 (pelvic clipped); Johnson (1982) reported no differences in the survival rates of pelvic- and pectoral-clipped muskellunge fingerlings.

However, McNeil and Crossman (1979) found that removal of a pectoral fin had a more adverse effect on subsequent survival than excision of a pelvic fin.

The short- and long-term survival rates of all three size-groups of muskellunge fingerlings stocked in Sparkling Lake were negligible; this could have been due to the paucity of rooted aquatic vegetation and lack of small forage fish in both 1976 and 1977. Northern pike predation did not contribute to the poor fingerling survival, since no northern pike were captured in fyke nets in the springs of 1981 and 1982 and none were observed in electrofishing runs conducted from 1976 to 1979. Other potential predators, walleye, smallmouth bass, and burbot were present and may have contributed to the low muskellunge survival.

Growth

Growth of muskellunge fingerlings, as indicated by mean lengths of recaptured fish collected at various intervals after stocking, appeared to be highest in Arrowhead Lake. Four-inch muskellunge stocked in this lake in 1976 and 1977 had consistently higher mean lengths when recaptured than their 8- and 12-in counterparts. Perhaps the high density of forage fish (mainly young-of-year yellow perch) allowed them to grow faster in the lake than the 8- and 12-in fish in the hatchery rearing ponds where there was probably more intense competition for food. Mean lengths at capture of 4-, 8-, and 12-in fish in the other three lakes did not appear to be consistently higher for one group when they were captured on subsequent sampling dates. The lack of small forage fish in each of these lakes may have contributed to slow growth.

MANAGEMENT IMPLICATIONS

1) Electrofishing surveys to assess the survival of stocked muskellunge fingerlings should probably not be initiated until mid to late fall. In this study the fingerlings were not evident inshore during late July and early September samplings but were captured in mid October-early November and were inshore even when ice was beginning to form along the shoreline.
2) Small muskellunge fingerlings can contribute to a fishery if stocked in lakes where numbers of large potential predators (particularly northern pike) are low, the abundance of small available forage is high, and there is adequate cover. Lakes that do not meet the criteria to stock with small fingerlings should be stocked with larger fingerlings.

ACKNOWLEDGEMENTS

We appreciate the assistance of H. Carlson, J. Smith, R. Steuck, and K. Arnold, who spent many hours in the field helping collect much of the data summarized in this report and assisted in marking and distributing the fingerlings to the study lakes. We also want to acknowledge E. Mertz, C. Gahler, and G. Lensert of the District Operations staff for their help in providing us with the muskellunge fingerlings. We are thankful to H.E. Snow, A.E. Ensign, L.D. Johnson, D.A. Hanson, and C.C. Krueger for their reviews of the manuscript and to C. Griesbach for typing the original and revised drafts.

REFERENCES

Belusz, L.C. 1978. An evaluation of the muskellunge fishery of Lake Pomme de Terre and efforts to improve stocking success. American Fisheries Society Special Publication 11:292-297.

Beyerle, G.G. 1981. Comparative survival and growth of 8.9- and 17.8- cm (3.5- and 7.0-inch) tiger muskellunge planted in a small lake with forage fishes. Michigan Department of Natural Resources Fisheries Research Report No. 1894. Ann Arbor, Michigan, USA.

Black, J.J., L.M. Andrews, and C.W. Threinen. 1963. Surface water resources of Vilas County. Wisconsin Conservation Department, Madison, Wisconsin, USA.

Caplan, D.L. 1982. An experimental study of interactions between young of the year pike (*Esox lucius*) and muskellunge (*Esox masquinongy*). Masters Thesis, University of Wisconsin, Madison, Wisconsin, USA.

Flickinger, S.A., and J.H. Clark. 1978. Management evaluation of stocked northern pike in Colorado's small irrigation reservoirs. American Fisheries Society Special Publication 11:284-291.

Gillen, A.L., R.A. Stein, and R.F. Carline. 1981. Predation by pellet-reared tiger muskellunge on minnows and bluegills in experimental systems. Transactions of the American Fisheries Society 110:197-209.

Johnson, L.D. 1981. Comparison of muskellunge (*Esox masquinongy*) populations in a stocked lake and unstocked lake in Wisconsin, with notes on the occurence of northern pike (*Esox lucius*). Wisconsin Department of Natural Resources Research Report No. 110. Madison, Wisconsin, USA.

Johnson, L.D. 1982. Factors affecting short-term survival of stocked muskellunge fingerlings in Wisconsin. Wisconsin Department of Natural Resources Research Report Number 117. Madison, Wisconsin, USA.

McNeil, F.I., and E.J. Crossman. 1979. Fin clips in the evaluation of stocking programs for muskellunge, *Esox masquinongy*. Transactions of the American Fisheries Society 108:335-343.

Oehmcke, A.A. 1951. Muskellunge yearling culture and its application to lake management. Progressive Fish-Culturist 13.63:70.

Ricker, W.E. 1975. Computation and interpretation of biological statistics of fish populations. Bulletin 191. Fisheries Research Board of Canada. Ottawa, Ontario, 382 pp.

Scott, W.B., E.J. Crossman. 1973. Freshwater fishes of Canada. Bulletin 184. Fisheries Research Board of Canada. Ottawa, Ontario, 966 pp.

Population Characteristics and Angler Use of Muskellunge in Eight Northern Wisconsin Lakes

DAVID A. HANSON

Wisconsin Department of Natural Resources
Box 309
Spooner, Wisconsin 54801

ABSTRACT

Prior to 1979, insufficient data existed on muskellunge populations and angler use to guide Wisconsin's muskellunge management plan. Consequently, a study was conducted from 1979 to 1983 to determine population characteristics and angler use of eight northern Wisconsin muskellunge lakes. Each lake was sampled with fyke nets for two consecutive springs, and a random-stratified roving creel census was conducted along with voluntary registration of angler-caught muskellunge for one open-water angling season.

Mean length-at-age of male muskellunge was shorter than for females. Males generally reached sexual maturity one or two years earlier than females and were shorter-lived. Because of the slower growth and higher mortality few males reached trophy size, and nearly all muskellunge larger than 40 inches were females. Growth of both sexes was related to the density of catostomids. Growth of males was inversely related to muskellunge density. Density of legal-size muskellunge (> 30 inches) averaged 0.33 fish/acre and ranged from 0.09 to 0.61 fish/acre; highest densities were found in dark, turbid waters.

Total angling pressure averaged 42.8 hours/acre. In five of the eight lakes muskellunge were the most sought after species. Overall, 42.2% of all angler trips were specifically for muskellunge. Muskellunge anglers fished an average of 16.8 hours/acre; exploitation rates averaged 27.5% and ranged from 13.8% to 42.0%. Quality of size structure of legal-sized populations was inversely related to angler exploitation rates. The 30-inch, size-limit regulation during this study failed to protect female muskellunge until their first spawning. In some lakes, high exploitation rates appeared to be limiting trophy muskellunge angling potential.

The muskellunge is Wisconsin's premier trophy sport fish. At the present time, fishable populations of muskellunge are found in 703 lakes and 48 streams covering 370,726 acres (not including the Great Lakes; WDNR 1982). Wisconsin's muskellunge waters have been classified into three categories for management and public information, with Class A waters being those premier waters considered to provide the best muskellunge angling. There are 317 Class A lakes and 17 Class A streams in Wisconsin, of which over 99% are located in the northwest and northcentral portions of the state.

In 1979, the Wisconsin Department of Natural Resources (WDNR) published strategic management plans for heavily-utilized sport and commercial species. These plans included information on management history, resource supply and user demand, program goals and objectives, and problems with consequent strategies in meeting goals and objectives. The muskellunge management plan predicted that angler use and harvest of this fish was expected to increase in Wisconsin at a time when populations were expected to decline (WDNR 1979); harvest of muskellunge was expected to be 10% higher than biologically sound by 1990. The plan, however, stated that there was insufficient data regarding the fishery to guide the management program. Because of the large number of muskellunge waters in the state and a paucity of information, this study was designed to determine muskellunge population characteristics, angler use, harvest and exploitation in nine representative Class A waters during the period 1979 to 1983. In addition, it was desired to determine characteristics of voluntary registration data of muskellunge catch and harvest in relation to creel census estimates. Results from one lake, Mud/Callahan Lake, were excluded from this report due to extreme differences in population characteristics and are summarized by Hanson in a separate WDNR research report.

STUDY SITES

In selecting the study lakes (Fig. 1) an attempt was made to utilize those with considerable variation in the following conditions; muskellunge growth rates and population density, impoundment versus natural lake, species composition of other game fish, water clarity, and depth. Very large and very small waters were deliberately not selected. Mean size of the eight lakes was 826 acres, somewhat larger than the average (591) of all Class A Wisconsin waters.

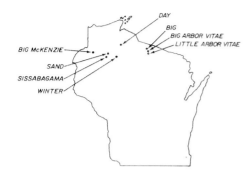

Figure 1. Location of study lakes.

All the lakes had mixed warm- and cool-water fish populations. Two lakes (Winter, and Day) had gamefish populations dominated by muskellunge and largemouth bass, the other six gamefish populations were predominantly muskellunge and walleye. Northern Pike were common in three lakes (Sand, Big McKenzie and Big). Other lake characteristics are listed in Table 1.

METHODS

Population Characteristics

Fyke-nets were fished from ice-out until after the peak spawning activity for two consecutive springs in each study lake. Winter and Mud/Callahan Lakes were netted in 1979 and 1980; Sissabagama and Little Arbor Vitae Lakes in 1980 and 1981; Sand, Big, Day and Big McKenzie Lakes in 1981 and 1982; and Big Arbor Vitae Lake in 1982 and 1983. From 8 to 20 fyke-nets were used in various combinations of 4x5 or 5x6 foot frames with 4 or 5 foot hoops, respectively, and mesh sizes ranging from 1/2 to 1 1/2 inches with leads from 35 to 100 feet long.

All captured muskellunge were measured to the nearest 0.1 inch (TL), weighed in a wetted net on a spring-loaded scale to the nearest 0.1 pound, sexed by presence of eggs or sperm, and tagged with a numbered metal tag placed around the preopercular bone or at the anterior base of the dorsal fin (the latter for large females, if the tag was too small for the preopercular bone; Johnson 1971). Approximately five scales were removed from the area below the anterior end of the dorsal fin and above the lateral line for aging. Occasional year-classes of known age fish (fin-clipped stocked fish) were present in some lakes and were used to verify scale-aging accuracy. Accuracy of aging agreed with that previously reported by Johnson (1971). Data from both years of spring netting were pooled within each lake for age and growth assessment. No mortality estimates were made due to small sample sizes of individual year-classes within each lake.

Population estimates of legal-size fish (30-in or longer TL) were determined by the Bailey modification of the Petersen estimate (Ricker 1975) of fish marked in the first spring-netting sample and recaptured by netting the following spring. Recruitment was accounted for by determining the one-year growth increment for a 30-inch muskellunge in each lake (mean of male and female from Table 2), which, when added to 30-in, was used as the minimum size in the recapture period. Density of muskellunge less than 30 inches could not be estimated by this method because of the systematic bias in capturing sexually-mature muskellunge in

Table 1. Characteristsics of eight study lakes in northern Wisconsin.

Lake	Surface acres	Maximum depth (feet)	Mean depth (feet)	Shoreline development factor	Total[a] alkalinity (mg/l)	Secchi disk (feet)	Color[a]	Turbidity[a] (NTU)
Winter	676	22	8.9	3.4	56	5	120	1.8
Sissabagama	719	48	15.5	2.2	32	8	10	0.6
Little Arbor Vitae	534	32	11.3	1.8	50	3	5	2.0
Big McKenzie	1185	71	18.7	1.5	70	10	5	1.2
Sand	928	50	21.1	1.2	34	11	25	1.2
Day	641	18	6.7	2.6	34	4	100	1.9
Big	835	61	20.5	2.3	56	13	15	0.9
Big Arbor Vitae	1090	41	18.1	2.4	50	4	10	1.9
Mean	826	43	15.1	2.2	47	7.3	36.3	1.4

[a] analysis by Wisconsin State Lab of Hygiene from a summer composite sample from 1 m below surface, mid-depth, and 1 m above bottom.

Table 2. Total length-age and weight-length relationships for eight northern Wisconsin lakes. All equations are of the form: Log Y = a+b Log X.

Lake	Sex	age (x); length (y)			length (x); weight (y)		
		a	b	R^2	a	b	R^2
Winter	M	1.22	0.36	0.72	-4.21	3.41	0.92
	F	1.20	0.43	0.80	-4.38	3.54	0.96
Sissabagama	M	1.19	0.34	0.55	-4.15	3.36	0.88
	F	1.08	0.52	0.65	-4.04	3.30	0.96
Little Arbor	M	1.20	0.34	0.59	-	-	-
Vitae	F	1.06	0.56	0.83	-	-	-
Big Mc-	M	1.14	0.46	0.81	-4.22	3.40	0.91
Kenzie	F	1.18	0.45	0.80	-4.64	3.69	0.93
Sand	M	1.17	0.40	0.83	-4.25	3.41	0.94
	F	1.10	0.53	0.79	-4.77	3.75	0.97
Day	M	1.27	0.24	0.45	-3.99	3.26	0.89
	F	1.28	0.27	0.30	-4.22	3.43	0.93
Big	M	1.19	0.37	0.83	-4.57	3.64	0.91
	F	1.23	0.38	0.79	-4.79	3.78	0.91
Big Arbor	M	1.25	0.29	0.66	-3.19	2.75	0.96
Vitae*	F	1.17	0.43	0.81	-4.16	3.39	0.97
Mean**	M	1.21	0.35	0.85	-4.08	3.32	0.98
	F	1.16	0.45	0.79	-4.43	3.56	0.99

*Weight-length relationship calculated from 1983 data only
**Calculated using estimated lengths and weights predicted from each lake's equations

the spring-spawning samples. Unbiased estimates of sub-legal muskellunge marked in the spawning run requires capture methods which do not contain the same bias as the method of capture in the marking period.

Potential forage for muskellunge was not systematically sampled, but it was suspected that muskellunge growth may have been related to the abundance of catostomids available. Correlation between catostomid density and muskellunge growth was examined using Spearman's rank correlation coefficient (Sokal and Rohlf 1969). Muskellunge growth was ranked separately for each sex by predicting the mean TL at age VI from the growth equations in Table 2. Catostomid density rankings were made independently by six field assistants and the final ranking was done by consensus of opinion from the individual rankings.

Angler Use

Angling regulations during this study included: an open season for muskellunge from the first Saturday in May through November 30, a 30-in minimum size limit and a daily bag limit of one fish. Motor trolling was prohibited on all lakes except Winter and Big McKenzie. Total fishing pressure on each study lake was estimated from a roving partial creel census during the angling season between the springs each lake was fyke-netted. Each creel census ended at freeze-up, usually 2 to 3 weeks prior to the official close of the muskellunge angling season. Little to no angling pressure occurred between freeze-up and the close of the muskellunge angling season, and muskellunge harvest in that period was assumed to be zero.

For 1980-83, a creel clerk was assigned to each lake, and they censused 1/2 the daylight hours on five days per week. Census shifts were defined as sunrise to mid-day or mid-day until 1/2 hour past sunset. One shift was scheduled each Saturday and Sunday with the remaining three scheduled randomly Monday through Friday (maximum of 1 shift per day). Five instantaneous counts were made during each shift on a stratified basis. Creel census methods for Lake Winter were that the lake was censused for the entire angling day, for seven consecutive days on an alternate week basis. Census data was processed by the statewide creel census program. Estimates of total angling effort are for daylight hours only. Estimates for lakes which had a significant night fishery (i.e., Sissabagama Lake) are minimum estimates of pressure and harvest.

Additional catch and harvest data were obtained from voluntary registration at cooperating resorts, taverns, and bait shops of legal-sized muskellunge caught (whether harvested or released) during the angling season. The voluntary information was solicited by news releases at the beginning of the angling season and by signs posted at the public access and resorts on each lake.

Three different estimates of exploitation rates were determined for each lake using data from; 1) voluntary registration, 2) projection of tagged fish observed by the creel census clerk to include uncensused time periods and 3) inclusion from all sources (voluntary registration, unprojected creel census, mail, phone, or office visitor). Estimates of exploitation rates from 1 and 3 should be considered minimum estimates.

Anderson and Weithman (1978) introduced proportional stock density (PSD) as an index of population size structure to evaluate problems of reproduction, growth, or mortality of cool water fish. Anderson and Gutreuter (1984) revised the definition of size categories of PSD to be used for muskellunge and discuss the use of relative stock density (RSD). They defined RSD as:

$$RSD(\%) = \frac{\text{quality size} \times 100}{\text{stock size}}$$

where quality size is 34 inches and longer and stock size is 20 inches and longer. However, PSD and RSD indices as previously defined were not used here due to systematic bias among sampling methods, (e.g., spring fyke-netting provided little data on sexually-immature, stock-size fish; no sub-legal, stock-size fish were reported in voluntary registration, etc.). I used RSD as an index of size structure but defined quality and stock sizes to be 34 and 30 inches and longer, respectively. I used both spring fyke-net length frequency data (RSD_{FN}) and population estimates (RSD_{PE}) as two separate methods to describe RSD for each lake.

RESULTS AND DISCUSSION

Population Characteristics

Fyke-net catch. A total of 2,762 muskellunge were captured during spring fyke-netting in the eight lakes; of these, 1,556 were males, 556 were females and the sex of the remainder was unknown (Table 3). Females were generally larger than males; 86% of the females captured were 30 in or larger and 25% exceeded 39 in, compared to 40% and <1% for males, respectively.

Growth. Growth and mean length-at-age of male muskellunge was slower than that of females (Tables 2 and 4). The estimated mean age for male and female muskellunge to reach 30.0 in was 5.8 and 5.1 years, respectively, although there was a considerable variability in growth among the lakes. Mean TL at each age from the eight study lakes is similar to data reported from other Wisconsin lakes (Johnson 1971 and 1981; Kempinger 1972; Schloemer 1938; and Wisconsin Department of Natural Resources file data), as well as reported data from Ontario (Harrison and Hadley 1979). However, mean TL at age from this study were less than those reported from southern or eastern states (Axon 1978; Belusz 1978; Bimber 1982; Haas 1978; Harrison and Hadley 1979; Houston 1952; Miles 1978; and Parsons 1959).

Weight of female muskellunge was significantly greater than equivalent size males ($P<0.01$) but again there was considerable variation in length-weight relationships among the lakes (Table 2). Mean length-weight relationships of muskellunge from this study were similar to data reported from other regions (Belusz 1978; Bimber 1982; Harrison and Hadley 1979; and Miles 1978).

Several types of growth equations were applied to age-length data to summarize growth for comparison among lakes. Growth was best described by $\log_{10}$ age-length regressions. The slope from the age-length equation (growth coefficient) was used as the dependent variable in correlation analysis to determine what factors affected growth. The accuracy of these equations in fitting the raw data (R^2 in Table 2) was correlated with the rate of growth (growth coefficient; $P=0.05$); lakes with the fastest growth had the best fit.

Table 4. Mean age frequency (%), and mean TL at age and their standard deviations (SD) of muskellunge from fyke-net catch of eight northern Wisconsin lakes.

| | Frequency (%) | | | | Mean length (in) | | | |
| | Overall mean | | SD | | Overall mean | | SD | |
Age	M	F	M	F	M	F	M	F
III	1.1	-	1.26	-	22.4	-	1.53	-
IV	17.9	2.6	9.33	5.01	25.9	26.7	0.66	1.61
V	32.4	16.7	8.77	7.45	28.3	29.9	0.69	1.34
VI	21.0	19.6	6.33	10.92	30.3	32.3	1.27	2.00
VII	10.3	17.8	5.18	9.62	32.1	34.9	2.01	2.98
VIII	7.4	15.8	5.06	8.90	33.3	36.6	1.87	2.38
IX	4.9	11.0	4.51	5.81	35.1	38.8	2.18	2.37
X	3.0	8.8	3.18	5.39	36.1	40.2	2.23	2.74
XI	1.4	3.8	1.25	2.14	37.4	43.4	2.00	2.91
XII	0.4	2.1	.60	1.64	38.2	44.7	1.67	2.33
XIII	0.1	1.8	.24	2.74	38.7	45.6	1.20	1.89
>XIII	0.0	0.4	0.00	0.42	-	-	-	-
N	1531	548	1531	548	1531	548	1531	548

Table 3. Mean length frequency distributions (%) and standard deviations (SD) of fyke-net catches of muskellunge from eight northern Wisconsin lakes.

| Total length (inches) | Male | | Female | |
	Mean	SD	Mean	SD
>21.0	0.2	0.29	0.7	0.95
21.0-23.9	2.6	1.90	0.1	0.35
24.0-26.9	18.8	5.58	3.5	2.37
27.0-29.9	38.0	7.30	10.0	6.46
30.0-32.9	25.6	6.36	21.9	8.21
33.0-35.9	9.8	4.34	21.7	9.78
36.0-38.9	4.3	3.86	16.8	8.36
39.0-41.9	0.7	0.97	14.4	7.06
42.0-44.9	0.0	0.00	7.4	6.05
45.0-47.9	0.0	0.00	3.1	2.80
>48.0	0.0	0.00	0.4	0.80
N	1556	-	556	-

Table 5. Correlation coefficients from analysis of factors affecting muskellunge population characteristics and angler use in Wisconsin lakes.

	Growth coefficient males	Growth coefficient females	Muskellunge density	Exploitation rate	Muskellunge hours/acre[a]	$RSD_{(FN)}$	$RSD_{(PE)}$
Mean depth	.60	.36	-.87**	-.86**	-.88**	.78*	.75*
S.D.F	-.52	-.52	.65	.58	.84*	-.27	-.51
Alkalinity	.56	-.03	-.02	-.39	.13	.40	.59
Seechi disk	.67	.07	-.81*	-.65	-.76	.90**	.69
Color	-.39	-.57	.63	.64	.63	-.43	-.72*
Turbidity	-.46	-.23	.81*	.35	.61	-.73*	-.56
Growth coefficient	---	.50	-.72*	-.50	-.60	.73*	.82*
Growth coefficient[m]	.50	---	-.50	.38	-.11	.22	.47
Muskellunge density[f]	-.72*	-.50	---	.61	.93**	-.76*	-.70
Exploitation rate[a]	-.50	.38	.61	---	.79*	-.62	-.79*

*P<.05
**P<.01

[a] does not include Day Lake (see text)

Growth coefficients for neither male nor female muskellunge were significantly related to mean depth, shoreline development factor, total alkalinity, secchi disk, color or turbidity data in Table 1 (P>0.05 for each singular analysis; Table 5). Growth coefficients for males were not correlated with growth coefficients for females, indicating that the well being or growth of one sex did not necessarily indicate similar growth of the other one. Growth of males was inversely related to the total density of legal-size muskellunge. Reasons why growth of one sex was not related to growth of the other, or why growth of females was not correlated with legal-size density, are unknown. However, there was a marked difference in the physical condition between males and females during the spawning season which may suggest behavioral difference between the sexes that could be related to growth. Male muskellunge during spawning showed recent cuts, scrapes, torn fins, etc. while females were generally in excellent condition.

While no abiotic factors were found to be related to muskellunge growth, a rank analysis of the density of adult catostomids (Table 6) was correlated to the mean length of age VI muskellunge (predicted from the growth equations; P<0.05 for females and P = 0.06 for males). It was expected that if actual catostomid densities had been estimated that a stronger statistical relationship would have been found, compared with the less sensitive analysis of rank data. Since muskellunge growth for males was found to be inversely related to adult muskellunge density, a rank analysis between muskellunge density and sucker density was done to provide insight if muskellunge populations would suppress the abundance of catostomids. No relationship was found between muskellunge density and catostomid density (r = 0.05, P>0.10).

Sexual maturity. Some individual male muskellunge first became sexually mature at ages III

Table 6. Population abundance, standing stock, and relative stock density (RSD) of legal-sized muskellunge and catostomid density rankings in eight northern Wisconsin lakes. RSD estimates are from fyke-net catch (FN) and population estimates (PE).

	P.E.	95% Confidence Interval	P.E./Acre	Lbs/Acre	$RSD_{(FN)}$	$RSD_{(PE)}$	Catostomid density rank
Winter	331	200-714	0.49	5.5	39.9	36.0	1
Sissabagama	116	78-218	0.16	1.8	43.7	44.8	6
Little Arbor Vitae	264	183-521	0.49	5.4	31.0	46.2	5
Big McKenzie	198	156-296	0.17	2.1	49.5	59.6	2
Sand	88	67-148	0.09	1.0	46.9	44.3	7
Day	394	276-720	0.61	6.0	21.3	19.8	8
Big	208	160-313	0.25	3.1	61.8	58.2	3
Big Arbor Vitae	411	292-691	0.38	3.9	35.1	38.9	4
Mean	251	-	0.33	3.6	41.2	43.5	-

Table 7. Estimates of angler use and catch rates of muskellunge from creel census data of eight northern Wisconsin lakes.

Lake	Year	Total hours per acre	Total angler trips	Muskellunge hours/acre	% Trips for muskellunge	Catch per hour[a]	Legal catch per hour[b]	Harvest per hour
Winter	1979	42.7	7,410	29.2	67.0	0.050	0.015	0.013
Sissabagama	1980	50.5	12,918	13.4	29.5	0.051	0.018	0.009
Little Arbor Vitae	1980	63.3	11,736	21.9	38.5	0.028	0.010	0.008
Big McKenzie	1981	33.5	9,506	5.8	11.8	0.044	0.012	0.005
Sand	1981	25.5	9,490	4.0	17.6	0.028	0.010	0.010
Day	1981	26.1	3,813	18.2	67.3	0.046	0.025	0.014
Big	1981	31.2	7,674	12.2	40.9	0.014	0.006	0.004
Big Arbor Vitae	1982	69.2	31,288	18.0	33.2	0.047	0.011	0.008
Mean	-	42.8	11,729	15.3	38.2	0.038	0.014	0.009

[a] Specific catch rate of all sizes of muskellunge caught
[b] Specific catch rate of only legal size muskellunge caught

and IV; however, the majority of individuals within a year-class matured at ages V or VI. Females matured later than males, some individuals at ages IV and V; but the majority of individuals within a year-class matured at ages VI - VIII.

Muskellunge density. Population densities of legal-size muskellunge averaged 0.33 fish/acre and ranged from 0.09 fish/acre to 0.61 fish/acre (Table 6). Mean standing stock of muskellunge 30 inches and longer was 3.6 lbs/acre and ranged from 1.0 to 6.0 lbs/acre. The density of legal-sized muskellunge in this study was associated with factors affecting water clarity with larger densities tending to occur in the shallower, more turbid waters. Density was inversely related to mean depth and seechi disk and had a positive relationship with turbidity (Table 5).

Angler Use

Angler characteristics. A total of 38,025 angler interviews and 4,941 instantaneous counts were made during creel censuses on the eight lakes. A typical angler from these eight lakes was male, between the ages of 16-64 and traveled more than 50 miles each way to fish; these groups comprised 84.6%, 83.0% and 81.7% of all anglers, respectively.

Angler fishing pressure. Total fishing pressure during the open water season (for all species) averaged 42.8 hours/acre and ranged from 25.5 to 69.2 hours/acre (Table 7). Angler pressure on these lakes was similar to slightly higher than that recorded on other lakes of similar size in northern Wisconsin (Department of Natural Resources, unpublished data). An average of 11,729 angler trips were made to each study lake. The percentage of angler trips specifically for muskellunge averaged 38.2% and ranged from 11.8 to 67.3%. On lakes without significant walleye populations (Winter and Day Lakes), the percent of anglers fishing for muskellunge averaged 67.2%. Each of the remaining lakes had a significant population of walleyes, and on these lakes the percentage of anglers fishing specifically for muskellunge was 28.6%, compared with 35.0% fishing for walleyes. In 5 of the 8 study lakes more angler trips were specifically for muskellunge than any other single species.

Muskellunge fishing pressure averaged 15.3 hours/acre and ranged from 4.3 to 29.2 hours/acre (Table 7). This pressure was significantly related to shoreline development factor and legal-size muskellunge density, as well as inversely related to mean depth and seechi disk when Day Lake data are excluded from analysis (an outlyer point). Day Lake receives less pressure than other lakes due to its remoteness from nearby communities and lack of available nearby lodging. The mean of 15.3 hours/acre observed in this study exceeded similar estimates from Cave Run Lake, Kentucky (4.7 in 1976 and 1977, Axon 1978) and Pomme de Terre Reservoir, Missouri (0.89 in 1976 and 1.9 in 1977; Belusz 1978).

Specific catch rates. Specific catch rates for all sizes of muskellunge averaged 0.038 fish/hour and ranged from 0.014 to 0.051 fish/hour (Table 7). Legal-sized specific catch rates averaged 0.014 fish/hour and ranged from 0.006 to 0.025 fish/hour. These catch rates were similar to those from Pomme

Table 8. Angler catch and harvest of muskellunge from creel census estimates and voluntary registration, eight northern Wisconsin lakes.

Lake	Creel census projections				Voluntary registration	
	Catch per acre[a]	Legal catch per acre[b]	Legal harvest per acre	Muskellunge released (%)	Legal catch per acre[b]	Harvest per acre
Winter	1.82	0.51[c]	0.44	11.4	0.33	0.29
Sissabagama	1.17	0.41	0.21	50.0	0.22	0.13
Little Arbor Vitae	0.99	0.36	0.28	21.6	0.32	0.30
Big McKenzie	0.37	0.13	0.05	58.0	0.01	0.01
Sand	0.23	0.07	0.07	4.5	0.05	0.05
Day	0.98	0.52	0.29	43.7	0.09	0.08
Big	0.24	0.11	0.07	38.2	0.07	0.04
Big Arbor Vitae	1.36	0.29	0.22	25.7	0.02	0.01
Mean	0.90	0.30	0.20	31.6	0.14	0.08

[a] includes all sizes of muskellunge
[b] includes only legal size muskellunge caught
[c] Not available from 1979 creel census records; estimated by multiplying harvest estimate from creel census times the ratio of legal-sized catch to harvest from voluntary registration.

de Terre Reservoir, Missouri (0.015; Belusz 1978), Chautauqua lake, New York (0.012; Mooradian and Shepherd 1973) and Cave Run Lake, Kentucky (0.017; Axon 1978).

Muskellunge catch and harvest. Total muskellunge catch (all-sizes) averaged 0.90 fish/acre but was highly variable, ranging from 0.23 to 2.09 fish/acre (Table 8). Legal-sized catch averaged 0.30 fish/acre and ranged from 0.07 to 0.52. The mean harvest rate from this study, 0.20 fish/acre, was less than the legal-sized catch rates due to catch and release fishing; which resulted in an average of 31.6% of the legal muskellunge caught being released. Harvest rates from this study were lower than that observed in Chautauqua Lake, New York (0.45 fish/acre; Mooradian and Shepherd 1973), but exceeded similar data from Cave Run Lake, Kentucky (0.09 fish/acre; Axon 1978) and Pomme de Terre Reservoir, Missouri (0.08 Fish/acre; Belusz 1978).

Size muskellunge caught. Overall, 62.4% of the muskellunge caught by anglers were of sub-legal size (Table 9). Over 60% of the legal-size catch was less than 34 in total length while less than 8% exceeded 40 in.

Exploitation rate estimates. Actual exploitation rates, using only harvested tagged fish from all sources of return information, averaged 25.9% and ranged from 13.8 to 42.0% (Table 10). Possible exploitation rates, including legal-size caught and released tagged fish, were slightly higher than actual exploitation rates. Since exploitation estimates from "all sources" and voluntary registration are known to be minimum estimates, and the "all sources" estimates always exceed the voluntary registration estimates, it was apparent that some angers were not cooperating with voluntary registration. Exploitation estimates from creel census averaged less than those from all sources and were therefore not as accurate. These actual exploitation rates were similar to average annual exploitation rates from: Escanaba Lake, Wisconsin (28%, $\geq$ age IV; Hoff and Serns 1986); Niagra River, New York-

Table 9. Mean total length frequency (%) and standard deviation (SD) of anglers catch of muskellunge from eight northern Wisconsin from creel census (all of total catch data plus CC of legal-size catch) and voluntary registration (VR). Voluntary registration data from 164 waters in 1981 in northern Wisconsin (NW) is also listed.

Total length	Total catch		Legal-size catch		
	Mean	SD	CC	VR	NW
<30	62.4	9.95	-	-	-
30.0-31.9	13.4	7.26	36.7	29.4	28.8
32.0-33.9	9.6	3.11	25.8	25.4	21.2
34.0-35.9	4.3	2.54	11.7	14.2	15.8
36.0-37.9	5.0	3.97	12.8	10.1	11.3
38.0-39.9	2.2	2.91	5.3	6.9	7.2
40.0-41.9	1.0	1.50	2.6	4.3	5.4
42.0-43.9	0.7	0.79	1.7	5.0	3.9
44.0-45.9	1.3	1.90	3.0	2.8	2.6
46.0-47.9	0.1	0.21	0.4	1.4	1.8
48.0-49.9	0.0	0.00	0.0	0.4	0.8
50.0-51.9	0.0	0.00	0.0	0.1	0.4
52.0-53.9	0.0	0.00	0.0	0.0	0.2
>54.0	0.0	0.00	0.0	0.0	0.1
N	1168	1168	711	559	3808

Table 10. Possible and actual exploitation rates of muskellunge from eight northern Wisconsin lakes.

	Possible			Actual		
Lake	Vol. reg.	Creel	All sources	Vol. reg.	Creel	All sources
Winter	36.0	43.1	44.0	34.0	38.8	42.0
Sissabagama	21.2	23.1	36.5	17.3	23.1	32.7
L. Arbor Vitae	35.0	24.7	38.1	32.0	19.6	35.1
Big McKenzie	5.0	31.5	21.2	5.0	17.5	13.8
Sand	20.5	7.2	25.6	20.5	7.2	25.6
Day	6.9	19.3	16.7	6.9	19.3	15.3
Big	--[a]	14.9	24.0	--[a]	14.9	18.7
B. Arbor Vitae	7.2	7.9	27.6	5.9	7.2	24.3
Mean	18.8	21.5	29.2	17.4	18.4	25.9

[a] Not available.

Ontario (13%, ≥ 30 in, Harrison and Hadley 1978); Chautauqua Lake, New York (12.2%, ≥ 30 inches, Bimber and Nicholson 1981); and Middle Island Creek, West Virginia (32%, ≥ 26 in, Miles 1978).

Factors Affecting Population Quality

Two parameters were used to define population quality—density of legal-size muskellunge and an index of the size structure of muskellunge at the beginning of the angling season. The density of adult muskellunge study was inversely related to mean depth and seechi disk and had a positive correlation with turbidity, i.e., greatest adult densities occurred in the more shallow dark water or turbid lakes. While some level of natural reproduction was observed in all lakes except Big McKenzie Lake, stocking of fingerling muskellunge was believed to contribute more to the adult populations than natural reproduction in these lakes. The turbid, more shallow lakes received the higher levels of stockings due to higher exploitation rates which occurred there, and this was suspected to be a more important influence in determining density than was some important but unknown habitat suitability. A negative association between northern pike presence and muskellunge density has been well documented (Dombeck, et al., 1986; Inskip, 1986; Inskip and Magnuson, 1986 Johnson 1981). The lakes with a greater mean depth had more abundant northern pike populations which may further explain the association of muskellunge density in relation to depth.

RSD values calculated from population estimates of stock and quality size, $RSD_{(PE)}$, were significantly correlated with similar estimates calculated from length frequency data from fyke-netting, $RSD_{(FN)}$, ($R^2=75.7\%$, $P<0.01$), although it was assumed the $RSD_{(PE)}$ estimates were more accurate in predicting true population quality. Both RSD indices were significantly related to mean depth, secchi disk, and growth coefficient of males (Table 5). $RSD_{(FN)}$ was significantly related to turbidity and legal-size density of muskellunge ($P=0.06$ for $RSD_{(PE)}$). RSD_{PE} was significantly related to exploitation rate ($P=0.10$ for $RSD_{(FN)}$). These correlations, plus the others previously mentioned, indicated that the muskellunge populations with the best quality in size structure occurred in deeper lakes with greater visibility. Angler exploitation was lowest in these lakes suggesting that angling was attributable in suppressing population quality in the shallow lakes.

Voluntary Registration Versus Creel Census

Estimates of muskellunge catch and harvest from voluntary registration averaged less than 50% of estimates from creel census (Table 8), although estimates for Little Arbor Vitae and Sand Lakes were in reasonable agreement with creel census estimates. Nevertheless, the poor disagreement with the other study lakes makes it difficult for a manager to accept voluntary registration estimates at face value without considerable knowledge of the circumstances from which the data were collected. Estimates from voluntary registration of angler harvest were correlated with similar estimates from creel census ($P<0.05$); however, the low coefficient of determination (57.8%) indicates a poor predictive relationship between the two methods of estimating harvest. Estimates of legal-size catch from the two methods were not correlated ($P>0.05$).

The length-frequency distribution of legal-size angler catch from voluntary registration indicated a larger size than from creel census data (Table 9). The catch of a few, extremely-large muskellunge in non-census hours would somewhat negatively bias creel census results; however, failure to voluntarily register small legal-sized muskellunge is believed to account for most of the differences in size observed.

The size of muskellunge from both the creel census and the voluntary registration was smaller than from voluntary registration data of 3,808 muskellunge caught from 164 waters in northern Wisconsin in 1981 (Table 9). There are two possible explanations for this difference; 1) the size of muskellunge caught in this study was actually smaller than the size caught in the 164 lakes; or 2) cooperation in registering smaller legal-sized muskellunge in the 164 water bodies, which did not have a creel census, was less than the cooperation in this study; the latter seems more probable. In other words, a more visible and active management program for muskellunge resulted in a greater amount of cooperation from the angling public with less bias.

By themselves, neither creel census or voluntary registration data provided consistently good estimates of exploitation rates compared with estimates from all sources (Table 10). This is not surprising considering muskellunge occur in low densities and exploitation rates are determined by marking only a portion of these fish. Best estimates would be expected from uniquely tagging a majority of the fish present and observing their catch from the broadest possible sampling method.

Since it was suspected that the degree of cooperation with voluntary registration was higher during this study than would normally occur due to high visibility of the census clerk, managers should utilize discretion when relying heavily on this type of data for making significant management decisions.

MANAGEMENT IMPLICATIONS

It is recognized that examination of interactions between biotic and abiotic factors in a study designed to determine population characteristics can lead to ambiguous interpretations and conclusions. The best studies to evaluate cause and effect relationships are those which were designed for that purpose; however, they are usually attempted after investigations such as this one provide evidence of important inter-relationships. While there were many intercorrelations in the analysis of these data, interpretation of the analysis provides some insight on the factors which affect population characteristics of muskellunge in Wisconsin.

No abiotic factors were identified as affecting the growth of muskellunge. Two biotic factors did show growth affects; the density of adults had an inverse affect on the growth of males, and the density of catostomids was positively correlated with the growth of both sexes. These findings suggest the following research studies might help enhance our understanding of muskellunge: 1) determine the behavioral differences between males and females and how these differences might be affected by density (i.e., do males exhibit a territorial behavior?) and 2) identify the food requirements of muskellunge and the role of catostomids in those requirements.

Tourism is a dominant economic industry in the region of Wisconsin which contained the study lakes (Fig. 1). Most large rivers and lakes within this region contain muskellunge. The high percentage of all anglers who were fishing specifically for muskellunge in the eight study lakes is a strong indication of the importance of the muskellunge in Wisconsin. Muskellunge anglers fished an average of 71 hours per legal-size fish caught. The fact that nearly 32% of those caught were voluntarily released testifies that many anglers regard the species as a trophy and are concerned about its possible over-exploitation.

Many factors were found to be significantly related to the quality of the size structure of the muskellunge population: mean depth, water transparency, and the growth coefficient of males had positive relationships with RSD indices, and color, turbidity, adult muskellunge density and exploitation rate had negative relationships. While these factors themselves are strongly inter-related, it is believed that exploitation rate and the growth of males are the two most important ones affecting RSD. The remaining factors are believed to be important in affecting growth and exploitation rates and thereby only indirectly important in affecting RSD or quality. Males grow slower, live shorter lives than females and rarely exceed 40 in TL, all of which would tend toward a lower RSD value.

Muskellunge are managed as "trophy" fish in Wisconsin. Prior to this study the acceptable level of angler exploitation of muskellunge was estimated to be 36% (WDNR 1979). Exploitation rates from these eight lakes averaged 25.9%. The finding that exploitation rates had a significant inverse relationship with RSD suggests that exploitation rates of 25 to 40% may be too high, if one of management's goals is to maintain population quality. Many anglers became concerned that muskellunge were being over-harvested in Wisconsin, because of a perceived decline in the annual catch of large fish, and they requested an increase in the minimum size limit from 30 to 36 in. While this study was not intended to evaluate the effects of various size limits, it did provide some biological information toward that end.

The goal when establishing size limits is generally to increase angler yield or increase the catch of large fish. In theory, size limits have been used to ensure recruitment by protecting individual fish until at least their first year of spawning, and/or delay angler exploitation until some later time in life. The statewide size limit at the time of this study (30 in) in general failed to protect females from harvest prior to their first spawning. The mean TL when the majority of individual muskellunge reached sexual maturity was 28 in for male and 32 in for females. Minimum size limits are often set higher than mean length at first spawning for development of trophy fisheries.

Unfortunately, when size limits are established for muskellunge, biological data on their populations are often scarce. Managers and anglers when considering possible effects of various size limits, tend to think in terms of changes in length or weight, and fail to consider changes in time required for fish to recruit to the new length. Using the data from this study, and assuming no change in age-length and length-weight relationships under various size limits, a theoretical change in size limit from 30 to 32, 34, and 36 in represents a small change in length (6.7%, 13.3% and 20.0%, respectively), but a large change in mean age to reach the new size limit (particularly for males; 20.7, 43.1 and 69.0%) and weight (particularly for females; 26.9, 56.7, and 92.5%, respectively). When a high minimum-size limit is chosen to restrict harvest and exploitation, the biological characteristics of the particular fishery should be carefully examined, since the higher size limit may create a "female muskellunge fishery" due to slower growth and higher mortality rates of males.

The definition of a "trophy muskellunge" is a sociological term and varies among anglers based on their previous experience and desires. Muskellunge have been managed primarily by uniform statewide regulations since 1903 (a minimum weight limit was first set at 4 lbs.). The great variability in population characteristics, particularly growth rates, in Wisconsin lakes suggests that optimum growth potential could be reached by managing each lake in accordance with its specific conditions, rather than by uniform management goals and regulations. Managers are also encouraged to obtain a better biological definition of the term, trophy management (if that is the program goal), prior to setting individual lake regulations.

Since the completion of this study, the statewide size limit in Wisconsin has been increased from 30 to 32 in; a few individual waters have a 34-in size limit which is being evaluated, and the opening of the muskellunge season has been delayed from the first Saturday in May until the Saturday nearest 30 May. If further restrictions are needed, those which reduce exploitation on all sizes of legal-size fish (i.e., further restrictions on season, bag and gear) should be considered.

ACKNOWLEDGEMENTS

Research technicians D. Stafford and R. Masterjohn were instrumental in coordinating field activities among the different study areas and assisting in data analysis. This project could not have proceeded so quickly without the cooperation and field assistance of Fish Management personnel and the U.S. Forest Service. Particular thanks go to; L. Andrews, T. Beard, M. Hoff, S. Johannes, F. Pratt, J. Roth, D. Scholl, S. Serns, J. Smith, R. Theis and R. Wendt and their assistants. J. Fabert, C. Goodman, and M. Staggs assisted with data processing. Finally, the cooperation of the resorts, taverns, and sporting goods stores who registered angler caught fish and assisted in other ways is greatly appreciated.

This research was supported, in part, by the Federal Aid in Fish Restoration Act under Wisconsin Dingell-Johnson Project F-83-R. The following muskie angling organizations donated funds for field activities; First Wisconsin Chapter-Muskies Inc., Hayward Lakes Chapter-Muskies Inc., Headwaters Chapter-Muskies Inc., Dave's Muskie Club, Alliance of Muskellunge Clubs of Wisconsin, and Chicagoland Chapter-Muskies Inc.

Thanks are due to L. Christenson, M. Dombeck, J. Klingbiel, E. Lange, S. Serns, H. Snow, and R. Strand for manuscript review. And to G. Hall for editorial assistance, and to M. Johnson for typing the manuscript.

REFERENCES

Anderson, R.O., and S.J. Gutreuter. 1984. Length, Weight, and Associated Structural Indices. Pages 283-300 *in* L. Nielson and D. Johnson, editors, Fisheries Techniques. American Fisheries Society, Bethesda, Maryland, USA.

Anderson, R.O., and A.S. Weithman, 1978. The concept of balance for coolwater fish populations. American Fisheries Society Special Publication 11:371-381.

Axon, J.R. 1978. An evaluation of the muskellunge fishery on Cave Run Lake, Kentucky. American Fisheries Society Special Publication 11:328-333.

Belusz, L.C. 1978. An evaluation of the muskellunge fishery of Lake Pomme de Terre and efforts to improve stocking success. American Fisheries Society Special Publication 11:292-297.

Bimber, D.L. 1982. Longevity, growth and mortality of muskellunge in Chautaugua Lake, New York. New York Fish and Game Journal 29:134-141.

Bimber, D.L., and S.A. Nicholson. 1981. Fluctuations in the muskellunge (*Esox masquinongy* Mitchell) population of Chautaugua Lake, New York. Environmental Biology of

Fishes. 6:207-211.

Dombeck, M.P., B.W. Menzel, and P.W. Hinz. 1986. Natural muskellunge reproduction in midwestern lakes. American Fisheries Society Special Publication No. 15: 122-134.

Haas, R.C. 1978. The muskellunge in Lake St. Clair. American Fisheries Society Special Publication 11:334-339.

Harrison, E.J. and W.F. Hadley. 1979. Biology of muskellunge (*Esox masquinongy*) in the Upper Niagara River. Transactions of the American Fisheries Society 108:444-451.

Hoff, M.H., and S.L. Serns. 1986. The muskellunge fishery of Escanaba Lake, Wisconsin, under liberalized angling regulations, 1946-1981. American Fisheries Society Special Publication 15:249-256.

Hourston, A.S. 1952. The food and growth of muskellunge (*Esoc masquinongy* Mitchell) in Canadian waters. Journal of Fisheries Research Board of Canada 8:247-268.

Inskip, P.D. 1986. Negative associations between abundances of muskellunge and northern pike: Evidence and possible explanations. American Fisheries Society Special Publication No. 15:135-150.

Inskip, P.D., and J.J. Magnuson. 1986. Fluctuations in growth rate and condition of muskellunge and northern pike in Escanaba Lake, Wisconsin. American Fisheries Special Publication No. 15:176-188.

Johnson, L.D. 1971. Growth of known-age muskellunge in Wisconsin and validation of age and growth determination methods. Wisconsin Department of Natural Resources Tech. Bull. 49. Madison, Wisconsin, USA.

Johnson, L.D. 1981. Comparison of muskellunge (*Esox masquinongy*) populations in a stocked and unstocked lake in Wisconsin, with notes on the occurrence of northern pike (*Esox lucius*). Wisconsin Department of Natural Resources Research Report 110. Madison, Wisconsin, USA.

Kempinger, J.J 1972. Muskellunge fishing on Escanaba Lake with liberalized fishing regulations, 1946-1971. Wisconsin Department of Natural Resources. Mimeograph. Madison, Wisconsin, USA.

McNeil, F.I. and E.J. Crossman. 1979. Fin clips in the evaluation of stocking programs for muskellunge (*Esox masquinongy*). Transactions of the American Fisheries Society. 108:335-343.

Miles, R.L. 1978. A life history study of the muskellunge in West Virginia. American Fisheries Society Special Publication 11:140-145.

Mooradian, S.R. and W.F. Shepherd. 1973. Management of muskellunge in Chautaugua Lake, New York. New York Fish and Game Journal 20:152-157.

Parsons, J.W. 1959. Muskellunge in Tennessee streams. Transactions of the American Fisheries Society. 88:136-140.

Ricker, W.E. 1975. Computation and interpretation of biological statistics of fish populations. Bulletin 191. Fisheries Research Board of Canada, Ottawa, 382 pp.

Schloemer, C.L. 1938. A second report on the growth of the muskellunge (*Esox masquinongy immaculatus* Garrard), in Wisconsin waters. Transactions of Wisconsin Academy of Sciences, Arts and Letters 31:507-512.

Sokal, R.R. and F.J. Rohof. 1969. Biometry. W.H. Freeman and Company, San Francisco, California, USA.

Wisconsin Department of Natural Resources (WDNR). 1979. Muskellunge management plan (sport: all waters). Pages 7-1 and 7-7 in Fish and Wildlife Comprehensive Plan, Part I: Management strategies 1979-1985. Wisconsin Department of Natural Resources, Madison, Wisconsin, USA.

Wisconsin Department of Natural Resources (WDNR). 1982. Wisconsin muskellunge waters. Wisconsin Department of Natural Resources Publication 1-3600(32). Madison, Wisconsin, USA.

The Muskellunge Fishery of Escanaba Lake, Wisconsin Under Liberalized Angling Regulations, 1946-1981

MICHAEL H. HOFF AND STEVEN L. SERNS

Wisconsin Department of Natural Resources
Box 440
Woodruff, WI 54568

ABSTRACT

The muskellunge population in Escanaba Lake has been unregulated by angling size, season, or bag limits from 1946 through 1981. To evaluate changes in the population under those conditions, the harvest, yield, and exploitation of wild and stocked muskellunge were analyzed for the period 1946-1981. Collection of pertinent data was facilitated by a compulsory, permit-type creel census in effect during the entire 36-year period. Harvest for fish age I and older averaged 25 fish annually over the entire period. From 1956 through 1981 the estimated population and exploitation rate for age II and older muskellunge averaged 47 fish and 29%, respectively. For age IV and older fish during this same period the mean annual harvest was 6 fish and the yield 70 lb; the population estimate averaged 25 fish and the exploitation rate 29%. Yearling stockings were nine times more effective at adding fish to the creel than fingerling stockings. Thirteen percent of the harvested muskellunge were creeled during the closed season for muskellunge on other Wisconsin waters, and only 5% of the muskellunge were part of a bag containing more than one muskellunge. Only 23% of the total number of muskellunge harvested in this period were 30.0 in or longer. The Escanaba Lake muskellunge population did not exhibit reductions in population size or annual harvest during the 36 years of angling without restrictions on size, season, or bag.

The muskellunge is presently managed as a trophy species in Wisconsin. While the number of legal-sized muskellunge is expected to decrease in the future as suitable habitat diminishes, angling pressure for this species is expected to increase (Klingbiel 1981). During the study period, harvest of muskellunge from most Wisconsin waters other than Escanaba Lake were regulated by a minimum length limit of 30 in, a bag limit of one fish, and a closed season from December 1 through the first Friday in May. To provide sustained or increased numbers of quality-sized muskellunge for future anglers, Klingbiel (1981) suggested that increasingly-restrictive regulations should be preceded by more extensive quantitative data on angling pressure, harvest, and exploitation. This report provides such information for Escanaba Lake for the period 1946-81, when there were no size, season, or bag limits on muskellunge.

STUDY SITE

Escanaba Lake is a 293-acre drainage lake in north-central Wisconsin. Maximum and mean depths are 26 ft and 14 ft, respectively. Although the lake has an inlet and two outlets, their intermittent flow and a fish barrier screen on the major outlet minimizes fish migration. The shoreline surrounding Escanaba Lake is completely state-owned and undeveloped except for one boat landing.

The lake is comparatively fertile for this region, with a total alkalinity of 16 ppm and a specific conductance of 46 umhos/cm (Kempinger and Carline 1977). Dissolved oxygen concentrations have not been a limiting factor for fish, since no summer or winter fish kills have been documented since 1946. Ice generally covers Escanaba Lake from late November to late April.

Twenty-four fish species have been identified from Escanaba Lake (Kempinger et al. 1975). Of these, muskellunge, walleye, northern pike, yellow perch, rock bass, and black crappie have been most important in the recent sport harvest. Only walleye and northern pike are not considered native to Escanaba Lake. Muskellunge are thought to be native to this lake, but records show a total of 309,000 muskellunge fry were stocked during the years 1937-41 (Kempinger et al. 1975). Muskellunge x northern pike hybrids have occurred naturally in the lake and were first noted in the angler catch during 1947.

From 1964-72, an experimental 22-in minimum length limit was imposed on northern pike from Escanaba Lake. No other size or bag limits or closed seasons have regulated any sport fish harvest since 1946.

249

METHODS

Muskellunge were captured with fyke nets during spring spawning seasons in 1953-72, 1974, 1977, and 1979-81. These fish were measured (total length) to the nearest 0.1-in, and a scale sample was removed from the nape above the lateral line for age determination. Fish were sexed, if possible, by the extrusion of gametes; with the exception of yearlings, they were tagged with a numbered, aluminum or monel preopercular strap tag or a monel jaw tag.

A complete creel census, under a compulsory, free permit system, has been in effect on Escanaba Lake since 1946. These permits were issued to anglers, prior to fishing, at the Northern Highland Fishery Research Area headquarters at the Escanaba Lake boat landing. All anglers were required by law to return the permits upon completion of their trip and to submit all creeled fish to a census clerk for inspection. Clerks recorded total hours fished (for all species to the nearest 0.5-hour), and for all fish measured (total length to the nearest 0.1-in) and weighed (nearest 0.01-lb for fish 0-3 lb, nearest ounce for fish 3-10 lb, and nearest 0.1-lb for fish exceeding 10 lb), recorded fin clips and tag numbers and collected a scale sample. When clerks were not available, anglers were required to record on their permit the number of each species harvested and the total hours fished.

Total annual yield (lb) was estimated for the years when weight data of harvested fish were in-

Table 1. Fishing pressure, harvest, and yield of Escanaba Lake muskellunge, 1946-1981.

Year	Angling hours[a]	Harvest	Yield (lb)	% of Harvest that was stocked	Mean total length (in) ± SD (N)		% of measured fish ≥ 30 in (N)		Harvest from Dec 1- 1st Friday in May
1946	9,386	14	91[b]	-	28.6±4.4	(12)	25	(3)	d
1947	6,295	7	68[b]	-	34.8±5.5	(7)	71	(5)	-
1948	12,776	5	38[b]	-	31.1±1.7	(5)	40	(2)	-
1949	19,332	11	74[b]	-	29.6±8.0	(8)	50[c]	(4)	d
1950	35,862	34	257[b]	-	29.0±8.1	(22)	50[c]	(11)	d
1951	21,874	21	85[b]	-	d		d		d
1952	25,702	52	200[b]	-	d		d		d
1953	28,862	20	133[b]	-	d		d		d
1954	17,965	14	117[b]	-	29.1±9.0	(13)	46[c]	(6)	-
1955	23,403	9	32[b]	-	27.3±4.8	(7)	14[c]	(1)	1
1956	24,412	6	30	-	28.1±6.1	(6)	33	(2)	-
1957	24,775	13	78	-	26.7±9.5	(13)	38	(5)	1
1958	26,368	16	80	-	24.8±8.7	(16)	31	(5)	1
1959	21,979	11	45	-	25.0±6.1	(11)	18	(2)	2
1960	22,214	18	90	-	26.4±6.9	(18)	28	(5)	4
1961	18,497	24	86	17	21.5±9.5	(24)	21	(5)	6
1962	22,367	74	207[b]	18[c]	22.3±6.1	(73)	12	(9)	3
1963	20,796	57	173	35	23.0±5.8	(57)	11	(6)	13
1964	12,769	22	131	25	28.7±6.0	(22)	36	(8)	2
1965	10,775	23	112	26	25.6±7.4	(23)	35	(8)	2
1966	13,716	22	153	27	29.2±7.9	(22)	50	(11)	1
1967	14,437	21	105	14	25.3±9.0	(21)	33	(7)	1
1968	9,898	24	76	8	23.7±4.9	(24)	8	(2)	4
1969	11,150	32	143[b]	19[c]	25.9±7.0	(31)	23[c]	(7)	4
1970	14,695	33	101	3	21.7±7.3	(33)	9	(3)	6
1971	16,246	25	74[b]	8[c]	22.4±6.2	(24)	13[c]	(3)	10
1972	11,271	39	143	5	24.0±5.7	(39)	8	(3)	4
1973	13,445	25	109	4	25.3±6.8	(25)	12	(3)	5
1974	11,107	13	64[b]	-[c]	25.6±7.3	(12)	17[c]	(2)	3
1975	17,883	25	108	12	24.8±8.0	(26)	16	(4)	3
1976	15,820	25	193	12	29.3±8.7	(25)	36	(9)	4
1977	18,247	22	91	-	25.1±5.7	(22)	9	(2)	9
1978	17,067	43	187	-	25.3±5.6	(43)	14	(6)	2
1979	12,892	53	267[b]	-[c]	27.2±6.5	(51)	16[c]	(8)	3
1980	14,023	28	169	4	29.2±5.5	(28)	43	(12)	2
1981	11,781	26	186	4	30.5±5.7	(26)	50	(13)	1
Total (N)	630,087	907	4,296	-	-		23	(182)	97
Mean±SD(N)	17,502±6,385 (36)	25±15 (36)	119±60(36)	-	25.6±7.2	(789)	28±16	(33)	4±3 (26)

[a]For all species. [b]Estimate. [c]Data incomplete. [d]No data.

complete by calculating the mean weight of muskellunge harvested during that year, multiplying that value by the number of harvested fish not weighed, and then summing the weights. Scale samples and length-weight data of creeled fish were incomplete prior to 1954. Muskellunge harvested prior to 1958 were not cataloged by angler file numbers; thus, composition of each angler's bag during this period was not analyzed. The angling year referred to in this report is defined as the open water and subsequent ice fishing season.

Muskellunge were aged using a microfiche reader equipped with a 36-43X objective. Scale ages of muskellunge caught by anglers from 1979 to 1981 were corroborated by age determination from cleithra (Casselman 1979). Asymptotic lengths (L_∞) of male and female muskellunge were computed from the Walford line (Ricker 1975).

The Bailey Modification of the Petersen formula (Ricker 1975) was used to estimate population size within the age range of fish tagged, based on the proportion of tagged to untagged fish harvested. Annual exploitation rate (μ) was determined as the percentage of fish tagged during the spring netting period that were harvested by anglers during the subsequent angling year. Instaneous rates of total mortality (Z) and annual rates of survival (S) and total mortality (A) were computed from the catch curve (Ricker 1975) of muskellunge captured in fyke nets from 1953-81 and angler-caught fish from 1946-81 (all years combined due to small sample sizes.

Muskellunge fingerlings stocked at the approximate rate of 1 fish/acre in 1961 and 1965 were marked by clipping the right pectoral fin; yearlings stocked at the rate of 0.7 fish/acre in 1961 and 0.5 fish/acre in 1966 were marked by excising the left pectoral fin. A chi-square test was used to test for differences in return to the creel of stocked yearlings versus fingerlings.

Stocked and hybrid muskellunge were excluded from mean size-at-age determinations. Mean size-at-age was computed only for fish fyke-netted in April and early May and fish angler-caught in April during the years 1953-81.

RESULTS

Fishing Pressure, Population Density, Harvest, and Exploitation

From 1946-81, an average 695 angling hours (for all species) was required to creel a muskellunge from Escanaba Lake (0.0014 fish/hour),

Table 2. Annual population estimates ($\hat{P}$) and rates of exploitation (μ) of Escanaba Lake muskellunge, 1956-1981.

Year	No. Marked	No. in recapture sample	No. of recaptures	$\hat{P}$	95% C.I.	Age range of tagged fish	μ(%)
1956	4	5	2	8	3-20	III-X	50
1957	9	7	2	24	9-60	III-XI	22
1958	8	7	1	32	10-58	III-IX	13
1959	5	3	1	10	3-18	IV-IX	20
1960	7	4	2	12	4-29	V-IX	29
1961	16	4	4	16	7-40	III-XI	25
1962	12	43	7	66	34-139	II-X	58
1963	3	50	1	77	23-139	II-IV	33
1964	27	22	4	124	55-311	I-X	15
1965	34	9	5	57	27-131	IV-XI	15
1966	16	12	2	69	25-173	IV-VIII	13
1967	7	8	2	21	8-53	V-X	29
1968	11	18	2	70	25-174	II-XII	18
1969	10	28	3	73	30-181	II-IX	33
1970	9	7	3	18	7-45	III-XIII	33
1971	1	3	0	-	-	-	-
1972	9	32	3	74	30-186	II-XII	33
1973[a]	-	-	-	-	-	-	-
1974	12	8	5	18	9-42	II-IX	42
1975[a]	-	-	-	-	-	-	-
1976[a]	-	-	-	-	-	-	-
1977	9	14	2	45	16-113	II-XII	22
1978[a]	-	-	-	-	-	-	-
1979	4	46	2	63	23-157	II-III	50
1980	11	25	3	72	29-179	III-XI	27
1981	13	16	4	44	20-111	IV-X	31
Mean				47	19-112		29
SD				31	13-75		13
N				21	21		21

[a]No fyke netting in these years.

while 3,042 hours were required to creel one 30 in or larger (0.0003 fish/hour) (Table 1). An estimated maximum of 5% of the total angling hours during the study period were spent fishing solely for muskellunge. Thus, from 1946-81 a minimum of 152 hours was required to creel a muskellunge 30 in and larger (0.0066/hour) for those anglers fishing specifically for muskellunge.

The average annual harvest and yield of Escanaba Lake muskellunge from 1946-81 was 25 fish (0.1 fish/acre) and 119 lb (0.4 lb/acre), respectively (Table 1). From 1956-81, estimates of the Escanaba Lake muskellunge population (age II and older) averaged 47 fish (0.2 fish/acre), while the exploitation rate of these fish averaged 29% (Table 2).

The mean annual harvest and yield of Escanaba Lake muskellunge age IV and older from 1956-81 was 6 fish (<0.1 fish/acre) and 70 lb (0.2 lb/acre), respectively (Table 3). During this same period, the average estimated number of fish age IV and older and their average exploitation rate was 25 fish (0.1 fish/acre) and 29%, respectively (Table 3).

Mortality and Mean Size-at-Age

Rates of instantaneous total mortality (Z), total annual mortality (A), and survival (S), estimated from the catch curve of ages V-XIII muskellunge fyke-netted from 1953-81 (Table 4), were 0.396, 0.327, and 0.673, respectively (r = -0.90). Values of Z, A, and S, computed from the catch curve of muskellunge harvested from 1946 to 1981 (Table 5) aged II-XIII and aged V-XIII, were 0.423, 0.345, and 0.655 (r = -0.97), and 0.385, 0.320, and 0.680 (r = -0.93), respectively. Thus, the average exploitation rate of age II and older Escanaba Lake muskellunge (29%) accounted for over 84% of the total annual mortality rate of fish age II and older. The average exploitation rate of muskellunge age IV and older (29%) comprised 89% and 91%, respectively, of the total annual mortality rate of age V and older fish computed from the catch curves of fyke-net catch and angler catch.

Mean size-at-age data indicate that Escanaba Lake female muskellunge reach a total length of 30 in between ages III and IV while males do not attain 30 in until between ages IV and V (Table 6). The asymptotic length (L ∞), or maximum length theoretically attainable in Escanaba Lake, of male and female muskellunge were estimated to be 43.9 in and 51.3 in, respectively. The L ∞ of male Escanaba Lake muskellunge is nearly identical to that in Chautauqua Lake, New York (43.3 in), although Escanaba Lake females have a much lower L ∞ when compared with Chautauqua Lake females (60.6 in) (Bimber 1982).

Table 3. Annual harvest, yield, population estimates ($\hat{P}$), and rates of exploitation (μ) of age IV and older Escanaba Lake muskellunge, 1956-1981.

Year	Harvest	Yield (lb)	No. Marked	No. in recapture sample	Recaptures	$\hat{P}$	95% C.I.	μ (%)
1956	2	17	2	2	1	3	1- 60	50
1957	6	65	8	6	2	19	8- 280	25
1958	5	59	7	5	1	21	8- 420	14
1959	3	25	5	3	1	10	4- 200	20
1960	7	61	7	7	2	19	8- 280	29
1961	5	68	15	5	4	18	9- 90	27
1962	8	88	6	8	2	18	8- 280	33
1963	4	43	1	4	1	3	1- 50	100
1964	9	73	19	9	4	38	19- 190	21
1965	10	86	34	10	5	62	32- 234	15
1966	12	129	16	12	2	69	29- 1,040	13
1967	8	84	7	8	2	21	9- 315	29
1968	2	20	8	2	-	-	-	-
1969	8	92	7	8	1	32	11- 630	14
1970	3	48	8	3	2	11	4- 27	25
1971	3	37	1	3	-	-	-	-
1972	3	60	5	3	-	-	-	-
1973[a]	3	49[c]	-	-	-	-	-	-
1974	2	33	6	2	2	6	3- 90	33
1975[a]	3	66	-	-	-	-	-	-
1976[a]	10	148	-	-	-	-	-	-
1977	3	33	8	3	1	16	6- 320	13
1978[a]	3	71	-	-	-	-	-	-
1979[b]	4	86[d]	-	-	-	-	-	-
1980	14	122	8	14	2	40	15- 100	25
1981	16	151	13	16	4	44	20- 111	31
Mean	6	70				25	11- 262	29
SD	4	37				19	9- 246	20
N	26	26				18	18	18

[a] No fyke netting in these years.
[b] No age IV and older fish tagged.
[c] Data incomplete - weight of one harvested fish unknown.
[d] Data incomplete - age and weight of one harvested fish unknown.

Table 4. Age-frequency of fyke-netted Escanaba Lake muskellunge, 1953-1981.

Year	I	II	III	IV	V	VI	VII	VIII	IX	X	XI	XII	XIII
1953			2	3	2			1	5	1	1		
1954			2		1	1							
1955			1			1							
1956			2	1					1				
1957	1		1	1		4	2			2			
1958			1		2	1	1	1	2				
1959	2	1	2	3	2	1	1	1	2				
1960		2		3	4	5	1	1	2				
1961	10		2		5	2	2	1	2	1	2		
1962	3	7				2	2	1		1			
1963	1	1	1	1									
1964	2	1	6	14		2		1		2			
1965	16		6	23				2	1	1	1		
1966		1		1	7	7		1					
1967	2	2		1		5			1				
1968		3		1	1		2	3				1	
1969	4	3		1	1	1			4				
1970	1	1	1		3	1	1			2			1
1971		1				1							
1972	3	4	1	1		1	1				2		
1973[a]													
1974	4	1	4				1		5				
1975[a]													
1976[a]													
1977	1	1	1	1		1			2	1		2	
1978[a]													
1979	1	2	2										
1980	1	1	3	4				1	1	1	1		
1981				5	6				1	1			
Total	52	32	32	46	59	28	21	14	27	13	7	5	1

[a] No fyke netting in these years.

Table 5. Age-frequency of angler-harvested Escanaba Lake muskellunge, 1946-1981.

Year	0	I	II	III	IV	V	VI	VII	VIII	IX	X	XI	XII	XIII	XV
1946			4	4	1		1								
1947					4	2	1								
1948			1	4											
1949		2		1	1	1	3								
1950		5	4	2	2	1	2	5	1						
1951[b]															
1952[b]															
1953[b]															
1954		1	2	4	1	2		1	1	1					
1955			4	1	1	1									
1956		1	1	2	1	1									
1957		4	1	1	1	2	1	1			1				
1958		6	3	2	2	1	1				1				
1959		5	1	2	1	1		1							
1960	1	1	8	1	3	1		1	2						
1961	1	16	1		1	2	1				2				
1962		31	33	2	1	3	1	1	2						
1963	1	6	27	19	1		2	1							
1964		1	4	8	8						1				
1965		5	4	2	3	5	1		1						
1966		1	6	3		3	7		1	1					
1967		4	6	3	1	2	3	1	1						
1968		6	9	7				1	1						
1969		3	17	2	1	1		1	5						
1970		15	9	4	1				1		1				
1971		5	13	3	1					2					
1972		7	22	7			1			2					
1973		4	13	4	1			1				1			
1974		4	1	4		1		1							
1975		5	15	1					2						1
1976		2	9	4	3	2			4						
1977		7	7	3	1	1		1							
1978		2	32	4	2	1			1						
1979		5	10	32			2		2						
1980			3	11	10	2		1					1		
1981			4	6	6	8		1	1						
Total	3	154	273	150	60	44	22	19	15	11	6	11	3	1	2
% of Total	1	20	35	19	8	6	3	2	2	1	1	1	1	1	1

[a] No XIV-year-olds were recorded.
[b] No data.

Fingerling and Yearling Stocking

Harvest of muskellunge stocked as fingerlings in Escanaba Lake during 1961 and 1965 constituted only 4% and 1% ($\overline{X} = 2\%$), respectively, of the total number stocked (Table 7). Total yield of these fingerling stockings, however, was 212% and 33% ($\overline{X} = 123\%$), respectively, of the total weight stocked. Harvest of muskellunge yearlings stocked in 1961 and 1966 was 31% and 5% ($\overline{X} = 18\%$), respectively, of the total number stocked, while yields of these stockings were 442% and 253% ($\overline{X} = 347\%$), respectively, which indicates a substantial contribution to the fishery from the perspective of yield. A chi-square test showed that, in proportion to numbers stocked, muskellunge stocked as yearlings contributed significantly ($P<0.001$) more to the creel than stocked fingerlings.

Eight (67%) of the 1961- and one (50%) of the 1965-stocked fingerling muskellunge harvested were creeled prior to their attaining a length of 30 in. Of the 1961- and 1966-stocked yearlings harvested, 36 (60%) and one (14%), respectively, were under 30 in when creeled. Since 46 of the 81 (57%) muskellunge harvested from the four stockings were creeled before reaching 30 in, contribution to the creel of these stockings was higher than it would have been under a 30-in minimum-size limit. However, yield was probably lower than if the fish had been protected from harvest until they attained 30 in.

Estimated rates of instantaneous total mortality (Z), total annual mortality (A), and survival (S) of the 1961 stocked yearlings, based on fyke-net catches of fish ages V-XII (Table 8), were 0.257, 0.226, and 0.774 ($r = -0.82$), respectively, which compared favorably to the mortality and survival rates of wild fish. Mortality and survival rate estimates

Table 6. Mean total length-at-age ($\pm$ SD, N in parenthesis) of fyke-netted and April angler-caught Escanaba Lake muskellunge, 1953-1981.

					Age					
I	II	III	IV	V	VI	VII	VIII	IX	X	XI
					Males					
-	-	25.2±1.5 (8)	28.6±1.3 (20)	31.7±1.3 (11)	33.4±1.3 (7)	35.1±1.5 (4)	36.6±0.5 (3)	38.3±2.2 (5)	36.5 (1)	
					Females					
-	-	27.4±2.0 (2)	31.8±0.1 (2)	34.2±1.7 (12)	35.8±1.0 (8)	37.4±0.4 (3)	39.9±2.7 (3)	42.4±1.1 (11)	43.7±2.3 (7)	44.8±2.4 (4)
				Males, Females and Undetermined Sex						
14.2±1.1 (25)	21.0±2.3 (37)	25.4±2.0 (43)	29.0±1.5 (33)	32.6±2.0 (36)	34.8±1.9 (23)	35.8±1.7 (14)	38.8±2.1 (11)	41.5±2.4 (19)	43.3±2.8 (11)	44.9±1.9 (6)

Table 7. Harvest and yield (lb) of stocked fingerling and yearling Escanaba Lake muskellunge, 1961-1981.

Year	Year of Stocking/Age at Stocking							
	1961				1965		1966	
	Fingerlings		Yearlings		Fingerlings		Yearlings	
	Hvst	Yield	Hvst	Yield	Hvst	Yield	Hvst	Yield
1961	-	-	5	3.4	-	-	-	-
1962	2	1.7	11	19.5	-	-	-	-
1963	4	4.3	15	45.4	-	-	-	-
1964	1	3.9	6	39.3	-	-	-	-
1965	1	7.0	5	42.5	-	-	-	-
1966	1	8.3	5	49.6	-	-	-	-
1967	1	5.3	2	21.0	-	-	-	-
1968	-	-	1	9.5	1	2.9	-	-
1969	1	19.1	5	61.6	-	-	-	-
1970	-	-	1	19.9	-	-	1	6.3
1971	1	12.0	1	16.3	-	-	-	-
1972	-	-	2	43.8	-	-	-	-
1973	-	-	1	22.8	-	-	-	-
1974	-	-	-	-	-	-	1	22.0
1975	-	-	1	26.5	1	20.6	1	18.8
1976	-	-	-	-	-	-	3	59.3
1977	-	-	-	-	-	-	-	-
1978	-	-	-	-	-	-	-	-
1979	-	-	-	-	-	-	-	-
1980	-	-	-	-	-	-	1	22.8
1981	-	-	-	-	-	-	-	-
Totals	12	61.6	61	421.1	2	23.5	7	129.2
No. Stocked	291	-	197	-	301	-	155	-
Pounds Stocked	-	29.0	-	95.3	-	71.0	-	51.0
Percent of Stocking	4	212	31	442	1	33	5	253

*Hvst = Harvest

were not possible for the other three stockings due to small sample sizes, although small numbers caught in fyke nets suggest high mortality rates for these stockings.

DISCUSSION

Based on this 36-year data base for Escanaba Lake, size limits would appear to be a more effective way to regulate muskellunge harvest than a daily bag limit of one fish or a December 1 to early May closed season. Also, size limits would tend to maintain higher yields of this species than would be possible without any limit, and only allow the harvest of fish which are trophy-sized.

During most years, muskellunge under 30 in comprised at least 60% of the harvest under the liberalized regulations on Escanaba Lake, and only 23% of the total muskellunge harvested from 1946 to 1981 were 30 in or longer (Table 1). The average length of creeled muskellunge did not exceed 26 in during 16 of the 28 years from 1954-81. Since 1946, 74% of the muskellunge harvest were age III or younger and 82% were age IV or younger.

Under liberalized regulations, the harvest of muskellunge under 30 in had a greater effect on the total muskellunge harvest than the impact of anglers keeping more than 1 fish/day, or harvesting fish during the time of year that muskellunge could not be legally harvested on other waters. During the period from 1958 to 1981, only 15 anglers exceeded the present Wisconsin daily bag limit of one muskellunge. These anglers harvested 33 fish (5% of harvest during this period), of which 27 were less than 30 in, 5 were 32 in or larger, and one was of unknown length. The most muskellunge creeled by any of these anglers was four, while most anglers creeled two fish. Two of these 33 fish were taken during the December-early May closed season on other Wisconsin waters. Although the timing and duration of the muskellunge closed season on most Wisconsin waters has changed since 1983, 13% (97 fish) of the Escanaba Lake muskellunge harvested since 1954 would not have been allowed under the previous December 1-early May closed season for most other Wisconsin waters. Of these 97 muskellunge, only four (0.4% of total harvest from 1954 to 1981) would have exceeded 30 in.

Even though the average exploitation rate has not been excessive compared to the maximum rate of 36% recommended to maintain fishing quality (Klingbiel 1981), the Escanaba Lake muskellunge population has not attained high average densities. From 1956 to 1981 the estimated annual population of Escanaba Lake muskellunge, age IV and older, averaged only 25 fish, which is much lower than the estimated state-wide density of one muskellunge greater than 30 in for every two acres of muskellunge water (WDNR 1982). However, mus-

Table 8. Fyke net catches of muskellunge stocked as yearlings in 1961 in Escanaba Lake, 1962-1972.[a]

Year	Age									
	II	III	IV	V	VI	VII	VIII	IX	X	XI	XII
1962	2									
1963		1								
1964			10							
1965				16						
1966					5					
1967						5				
1968							2			
1969								4		
1970									2	
1972											2

[a] None captured in 1971.

kellunge stocking has been infrequent in Escanaba Lake, and the average density there may be lower than in those lakes which are stocked regularly. Since 1946, a total of only 352 yearling and 592 fingerling muskellunge were stocked in Escanaba Lake; elsewhere in Wisconsin muskellunge stocking is considered an important management technique since natural reproduction is considered to be low among most populations (WDNR 1979).

These population estimates were not free of statistical biases, however, and these biases probably caused a consistent underestimate of the size of the Escanaba Lake muskellunge population. When using the Petersen Method to estimate the size of a fish population without a substantial probability that the estimate is negatively biased, Robson and Regier (1964) suggest that the number of marked fish times the number of fish examined for marks in the recapture sample should exceed four times the true population size. Since the true population size is rarely known, however, Ricker (1975) stated that the probability of statistical bias can be ignored, if tag recaptures number three to four more. In most cases these data failed to meet this criterion. Thus, substantial probability exists that many of the annual population estimates of Escanaba Lake muskellunge were negatively biased. A similar statistical bias probably also affected many of the estimates of annual exploitation rate. For example, the 1963 exploitation rate estimate of 100 was an obviously high estimate biased on only one tagged and recaptured muskellunge age IV and older.

With no size limit, bag limit, or closed season restricting harvest of Escanaba Lake muskellunge and no regular stocking program since 1946, trends in reduced harvest or population size were not evident. Even though exploitation rate composed a large percentage of the total mortality rate, the Escanaba Lake muskellunge population appears to have been able to sustain itself at a fairly constant level. In the four years immediately following removal of the minimum length limit, bag limit, and closed season on Escanaba Lake (1946 through 1949), muskellunge harvest/angler-hr was lower than the average of succeeding years. Although no data exists on the muskellunge population prior to 1946, this suggests that the restrictions regulating harvest prior to 1946 had not maintained substantially greater muskellunge densities in Escanaba Lake than were estimated for years after 1946.

Stocking of muskellunge yearlings was significantly more effective at adding fish to the creel than the fingerling stockings. In only one of the stockings did anglers harvest over 5% of the stocked fish, when 31% were harvested. However, 57% of that harvest was composed of fish less than 30 in and there is no way of determining how many of these fish would have been harvested under a 30-in minimum length limit. Yield may be almost as important an indicator of stocking success as harvest, and the average yield from the four stockings of fingerlings and yearlings was 235%. This mean yield value would likely have been even greater under a 30-in minimum length limit. The yield from three of the four stockings was greater than 200% of the pounds stocked.

CONCLUSIONS

A 30-in minimum length limit on Escanaba Lake muskellunge from 1946-81 would have theoretically reduced the harvest by 77% during this period, assuming no resultant changes in growth, mortality, and recruitment as a result of the harvest regulation. A daily bag limit of one fish and a closed season from December 1 through early May would have each theoretically reduced muskellunge harvest by only an additional 0.4%. The Escanaba Lake muskellunge population did not exhibit trends in reduced annual harvest or population density during the years of liberalized angling regulations and no maintenance stocking, which suggests that muskellunge harvest on some waters is being conservatively regulated by state-wide or region-wide regulations, particularly those which receive low angling pressure specifically for this species. However, until muskellunge populations can be managed on a lake by lake basis, some populations probably need to be conservatively regulated by state-wide or region-wide regulations so that populations which depend upon these regulations to maintain high densities and yields of trophy-sized fish will not be overharvested. It is possible that a minimum size limit of 30 or 32 in on Escanaba Lake muskellunge during the study period could have increased the yield and the number of quality- or trophy-sized fish in the harvest.

ACKNOWLEDGEMENTS

Most of these data were collected under the supervision of W.S. Churchill and J.J. Kempinger with assistance from O.C. Baccus. In recent years, seasonal personnel P. Kanehl and A. Braun assisted in data collection. Critical review of the manuscript was provided by L.M. Christenson, H.E. Snow, R. Hine, E.L. Lange, and three anonymous reviewers. E.L. Lange also provided statistical advice. C.

Griesbach typed the original manuscript and revisions. This research was supported, in part, by the Federal Aid in Fish Restoration Act under Wisconsin Dingell-Johnson Project F-83-R.

REFERENCES

Bimber, D.L. 1982. Longevity, growth and mortality of muskellunge in Chautauqua Lake, New York. New York Fish and Game Journal 29:134-141.

Casselman, J.M. 1979. The esocid cleithrum as an indicator calcified structure. Pages 249-72 *in* J. Dube and Y. Gravel, editors. Proceedings of the 10th Warmwater Workshop, Northeast Division American Fisheries Society. Ministère du Loisir, de la Chasse et de la Pêche du Québec, Montréal, Québec, Canada.

Kempinger, J.J. and R.F. Carline. 1977. Dynamics of the walleye (*Stizostedion vitreum vitreum*) population in Escanaba Lake, Wisconsin, 1955-72. Journal of the Fisheries Research Board of Canada 34:1800-1811.

Kempinger, J.J., W.S. Churchill, G.R. Priegel, and L.M. Christenson. 1975. Estimate of abundance, harvest, and exploitation of the fish population of Escanaba Lake, Wisconsin, 1946-69. Wisconsin Department of Natural Resources Technical Bulletin 84, Madison, Wisconsin, USA.

Klingbiel, J. 1981. The status of the muskellunge program. Wisconsin Department of Natural Resources Bureau of Fish Management, Administrative Report 11, Madison, Wisconsin, USA.

Ricker, W.E. 1975. Computation and interpretation of biological statistics of fish populations. Bulletin 191. Fisheries Research Board of Canada, Ottawa, 382 pp.

Robson, D.S., and H.A. Regier. 1964. Sample size in Petersen mark-recapture experiments. Transactions of the American Fisheries Society. 93:215-226.

Wisconsin Department of Natural Resources. 1979. Muskellunge management plan (sport: all waters). Pages 7-1 through 7-7 *in* Fish and Wildlife Comprehensive Plan Part I: Management Strategies 1979-1985. Wisconsin Department of Natural Resources, Madison, Wisconsin, USA.

Wisconsin Department of Natural Resources. 1982. Wisconsin muskellunge waters. Wisconsin Department of Natural Resources, Publication 1-3600 (82), Madison, Wisconsin, USA.

Introduction and Management of Northern Muskellunge in Iron Lake, Michigan

DELL H. SILER

Michigan Department of Natural Resources
P.O. Box 300
Crystal Falls, Michigan 49920

GEORGE B. BEYERLE

Michigan Department of Natural Resources
Institute for Fisheries Research
212 Museums Annex Building
Ann Arbor, Michigan 48109

ABSTRACT

The Michigan Department of Natural Resources (DNR), in 1962, introduced northern (Wisconsin variety) muskellunge into 396-acre Iron Lake with annual fingerling plants. By 1966, high survival of initial stocks had produced a substantial population of harvestable adult muskellunge.

In 1968, Iron Lake was designated a muskellunge brood stock lake and closed to winter angling. Angler harvest of muskellunge was further restricted in 1972 by imposition of an artificial lures only, 36-in minimum size limit for summer anglers. The resulting abnormally high population density of 30-in and larger muskellunge (0.84 per acre) occurred concurrently with a severe decline in previously substantial populations of black crappie and common sucker. In addition, some muskellunge emigrated into two popular brook trout streams. By 1978 public displeasure with the Iron Lake fishery stimulated the DNR to net and remove adult male muskellunge, move the muskellunge egg-collection operation elsewhere, and liberalize angling regulations (1981).

By spring 1982, at least 381 muskellunge (0.96 per acre) had been removed by netting and angling, and the catch-per-effort (CPE) of muskellunge in nets had declined 93%. Mean annual growth of adults was 0.98 in greater in 1981 than in 1979, and survival of naturally produced young was much higher. Northern pike was threatening to become the dominant large predator. Populations of black crappie and common sucker were increasing again, but yellow perch had, temporarily, become the dominant panfish.

The fish management experience at Iron Lake stresses the importance of carefully examining all possible consequences of any major actions being considered and of intensely monitoring a fish population being subjected to novel management techniques.

The muskellunge is the largest member of the pike family, Esocidae, and is native only to the United States and Canada. Throughout its native range, the muskellunge has evolved into several strains with more or less unique characteristics, e.g., spawning habits, markings, etc. Classification of these strains as subspecies is debatable (Hourston 1955; Reeve M. Bailey, personal communication).

The three common strains are the Great Lakes muskellunge, found in the Great Lakes and St. Lawrence basins, southwestern Quebec and southern Ontario; the Ohio muskellunge, found in the Ohio-Tennessee River system, including western Pennsylvania and New York; and the northern (Wisconsin) muskellunge, found in the headwaters of the Mississippi River and in the western Ontario waters which drain into Hudson Bay (Hourston 1955; Trautman 1957; Hubbs and Lagler 1958).

Two of the muskellunge strains are native to Michigan. The Great Lakes muskellunge is found mainly in Lake St. Clair and in certain other lakes which connect (or were formerly connected) with one of the Great Lakes. The northern muskellunge is native only to a few waters of the Mississippi drainage located in the western Upper Peninsula along the Michigan-Wisconsin boundary in Gogebic County. In the early 1960's, the Michigan Department of Natural Resources (DNR) began attempts to expand the range of the northern muskellunge (MacGregor et al. 1960). One purpose of the introductions (Schrouder 1973) was to establish muskellunge brood stock lakes to provide eggs for hatchery production of the hybrid tiger muskellunge (muskellunge x northern pike).

Muskellunge fingerling stockings began in 1962. By 1968 the muskellunge population was so substantial that the DNR chose Iron Lake as a source of eggs for hatchery production of northern muskellunge. To protect the egg source, the DNR closed the lake to all winter fishing and spearing, which caused immediate resentment among the large contingent of local winter-spear fishermen and panfish anglers.

To further protect the large muskellunge spawners, a 36-in minimum-size limit, artificial-lures-only regulation was adopted in 1972. Black crappie that had dominated the anglers' catch through the 1960's were rarely taken in the early 1970's. Anglers were quick to jump to the conclusion that the substantial muskellunge population was responsible for the declining population of panfish. Muskellunge sightings in two popular brook trout streams downstream from the Iron Lake outlet caused more public relations problems. Despite the removal of 223 male muskellunge in 1978-79, anti-muskellunge public pressure continued. Finally, in 1981 the egg collection operation was moved to another lake. In the winter of 1980-81 the DNR eliminated the ban on winter fishing and spearing in Iron Lake and established a daily creel limit of five esocids of any size. After 2 years of liberalized regulations, standard angling regulations for muskellunge in Michigan (one fish per day at least 30 in long) were reimposed on the Iron Lake fishery.

The objective of this paper is to document the results of several annual plantings of hatchery-raised fingerling northern muskellunge, and to report observations pertaining to winter spearing of muskellunge and changes in size limit and creel limit regulations in Iron Lake, Iron County, Michigan.

STUDY SITE

Iron Lake, located in central Iron County in the Upper Peninsula of Michigan, has 396 surface acres and a maximum and mean depth of 43 and 10.9 ft., respectively. The water, stained light brown, is moderately soft with an average methyl orange alkalinity of 30 ppm. Littoral areas vary from large sandy flats to bays with loose organic substrates. The lake has two small inlet streams and an outlet stream (Iron Lake Creek) that drains into the Iron River. Iron River is a top quality brook trout stream. Little information existed about the fish population in Iron Lake prior to the introduction of muskellunge. However, both the available data and angler reports suggest the existence of substantial populations of black crappies, golden shiners, and white suckers. Largemouth bass and smallmouth bass were present but not abundant.

METHODS

Northern muskellunge fingerlings were stocked annually in Iron Lake from 1962 through 1972, except for 1968. The planting time varied from June to August. Stocking rates and size of fingerlings varied considerably (Table 1). Annual trap- and/or fyke-netting of muskellunge was conducted from 1966 through 1982, except for 1973 and 1981 (Table 2). The first netting surveys to check survival and growth were in August 1966 and May 1967. Trap nets used had 4-ft pots, and the fyke nets had 4-ft diameter hoops. Mesh size in both types of nets was 0.75-in bar. Total net numbers used in each survey and the combination of fykes and traps was highly variable. All comparisons were made on the basis of catch per unit of effort (CPUE), per net day, generally of 24 hrs. Beginning in 1968, the annual netting operations were oriented to muskellunge egg collection, although data were collected on muskellunge CPUE and mean lengths for spawned fish. Data on CPUE of native fishes were obtained in 1967 and 1979. Between those years, data on fishes other than muskellunge were not recorded.

During the 1970 and 1979 nettings, the population of mature muskellunge was estimated by mark and recapture, using the multiple census formula of Schumacher and Eschmeyer (Ricker 1975). During the 1978 and 1979 netting operations, all captured male muskies were removed from Iron Lake to reduce the overall population without jeopardizing egg-taking potential. Following the liberalization of fishing regulations in 1981, an intensive creel cen-

Table 1. Number of northern muskellunge fingerlings stocked in Iron Lake, Michigan, 1962-72. Mean length calculated from U.S. Department of Interior, Fish and Wildlife Service, Manual of Fish Culture.

Year	Fingerlings stocked		Size of fish	
	Total	Per acre	Number per pound	Mean length (in)
1962	1,500	3.8	214	3.1
1963	1,750	4.4	97	4.1
1964	3,300	8.3	329	2.7
1965	3,000	7.6	27	6.3
1966	1,833	4.6	4	11.9
1967	3,000	7.6	27	6.3
1969	214	0.5	4	11.0
1970	850	2.1	---	---
1971	1,985	5.0	14	7.8
1972	1,600	4.0	13	8.0
Mean per year	1,903	4.8	88	6.7

Table 2. Summary of netting operations for muskellunge on Iron Lake, Michigan, 1966-82. Number of spawned fish in parentheses.

Year	Netting date	Effort (net days)	Fish caught	Total CPUE	Mean length, in inches of spawned muskellunge	
					Male	Female
1966	8/25-26	10	8	0.8	---	---
1967	5/10-12	9	33	3.7	---	---
1968	4/10-15	---	212	---	28.9 (-)	37.1 (15)
1969	4/20-5/1	33	293	8.9	27.9 (33)	33.4 (23)
1970	4/27-5/2	36	276	7.7	27.7 (208)	34.4 (55)
1971	4/26-5/11	157	405	2.5	27.1 (34)	33.4 (23)
1972	5/8-14	75	252	3.4	---	---
1974	4/29-30	6	62	10.3	29.1 (47)	34.5 (15)
1975	5/4-13	104	412	4.0	---	32.9 (70)
1976	4/9-25	30	231	7.7	---	30.6 (45)
1977	4/17-23	74	586	7.9	---	30.2 (86)
1978	5/1-11	28	201	7.2	---	---
1979	4/30-5/9	87	314	3.6	30.3 (125)	31.6 (170)
1980	4/27-5/4	58	148	2.6	---	---
1982	5/2-8	90	41	0.5	29.2 (24)	33.4 (17)

sus was conducted during the 2-month (January and February) 1981 winter fishing season and again the next summer (May through July).

Catch estimates for the 59-day winter creel census (January 1-February 28) were based on the assumption that on each randomly selected day of the census all anglers (spearers) and all captured muskellunge were checked by the census clerks. Limited access to Iron Lake, and the fact that all spearing was done from shanties, theoretically made the assumption valid (James R. Ryckman, personal communication). The estimate of daily total anglers was a negative linear relationship with day of the fishery. Daily total hours and total muskellunge speared were negative curvilinear relationships with day of the fishery (Siler and Beyerle 1984).

The summer census was conducted using standard instantaneous counts and random interviews (Merna et al. 1981).

RESULTS

Netting Operations

The first post-stocking survey in August 1966 produced eight fish ranging in size from 18.3 to 35.6 in (Table 2). A similar survey in May 1967 captured 33 fish from 20.3 to 31.0 in (mean 27.1). Although ages were not determined, the wide variation in lengths suggest the presence of at least two year-classes.

In the 1967 survey, white suckers and black crappies dominated the catch of native fishes (Fig. 1). Pumpkinseed, yellow perch, and largemouth bass were captured in fewer numbers. This order of abundance was similar to a electroshocking survey in August of 1966. By early 1970's, white suckers and black crappies were only rarely collected during netting operations (visual observations).

From 1969 through 1978, the annual CPUE of netted muskellunge was relatively constant (7.2 to 10.3), except for 1971, 1972, and 1975, when it was only 2.5, 3.4, and 4.0, respectively (Table 2). In these 3 years, because of high egg quotas, nets were left in longer after the peak spawning period. The mean length of spawned female muskellunge through 1974 averaged 34.7 in, then decreased in 1975 to 32.9 in, in 1976 to 30.6 in, and in 1977 to 30.2 in (Table 2, Fig. 2).

During the 1978 and 1979 netting operations, 97 and 126 adult male muskellunge, respectively, were transferred to another lake with an established muskellunge population. The removal of these 223 fish, ranging in size from 23.7 to 35.4 in, was reflected in the subsequent netting CPUE for adults, which decreased from an average of 7.6 muskellunge in 1976-78, to 3.6 in 1979 and 2.6 in 1980 (Table 2). Also, 1979 was the only year that more mature females than males were taken in the annual netting operation, and it was the first year since 1974 in which the mean length increased for spawned females

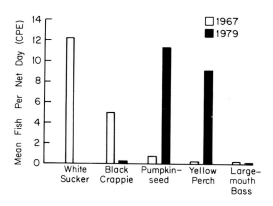

Figure 1. Net catches of native fishes in Iron Lake, Michigan, May 1967 and 1979.

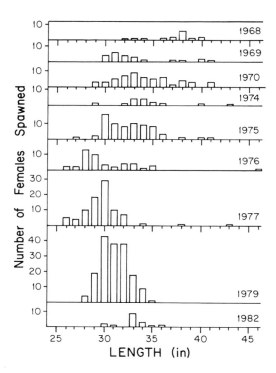

Figure 2. Variations in number and mean length of spawned female northern muskellunge from Iron Lake, Michigan, 1968-1982.

(Table 2, Fig. 2). In 1979, pumpkinseeds and yellow perch dominated the net catch of native fishes; black crappies and white suckers continued to be scarce (Fig. 1).

In 1982, the CPUE for muskellunge was only 0.5, a reduction of 79% from that in 1980 (following 2 years of manual removal) and 93% less than the mean CPUE for the 3 years previous to manual removal (Table 2). Evidence of better growth was indicated from four muskellunge (30.8 to 35.0 in) tagged in May 1982 and recaptured in May 1983. They had grown an average of 1.8 in, compared with a mean of 0.8 in for 33 female muskellunge (28.9 to 34.3 in), which had been tagged in May 1979 and recaptured in May 1980. In 1982, the mean length of female muskellunge increased to the levels of 1969-74 (Table 2).

The net catch of black crappies increased gradually between 1980 and 1982, then sharply in 1983 (visual observation). Meanwhile, a considerable increase in the yellow perch population was evident in 1979 (Fig. 1). By 1981 and continuing through 1983, yellow perch was the dominant panfish in Iron Lake (visual observation), but average growth was not as high as Michigan state average (Laarman 1963).

A small population of northern pike existed in Iron Lake at least since 1968. In 1980, coincident with the reduction of muskellunge, the northern pike population began a gradual increase, followed in 1982 by a sharp increase, when the CPUE for pike (0.87) exceeded the CPUE for muskellunge (0.46). No changes were noted in the populations of largemouth and smallmouth basses.

Population Estimates

In 1970 the population of sexually mature muskellunge was estimated to be 871 fish (95% confidence limits, 655-1,299), consisting of 79% males and 21% females. Ages of these fish were not determined, but mean length of males was 27.7 in (range 22.0 to 38.0), of females 34.4 in (range 29.0 to 41.0). Only 23% of the males but 96% of the females were legal size (30 in or larger). The estimated population density of legal-size muskellunge was 0.84 fish per acre (0.40 males and 0.44 females). In 1979, during the muskellunge egg collection operation, the population of mature females was estimated to be 732 fish (95% confidence limits, 433-2,367), or 1.85 females per acre.

Creel Census

The first creel census in 1981, during the months of January and February, estimated 2,485 (±113.0) hours of angler effort and a catch of 158 (±20) muskellunge (Table 3). In the second 1981 creel census, 15 May through July, the estimated catch was only 13 (±13) muskellunge for 1,161 (±272) angler hours (Table 4).

DISCUSSION

The 1966-67 nettings provided evidence that fingerling muskellunge from some of the earliest stockings in Iron Lake had survived in significant numbers. Considering the small size of the fingerlings stocked from 1962 through 1965 (Table 1), the relatively high survival was considered unusual (Johnson 1974, 1982). By 1970, the estimated population of legal-size muskellunge (30 in and larger) was 0.84 fish per acre, 2.5 times greater than the mean density (0.33 fish per acre) of 30-in and larger fish found in typical muskellunge lakes in Wisconsin (Hanson 1986).

One native fish species difficult to assess with

Table 3. Estimated harvest of northern muskellunge by spearers on Iron Lake, Michigan, January-February 1981.

Month	No. of anglers	Hours fished	Number of fish	Fish per hour	Percent of anglers interviewed
January	559	2,067.5	148	0.0716	37.39
95% C.L.	±13	±61.0	±11	±0.0056	±0.84
February	130	417.5	10	0.0240	16.92
95% C.L.	±20	±95.0	±17	±0.0409	±2.55
Total	689	2,485.0	158	0.0636	33.53
95% C.L.	±23	±113.0	±20	±0.0086	±1.13

Table 4. Estimated harvest of northern muskellunge by anglers on Iron Lake, Michigan, May-July 1981.

Month	No. of anglers	Hours fished	Number of fish	Fish per hour	Percent of anglers interviewed
May	21	165	3	0.0182	47.62
95% C.L.	±16	±68	±7	±0.0426	±36.28
June	229	609	7	0.0115	26.20
95% C.L.	±80	±200	±10	±0.0369	±9.15
July	117	387	3	0.0078	26.50
95% C.L.	±55	±171	±6	±0.0159	±12.46
Total	368	1,161	13	0.0112	27.45
95% C.L.	±98	±272	±13	±0.0118	±7.31

nets was the golden shiner, and it may have been important in influencing the population densities of both black crappies and muskellunge. The reduction of the muskellunge population density by manual removal (netting) and liberalized fishing would have resulted in less predation on adult golden shiners and white suckers, plus increased potential for reproduction of golden shiners. Thus, it is possible the golden shiners, white suckers, and black crappies all may have benefited from the reduction in the population density of muskellunge. When the dominant white suckers and black crappies were suppressed, a niche may have been provided for the yellow perch to become the dominant panfish in Iron Lake.

Annually, since 1974, a few yearling muskellunge were captured in survey nets, evidence that natural reproduction was occurring. In 1982, yearlings made up 26% of the total catch of muskellunge. This apparent increased survival of young muskellunge may have been one result of the decrease in numbers of larger individuals. Northern pike were first identified in the lake after the muskellunge population was well established. The substantial increase in northern pike numbers, following the 1981 reduction in the muskellunge population, further suggests that cannibalism may have been occurring in the muskie population.

The 1979 population estimate of 1.85 females per acre seems inordinately high when compared with Hanson's (1986) estimated 0.33 muskies (males and females) per acre for Wisconsin lakes. It is theorized that an overestimate occurred because the spawned (marked) females tended to leave the spawning area where the nets were set. This same phenomenon probably occurred during the 1970 population estimate but was much less a factor, because in 1970, 79% of the muskellunge captured (and marked) were males. Males would be more likely to remain in the spawning area after being marked and released.

CONCLUSIONS

The fisheries management experience at Iron Lake is worth examination. The original management goal was achieved with the successful introduction and subsequent development of a northern muskellunge population. However, the decision to make Iron Lake a muskellunge brood-stock lake, although a logical one for the DNR, turned out to be a problem. Because the DNR had relatively few dependable sources of northern muskellunge eggs, it was also logical to protect the potential new source of eggs with restrictive angling regulations. The adverse consequences of the resultant, and somewhat unexpected, expansion of the muskellunge population were not immediately obvious but occurred over a period of years. Intensive annual surveys of the fish population, which would probably have revealed the changing population structure, were not done. As is often the case under these circumstances, increasing complaints of anglers became the stimulus to corrective action by the DNR.

With the advantage of hindsight, two lessons to be learned from this management experience are: carefully consider beforehand all the possible consequences of any major change imposed on a fish population; and intensely monitor any fish population being subjected to novel management techniques.

REFERENCES

Hanson, D. 1986. Population characteristics and angler use of muskellunge in eight northern Wisconsin lakes. American Fisheries Society Special Publication 15:238-248.

Hourston, A.S. 1955. A study of variations in the maskinonge from three regions in Canada. Contributions of the Royal

Ontario Museum Division of Zoology and Palaeontology, No. 40, pp 1-13.

Hubbs, C.L., and K.F. Lagler. 1958. Fishes of the Great Lakes region. University of Michigan Press, Ann Arbor, Michigan, USA.

Johnson, L.D. 1971. Growth of known-age muskellunge in Wisconsin: and validation of age and growth determination methods. Wisconsin Department of Natural Resources, Technical Bulletin No. 49, Madison, Wisconsin, USA.

Johnson, L.D. 1974. Muskellunge survival in Wisconsin lakes. *In* Sixth Interstate Muskellunge Workshop, Morehead, Kentucky, USA.

Johnson, L.D. 1982. Factors affecting short-term survival of stocked muskellunge fingerlings in Wisconsin. Wisconsin Department of Natural Resources, Research Report 117, Madison, Wisconsin, USA.

Laarman, P.W. 1963. Average growth rates of fishes in Michigan. Michigan Department of Natural Resources, Fisheries Research Report 1675, Ann Arbor, Michigan, USA.

MacGregor, J.M., J.A. Scott, and B.C. Dean. 1960. A review of the life history and proposed management of the northern muskellunge, *Esox m. immaculatus*. Michigan Department of Conservation, Lansing, Michigan, USA.

Merna, J.W., J.C. Schneider, G.R. Alexander, W.D. Alward, and R.L. Eshenroder. 1981. Manual of fisheries survey methods. Michigan Department of Natural Resources, Fisheries Management Report 9, Lansing, Michigan, USA.

Oehmcke, A.A., L. Johnson, J. Klingbiel, and C. Wistrom. 1958. The Wisconsin muskellunge. Its life history, ecology, and management. Wisconsin Conservation Department, Publication 225, Madison, Wisconsin, USA.

Ricker, W.E. 1975. Computation and interpretation of biological statistics of fish populations. Bulletin No. 191. Fisheries Research Board of Canada, Ottawa, Ontario.

Schrouder, J.D. 1973. Muskellunge management in Michigan. Michigan Department of Natural Resources, Fisheries Division Technical Report 73-31, Lansing, Michigan, USA.

Siler, D.H., and G.B. Beyerle. 1984. Introduction and management of northern muskellunge in Iron Lake, Michigan. Michigan Department of Natural Resources, Fisheries Division Technical Report 84-12, Ann Arbor, Michigan, USA.

Trautman, M.B. 1957. The fishes of Ohio. Ohio State University Press, Columbus, Ohio, USA.

Characteristics of Native Muskellunge Streams In Eastern Kentucky

JAMES R. AXON AND LEWIS E. KORNMAN

Department of Fish and Wildlife Resources
Frankfort, Kentucky 40601

ABSTRACT

The need for more intensive management of the stream fisheries prompted the Kentucky Department of Fish and Wildlife Resources to initiate a muskellunge streams investigation. Fourteen streams were studied from 1980-1983, each during a 1- to 2-year period. Assessments included fish population, water quality, and morphology for each stream. The mean CPUE of muskellunge in the 14 streams was 0.6 fish per hour of electrofishing effort. Highest catches, up to 1.3 fish per hour, were in streams having the most fallen trees. Problems encountered in streams with a low or extirpated population of muskellunge were loss of spawning areas due to impoundment, sedimentation, or acidic water conditions. Golden redhorse, longear sunfish, and spotted bass were species found closely associated with muskellunge. Mean depth of pools, stream gradient, and fallen trees were the most important conditions identified with preferred muskellunge habitat. The supplemental stocking of 6- to 8-inch muskellunge should be continued in streams when benefits to the population and fishery are evident; streams having suitable muskellunge habitat should be managed for establishment of self-sustaining populations. Present and future concerns for Kentucky muskellunge streams are impacts from oil shale extraction, coal mining, oil-well drilling, clearing of trees in the riparian zone, impoundments, and illegal harvest of muskellunge.

The muskellunge ranges through more than 700 miles of 18 streams in Kentucky, primarily in the eastern part of the state. Four other streams contained native muskellunge, but the species has been extirpated because of habitat degradation resulting from coal-mining activities. Brewer (1980) expressed the need for habitat protection on Kentucky muskellunge streams following a study in 1969-72. He was most concerned about the potential impact of proposed flood-control projects on four streams. One of these projects was completed in 1974 with the impoundment of Cave Run Lake in the Licking River drainage; the dam impounded 58 miles of stream where muskellunge had previously ranged. However, an excellent muskellunge fishery has since developed in the lake from annual fingerling stockings; according to creel data, anglers harvest from 600 to 1,000 muskellunge from this lake annually. Impoundments already in existance on five streams were considered to be responsible for a decline in their muskellunge population, primarily because of the inundation of their spawning areas. Brewer (1980) recommended supplemental stockings of one large fingerling muskellunge per 2 acres of pool habitat in 17 streams to replace recruitment losses from adverse environmental conditions. Fingerling muskellunge (6-8 in) were stocked in 17 streams from 1973 through 1979. A D-J muskellunge stream investigation project was initiated in 1980 to evaluate current populations, stocking success, and environmental changes that may have occurred since the previous studies, and this paper reports on the results of those surveys.

Fourteen streams were examined from 1980-83; of these, two no longer support natural populations of muskellunge. It was considered that information gained from streams that formerly supported muskellunge would be just as important as that from native streams, if the causes that led to its elimination could be determined and a fishable population restored. North Fork Kentucky River is one of the former native muskellunge streams. The watershed of North Fork is the most heavily mined area for coal in Kentucky, which likely led to the extirpation of muskellunge in this stream.

STUDY AREA

Locations of the 14 streams, other streams that either have natural or extirpated populations of muskellunge, and related lakes are shown in Fig. 1. Kinniconick and Tygarts creeks are tributaries to the Ohio River; Red River, Station Camp Creek, and Sturgeon Creek are tributaries to the Kentucky River; Middle Fork, North Fork, and South Fork make up the headwater streams of the Kentucky River, while Collins Fork, Goose Creek, Little Goose Creek, Redbird River, and Sexton Creek are part of the South Fork Kentucky River drainage.

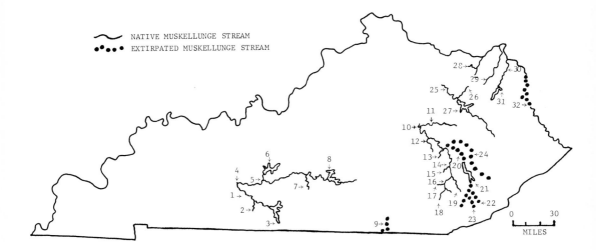

Figure 1. Location of muskellunge streams in Kentucky. Numbers identify the individual streams: 1 - Barren River, 2 - Drakes Creek, 3 - Barren River Lake, 4 - Green River, 5 - Nolin River, 6 - Nolin River Lake, 7 - Little Barren River, 8 - Green River Lake, 9 - South Fork Cumberland River, 10 - Kentucky River, 11 - Red River, 12 - Station Camp Creek, 13 - Sturgeon Creek, 14 - South Fork Kentucky River, 15 - Sexton Creek, 16 - Goose Creek, 17 - Little Goose Creek, 18 - Collins Fork, 19 - Redbird River, 20 - Middle Fork Kentucky River, 21 - Buckhorn Lake, 22 - Greasy Creek, 23 - Beech Fork, 24 - North Fork Kentucky River, 25 - Licking River, 26 - North Fork Triplett Creek, 27 - Cave Run Lake, 28 - Kinniconnick Creek, 29 - Tygarts Creek, 30 - Little Sandy River, 31 - Grayson Lake, 32 - Big Sandy River.

Many of these streams lie within the Appalachian Mountain Province, classified as the Eastern Coal Field. Stream lengths and portions inhabited by muskellunge are given in Table 1. Prior to these studies, 12 of the native muskellunge streams had been stocked periodically from 1973-83 (Table 2).

The physical dimensions of pools sampled for muskellunge were quite variable. Pool length and width varied from 0.1 to 2.9 mi and 20 to 140 ft, respectively; stream gradient per 5-mi section ranged from 1.0 to 8.6 ft/mi in sampled areas. The gradient for each stream varied from 3.0 to 25.3 ft/mi (Table 3). A greater percent of the streams was in pools, versus riffles, as would be expected in muskellunge streams.

Water quality characteristics are presented in Table 4. Temperature, dissolved oxygen content, and salinity were suitable for muskellunge, with the exception of summer temperatures in South Fork Kentucky River. Summer temperatures ranged from 72 to 88 F at all water quality sample stations. Mean summer temperatures for all streams were between 73 and 83 F. The only stream that had temperatures warm enough to be of concern for mus-

Table 1. Stream length, muskellunge range, and miles of suitable pool habitat for 14 streams in Kentucky.

Stream	Total length (mi)	General muskellunge range (stream mi)[b]	Miles of suitable pool habitat[c]
Kinniconick Creek	50.0	0-38	19.0
Tygarts Creek	85.5	0-75	43.0
Licking River[a]	82.0	215-280	--
Red River	96.1	0-60	35.0
Middle Fork Kentucky River	92.0	--	28.0
North Fork Kentucky River	161.0	--	--
Collins Fork	15.7	0-10	3.5
Goose Creek	44.5	0-30	8.8
Little Goose Creek	15.0	0-5	1.8
Redbird River	42.0	0-27	9.2
Sexton Creek	23.2	0-12	4.0
South Fork Kentucky River	44.6	0-45	32.0
Station Camp Creek	22.0	0-20	8.0
Sturgeon Creek	33.0	0-12	5.0

[a] Above Cave Run Lake.
[b] Middle Fork and North Fork Kentucky River do not support a natural population of muskellunge. Mile 0 represents the stream at its mouth.
[c] Miles of suitable pool habitat not determined for Licking River.

Table 2. Stockings of 6-8 inch muskellunge in eastern Kentucky streams since 1973.

Stream	Year stocked									
	1973	1974	1975	1976	1978	1979	1980	1981	1982	1983
Kinniconick Creek	212			390		203				
Tygarts Creek	400			833		414				
Licking River[a]			417	674	1,078	3,504	1,474		671	564
Red River	400			536		270				
Middle Fork Kentucky River				270	500	258				
Collins Fork					31	25	25	25	30	24
Goose Creek		130		224		215				
Redbird River										56
Sexton Creek		30		56		28				
South Fork Kentucky River		510		1,018		511				523
Station Camp Creek		50		98		49				
Sturgeon Creek		40		82		41				

[a] These stockings were at the headwater of Cave Run Lake.

kellunge was the lower section of South Fork Kentucky River, where the mean temperature was 80 and 83 F in the summers of 1982 and 1983, respectively; no muskellunge were sighted in this section. However, their absence may have been due to low gradient as well as high temperatures. Dissolved oxygen content ranged from 3.4 to 13.0 ppm from all streams. Mean specific conductance ranged from 104 to 442 µmhos. Salinity levels were high enough to be recorded in Licking River, Red River, Middle Fork Kentucky River, and North Fork Kentucky River; all but one were 0.2 ppt or less. A reading of 0.5 ppt was taken at the lower station in Licking River in the fall of 1983. A potential water quality problem of high salinity from brine, released from oil well operations, exists in most of the study streams. The highest salinity level (4.0 ppt) was recorded in South Fork of Red River in the fall of 1983.

Mean total alkalinity values ranged from 25 ppm in Kinniconick Creek to 150 ppm in North Fork Kentucky River. There was no apparent relationship between total alkalinity and populations of muskellunge, although Tygarts Creek had the sec-

Table 3. Physical characteristics of pools electrofished in each stream in eastern Kentucky. Gradient is for the entire stream.

Stream (no. of pools)	Mean depth (ft)	Mean maximum depth (ft)	Mean width (ft)	Mean length (mi)	Gradient (ft/mi)
Kinniconick Creek (10)	4.1	15.6	97.5	1.2	12.2
Tygarts Creek (15)	3.7	11.4	70.4	0.9	4.1
Licking River[a] (8)	2.2	4.2	46.9	0.7	3.3
Red River (14)	3.4	10.5	70.2	1.0	6.6
Middle Fork Kentucky River (12)	3.1	5.8	92.3	1.1	7.0
North Fork Kentucky River (10)	3.0	8.0	95.1	1.8	3.8
Collins Fork (11)	3.6	5.1	38.9	0.3	17.2
Goose Creek (10)	3.8	-	63.6	0.6	16.3
Little Goose Creek (2)	3.0	-	35.0	0.1	25.3
Redbird River (6)	3.5	-	58.3	0.7	22.5
Sexton Creek (3)	3.8	-	46.7	0.3	18.5
South Fork Kentucky River (9)	4.9	-	95.6	1.0	3.0
Station Camp Creek (5)	4.4	12.1	54.2	1.4	3.9
Sturgeon Creek (2)	4.3	11.5	63.8	1.3	17.6

[a] Above Cave Run Lake.

Table 4. Mean values for water quality parameters determined for 14 muskellunge streams in Kentucky.

Stream (year)	Summer temperature (F)	Dissolved oxygen (ppm)	Total alkalinity (ppm)	Turbidity (NTU)	pH	Specific conductance (umhos/cm)	Salinity (ppt)
Kinniconick Creek (1981)	77	8.1	30	19	7.0	104	0
Tygarts Creek (1981)	77	7.0	101	17	7.3	258	0
Licking River[a] (1982/1983)	74/78	10/8	88/99	35/63	7.5/7.5	-/369	0.02/0.07
Red River (1982/1983)	75/76	7.9/7.3	64/101	20/15	7.3/7.1	-/187	-/0.01
Middle Fork Kentucky River (1982/1983)	74/77	10/10	72/61	26/19	7.8/7.6	-/209	-/0.02
North Fork Kentucky River (1982/1983)	73/79	12/10	137/150	17/23	8.2/8.0	-/521	-/0.05
Collins Fork (1982/1983)	75/81	10.1/7.2	32/50	9/14	6.7/7.4	-/113	-/0.00
Goose Creek (1982/1983)	77/75	9.7/7.3	46/50	12/36	7.0/7.4	-/228	-/0.00
Little Goose Creek (1982/1983)	75/74	9.8/8.7	45/58	15/28	7.5/7.4	-/442	-/0.00
Redbird River (1982/1983)	77/80	9.7/8.0	47/68	5/19	6.7/7.7	-/440	-/0.00
Sexton Creek (1982/1983)	79/82	9.6/7.2	36/43	6/14	6.5/7.4	-/147	-/0.00
South Fork Kentucky River (1982/1983)	80/83	10.0/7.4	40/46	8/18	6.7/7.7	-/319	-/0.00
Station Camp Creek (1982/1983)	76/82	6.5/7.7	110/119	15/6	7.3/7.1	-/203	-/0.00
Sturgeon Creek (1982/1983)	72/80	6.9/8.0	50/58	10/8	7.0/7.2	-/159	-/0.00

[a] Above Cave Run Lake.

ond highest alkalinity and highest number of fish. Turbidity problems exist in Licking River, Middle Fork, and North Fork, where sedimentation is a severe problem. The mean depths of sediments not yet compacted in these streams were 1.4 and 0.3 ft at Licking River and North Fork, respectively, in 1983. The mean depth of sediment in Middle Fork was 0.3 ft above Buckhorn Lake and 0.01 ft below. Much of the instream cover from boulders and fallen trees in Licking River and North Fork did not provide habitat for fish due to being submerged in sediments.

Streams in the South Fork drainage were slightly acidic in 1982. They have a poor buffering capacity due to low total alkalinity. These streams are also susceptible to lowering of pH due to acid runoff from past deep coal-mining activities. However, the problem of acid pollution is not as apparent as it was in the period 1969-1973, when fish kills were common. The pH at Little Goose Creek, Goose Creek, and Collins Fork averaged 5.7, 6.5, and 6.6, respectively in 1969; in 1982 and 1983, the annual average for pH was 7.5-7.4, 7.0-7.4, and 6.7-7.4 in these streams, respectively. This improvement in acid conditions has resulted in increased and more varied fish populations in these streams.

METHODS

A total of 117 pools representing 106.2 stream miles were eventually sampled in the 14 streams. Pool area and accessibility were the most important considerations in determining sampling locations. Physical characteristics recorded for each study pool (Table 3) included width, depth, fish shelter, bottom type, pool-riffle ratio, vegetation, and land use. Lengths and gradients were determined by using topography maps. Water quality conditions were measured seasonally, except winter, at lower, middle, and upper stations within the larger streams, and at one or two stations on the smaller tributary streams (Table 4). Temperature and dissolved oxygen were measured with a YSI Model 54 oxygen meter. Total alkalinity was determined using bromcresol, green-methyl red as an indicator and titrating with 0.2N sulfuric acid. The pH and turbidity were measured with a Hach OR-E2-5 portable water analysis kit. A YSI Model 33 S-C-T meter was used to determine specific conductance and salinity.

Muskellunge populations were sampled for two consecutive years with a boat-type electrofishing unit utilizing AC output. Sampling was conducted when stream turbidity would not be a limiting factor to fish collection. The entire shoreline of each sample area was electrofished at least once; a second round of sampling was done when one round was thought to be insufficient. Captured fish from each pool were sorted to species, counted, measured to the nearest inch group, and released. Fish not readily identified were preserved and identified in the laboratory. Captured muskellunge were measured to the nearest 0.1 in and weighed to the nearest 0.01 lb; no attempt was made to try to determine their sex. Sighted muskellunge that escaped capture were also recorded. Scale samples were taken from each fish past the pectoral fin and above the lateral line. Legal-size muskellunge (≥ 30 in) were tagged on the anterior, basal edge of the dorsal fin with a numbered, monel, self-piercing jaw tag, with identification of the Kentucky Department of Fish and Wildlife Resources. Sub-legal muskellunge were marked by clipping one of their pelvic fins. In addition, self-addressed, mail-in survey envelopes were made available to fishermen at country stores and tackle shops within the area. A questionnaire was printed on the back of the envelopes for acquiring pertinent catch data.

Backcalculation of growth was determined from a modification of the Lee method (Everhart and Youngs 1981), and using a correction factor of 4.5 in determined by Brewer (1980). An estimate of the total muskellunge population was attempted for each stream using the Peterson method (Lagler 1956). According to Robson and Regier (1964), population estimates with this method were found to be highly inaccurate when there were fewer than four recaptures. Only in Tygarts Creek in 1981 were there enough recaptures to estimate the population (194 fish; Kornman 1983). Therefore, the catch per unit effort (CPUE) by electrofishing in each stream was utilized as an index of their abundance (Table 5). Success at capturing muskellunge depended on several factors such a fish size, mobility of fish, ability of the netters, proximity of fish to the boat, amount of stream cover, water depth, and turbidity.

RESULTS

Of a total of 270 muskellunge sighted at streams, 179 (66%) were actually boated. The CPUE for muskellunge at each stream ranged from 0.1 to 1.3 fish per hour; the mean CPUE was 0.6 fish per hour (Table 5). Fifteen muskellunge collected from the Middle Fork Kentucky River were not included, since they were all taken within 4.9 mi below Buckhorn Lake dam and assumed to have escaped from the lake, which has been stocked annually with muskellunge since 1979. The native population of muskellunge that earlier inhabited this stream declined

Table 5. Length frequency and catch rate of muskellunge captured from 2 years of electrofishing on Kentucky streams.

Stream	<10	10-12	13-15	16-18	19-21	22-24	25-27	28-30	31-33	34-36	37-39	Total	Fish per hour
Kinniconick Creek		8	4		2	5	2	4	6	4		35	0.5
Tygarts Creek		1	16	7	5	6	4	12	6	1	1	59	1.3
Licking River[a]					1	1		1				3	0.3
Red River			1		3		2	2				8	0.3
Middle Fork Kentucky River				2	3	1	4	3	1	1		15	0.3
North Fork Kentucky River												0	0.0
Collins Fork		5		2	2		2	1	1			13	0.9
Goose Creek		1	2	1	3	1	5	2	2	1		18	0.8
Little Goose Creek												0	0.0
Redbird River			1	1	2		1	2				7	0.4
Sexton Creek				1				1		1		3	0.7
South Fork Kentucky River			2	1		2	1					6	0.1
Station Camp Creek	1				2	2	1	3	2			11	0.8
Sturgeon Creek								1				1	0.2
Total												179	
Mean													0.6[b]

[a] Above Cave Run Lake.
[b] Results from Middle Fork Kentucky River and North Fork Kentucky River were not included in determining this value because these streams are not considered native muskellunge streams.

after Buckhorn Lake was impounded in 1961. Former spawning areas in Upper Middle Fork, Beech Fork, and Greasy Creek were separated from the population below the dam. Muskellunge were collected from Redbird River, which had not been stocked or previously considered a native muskellunge stream. No muskellunge were captured in Little Goose Creek, but muskellunge do frequent this stream, particularly in the spring during the spawning period.

The muskellunge CPUE was highest in Tygarts Creek, 1.3 fish per hour. Catch rates in Collins Fork, Goose Creek, Sexton Creek, and Station Camp creek were above the mean CPUE for all streams studied. South Fork Kentucky River is a better muskellunge fishing stream, particularly during the spring, than was indicated by the number of fish captured during the study. Conservation Officer Joe Burchell (personal communication) estimated 60 muskellunge were harvested in the South Fork during February-April 1983.

In Tygarts Creek, muskellunge were concentrated in the middle section, where 55 of the 59 captured fish were taken. Twenty-two of these were taken from a pool below Carter Caves State Park, an area where several spring-fed streams enter Tygarts Creek. Robert D. Hoyt, Western Kentucky University, (personal communication) observed a similar concentration of muskellunge in a spring-fed section of Barren River in Kentucky.

Muskellunge in Kentucky reach legal size (≥ 30 in) by their fifth or sixth year of life (Table 6); they grew slower through age IV in the South Fork Kentucky River drainage than in other streams. Similar growth rates for muskellunge were reported by Miles (1978) in West Virginia at Middle Island Creek and by Parsons (1959) in Cumberland Plateau streams in Tennessee.

Stocking may have had some part in strengthening certain year classes of muskellunge in some streams (Table 7). The majority of captured fish were from the 1979 and 1980 year-classes; most of the streams were stocked in 1979. Although no muskellunge were stocked in Tygarts Creek in 1980, the high numbers captured from the 1980 year class in this stream could be the progeny from fish of the 1976 stocking (Table 2).

Response by stream anglers to a mail-in survey was sporadic. The greatest number of returns was from Kinniconick Creek, Tygarts Creek, and Licking River, which can probably be attributed to the cooperation of several Kentucky Silver Muskie Club members who fish these streams. Based on tag returns from Kinniconick and Tygarts creeks in 1980-1981, the annual exploitation rate was 15 and 21 percent, respectively (Kornman 1983).

The three most abundant species collected from each stream are listed in Table 8. The golden redhorse ranked first in abundance in 9 to the 12 native muskellunge streams, second in two streams, and third in one stream; it comprised from 20 to 50 percent of the total population sample in 11 streams where it was the predominant species. Longear sunfish was among the three most numerous species in 9 of the 12 streams. The Upper Licking River was the only stream that did not have either the golden redhorse as the most abundant species, or longear sunfish as one of the three most numerous; gizzard shad were most abundant here, which is not unexpected because of the large shad population downstream in Cave Run Lake. Carp ranked second in abundance in Licking River, which is a reflection of

Table 6. Mean annual length (in) of muskellunge in 12 Kentucky streams.

Stream	Number of muskellunge	Age					
		I	II	III	IV	V	VI
Kinniconick Creek	34	11.4	17.6	22.2	26.3	30.1	33.1
Tygarts Creek	59	11.3	17.9	22.7	27.0	30.8	35.8
Licking River[a]	3	11.6	17.0	20.9	25.9		
Red River	8	10.8	17.2	23.1	28.0		
Middle Fork Kentucky River	15	11.8	16.0	21.5	27.5		
Collins Fork	11	9.8	14.4	19.2	23.1	27.0	30.7
Goose Creek	18	10.4	15.8	20.1	24.7	29.7	32.0
Redbird River	6	11.0	15.9	20.8	22.2		
Sexton Creek	3	9.1	13.2	19.6	23.8	30.4	
South Fork Kentucky River	6	10.6	16.6	20.4	24.0		
Station Camp Creek	10	11.4	18.0	23.2	27.9		
Sturgeon Creek	1	12.3	18.0	23.6	27.8		

[a] Above Cave Run Lake.

Table 7. Number of muskellunge from each year-class captured in each stream. The number in italics represent fish from a year-class when muskellunge were stocked in the stream.

Stream (Year sampled)	Year-Class								
	1974	1975	1976	1977	1978	1979	1980	1981	1982
Kinniconick Creek (80-81)	4	7	2	5	4	8	4		
Tygarts Creek (80-81)	1	2	*16*	5	4	*13*	18		
Licking River[a] (82-83)						*1*	2		
Red River (82-83)					2	*1*	4		1
Middle Fork Kentucky River (82-83)						7	6	1	1
Collins Fork (81-82)		1			*3*	4	3		
Goose Creek (81-82)		1	2	3	5	*4*	2	1	
Redbird River (81-82)					2	2	2		
Sexton Creek (81-82)			*1*	1			1		
South Fork Kentucky River (81-82)					*1*	*2*	3		
Station Camp Creek (82-83)						5	4	1	1
Sturgeon Creek (82-83)						*1*			
Total	5	11	21	14	21	48	49	3	3

[a] Above Cave Run Lake.

the severe sedimentation and turbidity problems here. Gizzard shad were second in abundance in Middle Fork Kentucky River, as a result of recruitment from Buckhorn Lake, and the most abundant species in Tygarts Creek. Striped shiner was one of the three most numerous species in six streams.

Table 8. The three most abundant fish species in 14 Kentucky streams, based on percent composition in each stream.

Stream[a]	Species (%)		
Kinniconick Creek	Striped shiner (33)	Golden redhorse (11)	Longear sunfish (9)
Tygarts Creek	Gizzard shad (31)	Golden redhorse (18)	Longear sunfish (8)
Licking River[b]	Gizzard shad (35)	Carp (16)	Golden redhorse (11)
Red River	Golden redhorse (31)	Longear sunfish (13)	Silver redhorse (12)
Middle Fork Kentucky River	Golden redhorse (39)	Gizzard shad (26)	Longear sunfish (4)
North Fork Kentucky River	Golden redhorse (33)	Spotted bass (19)	Channel catfish (5)
Collins Fork	Golden redhorse (20)	Striped shiner (12)	Longear sunfish (8)
Goose Creek	Golden redhorse (38)	Longear sunfish (13)	Striped shiner (8)
Little Goose Creek	Golden redhorse (31)	Spotted bass (6)	Striped shiner (4)
Redbird River	Golden redhorse (50)	Longear sunfish (11)	Striped shiner (8)
Sexton Creek	Golden redhorse (39)	Spotted bass (7)	Striped shiner (6)
South Fork Kentucky River	Golden redhorse (46)	Bigeye shiner (11)	Longear sunfish (10)
Station Camp Creek	Golden redhorse (29)	Longear sunfish (23)	Spotted sucker (6)
Sturgeon Creek	Golden redhorse (35)	Bluegill (10)	Longear sunfish (9)

[a] Middle Fork and North Fork Kentucky River do not support a natural population of muskellunge.
[b] Above Cave Run Lake.

Spotted bass was the dominant black bass species in 9 of the 12 native muskellunge streams. This species ranked second in abundance among all species in three streams. Smallmouth bass was equally abundant in Redbird River, and largemouth bass was just as abundant as spotted bass in Station Camp Creek. Largemouth bass was the most numerous black bass species in Collins Fork. Several pools in the upper section of this stream contained a dense stand of spatterdock *Nuphar* sp.; vegetation was not abundant in any other streams.

DISCUSSION

An examination of the physical features of pools where muskellunge were either captured or sighted provides an insight to preferred habitat of this species. Muskellunge were found in pools that had mean depths of 1.0 to 9.4 ft, maximum depths of 3 to 24 ft, mean widths of 30 to 128 ft, lengths of 0.1 to 2.3 mi, and gradients of 1.0 to 8.5 ft/mi. Although the range was wide for each of these parameters, muskellunge were found more frequently in pools having certain specific characteristics (Table 9).

Statistical interpretation of the pool characteristics where muskellunge were captured or sighted was not attempted due to known bias from less sampling efficiency as mean depth, width, and length of pools increased. Also, a few observed fish were later captured, so some fish could have been included twice in the calculations. In spite of these data weaknesses, mean depth and gradient are believed to be most critical for where muskellunge will be found, although all of these characteristics are interrelated.

Muskellunge occurred most frequently in pools with mean depths of 3.5 to 4.0 ft. When mean depth was less than 2.5 ft or greater than 5.4 ft, the rate of occurrence sharply declined. Muskellunge were found most often in pools with gradients of 3.0 to 6.9 ft/mi and infrequently below 2.0 ft/mi and above 7.9 ft/mi, which was also related to depth. Muskellunge appeared to prefer pools with a maximum depth of at least 5 ft, a mean width of more than 25 ft, and a total length of 1.8 mi or less. There was a gradual increase in fish captured and sighted per mile as pool length decreased. This relationship with pool length is probably more an indication of increased ease of

Table 9. Frequency of muskellunge captured and observed per mile in relation to mean and maximum depths, mean width, length, and gradient for the 93 pools sampled in 12 Kentucky streams.

	Fish per mile		Fish per mile
MEAN DEPTH (ft)		**POOL LENGTH (mi)**	
<2.0	0.88	<0.3	2.83
2.0-2.4	0.62	0.03-0.6	2.69
2.5-2.9	0.62	0.7-1.0	1.84
3.0-3.4	1.75	1.1-1.4	1.42
3.5-3.9	4.81	1.5-1.8	1.76
4.0-4.4	2.71	1.9-2.2	0.83
4.5-4.9	0.99	>2.2	0.60
5.0-5.4	1.60		
>5.4	0.76	**GRADIENT (ft/mi)**	
MAXIMUM DEPTH (ft)		>2.0	0.31
>5	0.45	2.0-2.9	1.16
5-6	1.71	3.0-3.9	3.13
7-8	1.61	4.0-4.9	3.41
9-10	3.42	5.0-5.9	1.30
11-12	1.67	6.0-6.9	3.20
13-14	2.02	7.0-7.9	1.88
15-16	2.85	>7.9	0.38
>16	1.32		
MEAN WIDTH (ft)			
>26	0		
26-35	2.50		
36-45	1.52		
46-55	1.68		
56-65	0.87		
66-75	2.75		
76-85	1.64		
86-95	2.97		
96-105	1.28		
>105	1.56		

escapement with increased pool length as it is for a preference for shorter pools.

Almost every captured muskellunge, age II or older, was near a fallen tree. Kornman (1983) noted that fallen trees were important as muskellunge habitat in Kinniconick and Tygarts creeks. Large woody debris in streams is evidently essential for muskellunge. In this study, the number of muskellunge sampled was directly related to number of fallen trees. Yearling muskellunge were most often captured along the shoreline in shallow water, where there was vegetation or tree roots. Kinniconick Creek and Collins Fork were the only streams that had aquatic vegetation of any significance.

CONCLUSIONS

Limiting factors to the muskellunge habitat and fisheries were identified and some of these can be resolved (Table 10). Strip mining for coal is currently a problem in the Upper Kentucky River drainage. Oil drilling has the potential for being a problem at Upper Licking River, Red River, Station Camp Creek, and Sturgeon Creek. Oil shale development is a future concern within Kinniconick Creek drainage and portions of Licking River, Red River, and Station Camp Creek. The greatest threat to muskellunge stream habitat is from the following authorized U.S. Army Corps of Engineers dams: Licking River (Falmouth Lake); Tygarts Creek (Kehoe Lake - deferred); South Fork Kentucky River (Bonneville Lake inactive); and Red River (Red River Lake - inactive). If constructed, these impoundments would eliminate most of the muskellunge habitat and populations in these streams. A total of 109 mi of native muskellunge streams has already been lost to impoundments.

Sedimentation is partly responsible for the loss for another 63 mi in North Fork Kentucky River, and it may soon cause the Upper Licking River to become unsuitable for muskellunge. Part of the answer to sedimentation is to maintain an adequate buffer zone along the muskellunge streams and their tributaries. Better enforcement of laws on proper surface mining and oil drilling operations would also serve to improve and maintain water quality in the Kentucky River and Licking River drainages. The streams in the South Fork Kentucky River drainage have the greatest number of management problems. One of the worst of these is from illegal harvest of

Table 10. Limiting factors to muskellunge habitat in 14 Kentucky streams during the 1980-1983 studies.

Stream	Loss of spawning area due to impoundment	Sedimentation	Need more fallen trees	Low total alkalinity	High turbidity	High salinity	Gravel dredging	Poor access	Illegal harvest
Kinniconick Creek				X			X		
Tygarts Creek		lower section					X	X	
Licking River[a]	X	X			X			X	
Red River		shifting sand			X			X	X
Middle Fork Kentucky River	X	above Buckhorn Lake			X				above Buckhorn Lake
North Fork Kentucky River	X	X			X				
Collins Fork				X					X
Goose Creek									X
Little Goose Creek			X	X					
Redbird River			X	X			X		X
Sexton Creek				X					
South Fork Kentucky River			X	X					X
Station Camp Creek								X	X
Sturgeon Creek									X

[a] Above Cave Run Lake.

sublegal fish and the use of illegal methods such as hoop nets and weirs. Collins Fork, with many narrow riffle areas, is particularly vulnerable to weirs.

The most important need in managing the native muskellunge fisheries in Kentucky streams is to assure the integrity of their habitat. Any drastic change in land use pattern within the drainages should be closely monitored to minimize any detrimental impact on the fisheries. The importance of mean and maximum depths to muskellunge reaffirms the concern for habitat loss due to sedimentation, gravel dredging, and locks and dams that alter pool depth. Stream banks and riparian zones need to be better protected to reduce sedimentation and provide instream cover from fallen trees. Stocking has a role in managing certain streams. Stocking native muskellunge streams is warranted in those streams where the population density is below a certain level and where the streams receive heavy fishing pressure. Other candidates for stocking are streams that have large populations of golden redhorse, longear sunfish and spotted bass and habitat suitable for muskellunge. Those which were native muskellunge streams should only be stocked to establish a self-sustaining population. The spawning areas in Middle Fork Kentucky River have been lost, so stockings below Buckhorn Lake dam would have to be on a maintenance basis to provide a put-grow-take fishery, as partial restoration for the fishery lost due to the impoundment.

ACKNOWLEDGEMENTS

This manuscript would not be possible without the diligent work by several District Fishery Biologists who investigated the 14 study streams, collected the data, and helped interpret the results. Our deepest appreciation goes to District Biologists Kerry Prather and Albert Jones, who along with assistant Albert Surmont, dedicated long hours to seeing that jobs were accomplished as planned.

REFERENCES

Brewer, D.L. 1980. A study of native muskellunge populations in eastern Kentucky Streams. Kentucky Department of Fish and Wildlife Resources. D-J Project F-31-R(1-6), Fisheries Bulletin 64, Frankfort, Kentucky, USA.

Everhart, H.W. and W.D. Youngs. 1981. Principles of fishery science, second edition. Comstock Publishing Associates, Cornell University Press, Ithaca, New York, USA.

Kornman, L.E. 1983. Muskellunge streams investigation at Kinniconick and Tygarts creeks. Kentucky Department of Fish and Wildlife Resources. D-J Project F-50, Fisheries Bulletin 68, Frankfort, Kentucky, USA.

Lagler, K.F. 1956. Freshwater fisheries biology, second edition. Wm. C. Brown Company, Dubuque, Iowa, USA.

Miles, R.L. 1978. A life history study of the muskellunge in West Virginia. American Fisheries Society, Special Publication 11: 140-145.

Parsons, J.W. 1959. Muskellunge in Tennessee streams. Transactions of the American Fisheries Society 88:136-140.

Robson, D.S. and H.A. Regier. 1964. Sample size in Petersen mark-recapture experiments. Transactions of the American Fisheries Society 93:215-226.

CULTURE

Culture of Purebred Muskellunge

JOHN H. KLINGBIEL

Wisconsin Department of Natural Resources
P.O. Box 7921, Madison, Wisconsin 53707

ABSTRACT

The use of conventional pond-rearing techniques for culture of purebred muskellunge has provided erratic and unpredictable results but the fingerlings produced are large enough to provide maximum survival after stocking. The most dependable phases of muskellunge culture are spawning, incubation and pond-rearing of fingerlings from 3 to 8 inches or larger. The phase of culture that needs the most improvement is rearing from swimming fry to fingerlings of about 3 inches. The ability to control environmental conditions and provide adequate forage is essential to stabilized production.

Intensive culture techniques using pelleted feed have revolutionized the rearing of hybrid muskellunge and northern pike; it has not been nearly as successful with purebred muskellunge. Rearing in tanks using live food has been successful but growth is somewhat slow. For some production programs it appears feasible to rear fingerlings to 3 inches in tanks using live forage and then transfer them to ponds to grow to stocking size. Solutions to production problems vary and are dependent upon local conditions and facilities as well as the size of fingerling needed for cost-effective stocking.

Natural reproduction of muskellunge is unable to maintain populations dense enough to meet angler demand in many waters. Fingerling stocking has compensated for this in native muskellunge waters and has also allowed the development of good fishable muskellunge populations in nonnative waters as well. Stocking programs are often limited, however, because of the high cost and difficulty of assuring adequate supplies of fingerlings.

Culture techniques have generally improved through the years but continue to give erratic results. The culturist's inability to adequately monitor or respond to critical environmental problems in ponds and provide adequate food supplies are the main problems. Intensive rearing in concrete tanks using pelleted food has revolutionized the rearing of northern pike and hybrid muskellunge. These methods have not been nearly as successful with purebred muskellunge. (Graff 1978; Bender et al. 1980; Jennings 1982; Bubnack 1982).

Purposes of this paper are to review current culture techniques, to help identify both strong and weak areas in cultural procedures, to suggest where intensive culture methods can best be used and where additional developmental efforts are necessary. The end products in culture programs are higher, more stable production and high quality fish that will survive and furnish the most cost-effective product for management.

MUSKELLUNGE CULTURE METHODS

Spawning and Incubation

Eggs are usually obtained from wild fish. Special angling regulations are often imposed on broodstock waters to help assure good populations of spawners. Sometimes captive broodstocks are held to assure an egg supply, but the number that can be held is often limited and they are susceptible to disease and food shortages.

Spawning normally occurs at temperatures of 45°F-56°F. Potential brood fish are usually captured with fyke or pound nets, but electro-shocking is at times equally effective and is not detrimental to egg quality. Many culturists emphasize gentle handling and minimizing stress to assure good egg quality. To facilitate handling, each female can be placed in a socklike bag of woven webbing, which gently restrains the fish while it is being checked for spawning condition. Although brood fish are commonly anesthesized, some culturists refrain from this practice believing it unnecessary and that it even hinders proper assessment of the fish's spawning condition. Proper assessment is essential to good fry production. Eggs should flow freely from the female if they are ready to be spawned; if they are watery or stringy, they are usually of inferior quality and are generally not used unless insufficient eggs are available.

Some culturists have found it necessary to use specialized spawning methods and equipment if ample brood stock are not available. Sorensen et al. (1966) reported that spawning of eggs with the aid of compressed air and collecting of sperm with a syringe has improved production in Pennsylvania. Egg production has occasionally been improved by injecting female muskellunge with carp pituitary to induce spawning; Jennings (1982) reported that of 17 females injected with 3 mg per pound of body weight, all matured and could be spawned within 96 hours.

Eggs are stripped into a dry pan and 1 or 2 drops of milt is added from each of 2 or 3 males. If it takes longer than 2 minutes to completely spawn a female, the eggs already taken are fertilized and processed before proceeding. Water is added immediately after the addition of milt and the mixture is gently stirred and set aside. After 1-3 minutes, the eggs are rinsed once or twice and placed in a pail to harden. Many culturists stress minimal handling throughout the entire egg processing procedure. Johnson (1958) reported that eggs are very soft when expelled and, if not allowed to harden, can be easily broken. After hardening for about an hour, eggs are carefully transferred to hatching jars for incubation. While eggs are hardening, water temperature changes are minimized; eggs are kept out of direct sunlight.

Muskellunge eggs are usually incubated in hatching jars. They incubate best at volumes of 1 to 2 1/2 quarts of eggs per jar. Numbers of eggs vary from 30,800 to 67,670 per quart (Johnson 1958). For the first 72 daily temperature units of incubation, water flows are maintained at approximately 1/2 gallon per minute per jar, and eggs are allowed to mat together lightly. Daily temperature units are the number of degrees Fahrenheit above freezing for a 24-hour period. Subsequently, flows are increased to 1 1/2 gallons per minute and eggs are separated and gently rolled. Hatchery water temperatures are regulated initially to those of the incoming eggs and then raised 1°F per day, usually until 55-58°F is reached. Final incubation temperatures are regulated so that hatching corresponds to optimum conditions in the rearing environment. Daily 1-hour prophylactic treatments of 1:4,000 or 1:6,000 formalin are often used; however, some culturists prefer two 15-minute treatments daily of 4 ppm diquat, because they believe formalin hardens the egg shells and can decrease hatchability. Treatments are administered by metering a constant flow of chemicals into the water supply for a specific treatment period.

Table 1. Time temperature relationship and daily temperature units required for hatching muskellunge eggs (from Piper et al. 1983).

Temperature F	Days to hatch	Daily temperature units to hatch (F)
45	21	273
47	20	300
49	19	323
51	18	342
53	16	336
55	14	322
57	12	300
59	10	270
61	9	261
63	8	248
65	7	231
67	6	210

Survival has been increased in some hatcheries by removing eyed eggs from the jars and completing the incubation on trays.

Hatching usually requires 300-330 daily temperature units depending on temperatures. Piper et al. (1983) provided a guide for hatching times at various stable temperatures (Table 1). Gala and Eipper (1969) indicated reduced hatchability at temperatures of 60°F or higher; Johnson (1958) recommended that incubation temperatures be above 55°F to maximize fry survival by assuring that yolk-sac materials contribute to growth, thus promoting early feeding.

Four or five days after hatching, retaining screens are removed from the hatching jars so the fry are able to swim out the jar overflow into the fry tank. Astroturf or other material is suspended in fry tanks to provide additional surface on which the fry can cling. In two to four more days they are actively swimming and ready to be stocked in rearing tanks or ponds. At this time they are approximately 0.5 in long; about 22,700 weigh 1 lb.

Those techniques currently in use for spawning and producing muskellunge fry are generally successful and production results are predictable. Fry production in Wisconsin from 1979-83 has ranged from 72-88% of the number of eggs collected. Although the success of incubating indivdual batches of eggs varies greatly, in most batches over 85% survive until eggs are eyed and dead eggs can be removed (Fig. 1).

Pond Rearing

Most large-scale rearing of muskellunge fingerlings is done in ponds. Pond preparation procedures vary depending upon physical conditions,

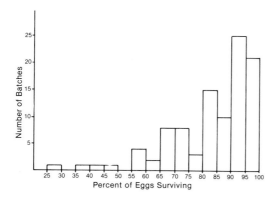

Figure 1. Frequency distribution of egg batches incubated in Wisconsin hatcheries from 1979-81 in relation to percent of eggs surviving until after eye-up.

water supply and fertility. Usually, ponds are kept dry in winter to reduce crayfish and macrophyte abundance. Ponds are filled early enough in the spring to encourage early zooplankton production. Contamination with unwanted fry can be minimized by screening inlets with saran socks and filling the ponds before any fry hatch in the water source. Ponds are usually fertilized to promote zooplankton pulses. Various materials and procedures are used for fertilization depending upon natural fertility and pond conditions. The use of alfalfa meal or pellets is increasing; rates as high as 389 pounds of pellets per acre have been used in Iowa (Jennings 1982). Torula yeast was reported to be an excellent fertilizer as well as an inhibitor of filamentous algae growth (Johnson 1958).

Monitoring of zooplankton abundance and size is important both before and after muskellunge fry are stocked. Standardized plankton tows taken twice daily are helpful in determining the condition of the food supply, assessing the effect of fertilization and determining when fry should be stocked.

Some ponds require that plankton be introduced in order to produce significant pulses; however, care must be taken to not introduce species that become so large that fry choke on them. Resulting mortalities from this can be significant (Walker et al. 1958). *Daphnia pulex* and *Daphnia longispina* are of-suitable size for muskellunge fry. According to Applegate (1981), the size of the food accepted is closely related to the width of the mouth of; he reported that when fry began feeding they were 0.55 in long and their mouths were 0.027 in wide. At that time the average size of zooplankton consumed was 0.02 in long, and 0.011 inches in diameter. When the fish were 23 days old and 1.42 in long, their mouth width had increased to 0.073 in; the mean length of the zooplankton consumed was somewhat larger than 0.039 in. He reported that *Moina brachiata*, often abundant in sewage treatment ponds, are eaten more readily than *Daphnia*, or a number of other zooplankters. Ideally, muskellunge fry should be stocked only when suitable zooplankton is available and danger of severe temperature drops are minimized. Analysis of records at Spooner Hatchery, Wisconsin, indicate that the best stocking temperature is 62°F, but subsequent temperatures are probably of even greater importance. Severely reduced feeding and subsequent mortality have occurred when water temperatures dropped into the high 40's a few days after fry stocking even though ample food was available. High temperatures have also been suspected to cause serious mortalities. Bonin and Spotila (1978) reported that muskellunge succumb to high temperatures more easily than hybrids and are particularly sensitive when 4 days old; they have a critical thermal maximum of 85.8°F at that age. Even though this is an extremely high temperature for that early in the rearing season, such temperatures may occur at the surface of ponds having extremely poor circulation during abnormally hot and sunny weather. Daily surface water temperature fluctuations as great as 17°F have occurred in rearing ponds at the Spooner, Wisconsin hatchery.

Fry are generally stocked in rearing ponds when they are developed enough to actively swim about, usually at 6-9 days after hatching. Stocking rates vary greatly depending upon the pond, the ability to control pond conditions and the length of the rearing period expected. At Woodruff Hatchery, Wisconsin, where pond conditions are closely monitored and maximum production is desired, stocking rates average 48,000 fry per acre-foot of water. All fry are stocked into a pond at one time and should be the same age and developmental state to minimize cannibalism.

Close monitoring of pond conditions is essential so the culturist can respond quickly to conditions that may endanger production. Factors such as pine pollen, which can cause gill abrasion, and predation by aquatic insects and birds present challenges. When predaceous insects appear, ponds are sprayed with 6 gallons of fuel oil per acre; trichodina and bacterial gill disease can be minimized with spray applications of 2 ppm potassium permanganate whenever necessary. Observing pond conditions as often as 4-5 times a day, particularly early mornings, may be necessary initially to assess abundance, type, and size of zooplankton and survival of muskellunge fry. Frequency of observations may be gradually reduced as the fish become larger.

Muskellunge can be reared for extended periods on zooplankton, but as they become older cannibalism increases and growth slows dramatically, unless larger food items are made available. Fry are ready to change from zooplankton to a fish diet after only a few days. When muskellunge are only 0.75 in long they can eat newly hatched white sucker fry; fathead minnow fry can be eaten by even smaller muskellunge. Various fish species are used for forage depending upon availability, size and phenology. Piper et al. (1983) included correct size, easy capture, attractive behavior, and long and slender body shape with good weight as requirements for suitable forage fish for muskellunge. White suckers are favored as initial forage fish by many fish culturists because of these characteristics and the phenology of their early life history. Fathead minnows are also commonly used; occasionally, fathead minnow brood stock is introduced to spawn directly in the muskellunge rearing ponds. Stocking fathead minnow brood stock too early may significantly reduce zooplankton numbers, thus reducing muskellunge growth and survival (Peterson 1984). Carp, goldfish and golden shiners are also commonly used as initial forage fish for muskellunge.

Timing of the first forage fish introduction is important because many small muskellunge fingerlings will choke if their food is too large. Even newly hatched white sucker fry should not be introduced until muskellunge have been feeding for 9 or 10 days. Culturists can determine when small muskellunge are able to eat available forage fish by observing them both in an aquarium. In some cases, it may be necessary to delay the hatching of forage fish by reducing incubation temperatures. Newly hatched white sucker fry are stocked daily at a rate of at least 10 per muskellunge fry. As the muskellunge grow they prefer larger forage fish, such as lake shiners, sucker fingerlings and various other species. When muskellunge become somewhat larger, forage fish of various sizes should be made available because muskellunge are seldom of uniform size. Forage fish must be in good health; if diseased, they will infect the muskellunge as well. As a precaution, all forage fish should be treated with diquat or other appropriate chemicals before they are stocked in the pond or other rearing facility.

Sometimes 1.5-3 in muskellunge fingerlings are transferred to ponds where forage fish are being reared. This assures that sufficient forage minnows are available initially and helps the culturist determine what auxiliary feeding program is necessary, because he knows the number of muskellunge he must feed. This rearing strategy also increases the quantities of forage fish reared in the pond because of the initial absence of predation from muskellunge. The size at which muskellunge fingerlings are harvested depends upon the survival expected in the waters to be stocked. In Wisconsin, 8 1/2 in has been established as a minimum size for stocking most waters. In ponds, a 10% survival of fry to stocking size is considered satisfactory. Survival from 2 in to stocking size averaged 59% at Woodruff Hatchery, Wisconsin, from 1979 through 1982 (Dimpfl 1983). It takes 2.75 lbs of forage fish to produce one lb of muskellunge fingerling at that hatchery; in less intensive pond culture, it is more commonly 3-3.5 lbs.

Growth rates of cultured muskellunge vary, depending on forage and water temperature. In 1982, the average fingerling reached 8.5 inches at Woodruff Hatchery by August 14 and at Spooner Hatchery by August 11 (David Hanson, pers. comm.). By the end of the fingerling harvest, usually in mid-September, most are 12-13 in long. Although production is somewhat lower at other Wisconsin hatcheries, at Woodruff 800 lbs of fingerling per acre are normally harvested. As many as 3,000-4,000 lbs/acre are harvested from individual ponds; the maximum was over 9,000 lbs/acre one year. Forage fish must be introduced frequently to reach these high production levels. Forage needs can be determined by observing their relative abundance on a daily basis.

Production costs vary greatly, depending not only on rearing success and feeding efficiency but also on the size at which fish are harvested and the items included in the cost. In Wisconsin, fry cost about $4.50 per 1,000. Costs of pond-reared fingerlings of stocking size are generally about $11 per lb or $2 each. Wisconsin's cost figures include considerable administrative overhead with no amortization of facilities, since virtually all its hatchery facilities have appreciated. Commercial prices may reflect maximum production costs since they include profit. Generally, muskellunge fingerlings are sold for $0.75 per in.

Tank Rearing

Tank-rearing procedures are used for some small-scale, purebred muskellunge production programs. Although many culturists are testing pellet feeding in conjunction with tank rearing, most actual production efforts depend upon natural food.

Swimming fry are stocked in tanks holding as little as 4 ft^3, or they are confined with screens to a small portion of a larger tank; crowding encourages initial feeding. *Daphnia* or other zooplankton are introduced. Brine shrimp are often used when other zooplankton become scarce. As muskellunge become larger, forage fish are introduced and more rearing space is required. Tanks or raceways sometimes holding well over 300 ft^3 are used; depths vary from 1.5-3 feet; water temperatures are usually kept at 68° to 72°F, if possible, to promote growth and minimize disease. Daily prophylactic treatments of various chemicals are administered, and tanks are cleaned whenever there are significant deposits on the tank bottom. If considerable size variation occurs the fish are graded to minimize cannibalism.

Success with rearing muskellunge fingerlings in tanks using natural foods has been quite variable; failures mostly occurred because of disease or food shortages. Bender (1982), in Pennsylvania, reported a 74% survival rate for the first 30 days, and a further 74% survival of those remaining from day 31 to 60 in 1982; he reported 44% survival for the first 30 days and 25% for the first 62 days in 1981 (Bender et al. 1981). Jennings (1982) reported that, in Iowa, 31% survived until they had grown to 33.3 fingerlings per lb.

Food conversion rates in tanks are generally higher than in ponds. At Wild Rose Hatchery in Wisconsin, conversion was 2.23 (AveLallemont & Batten 1983). Muskellunge growth rates in tanks are considerably less than in ponds, even when natural plankton and forage fish supplies are excellent (Fig. 2). Records of rearing in tanks and cage culture trials (Montz 1981) suggest that high loading rates may result in slower growth.

No accurate cost comparisons have been made between rearing in tanks and ponds because of the great differences in size of the fish raised, length of the rearing period and methods used to compute costs by different investigators. Jennings (1982) reported a production cost of $1.07 each for fingerling that were 24 per lb; 89% of these were reared in tanks, the remainder in ponds.

DISCUSSION AND CONCLUSIONS

Traditional culture techniques are usually successful when weather allows proper timing of plankton pulses, development of muskellunge and spawning and growth of forage species. Culturists must be able to compensate for variations in the normal phenology. They must not only know the relationships that exist at a particular time, but must somehow predict what they might be at a specific future time. To accomplish this, he/she must either be able to personally control rearing conditions or have a number of management options that can be used should the predictions be incorrect.

The value of controlled temperatures for muskellunge egg incubation has been widely recognized and placed in general use for a number of years (Oehmcke 1969). Although there have been difficulties at some hatcheries with spawning, incubation and production of healthy fry, adjustments to traditional procedures have largely resolved these problems. Likewise, there seem to be few technological problems with rearing fingerlings after they become about 3 in long, if sufficient forage is available.

Rearing problems appear to be greatest from the time muskellunge are swimming fry until they reach 3 in long. In some facilities, these problems can be addressed most easily by tank-rearing, where monitoring is easier. The disadvantage of slower growth in tanks is minimal at this stage; only an additional 5 days were necessary for the fingerlings to reach 3 inches in Wisconsin tests. The ability to control temperatures for at least 10 days after the swimming fry stage usually assures proper development for initial feeding and promotes good growth rates. This can be accomplished most easily in tanks, although the possibility of controlling temperatures in ponds should not be overlooked, particularly the ability to eliminate extreme highs and lows.

Perhaps the greatest problem in culturing large muskellunge fingerling is getting sufficient quantities of the proper size forage. It appears that more

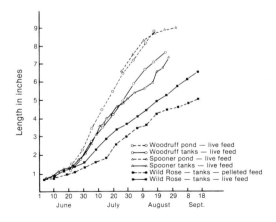

Figure 2. Length of muskellunge fingerlings reared in tanks or ponds at different Wisconsin hatcheries in 1983.

emphasis must be placed on this than on any other aspect of the rearing program. Although commercially available pellets could theoretically solve food shortages, better technology is necessary before high-quality fish can be produced using this type of food. Both better food quality and feeding techniques need to be addressed. Since relatively-low loading rates seem necessary for good growth, perhaps procedures for pellet feeding in ponds could be developed. This has been found practical and cost-efficient for northern pike and hybrid muskellunge. Until such technology is available for rearing large muskellunge fingerling, a number of approaches may be used to assure sufficient natural forage.

Daphnia pulex can be cultured in tanks for transfer to muskellunge-rearing units. Also, large quantities of zooplankton are often available in secondary sewage treatment ponds. The commercial availability of brine shrimp eggs, and the ability to control conditions under which they hatch, assures a timely source of that forage. The use of controlled temperatures may aid in the production of forage fish. Hatching dates of sucker eggs are commonly adjusted by controlling temperatures; use of this method for control of fathead minnow or carp spawning and/or growth may also be practical. The utilization of intensive, forage-production techniques, or species not normally used as forage, should not be overlooked.

Production of small fingerlings in tanks for transfer to ponds where they can grow faster appears to be practical; however, the choice of rearing technology used, and the cost-savings it may provide, must be evaluated in light of the size and quality of the fish it produces and the survival that can be expected after stocking. Different decisions are probably appropriate for different hatcheries and for the production scheduled for different waters.

Although muskellunge culturists await the "ultimate breakthrough" where high quality fingerlings can be reared intensively on pellets, those responsible for substantial production programs must be satisfied now by addressing the specific current needs of the production facilities available to them. However, relatively high production levels of good quality fingerlings are possible and can be predictably achieved with proper environmental controls.

REFERENCES

Applegate, R.L. 1981. Food selection of muskellunge fry. The Progressive Fish-Culturist 43:136-139.

AveLallemont, S. and D. Batten. 1983. Tank rearing of true muskellunge (*Esox masquinongy*) on live feed at Wild Rose State Fish Hatchery, 1983. Wisconsin Department of Natural Resources, Madison, Wisconsin, USA.

Bender, T.R., Jr. 1982. Intensive culture of Muskellunge (*Esox masquinongy*) using commercially prepared dry diets. Pennsylvania Fish Commission, Coolwater Diet Study 1982 Harrisburg, Pennsylvania, USA.

Bender, T.R., Jr., V.A. Mudrak, and S.E. Hood. 1980. Intensive culture of muskellunge (*Esox masquinongy*) using commercially prepared dry diets. Pennsylvania Fish Commission, Harrisburg, Pennsylvania, USA.

Bender, T.R., Jr., T.A. Wiggins, V.A. Mudrak, Z.B. Bean, C.R. Mann, and S.E. Hood. 1981. Intensive culture of muskellunge (*Esox masquinongy*) using commercially prepared dry diets. Pennsylvania Fish Commission, 1981. Project Report, Harrisburg, Pennsylvania, USA.

Bonin, J.D., and J.R. Spotila. 1978. Temperature tolerance of larval muskellunge (*Esox masquinongy* Mitchill) F1 hybrids reared under hatchery conditions. Comparative Biochemical Physiology. 59A:245-248.

Bubnack, J.R. 1982. 1982 muskellunge report Chautauqua Fish Hatchery. New York Department of Environmental Conservation, Albany, New York, USA.

Dimpfl, S. 1983. Muskellunge culture - intensive rearing for stocking. Wisconsin Department of Natural Resources, Madison, Wisconsin, USA.

Galat, D.L., and A.W. Eipper, 1969. Preliminary investigations on the effects of incubation temperature on the survival of muskellunge (*Esox masquinongy*) embryos. New York Cooperative Fishery Unit, Cornell University, Ithaca, New York, USA.

Graff, D.R. 1978. Intensive culture of esocids: the current state of the art. American Fisheries Society Special Publication 11: 195-201.

Jennings, T. 1982. 1982 hatchery branch production report. Iowa State Conservation Commission, Des Moines, Iowa, USA.

Johnson, L.D. 1958. Pond culture of muskellunge in Wisconsin. Wisconsin Conservation Department, Technical Bulletin 17, Madison, Wisconsin, USA.

Montz, G.R. 1981. Effects of stocking density on growth and feed conversion of muskellunge. Wisconsin Department of Natural Resources, Madison, Wisconsin, USA.

Oehmcke, A.A. 1969. Muskellunge management in Wisconsin. Wisconsin Department of Natural Resources, Bureau of Fish Management, Report 19, Madison, Wisconsin, USA.

Peterson, J., J. Bush and L. Sather. 1984. Adult fathead stocking prior to fry stocking. Wisconsin Department of Natural Resources, Madison, Wisconsin, USA.

Piper, R.G., I.B. McElwain, L.E. Orme, J.P. McCraren, L.G. Fowler, and J.R. Leonard. 1982. Fish hatchery management. U.S. Fish and Wildlife Service, Washington, DC, USA.

Sorenson, L., K. Buss, and A.D. Bradford. 1966. The artificial propagation of esocid fishes in Pennsylvania. The Progressive Fish Culturist. 28:133-141.

Walker, K., S. Kmiotek and A. Oehmcke. 1958. Muskellunge trough rearing experiment. Wisconsin Conservation Department, Northeast Area Investigation Memorandum 2, Madison, Wisconsin, USA.

Pennsylvania's Practices for Intensive Culture of Hybrid Muskellunge

THOMAS R. BENDER, JR. and DELANO R. GRAFF

Pennsylvania Fish Commission
450 Robinson Lane
Bellefonte, PA 16823-9616

ABSTRACT

Artificial propagation of esocids has been practiced since the late 1800's and hybrid muskellunge were first cultured in the 1930's. Intensive culture of hybrid muskellunge developed when it was discovered that they could be maintained on artificial diets. Present spawning, incubation and hatching techniques are essentially refinements of those developed for muskellunge. Development of new feeds and feeding techniques was a major factor in achieving the current state-of-the-art for culture of this fish. Cooperative efforts and interstate workshops have been instrumental in achieving widespread success with hybrid culture. Current research is directed at refinements in culture techniques and improvements in facility design equipment and feeds.

Esocids, specifically muskellunge, have been artificially propagated for nearly a century—New York started it in 1887, Wisconsin in 1899 (Buss 1961). Fry stocking was normal procedure in the early stages of muskellunge management, partly because hatcheries had great difficulty in producing larger fish. Gradually, techniques were developed that consistently provided fingerling-size muskellunge in ponds. Pond (extensive) culture was established as the standard approach to rearing esocids. Artificial propagation became a valuable tool in muskellunge management as a result of widespread adoption and refinement of techniques of pond culture developed in Wisconsin (Johnson 1958). The techniques used for culture of muskellunge were equally effective for production of northern pike and the hybrid between northern pike and muskellunge.

Hybrid muskellunge, also called tiger muskellunge, pike musky, or norlunge, were first artificially propagated in Minnesota. Mr. C. T. Falk successfully crossed muskellunge and northern pike at the Minnesota Department of Conservation's Nevis Hatchery in 1936 or 1937 (Eddy 1940). In 1939, Samuel Eddy repeated the cross and documented the results. This early work indicated that hybrid muskellunge might offer some advantages over regular species for artificial propagation. Eddy achieved about a 50% hatch, better than what was being accomplished with purebred muskellunge at that time; the fry also grew faster than purebred muskellunge. However, some disadvantages were also noted: the hybrids tended to have a higher percentage of stunted, deformed fish than purebreds. Despite the knowledge that hybrid muskellunge could be readily produced and grew well in a hatchery, it was to be some time before they would play a major role in esocid culture.

Extensive culture was successful, but it involved some inherent inefficiencies. Pond set-up and fingerling harvest were labor intensive, and production levels were unpredictable due to weather conditions and the fact that the fish could not be observed. In addition, large amounts of pond space had to be devoted to rearing forage. Some of these inefficiencies were eliminated by use of intensive or "tank" culture procedures developed in Pennsylvania (Sorenson et al. 1966). Those techniques provided the foundations for intensive culture of esocids as we know it today. Although the fish were reared in a controlled environment, primarily tanks, the rearing of forage still required ponds. Because intensive culture of esocids on live food involved a substantial amount of pond space for rearing forage and a great deal of labor to set up ponds, harvest, grade and transport forage, this approach could be considered "semi-intensive culture."

Intensive culture of hybrid muskellunge essentially became a major part of esocid culture with the discovery in 1968 that esocids, particularly the hybrid of northern pike and muskellunge, could be reared on commercially-available artificial diets (Graff and Sorenson 1970). The combination of tank culture techniques described by Sorenson and associates and the use of artificial diets made it possible to anticipate the mass production of esocids, much the same as trout. The expectation of an esocid that could be reared like trout was to be a major factor in enhancing the importance of hybrid muskellunge in hatchery programs.

SPAWNING AND HATCHING

Current practice among virtually all agencies working with hybrid muskellunge is the use of wild fish as brood stock. This can pose a problem if ripe fish of both species are not available for spawning at the same time. The general procedure is to use male northern pike and female muskellunge; however, no scientific information exists to demonstrate the superiority of either reciprocal cross—actually, availability of brood fish often determines how the hybrid cross is made. The preference for use of female muskellunge is based on the fact that brood muskellunge tend to be larger than northern pike and they tend to produce more and larger eggs. The perceived advantage is that the culturist has to spawn fewer females to produce a given number of eggs and the larger eggs produce larger fry (Shyrl Hood, pers. comm.).

Spawning techniques described by Sorenson et al. (1966) are still, for the most part, the state-of-the-art. Brood fish are anesthetized to permit ease of handling and to reduce the possibility of injury to the fish. The use of an inclined board or "rack" and an inflatable band or "blood-pressure cuff" greatly facilitates extrusion of eggs. Since adequate sperm can be a limiting factor, the use of a catheter and careful checking of sperm motility is still recommended procedure.

The potential difficulties posed by working with parent species that normally spawn at different water temperatures and, consequently, at different times, have resulted in growing interest in sperm storage. Currently, some culturists stockpile male northern pike in hatchery ponds or tanks in an effort to prolong "ripeness" and to assure their availability when the first ripe muskellunge females are captured. Obviously, it would be preferable if northern pike sperm could be taken when the gonads of male northerns were at an optimum stage of maturation and stored until needed. Limited success with short-term (7 days) cryostorage of northern pike sperm has been reported (de Montalembert et al. 1978), but there has been relatively little success to date with longer term storage (3-4 weeks). However, promising results have been achieved with salmonid sperm.

Incubation of any fish eggs requires an adequate supply of clean, disease-free water of the appropriate temperature, and hybrid muskellunge are no exception. Well or spring water sources are preferred since they are normally free of disease and turbidity and temperatures are generally stable. No matter what the source, the objective is always the same: a reliable flow of clear, disease-free water between 10-12 C.

The most common approach to incubation of hybrid muskellunge eggs is in jars, treated daily with formalin (40% solution of formaldehyde) at a concentration of 1,667 mg/l flow-through for 17 minutes. As eggs approach hatching, formalin treatments are stopped as the concentrations suitable as a fungicide on eggs are too high for fry. Not all culturists utilize jars for incubation. At Oneida Lake Hatchery in New York, Heath incubators (modified with smaller mesh screens) are the preferred method. The Heath incubator provides for better utilization of space than jars and requires less water. Utilization of less water can be a factor when one is concerned with significant heating and/or treatment costs. It is not recommended that hybrids be allowed to hatch in Heath trays (Richard T. Colesante, pers. comm.).

Hatching of hybrid muskellunge is handled in several ways. In many instances, eggs are hatched in incubation jars; and, over a 2- or 3-day period, fry are permitted to be carried with the discharge or effluent water into rearing tanks. Another widely accepted practice is to remove eggs from incubation jars or trays 1 or 2 days prior to hatching and place them on screens or screened trays suspended a few inches beneath the water's surface in rearing tanks. As eggs hatch, healthy fry can be encouraged to swim off the screens by periodic gentle jarring or bumping of the screen. Eggshells and dead, weak or deformed fry remain on the screen and can be easily removed and discarded after hatching is complete. In some situations, hatchery managers may prefer to "force hatch" hybrid muskellunge. This consists of removing eggs from the incubation unit as soon as the first few hatch and placing the entire lot in basins or other shallow containers with a relatively small amount of water. Care must be exercised at this point, since oxygen depletion can occur if too many eggs are placed in the basins. The containers are placed in a warm area, and as the water temperature increases, most of the eggs will hatch within several hours. After hatching, the fry can be decanted into rearing tanks. At Michigan's Wolf Lake Hatchery, newly hatched sac fry, which have been "cleaned up," are placed in Heath incubator trays (25,000-30,000 per tray), where they remain until swim-up. This method eliminates the need for daily formalin treatments to prevent the development of fungus (James Copeland, pers. comm.).

Active feeding by fry starts within 10 or 12 days following hatching at 10 C. The time to swim-up can be reduced considerably by increasing water temperatures one or two degrees per day to a maximum of 18 C (Thomas L. Clark, pers. comm.). Warming of rearing water is commonly practiced in modern facilities with heat exchangers incorporated in the hatch-house design.

FEED AND FEEDING

Hybrid muskellunge are normally started on artificial diets, but some culturists prefer to initially feed brine shrimp. The availability of canned brine shrimp eggs, and the use of incubation units to hatch them on a controlled schedule, provides a reliable and efficient approach to early life stage feeding of hybrid muskellunge. Brine shrimp are an adequate diet for fry to about 20 days; usually, at some point in this 20-day period artificial diet is introduced in addition to brine shrimp and a gradual transition is made to a 100% artificial diet.

The change from the long-standing practice of feeding esocids live food to the feeding of an artificial diet required substantial changes in fish culture techniques. When feeding live food, the major requirement was that food of the appropriate size be available at all times to avoid cannibalism and maintain good growth. While a great deal of effort went into producing, harvesting and grading forage to size, relatively little effort was involved in introducing the live food into rearing units. When feeding artificial food, no time was involved in producing or preparing the food, but a great deal of time was required to properly introduce the feed into rearing units. Hybrid muskellunge are reluctant to feed off the bottom of rearing tanks, preferring instead to strike food particles which are falling through the water column. Food that is not taken as it falls will accumulate on the bottom of the rearing unit. Successful feeding of hybrid muskellunge requires that small amounts of feed be introduced at frequent intervals. Hand-feeding hybrid muskellunge an artifical diet can be very labor intensive.

Modern techniques for culture of hybrid muskellunge universally include automatic, mechanical feeders in conjunction with electronic timers. Feeders are adjusted to dispense small amounts of feed, frequently, with a minimum of waste. Some speculation has arisen that the ideal system for feeding artificial diets to hybrid muskellunge would be demand feeders. To date no one has reported much, if any, success utilizing demand feeders for these hybrid fish. The current state-of-the-art remains mechanical feeders controlled by electronic timers. For fish up to 4 cm, a feeding schedule of 5 seconds every 2 minutes over a 24-hour period would be representative of practices in Pennsylvania. For larger fish, the time interval between feedings can be increased at the discretion of the culturist to reduce feed waste.

Currently, a number of feeder types are being used successfully. In Pennsylvania hatcheries, a basin-type feeder with adjustable openings and a time-adjustable traveling arm is employed. This feeder can either be suspended over or mounted on a rearing unit. Trough-type and pneumatic feeders are also in common use. Heat tapes can be utilized with all feeder types where moisture causes caking problems. Small pieces of wire can also be placed through openings to prevent plugging. No single design appears to be superior, and generally, because of ever-present budgetary problems, culturists use what they have available or what they can make or modify.

The biggest problem associated with use of prepared diets, particularly during the fry stage, is the accumulation of wasted feed, accompanied by water quality degradation. Initially, rearing units were drained and brushed on a daily basis as a means of removing accumulated wastes. This method generally resulted in turbid water and often injured fry, either because they were sucked against the screen or hit with the brush. Brushing was quickly replaced with siphoning devices and small pumps. Larger units such as swimming pool vacuums are now in common use for cleaning raceways. In Michigan (Pecor 1978), baffles are utilized in both tanks and raceways to accelerate flow and direct it along the bottom of rearing units, carrying with it accumulated wastes; with this method little additional cleaning is required.

Most commercially available trout or catfish diets have proved unsatisfactory for hybrid muskellunge. In 1970, the search for a satisfactory diet—both in terms of nutritional adequacy and acceptability to hybrids—led to the formation of a cooperative group of representatives from the U.S. Fish and Wildlife Service and pertinent state agencies. This group, later known as the Coolwater Diet-Testing Steering Committee, has focused on development of a suitable diet and best methodology for feeding hybrid muskellunge.

Results of a coordinated diet-testing program have been gratifying. The first generally acceptable artificial diet, known as W-7 (Orme 1978), provided a basis for reliable production of hybrid muskellunge under intensive culture. The W-7 formulation has since been improved, and refined versions, such as W-14 and W-16, are now available. The development of a satisfactory diet did not, however, automatically lead to easy commercial availability of that diet. Feed manufacturers have been reluctant to bid on the relatively small quantities of feed needed for hybrid muskellunge production. This problem has been reduced somewhat through the combining of all agency orders into a single purchase so that quantities involved are large enough to interest commercial feed vendors.

Another approach to assuring commercial availability of a satisfactory diet is to find a diet that is adequate for both salmonid and coolwater fish, one which is already commercially manufactured and which would be available from a number of vendors. In the search for such a diet, it was determined that Abernathy salmon diet—an open formula diet developed by the U.S. Fish and Wildlife Service—was quite satisfactory for hybrid muskellunge culture and available from commercial sources.

Experience and observation of problems with diets have caused fish culturists to be more exact in providing diet specifications. In effect, what is actually happening is that an acceptable diet for hybrid muskellunge is being developed through gradual modification of a proven salmonid diet. Work will continue in this area until sufficient nutritional information and performance evaluation relative to hybrid muskellunge is available to assure a diet that will consistently yield good survival and growth.

REARING

Hybrid muskellunge fry are generally started indoors in fiberglass tanks or larger concrete units. Although rearing unit dimensions vary, approximately 0.46 m x 0.6 m x 3 m seems average (not necessarily optimum) for fiberglass "starting units" and approximately 0.75 m x 0.76 m x 6 m is representative of the larger concrete units (Shyrl Hood, pers. comm.).

The first 20-30 days (after the initiation of feeding) are the most critical in tiger muskellunge culture. During that time the fish are "trained" to accept dry feed and non-feeders (fish which will not feed for whatever reason) drop out of the population. Work done in Pennsylvania hatcheries has shown that esocid swim-up fry which received no food can survive for as long as 14 days (Pennsylvania Fish Commission Hatchery Report, unpublished, 1983). However, it is not known when starving fry reach a "point of no return" after which the ingestion of feed will do no good. One can realistically expect 60-65% survival during the fry phase, and fish should be 3-4 cm in length 30 days after initial feeding. When fish are 30 days of age, the culturist can generally predict the success of the year's esocid production.

The early fry-rearing period is most critical, and meticulous care must be provided the fry at this time. Feeders are checked several times each day to be certain that holes are open. The high humidity found in most hatch-houses, as well as the high fat content of most starter diets, causes frequent blockage of feeder holes. Should this go unnoticed for any length of time, an increase in cannibalism will occur.

There is little agreement among culturists as to the optimum number of fry to start in a given amount of rearing area. Numbers from 100 to 1,200 fry per cubic foot of rearing area have been tried with varying results. Most culturists have individual preferences which are necessarily influenced by available facilities, such as water flow, number and size of rearing units, type of feeders, production quotas, etc. If fry are started at high densities, it will be necessary to thin them down sooner. Since hybrid muskellunge tend to suspend in the water column and remain stationary, some culturists feel that fry will begin feeding on dry feed better if crowded into a smaller area (approximately one-third of the tank) with a screen. Other workers feel that the fry will find the feed on their own and that crowding is unnecessary.

The high fat content and small size of the starter diet causes some of the feed to float temporarily before sinking. The floating feed tends to spread over the entire surface of the rearing unit causing a cleaning problem. Most of Pennsylvania's facilities utilize a thin strip of wood wedged across the width of the tank with one edge just below the water surface. The strips of wood are situated on either side of the feeder. In this way, floating starter diet is confined to the surface area within the wood strips and eventually sinks in one small area.

Frequent (daily, in many cases) constant-flow formalin treatments (167-250 mg/l for 1 hour) are administered prophylactically to fish at most facilities to control gill fungus. Gill fungus and systemic bacterial infections are the major disease problems associated with the intensive culture of esocids.

Cannibalism can be a problem, especially during the initial feeding period, and cannibals should be removed when observed. However, fish losses in the early fry-rearing stage cannot always be attributed to cannibalism. In 1979, a check of rearing units at one of Pennsylvania's hatcheries indicated that, despite the best efforts of experienced fish culturists, fry do escape from rearing units, and this number could be significant (20-30 per tank in 2-3 hours). Escapement can be detected by covering the effluent pipe with a piece of fine mesh material such as 200-300 micron saran screening (panty hose also works well) for 48 hours. Any fish which manage to get through the retaining screen will be prevented from escaping through the overflow pipe.

Those who culture hybrids differ in their opinions about water temperature. Although many facilities are unable to control water temperatures after incubation and early fry rearing, culturists in those that can seem to prefer temperatures in the 18 C to 20 C range. Meade et al. (1983) recommend 20-22 C for fish 3-4 cm long and 23 C for fish 12-13 cm long as optimum temperatures for maximum production. At Pennsylvania facilities attempts are made to maintain rearing water temperatures at about 20 C while fish are held indoors.

Most culturist agree that hybrid muskellunge prefer subdued rather than bright light. Many hatcheries cover rearing units, leaving an opening directly above the feeders. If larger fish (10 cm or more) are held indoors, some type of tank cover should be provided to prevent fish from jumping out of the tank when lights are turned on.

Hybrid muskellunge have been transferred to outdoor raceways at a length of 3.8 cm and successfully reared to a length of 10.6 cm on dry feed with a 91% survival (Pennsylvania Fish Commission Hatchery Report. unpublished, 1978). These fish were transferred outdoors at a smaller size than is the generally-accepted practice. From these tests, it appears that a great deal of time and effort could be saved by rearing hybrids in raceways rather than tanks, once they have attained a length of 3.8 cm.

For convenience, the remaining portion of the rearing period can be broken down into two parts: (1) the fingerling phase, a 30-day period during which the fish are generally maintained in the building at reduced densities; and (2) the advanced fingerling phase, a 30-60 day period during which the hybrids are reared in outdoor raceways. An 80-90% survival can be expected for both the fingerling and advanced-fingerling portions of the rearing period. Fish length will range from 7-10 cm at the end of the fingerling phase and 20-30 cm at the end of the advanced fingerling period. Overall, feed conversion for the entire rearing period will range from 1.5 to 2.0. It is possible to have feed conversions of less than 1.0, particularly during the fingerling phases (Pennsylvania Fish Commission Hatchery Report, unpublished, 1978).

Little information is available relative to the cost of producing tiger muskellunge fingerlings using live versus dry feed. A test conducted at a Pennsylvania Fish Commission hatchery in 1977 resulted in an estimated cost per pound for tiger muskellunge fingerlings, reared for 98 days on live versus dry feed (W-7), of $16.62 and $1.52, respectively (Pennsylvania Fish Commission Hatchery Report, unpublished, 1977).

HATCHERY DESIGN

Recent advances in hybrid muskellunge culture, most of which resulted from the use of formulated dry diets, have made it possible for many hatcheries to become dual-purpose facilities, capable of rearing both salmonids and coolwater fish.

A modern dual-purpose hatchery will have a device to heat water, accompanied by a temperature-control unit of some type, generally a type of heat exchange system with blending valves. A recent evaluation of a liquid-to-liquid heat pump shows promise (Fuss 1983a). Heating systems are used primarily during egg incubation and early fry rearing when temperatures of available water sources are less than optimum. There will generally be a degassing device of some sort since problems related to gas supersaturation, caused by heating and/or pumping, have become apparent in recent years. There will also be a warming pond, or some other warmwater source, which can be blended with coldwater sources (when necessary) to maintain temperature control flexibility during the latter part of the production cycle when warmwater sources are actually warm. Source water will be filtered or treated with ultraviolet light if necessary. Electrical systems will be adequate to permit the use of automatic feeders and timing devices. The capability to recondition water through aeration, so that recycling or reuse is possible, has been incorporated in today's hatcheries. Effluent treatment has become an important part of modern rearing facilities. A separate clean-out system, which carries cleaning water directly to a clarifier for primary solids removal, is currently in use in Pennsylvania.

Dual-purpose hatcheries tend to be more cost effective due to better utilization of facilities, equipment and manpower, and they appear to be the trend in hatchery development for the future.

FUTURE NEEDS AND DIRECTION

Hybrid muskellunge are an integral part of a modern revolution in the art and science of esocid culture. Hybrid muskellunge are reared in tanks and raceways solely on commercially prepared diets. Large-scale intensive culture of hybrid muskellunge in tanks and raceways, using artificial diets, is now a reality. Research efforts are no longer directed at developing the art and science of intensive culture; rather they are directed at refining techniques and improving facilities, equipment and diets utilized in intensive culture of hybrid muskellunge.

There are still some problem areas and it seems certain that advances will not be made as rapidly as in the past. The accumulation of more information relative to nutritional, physiological, physical and behavioral requirements of tiger muskellunge may some day permit the rearing of hybrids in "cookbook" fashion, using mathematical formulae which take all variables into consideration (Westers 1978). We are beginning to see the inclusion of computers as a tool in fish culture. Programs capable of predicting feed requirements, growth, survival, stocking dates, etc. are now available for use in home computers (Steven Pastor, pers. comm.). Microprocessors can now be utilized in hatchery situations to operate automatic feeders and monitor such things as water temperature, water levels and associated warning systems (Fuss 1983b). Not many years ago the intensive culture of esocids on artificial diet was regarded as a novelty with no practical application. Now fish culturists can provide hybrid muskellunge in the numbers and sizes requested by fishery managers, and they can do so efficiently. The state-of-the-art of intensive culture of hybrid muskellunge is that the art is becoming science.

REFERENCES

Buss, K. W. 1961. The northern pike. Pennsylvania Fish Commission Benner Spring Fish Research Station, Special Purpose Report. 58 pages.

de Montalembert, G., C. Bry, and R. Billard. 1978. Control of reproduction in northern pike. American Fisheries Society Special Publication 11:217-225.

Eddy, S. 1940. Do muskellunge and pickerel interbreed? Successful hybridizing in hatchery suggest they do. Progressive Fish-Culturist 48:25-27.

Fuss, J. R. 1983a. Evaluation of a heat pump for an aquacultural application. Progressive Fish-Culturist 45:121-123.

Fuss, J. R. 1983b. Using single board computers for control applications. Fisheries, Volume 8(5):8-13.

Graff, D. R. and L. Sorenson. 1970. The successful feeding of a dry diet to esocids. Progressive Fish-Culturist 32:31-35.

Johnson, L. D. 1958. Pond culture of muskellunge in Wisconsin. Wisconsin Conservation Department, Technical Bulletin 17. Madison, Wisconsin, USA.

Meade, J. W., W. F. Krise, and T. Ort. 1983. Effect of temperature on production of tiger muskellunge in intensive culture. Aquaculture 32:157-164.

Orme, L. E. 1978. The status of coolwater fish diets. American Fisheries Society Special Publication 11:167-171.

Pecor, C. H. 1978. Intensive culture of tiger muskellunge in Michigan during 1976 and 1977. American Fisheries Society Special Publication 11:202-209.

Sorenson, L., K. Buss, and A. D. Bradford. 1966. The artificial propagation of esocid fishes in Pennsylvania. Progressive Fish-Culturist 28:133-141.

Westers, H. 1978. Biological considerations in hatchery design for coolwater fishes. American Fisheries Society Special Publication 11:246-253.

Iowa Culture of Muskellunge on Artificial Diet

WALLACE D. JORGENSEN

R. R. Box 7722
Fish Hatchery
Spirit Lake, Iowa 51360

ABSTRACT

Natural food diets (zooplankton and minnows) and artifical dry diets were compared as a means of increasing fry survival during the intensive culture of muskellunge at the Iowa Spirit Lake Hatchery. Muskellunge fry fed the natural diets had a constant supply of zooplankton available; at about 1.5 inches, fathead minnows were added to their daily diet until stocking occurred. Survival to stocking size (5.3 in) was 18.0%. Artificial feeding of muskellunge was initiated on an experimental basis in 1982 and tested on a production scale in 1983. Brine shrimp and Abernathy feed were used for the initial training process, with brine shrimp withdrawn when the fry converted to dry feed. In 1982, only 8.0% of the initial total were reared to stocking size, because a malfunction in the water supply caused the death of 910 fish. In 1983, 53% of the artificially-fed musky survived to a mean stocking size of 5.4 inches, compared with 18.0% for those on the minnow diet. Cost comparisons for the two rearing methods in 1983 were $1.76/fish for the live-food diet and $0.29/fish for the artificial diet.

Spirit Lake Fish Hatchery is the only muskellunge-rearing facility in Iowa. Both intensive (tank-rearing) and extensive (pond-rearing) culture methods have been used in Iowa. The intensive culture diets are similar to those used in Pennsylvania (Sorenson et al. 1966), in which zooplankton and fathead minnows are fed as the principal forage. Extensive culture diets have also utilized zooplankton and fathead minnows and parallels some of the culture work reported in Wisconsin (Johnson 1958).

Problems in Iowa associated with intensive culture of muskellunge have been 1) collecting adequate quantities of zooplankton, 2) capturing sufficient numbers of acceptable-size minnows, and 3) poor musky survival. Problems with extensive culture of these fish have been 1) controlling disease, 2) cannibalism, 3) erratic production numbers, and 4) removing musky from rearing areas under less than ideal conditions, which results in loss of fish. All of these factors have made musky production in Iowa inconsistent and expensive.

Graff and Sorenson (1970), reported encouraging results from feeding musky a dry diet. Iowa's objectives in 1982 and 1983 were to explore the feasibility and develop techniques for rearing muskellunge on an artificial diet. If successful, this would provide a consistent food source, reduce production costs and, ultimately, make musky production more predictable.

METHODS

Artificial Feeding

In May 1982, 2,000, 10-day old, swim-up fry were placed in a concrete rectangular tank measuring 15.3 ft x 4.3 ft x 1.17 ft deep; total tank volume was 77 ft^3. The fry were initially confined in a 4.3 ft x 4.8 ft (22 ft^3) area at the water inlet end of the tank for diet-training purposes. Fresh water flowed through the tanks at approximately 20 gpm which provided an exchange rate of 2.0 times each hour. Water temperatures were uncontrolled and ranged from 54° F at the beginning of the project to 80° F during the midsummer.

Training began immediately after the fish were placed in the rearing unit. Brine shrimp nauplii, hatched from 0.04 lb of eggs, were fed once each morning. Immediately following the daily brine-shrimp introduction, Abernathy starter, S8-1(82) formulation, was scattered by hand on the water surface for 10 minutes. Dry food was also dispensed from a circular automatic feeder (Mo-Do Enterprises) for 30 seconds every 10 minutes, 24 hours per day throughout the rearing period. After 30 days, brine shrimp and hand feeding were discontinued. When fish outgrew the Abernathy starter, they were fed a W-14 diet, progressing gradually from number 1 to 4 granules. Feed size was changed when visual observaton indicated fish could consume larger particles. Feed-size transitions were done gradually over an 8-day period by mixing 75% small granules to 25% large granules, increasing to 50-50%, then to 100% large. Fecal material and uneaten food was siphoned from the tank once daily. Prophylactic one-hour, static disease treatments were administered throughout the growing period to combat bacterial gill disease, gill fungus and, columnaris.

In May 1983, 2,000, 10 day-old, swim-up fry

were placed in each of two concrete tanks. Tank volume, water exchange rate and training area were identical to 1982. Water temperature was 56° F at the outset of the project, reaching 82° F during midsummer and then decreasing to 74° F just before the fish were stocked. As fish grew in each tank and appeared crowded, their confinement was expanded from 22 ft^3 to 47^3 ft, then to 71^3 ft.

The fish training and feeding procedures used in 1983 were identical to those in 1982 with two exceptions: 1) the amount of brine shrimp fed daily in each tank was decreased to 0.02 lb per tank for 20 days; 2) standard 12-ft Louden feeders dispensed Abernathy starter food for about 0.5 seconds every 5 minutes, 24 hours per day for the first 24 days of the rearing period. During the last 68 days, food was dispensed for 0.5 seconds every 10 minutes, 24 hours per day. Fecal material and uneaten food was siphoned daily, with the exception of the last 14 days when they were disposed of by simultaneously brushing tank bottoms and flushing water through the tank and out the drain. As in 1982, prophylactic one-hour, static disease treatments were administered throughout the rearing period.

Live Food Rearing

In May 1983, 2,000, 10-day old, swim-up fry were placed in each of two concrete tanks. Water suply and temperature, tank volume, water exchange rate, training area, and confinements were identical to the 1983 artificial feeding procedures.

Zooplankton were collected from fertilized hatchery ponds and distributed daily to the musky tanks. After 20 days on zooplankton, fathead minnows were added to their daily diet. Minnow fry were netted from minnow-rearing ponds at the hatchery; if those minnow sizes or quantities were inadequate, others were seined from small lakes in the area. Prophylactic one-hour, static disease treatments were administered throughout the rearing period.

RESULTS

Artificial Feeding

In 1982, 160 musky, 8.0% of the initial 2,000, survived to a mean stocking length of 5.6 in and a weight of 33.6 fish/lb. Musky consumed 1.26 lb of brine shrimp, but no records were kept for artificial feed. A water-supply malfunction, 36 days into the study, resulted in the death of 1,000 musky. Examination of these fish revealed that 910 had consumed the dry diet.

In 1983, 2,103 musky, 52.6% of the 4,000 stocked, survived to a mean stocking length of 5.4 in

Table 1. Survival of muskellunge reared on live food and artificial diet in Iowa, 1983.

	Initial stock	Percent survival	Average length (in)	Number fish/lb	Number days old
Minnow diet	4,000	18	5.3	44	101
Artificial diet	4,000	53	5.4	46	102

and weight of 46 fish/lb (Table 1). During the production period, musky consumed the nauplii hatched from 0.84 lb of brine shrimp and 188 lb of dry feed. This amount of food produced 46 lb of fish for a food conversion of 4.08. These artificially fed musky cost $0.29 per fish (Table 2) to produce.

Live Food Rearing

In 1983, 722 muskies, 18% of the original total, survived to a mean stocking length of 5.3 in and a weight of 44 fish/lb (Table 1). These muskies cost $1.76 per fish (Table 2) to produce.

Two experiments were conducted to determine if musky would convert back to live from artificial food diet. 1) Three dozen fathead minnows were placed in a tank where the muskies had been fed artificially; within 2 hours all acceptable-size minnows were eaten. 2) Two hundred forty musky were fed only fathead minnows for 2 weeks with no musky loss. Both experiments were conducted after the 102 day, artificial-feeding period had ended.

SUMMARY

Completion of this project demonstrated that:
1. Musky can be successfully produced using brine shrimp and artificial feed. In 1983, 53% survived from swim-up fry to suitable stocking size of 5.4 in. A 50% survival to fingerlings is acceptable in Iowa.
2. Fry to fingerling survival was better with artificial food. The lower survival with live food can be

Table 2. Production costs in Iowa for intensively-reared muskellunge fed live food and artificial diet.

	Man-hour costs	Chemical costs	Fertilizer costs	Feed costs	Cost/fish
Minnow diet	$1170.40	$ 81.30	$26.00		$1.76
Artificial diet	$ 389.16	$109.89		$107.47	$0.29

attributed to inconsistent food quantities, unacceptable size of food, cannibalism, and the introduction of disease organisms with each feeding.
3. Musky can be trained to eat a dry diet at a density of 90 fish/ft^3.
4. Musky will survive and grow at a density of 13.7 fish/ft^3 or 0.31 lb/ft^3.
5. Production costs are considerably lower using artificial food than live organisms.
6. Grading of fish on the artificial diet is not neccessary, since growth rate is more consistent.
7. Grading of fish on a live food diet is generally necessary throughout the growing period to control cannibalism.

In 1984, Iowa planned to meet all its muskellunge-stocking commitments with fish reared on a dry diet. In addition, work continued on reducing the food-conversion rate and determining optimum tank-loading densities.

ACKNOWLEDGEMENTS

I would like to thank the Spirit Lake Hatchery staff for their dedication on this project and thank Terry Jennings and Alan Moore for reviewing this manuscript.

REFERENCES

Sorenson, L., K. Buss and A. D. Bradford. 1966. The artificial propagation of esocid fishes in Pennsylvania. Progressive Fish Culturist 28: 133-141.

Johnson, L. D. 1958. Pond culture of muskellunge in Wisconsin. Wisconsin Conservation Department Technical Bulletin 17. 54 pp.

Graff, D. R. and L. Sorenson. 1970. The successful feeding of a dry diet to esocids. Progressive Fish Culturist 32:31-35.

Disease Incidence and Management Implications of Cultured Esocids

JAMES E. HARVEY

Pennsylvania Fish Commission
Linesville Fish Cultural Station
Linesville, Pennsylvania 16424

ABSTRACT

Case histories of disease incidence were reviewed for the five-year period, 1978-82, at Pennsylvania Fish Commission hatcheries involved in the rearing of esocids. Bacterial columnaris, bacterial hemorrahagic septicemia, gill fungus, bacterial gill disease, external protozoans and monogenetic trematodes were the most prevalent disease agents diagnosed in conjunction with excessive fish mortalities. Gill-associated problems were most frequently diagnosed, followed by systemic bacterial infections and external skin problems. Various treatments used in controlling esocid diseases are discussed. Cultural practices are considered in their relationship to fish health and the major etiologic agents.

The Pennsylvania Fish Commission (PFC) is involved in rearing programs with muskellunge, tiger muskellunge (northern pike × muskellunge), amur hybrid pike (northern pike × amur pike), northern pike and chain pickerel. Six fish cultural stations, located throughout the commonwealth of Pennsylvania, annually produce in escess of 150,000 fingerling esocids. Health management of esocids at these facilities is of constant concern to biologists, fish culturists and administrators in their efforts to produce a quality product in sufficient numbers to meet annual production quotas.

Graff (1978) stated that "pathology as related to the intensive culture of esocids is a young and developing science. Much basic information is needed." Due to the nature of extensive pond culture, in that esocids cannot be observed closely for signs of disease problems till harvest, control of epizootics is difficult to achieve. Most published accounts of disease incidence at fish cultural facilities, such as Meyer (1970), involve ictalurids, cyprinids, salmonids and centrarchids. Reports of esocid diseases associated with fish cultural operations are usually found in unpublished reports of state and federal agencies and are not readily accessible. Management of fish health in natural brood populations of esocids is nearly impossible, yet serious disease problems can exist. The objective of this paper is to describe the shortcomings of the diagnostic pathology of cultured esocids, what has been observed in regards to esocid diseases in Pennsylvania, and how fish health problems are dealt with here. It is also hoped to provide a stimulus for continued development of the state-of-the-art in fish culture and pathology.

METHODS

Case histories involving diagnostic examinations of esocids at PFC hatcheries were reviewed for the period 1978-82. A case history is a pathological examination of moribund specimens from a production lot of fish. Such examinations are requested by hatchery personnel when there are high fish mortalities, or when there is a concern that a disease problem is about to occur due to increasing daily mortalities. All case histories involved microscopic examinations of wet mounts of scrapings from skin, gill and gut. Also included were macro- and microscopic examinations for external and internal signs of disease. Bacterial isolations from kidneys and lesions on tryptic soy agar were presumptively identified and drug sensitivities performed. No virological or histological work was routinely conducted. Production and lot history information was obtained from hatchery personnel for each case. Treatment and management recommendations were made and followed up to determine their efficacy. All lab procedures utilized in the diagnostic case histories were in accordance with "Procedures for the Detection and Identification of Certain Fish Pathogens," McDaniel (1979).

RESULTS

Table 1 reflects the diseases and etiologic agents identified in 64 case histories, the number of times they were diagnosed, percent occurence and percent of total pathogens found. The most prevalent diseases found, in decreasing order, were bacterial columnaris caused by *Flexibacter columnaris*, bacterial hemmorhagic septicemia caused by *Aero-*

Table 1. Diseases and etiological agents of esocids identified in sixty-four case histories from 1978-82 in Pennsylvania.

Gill diseases	Number of times identified	Percent of occurrence	Percent of total identified diseases
Bacterial columnaris	16	25	15
Fungus	14	22	13
Bacterial gill disease	10	16	10
Protozoan parasites and monogenetic trematodes	8	13	8
Environmental gill disease	1	2	1
Sub-total	49		47
Systemic bacterial infections			
Aeromonas hydrophila	18	28	17
Unidentified	6	9	6
Pseudomonas fluorescens	2	3	2
Aeromonas salmonicida	1	2	1
Sub-total	27		26
External skin problems			
Bacterial columnaris	9	14	9
Protozoan parasites and monogenetic trematodes	8	13	8
Fungus	4	7	4
Sub-total	21		20
Other			
Hexamita sp.	4	7	4
Suspect nutritional	2	3	2
Gas supersaturation	1	2	1
Sub-total	7		7
GRAND TOTAL	104		100

monas hydrophila and *Pseudomonas fluorescens*, gill fungus, bacterial gill disease caused by *Flexibacter* sp. and external protozoan parasite and monogenetic trematode infestations. There were 16 case histories (25%) where no disease agent could be identified. Of the total agents found, 47% were related to gill problems, 26% were diagnosed as a systemic bacterial infection, and 20% were problems associated with the external skin surfaces. Other problems, which included *Hexamita* sp. infestations, suspected nutritional disorders and instances of total dissolved gas supersaturations, comprised the remaining 7%. Most case histories identified multiple etiologic agents and/or diseases. In all cases, however, either control was achieved by the recommended treatment or mortalities decreased to a chronic level. The majority of the major etiologic agents or diseases identified caused high mortalities when the fish were subjected to a predisposing environment of stress, such as crowding, low dissolved oxygen, high un-ionized ammonia levels, or less than optimum water temperatures or pH levels.

DISCUSSIONS

Bacterial columnaris, gill fungus, bacterial gill disease and environmental gill disease were found most often on the gills of esocids less than twenty days old. Esocids being fed dry diets at this age receive starter mash type feeds that lower the water quality and coat the gills, thereby creating hyperplasia and an environment conducive to opportunistic pathogens. Tank hygiene is of utmost concern. Daily siphoning of waste feed and cleaning of tanks is essential. Prophylactic chemical treatments are required to prevent these diseases from becoming

Table 2. Standard drug and chemical treatments for various esocid diseases in Pennsylvania.

Drug or Chemical	Rate	Duration	Pathogens
Formalin	1:600 (eggs only)/ 15-17 minutes	daily	fungus
Formalin	1:4,000-1:8,000/ 30-60 minutes	2-5 days or daily	external protozoans and monogenetic trematodes, fungus, bacterial gill disease, columnaris
Hyamine 1622	1/2-2 PPM/ 30-60 minutes	2-5 days	bacterial gill disease, columnaris
Diquat	8-16 PPM/ 30-60 minutes	2-5 days	bacterial gill disease, columnaris
Roccal II	1/2-1 PPM/ 30-60 minutes	2-5 days	bacterial gill disease, columnaris
Terramycin, Neoterramycin, Furacin (water soluble)	1/2-5 PPM/ 30-60 minutes (bath)	3-15 days	systemic bacterial infections, bacterial gill disease, columnaris
Terramycin, Neoterramycin (oil soluble)	2-4 grams/ 100 lbs. fish/day (oral mixed in feed)	5-15 days	systemic bacterial infections
Furazolidone (oil soluble)	2-4 grams/ 100 lbs. fish/day (oral mixed in feed)	3-10 days	systemic bacterial infections, internal protozoans (Hexamita sp.)
Salt	0.5% by weight of water	1 hour/day in rearing units; continuous in transportation units	external protozoan parasites, osmoregulatory stress

established. By the time a fish culturist observes obvious signs of gill problems, such as flared opercula and filamentous strands of growths on the gills, it is usually too late to prevent high mortalities. Gill-associated problems were usually treated with diquat, quaternary ammonium compounds (hyamine and roccal), water soluble terramycin baths, nitrofurans or formalin (Table 2). Ultraviolet treatment of source water shows promise for elimination of the etiologic agents of gill disease before they come in contact with fish.

Systemic bacterial pathogens comprised more than a quarter of the total identified etiologic agents with *A. hydrophila* being the most prevalent. A phenomenon called "red spot," as described by Amlacher (1970), has been identified in breeding adults in a natural brood population of muskellunge in the Pymatuning reservoir and sanctuary in northwestern Pennsylvania. Bothe *P. fluorescens* and *A. hydrophila* were presumptively identified in necrotic lesions and kidneys of adult muskellunge from these two water-bodies. The incidence of this disease has been increasing for several years and is thought to have contributed to a poor trapnetting season in the spring of 1983. Six adult muskellunge were trapnetted in 1983, as compared to catches in excess of 100 individuals per season in previous years. According to Dr. Thomas Wellborn, (pers. comm.), little can be done to control this disease in the wild. Factors such as high population densities, adverse environmental conditions, inbred brood stock, and velogenic strains of bacteria have been suggested, but no clear answers have emerged. *A. hydrophila* and *P. fluorescens* are considered common water and soil bacteria that are universal in distribution and opportunistic pathogens when fish are stressed.

Furunculosis (*A. salmonicida*) was identified once and is not widespread in esocids reared in Pennsylvania. The route of infection of this one case came from a carrier-population of salmonids. Water-soluble baths of terramycin, neo-terramycin, and nitrofurans, along with the same drugs mixed in the feed, have given acceptable control for systemic bacterial infections in intensive culture. Drug-resistant bacteria have not been encountered to the degree that they have been found in salmonid culture in Pennsylvania. Ultraviolet treatment of influent water is advantageous in cutting down the numbers of potentially pathogenic bacteria coming in contact with esocids in culture situations. Systemic bacterial infections in pond culture and in larger bodies of water are not as common as they are in intensive culture, but they are more difficult to control when they do occur. The unidentified systemic bacterial infections that were found (Table 1) could be controlled with drugs. Drug sensitivity testing, therefore, is essential.

Pathogens associated with the external skin comprised one-fifth of the total agents found (Table 1).

Bacterial columnaris was usually treated with flow-through diquat treatments or water-soluble drug baths (Table 2). Protozoan parasites and monogenetic trematodes included *Scyphidia* sp., *Glossatella* sp., *Trichodina* sp., *Epistylis* sp., *Costia* sp., *Ichthyopthirius multifilis*, *Chilodonella* sp., and *Gyrodactylus* sp. All of these parasites, except *I. multifilis*, were effectively controlled with formalin. It is very important to be able to diagnose the tomite stage of *I. multifilis* before the obvious adult stage is observed. The tomite stage can be treated with formalin, but the adult stage cannot. Fungus, including *Saprolegnia* sp., was easily controlled with formalin. However, a dual infection of massive amounts of fungus and protozoan parasites was very hard to control with flow treatment, because it did not penetrate the fungal masses where protozoan parasites sought refuge.

In 16 case histories no diseases or etiologic agents could be identified, even though abnormal mortalities were occurring. Viral infection cannot be ruled out. Only a few case histories involved occasional checks for virus isolation; this is not a standard protocol as it is in salmonid diagnostic pathology. The high percentage of unidentified mortalities (25%) would be unusual in salmonid diagnostic pathology, where normally the etiologic agent can be readily identified. It is difficult to identify environmental problems because loading-rate information is limited for esocids. Most fish culturists are aware of oxygen requirements of esocids, consequently, oxygen deficiency problems were never diagnosed. Nutritional problems were diagnosed when obvious signs of scoliosis and lordosis were present. Gas supersaturation problems have been kept to a minimum by timely checks for water sources by hatchery managers. A saturometer is a must for esocid culturists, especially when water is being heated; this creates a potential gas supersaturation situation.

The need for improved and specific diagnostic techniques for esocids is evident in the high number of unaccountable mortalities and the relatively numerous, unidentified systemic-bacterial infections encountered in these case history files. A substitute for malachite green should be found. Combination treatments with malachite green and formalin have historically been effective in controlling gill monogenetic trematode infestations; however, the PFC does not presently use malachite green due to possible health hazards to humans. Also, until more control over natural brood populations occurs, better artificial diets are developed, extensive culture is phased out completely and loading-rate/ feeding-rate formulas are perfected, it is recommended that the esocid culturist follow these rules:

1) Keep the dissolved oxygen in all rearing units as close to saturation as possible.
2) Maintain tank hygiene in intensive culture.
3) Suppress supersaturated gas levels as close to saturation as possible.
4) Maintain and closely monitor optimal water temperatures.
5) Prophylactically treat for gill diseases on a regular basis.
6) Utilize healthy forage fish or feed a high-quality dry diet.
7) Observe the fish as often as possible for abnormal or changing signs.

There are too many cases in which an etiologic agent or disease cannot be found in conjunction with high mortalities. Until improved diagnostic techniques decrease that number, the health management of esocids will be an art and not a science.

ACKNOWLEDGEMENTS

The author acknowledges Delano Graff and Shyrl Hood, of the Pennsylvania Fish Commission, who critically reviewed the manuscript. Many thanks to the other Pennsylvania Fish Commission personnel who assisted me in this endeavor.

REFERENCES

Amlacher, E. 1970. Textbook of Fish Diseases. (translated and updated by D.A. Conroy and R.L. Herman) T.F.H. Publications, Inc., Neptune City, N.J. 302 pp.

Graff, D.R. 1978. Intensive culture of esocids: the current state of the art. American Fisheries Society Special Publication. 11:195-201.

McDaniel, D. (Editor). 1979. Procedures for the detection and identification of certain fish pathogens. American Fisheries Society, Fish Health Section. Bethesda, Maryland USA. 118 pp.

Meyer, F.P. 1970. Seasonal fluctuations in the incidence of disease on fish farms. American Fisheries Society Special Publication. 5:21-29.

Effects Of Temperature, Diet Composition, Feeding Rate, And Cumulative Loading Level On Production Of Tiger Muskellunge

JAMES W. MEADE

U.S. Fish and Wildlife Service
National Fishery Research and Development Laboratory
Wellsboro, Pennsylvania 16901

CAROL A. LEMM

U.S. Fish and Wildlife Service
Office of Cooperative Research Units
Washington, D.C. 20204

ABSTRACT

Tiger muskellunge (northern pike × muskellunge) were reared to determine effects of temperature, diet composition, feeding rates, and loading levels (weight of fish of a specified length per unit water flow) on growth, production and tissue composition. Optimum temperature for production was 20-23 C, depending on fish size. Protein requirement at 20 C was at least 45% of the diet. Protein level effect on growth was greater at lower temperatures. Recommended feeding rate is 13% of body weight per day for 7 cm fish reared at 20 C. Loading at less than 4.5 kilograms of fish/liter/minute of water flow produced fish of high quality; at higher loading rates, up to 12 kilograms/liter/minute, liver vacuolation increased but growth did not decrease significantly.

Temperature, diet composition, feeding regime, and loading level (weight of fish of specified length per unit flow of water) are critical aspects affecting the success and efficiency of intensive fish culture. During the 1970's the intensive, flowing-water culture of tiger muskellunge (hybrids of northern pike and muskellunge) was developed and became an integral part of several state fish culture programs, especially in the northern U.S. Westers (1978) discussed intensive culture of tiger muskellunge and other coolwater fishes, but little quantitative information on the production of any specific coolwater fish was available.

For tiger muskellunge, neither the optimum production temperature nor the preferred or selected temperature was well established. Little was known of the nutritional requirements, but it was evident that, for coolwater fishes, protein requirement for optimum growth varied with species (Ketola 1978) and temperature (Satia 1974; Page and Andrews 1973). For tiger muskellunge, a feeding rate estimate had been made (Pecor 1978), and rates of oxygen consumption and ammonia production were reported to be less than those for salmonids (Pecor 1979).

In 1980, we began a two-year series of experiments on tiger muskellunge culture, with the objective of establishing production-rearing guidelines for temperature, diet protein, feeding regime, and loading.

METHODS

All tests were conducted at the National Fishery Research and Development Laboratory, Wellsboro, PA. Fish were reared in heated, aerated well water of pH 6.8-7.0, total hardness 30 mg/l (as $CaCO_3$), total gas saturation 103%, and incoming dissolved oxygen (DO) at 90% of the saturation. Fish were supplied by the Pennsylvania Fish Commission from northwestern Pennsylvania stocks. Diets were dispensed by automatic feeders, and unless otherwise noted, the diet was W-7 (Orme 1978).

Analysis of variance and Duncan's multiple range test (Duncan 1955) were used to analyze growth (calculated as mean wet-weight gain and mean length increment per fish per day) and production (as net weight increase for the entire group of fish in one rearing unit). Materials and methods used in the study of specific factors are given in later sections.

RESULTS

Effect of Water Temperature on Production

Two temperature tests were conducted. In the first, 3-4-cm fish were reared at 14, 16, 18, 19, 20, 21, 22, and 24 C, and fed at 23% of initial body weight (BW) per day. The hatchery constant equivalent (HC), determined according to Westers

(1981), was 80 to 115, from start to finish of the 22-day test. Food was dispensed every 5 min from 1400 to 0800 hours. For the second test, fish 12-13-cm long were reared at 18, 20, 21, 22, 23, 24, 26, and 28 C. Food was offered at 10% BW per day, or HC 125 to 165, from start to finish of the 28-day test, and dispensed every 30 min, from 1600 to 0800 hours.

For 3-4-cm fish, survival at 24 C was reduced by cannibalism, probably the result of insufficient food. Production at 24 C was also reduced (Fig. 1). Mean length, or growth, increased directly with temperature and was linear when plotted against temperature units (TU), where one TU = 1 C/day (Fig. 2). Maximum production occurred at 20 C (Fig. 2).

For 12-13-cm fish, survival was not significantly affected by temperature, though survival tended to increase with temperature. Both mean growth (Fig. 3) and production, and the growth and production efficiencies were greatest at 23 C. Additional data for these tests are available in Meade et al. (1983).

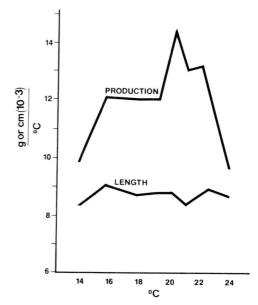

Figure 2. Mean daily production and length increase as functions of temperature units for tiger muskellunge 3-4 cm long reared for 22 days.

Dietary Protein Level

Knowledge of the protein requirements of cultured fish is esential for the development of economical, nutritionally-complete diets, since protein is the most expensive component. Tiger muskellunge 2.5-cm ($\bar{x}$ TL) were reared on diets containing 35, 45, 55, or 65% protein at each of three temperatures -- 17, 20, and 23 C. A fifth diet, W-7, containing 50% protein, was used as a reference during the 7-week test. Details of methods and a related discussion of results are available in Lemm and Rottiers (1986).

Cannibalism was severe during the first two weeks, which probably indicates underfeeding or problems in system design. The 65% protein diet test was discontinued, because the two-week survival was only 8% among those fish -- presumably because the food floated and was relatively unavailable. After two weeks, all fish from the replicate groups within each test set were pooled, inventoried and redistributed at lower densities.

Hematocrit and hemoglobin mean values were not significantly different among fish on different diets. Tissues of fish sampled from each 20 C group, showed no evidence of gross or histological pathology. Proximate analysis at the beginning and end of

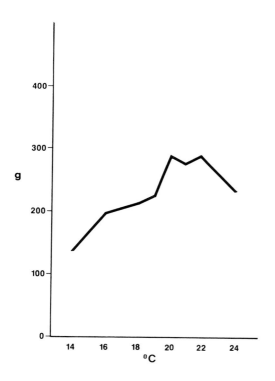

Figure 1. Production (weight increase in g) of tiger muskellunge 3-4 cm long reared for 22 days at various temperatures.

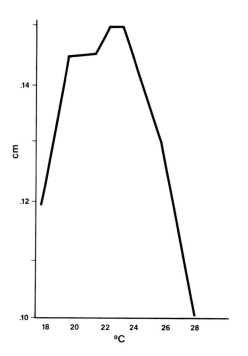

Figure 3. Computed length gain of tiger muskellunge 12-13 cm long reared for 28 days at different temperatures.

the study showed body composition to be similar for all fish (14.6% protein, 3.6% lipid, 2.6% ash, and 78.8% moisture, as means of wet weight). Correlations between water temperature and growth ($r = 0.82$ to 0.96) were significant for fish on all diets.

Growth of fish increased with protein levels of 35 and 45% at 17 and 20 C; however, an additional 10% protein content in the diet did not result in further significant increases in weight at these temperatures. At 23 C, there was no significant difference in weight gains of fish fed any of the diets. After week 2, fish fed the reference diet were significantly heavier than the other fish; however, their lipid content was also significantly higher.

Growth of tiger muskellunge 3-13-cm, reared at 20 C or less, was highest among those fed 45-55% protein diets (Fig. 4). The protein requirements for tiger muskellunge were similar to those reported for smallmouth bass (45% Anderson et al. 1981), chinook salmon (40-55% Delong et al. 1958), plaice (50% Cowery et al. 1972), grouper (40-50% Teng et al. 1978), and puffer fish (50% Kanazawa et al. 1980); and higher than those for rainbow trout (40% Satia 1974), and for redbelly tilapia (35% Mazid et al. 1979).

Livers of fish fed test diets showed more abundant vacuolization than those of fish fed reference diet. The vacuoles were probably filled with glycogen derived from dextrin. Cowey et al. (1975) found that liver glycogen levels were higher in plaice fed 20% carbohydrate (dextrin—glucose) diets than in those not fed carbohydrate. Refstie and Austreng (1981) reported that total carbohydrate in rainbow trout livers increased with dietary glucose. No mortality or liver pathology attributed to high dietary carbohydrate was reported by those authors. Hilton et al. (1981) suggested that in trout elevated liver glycogen, which may result in impaired liver function and lower survival, resulted from high digestible-carbohydrate diets.

Feeding Rate

Efficiencies and inefficiencies in feeding rates are translated directly and immediately into changes in operating costs. Two tests were designed to determine effects on growth of amount of food and frequency of feeding. In the first test, 50 fish, 6.3-cm ($\bar{x}$ TL), were reared in each of 22 tubs (Meade and Fuss 1983). Flow was 4 L/min. Food was dispensed daily by automatic feeders at intervals of 5, 15, 30, or 120 minutes from 1200 to 0800 h; amounts of diet were 5, 10, or 20% BW, dispensed to fish in two tubs at each of the four time intervals. The 5-min interval for 5% BW was not used because of the small amount of food involved per feeding. Fish were inventoried and diet amounts adjusted weekly. Because of mechanical problems the

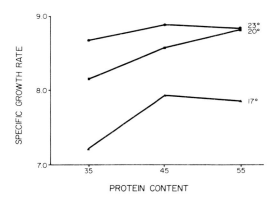

Figure 4. Specific growth rate of tiger muskellunge reared at three temperatures and fed four diets, each with different protein content.

Table 1. Growth, survival, and food conversion for tiger muskellunge reared at 20 C for 21 days and fed at 5, 10 and 20% of body weight at four feeding intervals. Values are means for two groups, each containing 50 fish, 6.3 cm ($\bar{x}$ TL) and 1.2 g ($\bar{x}$ weight) initial size.

Feeding rate (% body weight)	Interval between feeding (min.)	Feedings per day	Weight increase (g)	Length increase (cm)	Food conversion
5	5	241	(a)	(a)	(a)
	15	81	42.9	2.36	1.52
	30	41	17.0	2.13	3.97
	120	11	-15.7	1.64	(b)
10	5	241	115.4	3.94	1.42
	15	81	68.5	3.10	2.10
	30	41	56.0	3.16	2.44
	120	11	43.8	2.40	3.22
20	5	241	125.3	4.24	3.05
	15	81	121.3	4.19	2.81
	30	41	110.8	4.07	3.03
	120	11	89.1	3.42	3.53

a. The 5-minute interval for 5% body weight was not included due to the small amount of food provided per feeding
b. Negative value.

fish received half rations and irregular feedings during the first two test days, and during the first three test days food was offered only 18 hours per day. Feeding durations were altered slightly to make adjustments in the quantity of food dispensed. Feeding durations were 3 to 7 seconds, and test duration was three weeks. In the second test, 595 liter rearing units were stocked with 150 fish, 12.1-cm $\bar{x}$ TL. Food was dispensed every 30 minutes, 24 hours per day. Feeding durations of 7.7 to 30 seconds provided differences in amounts of food dispensed to triplicate lots of fish fed 5, 10, 15, and 20% BW. Food amount was adjusted weekly, and test duration was four weeks.

Production of the 6.3-cm fish increased with amount of food (Table 1, P≤0.005). As feeding frequency increased, or the interval between feedings decreased, production, survival and gain in length seemingly increased, though not significantly. Production differences due to food amount ranged from a high of 118% increase for fish fed 20% vs 5% BW at 120-minute intervals, to a low of 8% increase for fish fed 20% vs 10% BW at 5-minute intervals.

Feeding interval results can be thought of as reflecting the effects of number of feedings, rather than of time between feedings (Fig. 5). Among fish fed 5% BW, those fed at 15-minute intervals gained 150% more weight than those fed at 30-minute intervals (Table 1). Fish fed 5% BW at 120-minute intervals lost weight. Fish fed 10% BW at 5-minute intervals gained 68%, 106%, and 163% more weight than those fed at 15-, 30-, and 120-minute intervals, respectively. Fish fed 20% BW were least affected by feeding frequency; those fed every 5 minutes gained only 3, 13, and 41% more than those fed at 15-, 30-, and 120-minute intervals. Frequency of feeding became less important as food amount increased (Fig. 5).

For 12.1-cm fish, the 5% BW feeding level resulted in poor production (Fig. 6) and growth, and the 15% and 20% BW feeding rates produced little growth advantage over the 10% BW level. Food conversion rate increased with food amount, or food-to-flesh conversion efficiency decreased. Overall, food conversions were not good, which may have been due to low rearing densities.

Growth variation among individual fish within each group was not measured, but no obvious differences were noticed. For some species, particularly channel catfish (Randolph and Clemens 1976) and striped bass (Nick Parker, pers. comm.), frequent feeding of small amounts of food sometimes results in increased size variation.

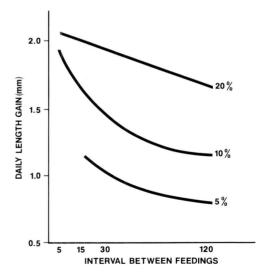

Figure 5. Growth of 6.3-cm ($\bar{x}$ TL) tiger muskellunge fed at 5, 10 and 20% of their body weight per day at four feeding intervals for 3 weeks at 20 C.

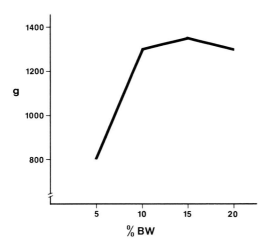

Figure 6. Production of tiger muskellunge 12-13 cm long fed at 4, 10, 15 and 20% of their body weight daily for 4 weeks at 20 C.

Loading

The need to determine the effects of loading, defined as weight of food fed daily, or the weight of fish of a specified size per unit flow of water, on production and quality of cultured fish is becoming increasingly important as intensive culture systems are developed. Qualitative standards are necessary for efficient production management. We determined the effects of reused, aerated water on growth, survival, feeding efficiency and quality (based on tissue histology of tiger muskellunge, 18-cm long.

Rectangular fiberglass 750 liter rearing units were divided into three 250 liter sections, with baffles allowing only a one-way flow. The first and third sections of each unit were used as rearing containers and the center section was used for aeration (Fig. 7). A series of four divided tanks provided a total of eight in-series rearing areas (sections). Water flow was 4 L/min. Water from overflow drains at the lower end of each 750 liter tank flowed into 75 liter tubs for aeration. The series was set-up in duplicate.

Six kgs of fish were placed in each of the 16 rearing sections, where the fish were acclimatized for 12 days. During the 8-week test, freshwater inflow was provided to only the first section of each of the two series; inflow to each of the other sections was used, reaerated water from the upstream section. The fish load on the water flow is referred to as cumulative load (CL), expressed as kilograms (of fish) per liter of water flow per minute (kg/L/min).

Cumulative load should not be confused with fish density, a term not used in this discussion and an expression of fish weight per unit volume. The relation of CL to water use and reuse is shown in Table 2.

Growth estimates were made from weight of sampled fish and total fish counts on odd-numbered weeks and from complete weight inventories on even-numbered weeks. The weight of fish in each rearing section was adjusted to 6 kg at each inventory. Daily feeding rate for all fish was 5% BW of fish in the rearing section having the greatest total weight of fish. Diet was dispensed every half hour, 23 hours per day (47 daily feedings). Rearing sections were cleaned with a siphon twice per day. About 80 liters of fresh water was metered into the section being cleaned to replace water lost through the siphon. The tanks, screens and baffles were scrubbed twice weekly with a brush and granular salt. Flow rate, DO and temperature were checked daily, and conductivity, ammonia and turbidity periodically. Concentration of un-iodized ammonia was calculated according to Emerson et al. (1975). Water samples were taken before the 0800-h cleaning, and at other times for comparison. At the end of the test, two fish from each unit were taken at random for cursory histological examination.

Water quality deteriorated with cumulative load-

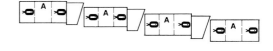

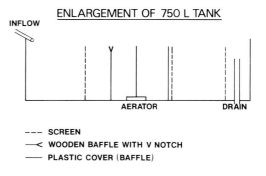

Figure 7. Schematic diagram of rearing and aeration unit arrangement for study of the effects of cumulative load on tiger muskellunge.

Table 2. Means of water chemistry values collected during a 56-day study of the effects of cumulative load on tiger muskellunge. Data for two identical systems at each cumulative load were averaged. Conductivity, turbidity, pH and ammonia values were for water from the discharge end of the rearing unit.

Physical or chemical value	Cumulative load in kilograms per liter per minute							
	1.5	3.0	4.5	6.0	7.5	9.0	10.5	12.0
Rearing units[a]	1	2	3	4	5	6	7	8
Water Use	1st	2nd	3rd	4th	5th	6th	7th	8th
Temperature (°C)	19.0	19.0	18.5	18.5	18.0	18.0	17.6	17.6
Oxygen (DO in mg/L)								
Mean								
In	8.4	8.3	8.5	7.4	8.2	7.7	8.3	7.9
Out	6.4	5.4	4.5	4.6	4.8	4.8	4.7	4.0
Outflow range								
Minimum	4.0	2.2	3.0	2.0	2.8	2.2	3.5	1.2
Maximum	8.2	8.4	7.9	8.0	9.0	8.2	8.0	7.4
Conductivity (μmhos)	76	80	90	95	85	102	108	123
Percent increase in turbidity over source water	3		5		8		9	
pH	6.6	6.8	6.5	6.7	6.5	6.7	6.5	6.6
Ammonia (mg/L)								
Total	0.15	1.20	1.20	1.80	3.40	3.80	5.00	4.60
Un-ionized	0.001	0.003	0.001	0.003	0.004	0.006	0.006	0.006

[a] Each unit was duplicated (1a and b, 2a and b, etc.)

ing (Table 2). Dissolved oxygen ranged from 9.0 to 7.4 mg/1 in unit inflows and from 9.0 to 1.2 mg/1 in outflows. Although incoming water was about 90% saturated with DO, mean DO concentrations in outflows of all reuse units were somewhat reduced. Conductivity increased with CL, from 76 μmhos to 124 μmhos. Total ammonia and un-ionized ammonia concentration increased with CL and with test duration. Ammonia concentration fluctuated diurnally, and was generally higher at 1100 hours than at 0730 or at 1500 hours (not shown in table). Ammonia production averaged about 9.5 g/kg of food. A poor food conversion rate (4.9) indicated that nearly 60% of the food may have been uneaten. Adjusted for estimated food wastage the production of ammonia was about 24 g/kg of food. A typical ammonia production rate in salmonid culture is about 35 g/kg of food, with a conversion rate of 1.5 (equivalent to an ammonia production rate of about 26 g/kg at a food conversion of 2.0). However, the salmonid diet has only about 65% of the protein level of the tiger muskellunge diet, indicating that the ammonia production rate per kilogram of crude protein for tiger muskellunge, about 48 g/kg protein, was somewhat lower than might be expected based on salmonid metabolism (about 80 g/kg protein).

Growth was slightly reduced as CL increased (Table 3), but concentrations of un-ionized ammonia were low (Table 2). The apparent growth reduction at CL's of 4.5, 7.5, and 10.5 are not understood. Rearing sections with these CL's were directly below tub aerators, which provided water with high incoming DO, even though water retention time in those sections was only 1/3 that of tank midsection aerators. The aerators did not cause gas supersaturation as determined with a Weiss saturometer. Survival was poor at CL of 12; a DO depletion, caused by mechanical failure, resulted in a nearly-complete kill. Production was variable after the second water use and followed the same pattern as growth. Fish production at CL 9.0, or in the sixth-use water, was equal to production in the first- and second-use units. Differences in net-weight gain were significant ($P \leq 0.05$) only between fish reared in the initial unit (CL 1.5) and those reared in the eighth-use unit (CL 12.0). Food conversion was inversely related to production. In general, tissue damage appeared to begin at CL 3.0 and became progressively more severe as CL increased.

The results indicated that, in a serial reuse system with aeration, production can remain high, but fish quality may deteriorate. Differences in fish quality occurred in fish reared in water that was passed (reused) two times (CL 4.5) or more.

Table 3. Production and mean length gain for tiger muskellunge reared 56 days at various cumulative loading rates.

Cumulative load (kg/L/min)	Production (kg)	Daily length increase (cm)
1.5	5.1	0.14
3.0	5.0	0.12
4.5	1.6	0.08
6.0	4.2	0.12
7.5	1.6	0.07
9.0	5.0	0.10
10.5	2.5	0.07
12.0	2.3	0.10

APPLICATIONS TO PRODUCTION

Water temperature may be the most important factor accounting for differences in growth rate of cultured fishes. Heating water for a short period in the spring may be an efficient use of resources, since additional length increments can be added with relatively little heated water. Small tiger muskellunge required the fewest temperature units per unit of length increment. Fig. 2 reflects the efficiency of heating for production and could be used to model economic break-even points. We consider 23 C to be the optimum temperature for maximum production of fish larger than 8 cm (TL), and 20 to 22 C as the optimum range for production of fish 3-8 cm long. However, lower temperatures are used in some facilities to reduce disease problems. Prediction of an optimum temperature for all sizes and all strains at all production facilities is unrealistic. (See Meade et al. 1983, for a pertinent discussion.)

If tiger muskellunge are reared at less than their maximum production temperature of 23 C, the level of protein in the ration can have an important influence on growth. At lower temperatures, tiger muskellunge -- and possibly other fish -- require a higher amount of protein in the diet to achieve maximum growth. Therefore, one could reduce the detrimental effects of a lower-than-desired temperature by using a higher protein diet.

There is always an optimum feeding level, depending on how the feeding objectives are defined and weighted. The optimum level is usually defined in terms of growth rate (near maximum), feed conversion rate (near minimum), and quality of fish produced (measured in such terms as external appearance, including absence of deformities and integrity of fins and skin; health of tissues; hardiness; survivability; and palatability of flesh). We backcalculated hatchery constants (HC) by using data from the feeding rate test conducted at 20 C; from these data, tiger muskellunge should be fed at a hatchery constant of about 90, and fish of 7 cm TL should receive about 13% BW per day.

Growth-rate differences attributable to feeding regime attenuate with increasing fish size. Conversely, the smaller the fish, the more critical it is to establish and use a high-frequency, or short-interval, feeding regimen. We recommend that fish less than 8 cm TL be fed at least every 5 minutes and larger fish every 15 minutes, until other methods such as demand-feeding technqiues are developed for these hybrids.

Judging from these results, the production efficiency of tiger muskellunge in intensive flowing-water systems can be improved in several ways. Growth can be increased by limited water heating in early spring; by feeding high protein diets to fish cultured at or below 20 C; and by more frequent feeding, especially for small fish. Limiting loading levels may offer a quality advantage. These advantages must be balanced against the costs of water heating, higher dietary protein and installation and maintenance of automatic feeders, as well as against the efficiency of water use. The heating, protein, and feeder cost-benefits can be easily determined, if a monetary value is placed on the weight of stocked fish. However, the desired balance between production costs and fish quality remains to be determined. It may be most economical to manage production at the fourth-use level (CL 6.0 in the present study), if fish grown in frequently reused water can be easily reconditioned several weeks before they are stocked. Conversely, it may be wasteful to attempt production rearing at CL 10.5-12.0, because of increased food costs, severe organ damage and probable reduced survival potential.

Hatchery production standards for tiger muskellunge (size, growth rate, number and quality) should be established from the administrator's point of view and developed for specific management needs. Current production standards are often based on false economics, generally reflecting only total fish produced, whereas the determining factor should be survival of stocked fish in the wild. From field studies, management biologists should determine appropriate size and number requirements for stocked fish. The hatchery manager should determine the most efficient production schedule to meet those requests. To do that requires knowledge of the effect of size and number of stocked fish and dates of stocking on survival. Until such field data are available, managers can only attempt to optimize hatchery production.

Toward the goal of optimizing hatchery production of tiger muskellunge, the following recommendations are made:

Rearing Temperature
    fish over 8 cm . . . . 23 C
    smaller fish . . . . 20 C
Feeding Rate
    hatchery constant (20 C) 90
Diet
    Protein level (20 C) 45%
Water Uses[1/]
    rearing groups in series and each
    depleting available DO before
    outflow to reaeration. 3

[1/] probable safe minimum number of uses, given neutral pH, DO at 90% of saturation at each inflow, no significant toxicants in source water, and moderate buffering capacity)

ACKNOWLEDGEMENTS

We thank W. Krise for his help throughout the study; T. Ort, S. Berryman, D. Shellman, W. Bump, D. Rottiers, R. Herman, D. Tandjung, C. Coutant, and R. Soderberg, made important contributions.

REFERENCES

Anderson, R. J., E. W. Kienholz, and S. A. Flickinger. 1981. Protein requirements of smallmouth bass and largemouth bass. Journal of Nutrition 111:1085-1097.

Brown, M. E. 1957. The physiology of fishes. Vol. 1, Academic Press, New York, NY, USA.

Cowey, C. B., J. W. Adron, D. A. Brown, and A. M. Shanks. 1975. Studies on the nutrition of marine flatfish: the metabolism of glucose by plaice (*Pleuronectes platessa*) and the effect of dietary energy source on protein utilization in plaice. British Journal of Nutrition 33:219-231.

Cowey, C. B., J. A. Pope, J. W. Adron, and A. Blair. 1972. Studies on the nutrition of marine flatfish: the protein requirement of plaice (*Pleuronectes platessa*). Bristish Journal of Nutrition 28:447-456.

DeLong, D. C., J. E. Halver, E. T. Mertz. 1958. Nutrition of salmonid fishes. VI. Protein requirements of chinook salmon at two water temperatures. Journal of Nutrition 65:589-599.

Duncan, D. B. 1955. Multiple range and multiple F tests. Biometrics 11:1-42.

Emerson, K., R. C. Russo, R. E. Lund, and R. V. Thurston. 1975. Aqueous ammonia equilibrium calculations: effect of pH and temperature. Journal of the Fisheries Research Board of Canada 32:2379-2383.

Hilton, J.W., C.Y. Cho, and S.J. Slinger. 1981. Effect of extrusion processing and steam pelleting diets on pellet durability, pellet water absorption, and the physiological response of rainbow trout (*Salmo gairdneri* R.). Aquaculture 25:185-194.

Hilton, J. W. 1982. The effect of pre-fasting diet and water temperature on liver glycogen and liver weight in rainbow trout, *Salmo gairdneri* Richardson, during fasting. Journal of Fish Biology 20:69-78.

Johnson, L. D. 1978. Evaluation of esocid stocking program in Wisconsin. American Fisheries Society Special Publication 11:298-301.

Kanazawa, A., S. Teshima, M. Sakamoto, and A. Shinomiya. 1980. Nutritional requirements of the puffer fish: purified test diet and the optimum protein level. Bulletin of the Japanese Society of Scientific Fisheries 46:1357-1361.

Ketola, H. G. 1978. Nutritional requirements and feeding of selected coolwater fishes: a review. Progressive Fish-Culturist 40:127-132.

Lemm, C.A., and D.V. Rottiers. 1986. Growth of tiger muskellunge reared at three temperatures and fed different amounts of protein. Progressive Fish-Culturist 48:101-106.

Mazid, M. A., Y. Tanaka, T. Katayama, M. A. Rahman, K. L. Simpson, and C. O. Chichester. 1979. Growth response of *Tilapia zillii* fingerlings fed isocaloric diets with variable protein levels. Aquaculture 18:115-122.

Meade, J. W., and J. T. Fuss. 1983. Design of small, low-cost rearing unit. Progressive Fish-Culturist 45:181-182.

Meade, J. W., W. F. Krise, and T. Ort. 1983. Effect of temperature on production of tiger muskellunge in intensive culture. Aquaculture 32:157-164.

Orme, L. E. 1978. The status of coolwater fish diets. American Fisheries Society Special Publication 11:167-171.

Page, J. W., and J. W. Andrews. 1973. Interactions of dietary levels of protein and energy on channel catfish (*Ictalurus punctatus*). Journal of Nutrition 103:1339-1346.

Pecor, C. H. 1978. Intensive culture of tiger muskellunge in Michigan during 1976 and 1977. American Fisheries Society Special Publication 11:202-209.

Pecor, C. H. 1979. Experimental intensive culture of tiger muskellunge in a water reuse system. Progressive Fish-Culturist 41:103-108.

Randolph, K. N., and H. P. Clemens. 1976. Some factors influencing the feeding behavior of channel catfish in culture ponds. Transactions of the American Fisheries Society 105:718-724.

Refstie, T., and E. Austreng. 1981. Carbohydrate in rainbow trout diets. III. Growth and chemical composition of fish from different families fed four levels of carbohydrate in the diet. Aquaculture 25:35-49.

Satia, B. P. 1974. Quantitative protein requirements of rainbow trout. Progressive Fish-Culturist 36:80-85.

Stuntz, W. E., and J. J. Magnuson. 1976. Daily ration, temperature selection, and activity of bluegill. Pages 180-185 *in* G. W. Esch and R. W. McFarlane, editors, Thermal Ecology II, CONF 750425. NTIS, Springfield, VA, USA.

Teng, S. K., T. E. Chua, and P. E. Lim. 1978. Preliminary observation on the dietary protein requirements of estuary grouper, *Epinephelus salmoides* Maxwell, cultured in floating net cages. Aquaculture 15:257-271.

Westers, H. 1978. Biological Considerations in hatchery design for coolwater fishes. American Fisheries Society Special Publication 11:248-253.

Westers, H. 1981. Fish culture manual for the state of Michigan (Principles of intensive fish-culture). Lansing, MI, USA.

SPECIAL PANELS ON MUSKELLUNGE MANAGEMENT

PANEL 1 - CATCH AND RELEASE

Catch and Release of Muskellunge - Philosophy and Methods

PAUL L. GASBARINO

Muskies Canada
P.O. Box 309, Station K
Toronto, Ontario M4P 2G7

ABSTRACT

Organized muskellunge fishing clubs, such as Muskies Canada and Muskies, Inc., feel very strongly that the future of the muskellunge is in doubt. Serious muskellunge anglers feel an obligation to help protect and stsudy this valuable predator, so they are concerned with environmental changes that affect its biology and habits. The dedicated muskellunge angler is also concerned that trophy fish remain available. Since it is the largest fresh water predator, it is hoped there will be a new world record. Current conditions indicate that, if trophies of more than 35 pounds are to be expected, most of the smaller fish must be released. Voluntary live release is the most direct and unselfish act an angler can practice to aid this species.

Muskellunge fishing clubs practice and promote live release extensively. Annual records of release and related statistics are shared with resource management authorities, who in turn share their biological expertise with anglers. Anglers are being taught the sensitivity of the muskellunge and the need to handle it correctly. The fishing public is being made aware of the release concept and proper methods of live release through the media, sports shows and direct demonstration. Achievements by fishermen in this program are publicly acknowledged.

At the present time, anglers have shown a willingness to practice "catch and release," and to use one or more of ten general release methods which are best described by the handling technique or equipment employed in each. The ten categories are: net, stretchers, hand, lip-lock, pistol grip, vise grip, tailing, stunning and two gaffing methods. Each technique requires special care, consideration and practice. To ensure future survival of the muskellunge, it is the hope of Muskies Canada that fishery agencies in Canada and the U.S., undertake studies to document the best release-handling methods.

Muskies Canada, Muskies, Inc. and other non-aligned muskie clubs, are dedicated to the preservation of the muskellunge. Members believe that the muskellunge is a valuable predator and that everything possible should be done to ensure the species survival. They are concerned about the effect on the natural distribution of muskellunge of such activities as habitat alteration, population growth, competition with northern pike and other fishes, poaching, angler exploitation and commercial fishing.

Changing times have not been good to the muskellunge, and musky anglers believe that, if everything remains as it is today, this species' future is bleak indeed. Concerns are that certain environmental changes will accelerate, making the attempts to protect this fish even more difficult. Anglers must anticipate and solve those problems, be dynamic in their approach to research and management, and be completely knowledgeable about muskellunge biology, habits and habitat. Hatcheries are necessary to insure a stable population of muskellunge.

Biologists have taught that even though the muskellunge is a tremendous fighter, it is also sensitive and has to be treated with respect (Casselman, 1975). Various studies have indicated that a muskellunge undergoes a constant struggle for survival from egg to maturity; the general rule of thumb is that of the eggs laid by one muskellunge in spring, one pair is all that can be expected to live to adulthood (pers. comm. E.J. Crossman). Serious muskellunge anglers have thus adopted live release habits to ensure future spawners and trophies.

For the past five years, many organized musky clubs have promoted and practiced live releases; they have released over 90% of all the muskellunge caught by their members. Of course, some individual fish are kept as trophies. A trophy fish is in the eyes of the beholder; one person's 15-pounder may be the biggest he will ever get, while someone else may not be satisfied unless his quarry is over 30 pounds.

In furthering the belief that live release can help the overall muskellunge stocks, Muskies Canada and Muskies, Inc. have also instituted public-release award programs, whereby a live-release certificate is presented to all anglers who submit an affidavit indicating the release of a legal-size muskellunge. We are confident that some of these releases will ensure future trophies that could exceed 50 lb. Statistics of releases by club members and by non-member anglers are available through Muskies Canada and Muskies, Inc. Release is stressed through sports shows, the media, and demonstration.

Proper handling of muskellunge is the most important aspect of live release. It should be recognized that putting a muskie back correctly goes a long way toward support of a continued trophy fishery. Once this concept is accepted, then the best ways to release muskellunge have to be studied. Also, in support of the release concept, there is a pressing need to study and understand the effects of live release. Concepts of monitoring release, such as tagging, fin clipping, angler-log sheets and possible survival rates require constant investigation and expansion (Richards and Ramsel, 1986).

Following is an outline of eight acceptable release methods and the tools required for successful release. It is recommended that each angler choose methods that are practical and with which they are comfortable. Two or three methods should be used regularly, wherever state or provincial laws allow. Three other methods are described which are practiced by some anglers, but are not recommended because of probable injury to the fish.

RECOMMENDED RELEASE METHODS

(1) Hand Release

At boatside, the fish is gripped behind the head and gill opening by one hand (Fig. 1). The angler places his thumb over the cleithrum on one side and his forefinger over that on the other side; the fish is carefully gripped without squeezing it and the hook is removed with the free hand.

Special care is needed to prevent crushing or damaging the gills. Some anglers recommend hold-

Figure 1. Hand release of muskellunge at boatside.

Figure 2. Net release of muskellunge at boat.

ing the fish across the bridge of its head by the edge of the eye sockets, but this method can be injurious; either the eyes are pushed in and injured, or the skin on the eye sockets is badly bruised, leaving the fish subject to infection and blindness. Hand release is best used with fish under 10 lb. and can be done in the water quickly. With lightly hooked fish, needlenose pliers should be used; often in this way the fish is never touched; other fish can be shaken off the hooks at boatside.

(2) Net Release

This is the most-often utilized form of landing, and the method with which most anglers are comfortable (Fig. 2). Once netted, the fish can be left in the water with the mesh gathered around it so it can be held firm while removing the hooks. If the fish has to be brought aboard, it should be lifted very slowly by the net hoops, not the handle which will often bend, and the muskellunge laid gently in the bottom of the boat. The fish is held down carefully, the hooks removed and then it is returned to the water. While abroad the boat, the fish can be restrained by one palm pressed down behind the head and the other holding the fish around the tail. It is helpful to have a piece of material, either cloth or plastic, in which the muskellunge can be folded, or it could be placed over the fish with the angler straddling it. The hooks can be removed by the aid of needlenose pliers, or if necessary, cut with side or bolt cutters. The muskellunge is then lifted into the water with as much of its body supported as possible. If it requires resuscitation, support it under the stomach with one hand and one around the tail and "walk it" through the water as the boat drifts. Pulling it backwards should be avoided, as this movement is unnatural to the respiration of the fish; it could allow debris to enter and damage the sensitive gills, or simply not aerate the gills adequately. The size of net and its quality should also be considered. The net should be large enough to engulf the entire fish. As there is always the chance of taking a 45 to 50-in muskellunge the net should be at least that deep. All nets should have a sturdy frame. The kind of mesh is also important; cotton is preferable to nylon which is coarse. Damage can occur to netted fish, so gentle net handling is mandatory.

(3) Stretcher or Cradle (pers. comm. Robert Strand.)

Knotless net mesh or porous cloth about 24 to 36 inches wide is stretched between two wooden poles

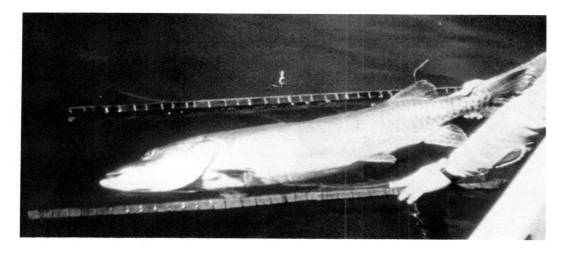

Figure 3. Stretcher method of release of muskellunge at boat.

about 4 to 5 ft long. The cradle can be attached directly to the boatside or be completely free (Fig. 3). The muskellunge is drawn into the netting and folded in length-wise, so that it is held in envelope fashion while the hooks are removed. It is a good way to hold fish securely without injury, as it is totally immobilized. This method is excellent for research work and tagging as the entire fish can be examined in the water. This procedure should be used only when two or more anglers are present; it is effective and highly recommended by experienced anglers.

(4) Lip Lock (Fig. 4)

This is a sound method commonly attributable to the Leech Lake area of Minnesota and often referred to as the Leech Lake lip-lock.

The lip-lock is a hand grip, where the first three fingers of one hand are slid inside the membranes (between the halves of the lower jaw) connecting the gill cover and jaw; this area is expandable and there is room for the fingers without danger of insertion in the gills. The thumb is clamped on the outside of the jaw and the muskellunge held firm. Anglers wishing

Figure 4. Lip Lock method of release of muskellunge.

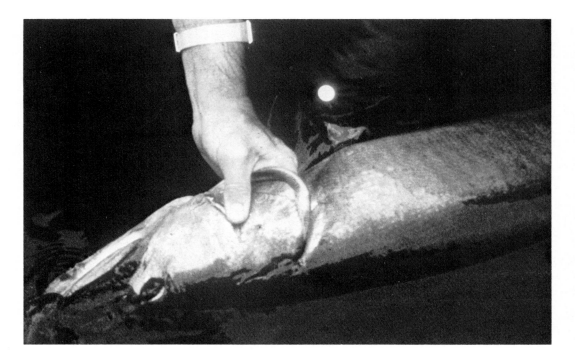

Figure 5. Gill-cover grip method of release.

to attempt this method should ask for a demonstration from an experienced angler.

(5) Gill-Cover Grip (Variation of No. 4)

The safest way to do this is to grip the gill cover between thumb and fingers, 2 or 3 in up from the jaw, and gently turn the fish on its side. The hooks can then be safely removed with a free hand (Fig. 5). This method is favored by many highly experienced anglers. Twenty-pound or heavier muskellunge can be handled successfully. However, anglers should be aware of the potential hazards of employing this method. The gills or gill cover could be damaged; also the anglers' fingers can slip into the gill rakers. In addition, free hooks on the lure could become embedded in the anglers wrist.

(6) Gaff

Gaffing is in common use in some areas (Fig. 6), and an accepted procedure. The point of the gaff is inserted down through the lower jaw as close to the tip as possible, missing the tongue and only piercing the lower jaw membrane. The fish is thus secured by a firm grip on the shank of the gaff hook by either one or two hands. This allows limited mobility of the fish and a strong hold on the head, enabling safer dehooking and more control where hook removal is troublesome, such as deeply hooked, multiple-hooked or fish rolled up in lure, line or leader. When gaffed, the fish is held beside the boat and most often does not have to be boated.

This is a good method for large muskellunge, especially if hooks have to be cut out with a knife, or cut off with side cutters. However, this requires skill so that the gaff is used as a tool and not as a weapon. Care must be taken not to pierce anything except the lower jaw membrane, which will heal quickly with no permanent injury. When gaffed, the muskellunge should not be lifted out of the water, which is supporting the fish's weight. If a large fish is lifted by a gaff inserted in this way, it is quite likely that the connection between the two halves of the lower jaw will separate and inhibit capture of food later.

Figure 6. Example of proper gaffing method of release.

(7) Tailing

This is a relatively new and extremely effective procedure; it has been adapted from a common practice in use with salmon-landing methods. A plastic-covered loop over a steel cable at the end of a long handle is slipped over the muskellunge's tail and pulled firm (Fig. 7). The fish is then held firmly and can be netted, gaffed or stretchered. The tailer is easy to use but with the head free, caution is advised. However, often the muskellunge calms down when tailed. This is a very good method on large boats, or with fish over 25-30 lb. Hand-tailing and turning the fish on its side reportedly causes the fish to become dormant and is an effective release method for small muskellunge.

Figure 7. Tailing release method of muskellunge.

(8) Long-Handle, Pistol-Grip Hook-Remover

This device has a squeeze handle, a six-inch shank and alligator clamp at the end. It is clamped on a hook while the fish is in the water and the fish is released without being touched. Depending on how the fish is hooked, this can be effective with small to medium fish. It keeps the hands away from the fish's mouth and gives leverage for hook removal. However, equipment quality is essential; avoid tools that are cheaply made. Needle-nosed pliers also allow a tight grip, and the hooks can be pulled straight out with no twisting.

METHODS PRESENTLY IN USE BUT NOT RECOMMENDED

(9) Vise Grips

These are effective when used properly. A modified vise grip, with half-inch lugs welded to the inner surface of each jaw, is slid open around the lower jaws of the fish and clamped shut. The muskellunge is locked on and held tight. This procedure requires a strong wrist and a firm grip. Here again, the fish can be held beside the boat and secured while hooks are removed without permanent injury. A drawback is the chance of the fish pulling the vise grips out of the hand, breaking free and swimming around with a heavy tool locked on the side of his mouth. A strong cord could be attached to the grips, but this could cause anglers injury, if the muskellunge starts twisting around out of control.

(10) Stunning

This procedure is used by a few anglers, but it is discouraged. At boatside, the fish is rapped over the head and then hand-held or boated. After the hook removal, it is released and swims off, supposedly unharmed. There is no way to judge the amount of force necessary to render the fish unconscious, or if the fish is seriously injured.

(11) Poor Gaffing Method

A second gaff method is to insert the hook through the gill plate and out the mouth, without striking the gills, tongue, eyes or head. Only the most skilled anglers should use this method. It is not recommended because of the probabilities of causing serious injury to the fish.

DISCUSSION

Anglers need to understand what to do from the time the muskellunge is hooked until after its release. Tackle should match possible size of muskellunge expected. While angling in an area known to have fish around 20 lb, but few over, then 15 to 20-lb line is appropriate; when 40 to 50-pounders are the quarry, then heavier line should be used.

A fish should be brought to the boat quickly and left in the water whenever possible; it should not be handled unless necessary, or handled gently. The entire body weight should not be supported by holding a muskellunge by the head (Fig. 8). The angler must

Figure 8. Example of proper method of supporting the body weight of the fish.

learn how to coddle a muskellunge, resuscitate it, and remove hooks, as well as how it's affected by stress (Beggs, et al., 1980). A fighting fish builds up lactic acid. The more the muskellunge struggles, the less intake of oxygen and the more often it needs resuscitation; also the cooler the water, the better the muskellunge maintains its resistance to stress. It is important to minimize landing time and reduce stress factors. Sometimes fish roll in the line and leader, or hooks are imbedded to an extent that the fish has to be restrained while hooks and line are cut to free it. Here, fast action is the order of the day!

Fish should not be left out of the water more than two minutes. There are also occasions when a legal-sized muskellunge may be so badly injured that it would not survive release; if it is gill-hooked and bleeding very badly or hooked in the eyes, it is more humane to kill it quickly. Kept fish should not be wasted, however, they can be mounted, eaten, or retained for research. Any captured fish should be examined, measured, and observed for general physical appearance, tags, injuries or disease such as lymphosarcoma (Fig. 9). Written records of the observation should be made. If the fish is to be kept for scientific purposes, it should be frozen and a resource agency biologist notified of its whereabouts.

Proper handling of muskellunge is the most important aspect of live release. Handling should be limited and gentle. If stressed, it should be cradled over the crook of one arm and held around the tail by the other and moved in a slow S-fashion until it can swim free (Fig. 10). It can be taken to shallow water where the angler can wade along with the fish holding it as previously described, until hopefully the angler is left wading alone.

Muskellunge are unpredictable; each time one is hooked, it reacts differently. Each release is unique and only experience with fish of many sizes and varied circumstances will improve technique. It is imperative that experiments to monitor release successes be planned and implemented so that the angling public will become aware that live release is a creditable practice.

Prior to extensive use and experimentation with live release by organized muskellunge anglers, there was concern about the value of the concept. Evaluation has indicated that use of proper release methods could lead to survival of 70-100% of fish, which could be caught by another angler, or at least to increase to greater trophy size. It is the responsibility of experienced anglers to pioneer release experiments to improve that survival ratio (Dent, 1986).

Angler concern for the musky must be communicated and proven release methods demonstrated to the public while the search for overall improvement in management of the muskellunge continues.

ACKNOWLEDGEMENTS

This work required the input of a large number of people whose assistance I would like to acknowledge, as follows:

John Anderson; Jacqueline Brand; Craig Birchall; Canadian National Sportsmen Shows; Gen Carr; Dr. John Casselman; Dr. Ed Crossman; Paul Curran; John Damman, Jim Davis; Ron Dent; Barry Diebold; Mark Falco; Carl Fochs; Gil Hamm, Jim Hutchings; Al Jessop; James Kelly; George Migas; Ministry of Natural Resources

Figure 9. An example of Lymphosarcoma in muskellunge.

Figure 10. Shallow water resuscitation of muskellunge.

(Ontario); John Parry; John Power; Larry Ramsell, Cecil Ray; Brendan Reid; Kevin Richards; Herb Rieneke; David Sault; Larry Slogoski; Robert Strand; Peter Wilkinson; Sam and Fran Wilson.

REFERENCES

Beggs, G.L., G.F. Holeton, E.J. Crossman, 1980. Some physiological consequences of angling stress in muskellunge (*esox masquinongy*) Mitchell. Journal of Fish Biology. 17(649-659)

Casselman, J.M. 1975. Age and growth of muskellunge from the upper Niagra River with reference to over-exploitation. Report prepared for the Ontario Ministry of Natural Resources, Niagra District, Fonthill, Ontario.

Dent, R.J. 1986. A case history of voluntary catch and release of muskellunge with management implications for fishery programs. American Fisheries Society Special Publication Number 15:316-322.

Richards, K., and R. Ramsell. 1986. Quantifying the success of muskellunge catch and release programs. A summary of cooperative angler tagging studies. American Fisheries Society Special Publication 15:309-315.

Quantifying the Success of Muskellunge Catch and Release Programs: A Summary of Cooperative Angler-Tagging Studies

KEVIN RICHARDS

Missouri Department of Conservation
1104 South Grand
Sedalia, MO 65301

ROD RAMSELL

Muskies, Inc.
6048 51st St. North
Oakdale, MN 55109

ABSTRACT

Catch and release fishing for muskellunge is one direct method by which anglers can contribute to the future of quality of their fishing experience and enhance the survival of these trophy fish. Muskellunge fishing-club members from seven states participated in tagging studies in an effort to quantify survival of angler-caught and released muskellunge. Data from over 1,600 fish tagged and released since 1976 were examined for this summary. Approximately 17% of these angler-tagged muskellunge were recaptured by anglers. Recapture rates for 578 muskellunge caught, tagged and released by 16 experienced anglers ranged from 22% - 44%, and averaged 28%. These rates equal, or exceed, most muskellunge-exploitation extimates reported by fisheries agencies, which indicates excellent survival and tag retention for the fish handled by these anglers. In the future, more extensive tagging studies on fewer bodies of water may provide even more useful information on the positive effects of muskellunge catch and release fishing.

In the management of muskellunge populations fishery agencies commonly utilize minimum-length limits and daily creel limits to restrict harvest. Minimum lengths range from 24 to 36 in, with 30 in used most frequently; creel limits normally allow the harvest of only one or two muskellunge per day. The rationales for these regulations are to enhance the trophy potential of the muskellunge fishery, to reduce exploitation, and to protect immature fish (Miller 1983). Despite the implementation of these restrictive measures, muskellunge populations are typically characterized by low densities, high exploitation rates, and limited trophy potential.

Fishery agencies realize that many factors other than harvest affect the structure of muskellunge populations; these include habitat requirements, predator-prey relationships, disease, genetics, and stocking techniques. All of these factors are being researched, as exemplified by the content of this International Symposium on Muskellunge. In addition, clubs of dedicated muskellunge anglers have offered to assist fishery agencies with financial support and volunteer labor. Many have also shown a willingness to adopt self-imposed harvest regulations, more restrictive than those enforced by fishery agencies.

Since 1970, voluntary catch and release fishing for muskellunge has grown in popularity as a direct method by which anglers attempt to enhance the future quality of their fishing experience. Voluntary catch and release fishing can increase survival of muskellunge and benefit the fishery, if hooking and handling mortality of released fish is not excessive. Porter (1977) stated that this was one of the more effective means of ensuring more legal muskellunge in a population. Dent (1986) presented a case history of voluntary catch and release of muskellunge, including application of a catch and release model developed by Clark (1983).

The first voluntary release award program for muskellunge was initiated in 1954 by The Musky Club Alliance of Wisconsin (Ogden, 1986). Currently, Muskies, Inc. (MI) is one of the largest organized muskellunge fishing organizations, with over 3000 members and 37 chapters in the United States. According to MI records, from 1970 to 1975, members of organized muskellunge fishing clubs began releasing an increased percentage of the legal muskellunge they caught. MI members released 19% of the legal muskellunge they caught in 1970, 59% in 1972 and 85% in 1975 (Davis 1983). While anglers were releasing legal muskellunge in record numbers there was little evidence that these released fish were surviving and contributing to the quality of muskellunge fisheries.

Since 1975, results of two studies have indicated that release survival of angler-caught muskellunge, under research conditions, is relatively high. Beggs et al. (1980) reported a 70% survival rate for angler-caught muskellunge which were anesthetized, transported to a laboratory, and subjected to physiological testing; this study was conducted during the mid-summer when angling stress from high water temperatures could be greatest. Strand (1986) reported 100% survival for 14 adult muskellunge which were angler-caught, surgically implanted with radio transmitters and released.

In an effort to document the survival of angler-caught and released muskellunge under typical angling conditions, MI members initiated a tagging program in 1976. This program evolved into six separate studies. The purposes of this paper are to 1) summarize the data from over 1,600 angler-caught muskellunge tagged during the six studies and 2) make recommendations for the design and direction of future studies.

METHODS

The six separate studies each maintained their own set of tagging records (Table 1). Each of the studies was administered somewhat differently, but, in most cases, they were coordinated with local representatives of state fishery agencies; posters explaining the studies were distributed locally, and cooperators were instructed not to tag muskellunge which had been injured or appeared to suffer severe physiological stress. Specifics of each study will be discussed with results, as study design is an important aspect of this paper.

For this summary, the tagging coordinators of the six separately administered tagging studies were contacted, and agreed to share available information. Follow-up contacts were made with the coordinators and local fishery biologists. Each of the six studies will be summarized separately, followed by a general discussion of data which are comparable between studies.

Table 1. Pertinent data for six muskellunge tagging studies conducted by Muskies, Inc. chapters.[a]

Study	Participating chapter(s)	Number of lakes/streams	Tagging initiated	Tagging ended[b]	Tag type/ placement	Number tagged	Number recaptured	Recapture rate
Minnesota	Twin Cities, MN	1	1976	1979	Floy FD-68B (T-bar)/Dorsal	220	33	15.0%
New York	Niagara River, NY	1	1979	1982	National-monel band #4/Dorsal	276	29	10.5%
Missouri	Pomme de Terre, MO	1	1980	1981	Floy FD-68B (T-bar)/Dorsal	49	5	10.2%
Ohio	Cleveland, OH	1	1982	1982	National-monel band #4/Opercle	43	4	9.3%
Pilot-ITP	West Virginia, WV Headwaters, WI Hayward Lakes, WI Niagara River, NY Pymatuning, PA	16	1978	1978	Floy FT-2 (Dart)/Dorsal	60	4	6.7%
International Tagging Program (ITP)	West Virginia, WV Headwaters, WI Hayward Lakes, WI God's Country, WI Lakeland, WI	115	1979	1982	National-monel band #4/Dorsal	1016	203	20.0%
Total	--	--	--	--	--	1664	278	16.7%

[a] Additional information for each study is available from the appropriate chapter or state fishery agency.

[b] Last tagging year for which at least one full year of recapture information was available.

RESULTS

Minnesota

In 1976, the Twin Cities Chapter initiated a tagging study in cooperation with the Minnesota Department of Natural Resources. Tagging was conducted on Leech Lake, a 110,000-acre lake in northcentral Minnesota. Floy FD-68B T-bar anchor tags were inserted at the base of the dorsal fin. Participating anglers were given a set of guidelines and separate forms to use with each tagged muskellunge. Cooperators were authorized to tag muskellunge over 20 in, yet most of those tagged exceeded the 30 in legal-size limit which was in effect through 1982. A total of 220 muskellunge were caught, tagged and released by cooperating anglers (Table 1). The 33 recaptures reported through 1980 included several multiple recaptures. Three recaptures represented trophy-sized muskellunge; these fish measured 35, 43, and 45 in when released and 48, 48, and 51 in, respectively, when recaptured. Through 1980, the overall recapture rate was 15.0%.

New York

In 1979, the Niagara River Chapter initiated a tagging study in cooperation with the New York Department of Environmental Conservation. Tagging was conducted on the upper Niagara River between Lake Erie and Niagara Falls. This area covers approximately 33 river miles and 12,000 acres. National monel band tags, size 4, were attached to the anterior base of the dorsal fin of muskellunge 20 in and larger. The minimum legal-length limit for this area is 30 in. Cooperators were given a detailed set of guildelines and diaries to be used to record tagging and other information. At the end of each year diaries were collected and tagging information summarized by state fishery biologists. A copy of the diary summary, along with recommendations for the next year, were given to cooperators.

Through 1982, cooperating anglers tagged and released 276 muskellunge (Table 1). Of the 29 reported recaptures through 1983 (10.5%), 12 were re-released and one of these was recaptured a second time. A 48 in muskellunge released during this study was recaptured eight days later, thus providing two anglers the thrill of a trophy catch.

Missouri

In 1980 and 1981, the Pomme de Terre Chapter conducted a tagging study in cooperation with the Missouri Department of Conservation on 7,800-acre Pomme de Terre Lake. Participating anglers were given individual tagging demonstrations prior to issuance of tags. Cooperators were authorized to tag muskellunge over 23 in with Floy FD-68B T-bar anchor tags inserted at the base of the dorsal fin. The same type of tags were being used concurrently for a tagging study conducted by state biologists; therefore, special posters explaining the program were not necessary. Muskellunge harvest in Missouri is regulated by a 30-in minimum-size limit.

Through 1981, 49 muskellunge were tagged and released by cooperating anglers (Table 1). Five recaptures were reported through 1983 (10.2%); three were re-released but none were caught a second time. The time span between tagging and recapture for these five fish ranged from 11 to 837 days. The one which was at large for 837 days was first recaptured by state biologists during electrofishing operations, 703 days after it had been tagged and released; at that time the Floy tag was still in place and a dorsal band tag was also applied. When re-caught by an angler, 134 days later, only the band tag was in place; the original Floy tag had been lost. Four additional muskellunge, which had been tagged with Floy tags during electrofishing operations in previous years, were recaptured by electrofishing in 1983, and a dorsal band tag was applied. Two of these fish were recaptured by anglers in 1983; one had lost its original Floy tag, and the other Floy tag broke apart while being examined by a state biologist (Ron Dent, pers. comm.).

Ohio

The Ohio tagging study was initiated by the Cleveland Chapter in 1982, in cooperation with the Ohio Department of Natural Resources. Tagging was conducted on 3,000-acre West Branch Reservoir. Participating anglers were given tagging instructions, and tagging records were maintained by the Ohio Department of Natural Resources. Fish from 21.5 to 42.0 in were tagged with National monel band tags, size 4, attached to the opercle. Most of the fish tagged were hybrid muskellunge, although a few true muskellunge were also tagged. There is no minimum-length limit for muskellunge in Ohio.

A total of 43 muskellunge were tagged and released by cooperating anglers in 1982 (Table 1); through 1983, there were four reported recaptures (9.3%).

Pilot Intranational Tagging Program

In 1978, five MI chapters participated in this pilot study. Tagging was conducted on 16 bodies of water in four states. Floy FT-2 dart tags were inserted at the base of the dorsal fin of muskellunge 28 in or larger. Participating anglers were given a set of guidelines and separate tagging forms to be used for each fish.

During 1978, 60 muskellunge were tagged and released by cooperating anglers (Table 1). Four tags (6.7%) were returned through 1983. Ten of the 60 fish were tagged in rivers in West Virginia, and three of the four recaptures were West Virginia fish, all of which had been tagged by the same angler.

Intranational Tagging Program

In 1979, the Intranational Tagging Program (ITP) was initiated under its present format. Five chapters from two states are currently active in this study. Muskellunge, 28 in or larger, have been tagged on 115 bodies of water with National monel band tags, size 4, attached to the anterior base of the dorsal fin. Participating anglers were given a detailed set of guidelines and separate forms to be used for each tagged muskellunge.

The ITP has been the most extensive of the tagging studies. From 1979 through 1982, 1,016 muskellunge were tagged and released by cooperating anglers (Table 1). Of the 203 reported captures through 1983 (20.0%), 61 were re-released and 22 of these (18.0%) were caught a second time. Tagging continued through 1983, with an additional 436 muskellunge tagged and released that year.

Table 2. Summary of ITP muskellunge tag returns by body of water, 1979-1983.

Lake or Area, State	Number tagged	Number recaptured	Recapture rate
Chippewa Flowage, WI	451	119	26.4%
Eagle River Chain, WI (10 lakes)	72	3	4.2%
North and South Twin Lakes, WI (2 lakes)	63	4	6.4%
Lake Wissota, WI	61	13	21.3%
Minocqua Chain, WI (5 lakes)	53	11	20.8%
Cisco Chain, WI/MI (15 lakes)	34	9	26.5%
West Virginia (12 creeks)	33	8	24.2%
Palmer and Tenderfoot lakes, WI (2 lakes)	32	3	9.4%
Boulder Lake, WI	22	5	22.7%
Sub-total	821	175	21.3%
All other waters (64 lakes, 2 rivers)	195	28	14.4%
Total	1016	203	20.0%

Because of the large number of waters on which tagging took place, many have had few fish tagged or recaptured. On the other hand, there are noteworthy cases of intensive tagging efforts on nine individual bodies of water, or systems (Table 2). Six of the nine areas with over 20 tagged muskellunge had recapture rates greater than the ITP average.

The Chippewa Flowage is an excellent example of an intensive ITP tagging effort on a single body of water. This flowage is a 15,000-acre impoundment in northwest Wisconsin. From 1979 through 1982, 451 muskellunge were tagged and released by cooperating anglers. Of these, there were 119 reported recaptures through 1983, for an impressive recapture rate of 26.4% (Table 2). Tagging continued on the Chippewa Flowage in 1983 with an additional 153 muskellunge tagged and released, which brings the five-year total of tagged releases to 604 muskellunge. The only other study, in which a comparable number of angler-tagged muskellunge were released, was conducted on 274,000-acre Lake St. Clair, located on the Michigan-Ontario border. Haas (1978) reported that members of the Michigan-Ontario Muskie Club tagged 530 muskellunge between 1969 and 1975 using plastic dart tags (Floy FT-2). Through 1977, only 32 recaptures (6.0%) had been reported.

The detailed records which have been kept for the ITP allow for stratification of the results by angler, size at tagging, month of tagging and bait used for original capture. Approximately 100 anglers have participated in the ITP to date. Sixteen anglers have each tagged over 10 muskellunge and have recapture rates above the ITP average (20%); they account for 578 of the 1,016 tagged muskellunge. Their mean recapture rate is 27.5%; range is from 21.7 to 44.4%. The angler who had the second best rate of return (42.9%) was the same individual who had tagged three of the four recaptured muskellunge during the Pilot ITP.

Each of these 16 anglers was sent a questionnaire concerning handling, tagging and release techniques in an effort to document successful handling methods; responses were received from 14. Five of the respondents land the fish as soon as possible to avoid stressing the fish; the other nine land the fish when the opportunity presents itself and they can feel comfortable with handling it. None of the anglers play the fish to exhaustion. All use a net, at least some of the time; alternative landing methods included hand grabbing over the back behind the gills, unhooking in the water without touching the fish and tailing. Eleven respondents handle their catch in the boat, two handle fish in the water, and one does both. When a muskellunge is brought into the boat it is normally restrained by keeping it in the net, straddling it, or wrapping it in some type of cloth or carpet. When releasing a muskellunge most respondents hold the fish by the tail, support it with the other hand and move it back and forth in the water. Twelve respondents prefer to release muskellunge in water less than five feet deep. While all of the respondents examine muskellunge for serious injury or bleeding, the final decision on whether or not to release an injured fish is an intuitive judgement; slight bleeding was of little concern to most of the anglers. The variation in handling methods utilized may indicate that the best methods for handling a muskellunge which is to be released are those which individual anglers can practice with speed and confidence.

The size of a muskellunge at the time of capture and release may affect its survival chances. Stratification of the ITP results by size of fish indicates higher recapture rates for muskellunge which were 30 to 39 in when tagged, compared with those which were over 40 in when tagged (Table 3). Future studies could test this observation by tagging equal numbers of 30 and 40 in muskellunge in the same body of water.

Stratification of ITP muskellunge tag returns by month of tagging indicates consistently high recapture rates, and thus high survival, for fish tagged in the months of June, July, and August (Table 4). This is encouraging as high summer water temperatures are often mentioned as a handling concern. The relationship between release survival and water temperature could be determined through future studies in which temperature is used as the primary variable. Considering the large sample size, the relatively low recapture rate for ITP muskellunge tagged and released during September is unexplainable.

Summary of ITP muskellunge tag returns by bait used for original capture also yields some interesting information (Table 5). The most commonly-used lures (bucktails) have a recapture rate approximately equal to the ITP average. Topwater lures have a high recapture rate of nearly 31%. All other bait types, including live bait, have recapture rates which are lower, but consistent, ranging from 14.1 to 16.7%.

The time span between release and recapture of ITP muskellunge has ranged from 15 minutes to

Table 4. Summary of ITP muskellunge tag returns by month of tagging, 1979-1983.

Month of tagging	Number tagged	Number recaptured	Recapture rate
January	0	0	--
February	0	0	--
March	3	0	0.0%
April	10	3	30.0%
May	66	12	18.2%
June	199	46	23.1%
July	181	44	24.3%
August	144	34	23.6%
September	246	33	13.4%
October	129	27	20.9%
November	22	3	13.6%
December	8	1	12.5%
Unknown	8	0	0.0%
Total	1016	203	20.0%

Table 3. Summary of ITP muskellunge tag returns by size at tagging, 1979-1983.

Length when tagged (in)	Number tagged	Number recaptured	Recapture rate
<30	9	1	11.1%
30-31	330	67	20.3%
32-33	248	51	20.6%
34-35	158	33	20.9%
36-37	112	22	19.6%
38-39	74	18	24.3%
40-41	46	7	15.2%
42-43	23	3	13.0%
44-45	8	0	0.0%
≥46	8	1	12.5%
Total	1016	203	20.0%

Table 5. Summary of ITP muskellunge tag returns by bait used for original capture, 1979-1983.

Bait type	Number tagged	Number recaptured	Recapture rate
Bucktail	346	72	20.8%
Topwater	190	58	30.5%
Jerk Bait	166	25	15.1%
Crankbait-deep	117	19	16.2%
Live	64	9	14.1%
Crankbait-shallow	61	9	14.8%
Jig	30	5	16.7%
Spoon	2	0	0.0%
Unknown	40	6	15.0%
Total	1016	203	20.0%

1,440 days. Fifty-one percent of the recaptured muskellunge were caught within 300 days of release. An additional 32% were caught between 301 and 600 days after release, and the remaining 17% were caught over 600 days after release.

DISCUSSION

The six cooperative muskellunge-tagging studies included in this review represent a significant volunteer effort by dedicated muskellunge anglers. The basic purpose behind these studies was to gather proof that angler-caught and released muskellunge survive and contribute to the future quality of the muskellunge fishery. The purpose was achieved, and the proof has been used to encourage the catch and release ethic. Since these studies were first initiated the catch and release philosophy has gained additional acceptance. MI members reported a 94% release rate for legal muskellunge they caught in 1984 (Davis 1984), and other, non-affiliated muskellunge clubs seem to be following suit.

Four different tag applications were used in these studies. Highest recapture rates were observed for monel-band tags attached to the anterior base of the dorsal fin, lowest rates for dart tags inserted at the base of the dorsal fin. Similiar low-recapture rates were reported for dart-tagged muskellunge by Haas (1978). T-bar tags inserted at the base of the dorsal fin and monel-band tags attached to the opercle were associated with intermediate return rates. Tag loss and deterioration were reported for T-bar tags, and skin growth over monel tags was a reported problem.

Sample size is an important consideration of any mark and recapture study. The three studies with the largest sample sizes also had the highest recapture rates (Table 1).

Reported angler-recapture rates for muskellunge caught, tagged and released by anglers in these six studies approximate recapture rates by anglers in many studies where muskellunge were tagged by state fishery crews during netting or electrofishing operations (Table 6). This indicates that properly-handled and released muskellunge experience survival rates similar to those which have not been previously captured by angling.

Future studies of this type should be even more closely coordinated with local fishery agencies and some, at least, designed to statistically validate some of the correlations and observations which have been discussed in this summary. Indiscriminant tagging of muskellunge should be discouraged, as it increases the handling time between hooking and release and provides only limited data. Continued cooperation between muskellunge anglers, organized muskellunge fishing clubs, and resource mangers can result in even more improved muskellunge fisheries.

ACKNOWLEDGEMENTS

Special thanks go to all the cooperating members of Muskies, Inc., tagging coordinators, and biologists who shared their data for this summary. Ronald J. Dent assisted in manuscript review. Melanie A. Benitz typed the manuscript and tables and deserves credit for her patience.

REFERENCES

Beggs, G.L., G.F. Holeton, and E.J. Crossman. 1980. Some physiological consequences of angling stress in muskellunge. Journal of Fish Biology 17:649-659.

Clark, R.D. 1983. Potential effects of voluntary catch and release of fish on recreational fisheries. North American Journal of Fisheries Management 3:306-314.

Davis, B. 1983. Muskies, Inc. members only fishing contest results, 1970-1982. Muskie 17:1-36. St. Paul, Minnesota.

Davis, B. 1984. 'Lunge Log. Muskie 18:4-8. St. Paul, Minnesota.

Dent, R.J. 1986. A case history of voluntary catch and release of muskellunge with management implications for fishery programs. American Fisheries Society Special Publication 15:316-322.

Haas, Robert C. 1978. The muskellunge in Lake St. Clair. American Fisheries Society Special Publication 11:334-339.

Table 6. Summary of tag return rates by anglers for muskellunge in studies conducted by state fishery agencies.

Study	Location	Tag type	Recapture rate
Spangler 1968	Pigeon and Sturgeon Lakes, Ont.	Disc	13%
Mooradian and Sheperd 1973	Chautauqua Lake, NY (1941-1960)	Monel Band	10%
	Chautauqua Lake, NY (1961-1971)	Monel Band	18%
Hacker 1973	Little Green Lake, WI	Disc	30%
Haas 1978	Lake St. Clair, MI/ONT	Dart	12%
Miles 1978	Middle Island Creek, WV (Total)	Monel Band	35%
	Middle Island Creek, WV (≥28 in)	Monel Band	49%
	Middle Island Creek, WV (<28 in)	Monel Band	15%
Dent 1986	Pomme de Terre, MO	T-bar	15%
	Pomme de Terre, MO	Monel Band	26%

Hacker, Vernon A. 1973. The results of a ten year voluntary muskellunge creel census at Little Green Lake, Green Lake County, Wisconsin, 1963-1972. Fisheries Management Division Management Report No. 58. Wisconsin Conservation Department, Madison.

Miles, Robert L. 1978. A life history study of the muskellunge in West Virginia. American Fisheries Society Special Publication 11:140-145.

Miller, M.L. 1983. The status of muskellunge management in North America (results of a mail survey conducted during fall 1981). Division of Wildlife, Ohio Department of Natural Resources, Columbus.

Mooradian, S.R. and W.F. Sheperd. 1973. Management of muskellunge in Chautauqua Lake. New York Fish and Game 20:152-157.

Ogden, K. 1986. Musky/Club contributions to muskellunge management and the sport fishery. American Fisheries Society Special Publication 15:326-328.

Porter, L.R. 1977. Review of selected literature on muskellunge life history, ecology and management. Minnesota Department of Natural Resources, Special Publication 119. St. Paul.

Spangler, G.R. 1968. Angler harvest and mortality of *Esox masquinongy* in Pigeon and Sturgeon Lakes, Ontario. Journal of the Fisheries Research Board of Canada 25:1145-1154.

Strand, R.F. 1986. Identification of principal spawning areas and seasonal distribution and movements of muskellunge in Leech Lake, Minnesota. American Fisheries Society Special Publication 15:62-73.

A Case History of Voluntary Catch and Release of Muskellunge with Management Implications for Fishery Programs

RONALD J. DENT, JR.

Missouri Department of Conservation
1104 S. Grand St.
Sedalia, MO 65301

ABSTRACT

Many muskellunge fishing clubs, realizing potential overharvest of muskellunge, have promoted voluntary catch and release of these fish. At Pomme de Terre Lake, Missouri a local chapter of Muskies, Inc. (MI) established and promoted a voluntary release program for legal-size muskellunge. Their efforts, in cooperation with the Missouri Department of Conservation, resulted in the release of a substantial percentage of caught muskellunge.

This voluntary release of muskellunge contributed to a sustained quality fishery in the face of increased fishing pressure and a limited stocking program. The promotion of such release programs, in cooperation with muskellunge fishing clubs, is recommended as a supplemental management tool which can improve the quality of a muskellunge fishery by reducing harvest of legal fish.

Pomme de Terre Lake is the only large reservoir in Missouri which has a muskellunge stocking program in effect. This reservoir was chosen as the site for a muskellunge program because of its relatively small size and good habitat (flooded timber and aquatic vegetation), factors that should contribute to a reasonable return to the creel of stocked fish. The objective of the program was to provide Missourians with a trophy fishery utilizing a predator that could prey upon large non-game fishes such as gizzard shad, carp and various redhorses.

Muskellunge were first stocked into Pomme de Terre Lake in 1966. Since this time a total of 18,142 ten-inch fingerlings have been stocked. The stocking rate has fluctuated between 0 and 3,370 fingerlings and has averaged 1,007 fingerlings annually since 1966 (Table 1). Fry stockings were attempted from 1966 to 1969; however, these releases were discontinued after 1969 due to poor survival (Belusz 1978). Natural reproduction is extremely limited and stocking is needed to maintain the population. This lake is near the extreme southern range of the muskellunge (Crossman 1978). A 30-in minimum length limit, with a daily limit of one and a possession limit of two legal muskellunge, was implemented in 1967.

In 1975, the Pomme de Terre Chapter of Muskies, Inc. (MI) was formed. Membership has averaged 200 anglers since 1981. The chapter has promoted the release of muskellunge by awarding plaques at the end of the year to anglers releasing the most and the largest fish. Members fishing contests and an annual open tournament support the catch and release philosophy by offering the largest prize to the angler releasing the largest or most muskellunge during each event.

The objectives of this paper are to provide an example of a successful voluntary-release program and to show how the release of legal muskellunge can be a valuable addition to management programs, in which reducing overharvest is a principal goal.

Table 1. Stocking history of muskellunge in Pomme de Terre Lake, Missouri, 1966-1983.

Year	No. fingerlings stocked	Total length range (in)
1966	1,507	7.0-15.0
1967	835	7.0-15.0
1968	798	9.0-11.0
1969	370	9.0-11.0
1970	298	9.0-11.0
1971	0	--
1972	419	9.0-11.0
1973	0	--
1974	207	9.0-11.0
1975	298	9.0-11.0
1976	1,225	10.0-13.0
1977	2,393	11.0-12.0
1978	3,370	12.0-15.0
1979	1,186	12.0-14.0
1980	150	12.0-14.0
1981	835	12.0-14.0
1982	1,676	12.0-14.0
1983	2,575	9.0-12.0
Total	18,142	

STUDY SITE

Pomme de Terre Lake, a 7,800-acre flood control and recreational impoundment in west central Missouri, was built in 1961 by the U.S. Army Corps of Engineers. The dam impounds the Pomme de Terre River and Lindley Creek (Fig. 1). Physical and chemical descriptions of the lake are available in Hanson (1977), Belusz (1978) and Dent (1986).

METHODS

From 1976 to 1983, a roving creel survey was conducted on both arms of the lake from March 1 through November 30. Sampling methods and data calculations were made as described by Hanson (1977) and revised by Dent (1986). In 1983, the creel clerk was instructed to also ask all anglers fishing for muskellunge if they belonged to MI. This information was used to compare the numbers and sizes of muskellunge caught and released by members and non-members of MI.

From 1966 to 1975, most marinas kept records of the number of legal-size muskellunge caught by anglers. From 1976 through 1983, one marina (State Park Marina) also kept records of the number and length of legal muskellunge caught and released. No differentiation was made during these years for MI and non-MI members. From 1980 to 1983, this marina was also designated as the headquarters for all MI muskellunge catch and release records. These records, which included the angler's name, size of muskellunge caught, location of catch and lure type, and weather conditions were made available to the Missouri Department of Conservation (MDC) and supplemented the creel survey data. Data on released fish are expressed as a percentage of the fish caught that were released.

For three days after the 1982 and 1983 tournaments, a thorough search of the lake was made in an effort to find any muskellunge that had died after being released.

Modeling of the muskellunge population, catch and harvest was made to compare the effects of a fishery without a release program and one with a high release in order to demonstrate to fishery agencies the impact a release program can have on a muskellunge fishery. The model used was proposed by Clark (1983). Assumed values of annual survival for Pomme de Terre muskellunge in the model were Age I and II .85, Age III .80 and Age IV-XIII .50. The survival rates for Age I were similar to those reported by Belusz (1978). The annual survival rates for Age II-XIII were similar to those reported by Johnson (1975).

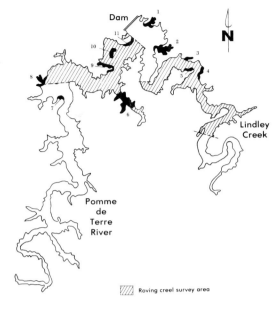

Figure 1. Roving creel survey areas and the 11 areas (numbered black spots) where 75% of the legal muskellunge were caught on Pomme de Terre Lake, Missouri, during 1981 to 1983.

The annual rate of exploitation of legal fish was 0.29, which was a seven-year average from Pomme de Terre Lake creel returns and data supported by spring electrofishing tagging studies conducted by MDC personnel from 1976 through 1983. FD68B Floy tags were used from 1976 to 1982 and monel-band dorsal tags were used in 1983. No rewards were given for returned tags. An assumed stocking of 1,000 Age 0 muskellunge each year was used, which approximates the 18-year average for Pomme de Terre Lake. The 78% release rate was the four-year (1980-83) mean release rate for Pomme de Terre Lake anglers. A survival rate of 0.80 for fish caught and released was assumed and approximates data reported by Beggs (1977) and Strand (1986) for angler-caught and released muskellunge. The model computed the number of legal muskellunge released, the number in the population, the catch and harvest of legal muskellunge and of trophy size fish, which was assumed to be 40 in and larger. The population, catch and harvest of legal muskellunge reported are for fish between the ages of four and thirteen.

RESULTS

The Muskellunge Fishery

Fishing pressure for muskellunge in Pomme de Terre Lake has increased steadily since 1976. Average fishing hours and trips from 1981 through 1983 were more than triple the pressure estimates made in 1976-80 (Table 2). This increase in muskellunge fishing pressure was related to the promotion of the fishery by MI and to increased fishing success which was evident in the catch rate from 0.008 in 1976 to a high of 0.036 in 1981. Increases in the estimated catch of legal muskellunge also occurred in the 1981-83 period averaging 0.15/acre, compared with the 1976-80 average of 0.05/acre (Table 2). The catch of legal muskellunge is usually dominated by three and four-year old fish, since they enter the 30-in, legal-size limit at that age.

Even though the legal catch has increased, the harvest of legal fish has remained low since 1979. Harvest from 1979 to 1983 averaged 0.03 muskellunge/acre/year, a 75% reduction in harvest compared to the legal catch during this same period. The increase in MI membership and the promotional program of the Pomme de Terre Chapter anglers contributed to the release of more legal muskellunge and the decline in harvest.

The catch rate of legal fish has increased substantially, as anglers learn more about the favorite feeding areas and seasonal movement patterns of muskellunge (Table 2). MI catch data from 1981 to 1983 revealed that approximately 75% of the recorded legal muskellunge caught for these years were caught at 11 locations on the lake (Fig. 1) According to creel survey estimates the return to the creel of stocked muskellunge caught as legal fish has averaged 29% since the first stocking occurred in 1966. Exploitation estimates determined from returns of non-reward, tagged legal muskellunge averaged 15% (range 0 to 33%) each year from 1976 to 1982 for Floy tags and 26% in 1983 when monel tags were used.

One of the primary objectives of MI is to promote the release of nontrophy size legal muskellunge. The opinion of what size fish is considered a trophy varies among anglers; however, most would agree that a minimum-size trophy fish at Pome de Terre Lake would be 40 in. Few legal fish were entered on the catch records of State Park Marina for 1976 and 1977; the release rates were 61% and 20%, respectively for those years (Table 3). During 1979, the release percentage increased to 54%. Starting in 1980, catch records at the marina were separated for MI and non-MI members (Table 4). The release percentage for MI members was 73% in 1980; this increased to 94% in 1983. A corresponding increase in the release percentage occurred for non-MI members for the same years, from a low of 14% in 1980 to a high of 73% in 1982 (Table 4). The release percentages calculated for creel survey data in 1983 are similar to those calculated for the marina catch information, 95% for MI and 71% for non-MI members. The steady increase in the release percentage at Pomme de Terre Lake parallels that in the national Muskies, Inc. members-only contests from 1970-82 (Davis 1983). The 1982 release percentages for MI members from Pomme de Terre Lake and the national MI members were both 93%.

The average size of muskellunge caught and released by all fishermen according to State Park

Table 2. The muskellunge fishery of Pomme de Terre Lake, Missouri, as determined from a roving creel survey 1976-1983.

	1976	1977	1978	1979	1980	1981	1982	1983
Estimated fishing pressure (no/acre)								
hours	1.8	0.91	1.3	1.8	2.4	4.9	4.4	5.0
trips	0.46	0.23	0.28	0.44	0.41	1.13	1.08	1.11
Estimated catch (no/acre)								
legal	0.02	0.02	0.03	0.07	0.10	0.18	0.15	0.12
sublegal	0.007	0.009	0.04	0.09	0.08	0.02	0.02	0.02
Estimated harvest (no/acre)	0.02	0.02	0.03	0.04	0.04	0.04	0.02	0.03
Catch rate (fish/hour)								
legal	0.008	0.011	0.017	0.028	0.027	0.036	0.032	0.020
sublegal	0	0	0	0.016	0.019	0.002	0.003	0.002
Mean length caught (in)								
legal	37.3	34.9	38.3	35.8	34.8	35.6	36.4	37.4
sublegal	24.8	18.3	20.9	20.4	23.8	24.9	18.1	21.5

Marina catch records for 1976-83 fluctuated between 32.1 and 36.8 in and averaged 35.9 in (Table 3). Average sizes of released muskellunge at Pomme de Terre Lake were similar to those released in the national MI contests (Davis 1983). No significant differences were evident in the average size of muskellunge released by MI or non-MI members at Pomme de Terre Lake.

Table 3. Number and average length of legal muskellunge caught and released by all anglers entered on the catch records of State Park Marina, Pomme de Terre Lake, Missouri 1976-1983.

	Number caught	Number released	Release percentage	Total avg. length (in) of released muskellunge
1976	18	11	61	34.3
1977	10	2	20	36.8
1978	25	9	36	32.1
1979	54	29	54	33.7
1980	87	47	54	34.5
1981	237	196	83	35.8
1982	197	175	89	36.8
1983	125	111	89	36.5
Total	757	580	--	35.9

Table 4. Comparison of the release percentages for MI and non-MI members from Pomme de Terre Lake, Missouri, based on catch records of State Park Marina and creel survey data 1980-1983.

	Marina catch records			Creel survey
	MI (%)	Non-MI(%)	Combined (%)	Combined (%)
1980	73	14	54	68
1981	91	54	83	80
1982	93	73	89	95
1983	94(95)[a]	69(71)[a]	89	(86)[a]

[a](Roving creel survey data release rates)

Due to recruitment of fish into the 30-in legal-size limit a higher percentage of smaller than larger muskellunge are released. However, the multiple catch and release of all sizes of muskellunge has resulted in more bigger fish in the population. Multiple catches of tagged muskellunge reported to MDC by anglers support high survival rates of caught and released muskies. As a way of evaluating the effectiveness of a release program the data were separated by the percentage of fish released of the total number caught within individual size groups (Table 5). These data are especially useful in demonstrating the increase in the release rate of larger muskellunge, as anglers select increasingly larger fish for personal trophies.

A high percentage of the 30-32 in muskellunge were released each year, and for the last 7 years of records (1977-83) the release percentage averaged 83% (Table 5). In the last three years over 90% of all muskellunge caught in this size range were released. Similar increases in the release percentage of fish between 33 and 40-in size groups were evident. The most striking evidence of the esteem that Pomme de Terre anglers have for muskellunge is the fact that in 1982 and in 1983, 80% or better of the muskellunges caught over 40 in were released; this includes three muskellunge over 46 in. Apparently, as anglers become more experienced at catching muskellunge their minimum size for a personal trophy also increases.

Survival estimates of muskellunge caught and released by anglers are not prevalent in the literature. Beggs (1977) discussed the physiological changes in pH, oxygen consumption, carbon dioxide levels and lactic acid concentrations of 25 muskellunge caught by anglers in Ontario. She concluded that delayed mortalities of caught and released muskellunge could result from acidosis and oxygen starvation caused by high lactic-acid concentrations, when a fish is played too long. Nine of

Table 5. The percentage of legal muskellunge released based on the total number caught within each size group. These data are based on the Pomme de Terre Lake catch records kept by State Park Marina and include both MI and non-MI members.

Size Group (in)	1977	1978	1979	1980	1981	1982	1983	Weighted $\overline{X}$
30-32	25	63	70	65	93	90	100	83
33-34	33	50	50	47	82	91	84	74
35-36	0	33	40	75	90	82	87	87
37-38	0	0	0	40	87	91	100	85
39-40	50	0	67	20	50	94	86	72
>40	0	0	22	33	42	83	80	48
Total number caught	10	23	54	87	232	195	129	730

thirty (30%) angled-muskellunge died during the study, and two delayed mortalities occurred after the fish were released back into the lake. The 70% survival rate for these fish was considered good under stressful test conditions that included hourly cannulation of blood samples and holding for up to 87 hours. Strand (1986) reported a survival rate of 100% during a 17-day period for fourteen muskellunge caught by anglers and surgically implanted with radio-transmitters on Leech Lake, Minnesota.

Observations during a three-day search of Pomme de Terre Lake, following 2.5-day muskellunge tournaments in 1982 and 1983, support the high survival rates reported by Beggs and Strand. The total number of legal muskellunge caught and released during the tournaments were 22 (82%) in 1982 and 15 (82%) in 1983. Two dead muskellunge were found in 1982 and no dead fish were found in 1983, which resulted in observed survival rates for released fish of 91% and 100%, respectively.

In Pomme de Terre Lake, the majority (67%) of the legal muskellunge are caught during two periods, May 1-June 15 and September 15-October 31. The observations for survival of released fish from the 1982 and 1983 Pomme de Terre Lake tournaments, and those reported by Beggs and Strand, are probably indicative of average survival of angler-caught and released muskellunge during these fishing periods.

Based on modeling predictions, the number of legal muskellunge in the lake would increase by 549 fish (78%); the total muskellunge catch would increase by 311 fish (71%), but the legal estimated harvest would decrease by 271 fish (62%). The catch of 40-in and larger muskellunge would increase by 61 fish representing an eightfold increase in trophy-size fish with the 78% release rate (Table 6). The modeling predictions of catch correspond to increases in the catch of legal muskellunge, and of 40-in and larger fish, at Pomme de Terre Lake from 1981 to 1983. The magnitude of the decrease in the harvest and subsequent increases in the population and catch are significant and are indicative of the changes that should occur with a release program. These data should provide incentives to other fishery agencies to promote catch and release. The increase in the trophy catch should help promote the catch and release concept to muskellunge anglers.

MANAGEMENT IMPLICATIONS

The threefold increase in muskellunge fishing pressure noted in Missouri during the 1980's is similar to increases observed in other states including: Wisconsin, where the number of muskellunge anglers doubled to 220,000 in the 1980's (Anon. 1982); Lake St. Clair Michigan, which had estimated averages of 1.0 million angler days/year in the 1960's and 1.5 million in the 1970's (Porter 1977); and in Cave Run Lake Kentucky, where 0.1 muskellunge trips/acre in 1975 increased to 1.3 trips/acre for the years 1976-79 (Axon 1981).

Table 6. Comparisons of the predicted legal muskellunge population and fishery at Pomme de Terre Lake, Missouri, from a model with a 78% release rate and no release.

	Released %	
	78	0
Population	1256	707
Catch	746	435
Number released	582	0
Harvest	164	435
Catch ≥ 40"	69	8

Reduced numbers and sizes of fish were noted as symptoms of overharvest, which became prevalent in many muskellunge waters as fishing pressure increased, and lower rates of natural reproduction were documented (Porter 1977). Reductions in the harvest of muskellunge became an important tool for muskellunge management. Miller (1983) reported that, in a survey of 19 states and three Canadian provinces that had a muskellunge management program, 18 states and one province were using minimum-size limits ranging from 24 to 36 in. The rationale for harvest regulations indicated by more than 70% of the agencies responding to his survey were: prevention of overharvest, protection of immature fish and promotion of a trophy fishery. In addition, the collection of length-frequency data on released muskellunge could be a source of information previously overlooked by fishery agencies to supplement a lack of creel or population data. A voluntary release program meets all these objectives without penalizing the novice angler from keeping his "first legal," which many times may be the largest fish of his career. At the same time, the benefits of releasing nontrophy-size fish will be appreciated by the avid muskellunge angler. The trophy aspect of a muskellunge program should be promoted or publicized by encouraging release of all nontrophy-size fish (Porter 1977).

Improvements in handling techniques for angler-caught muskellunge are needed, along with more studies to better document survival of angler-released fish under various environmental conditions and handling techniques. These data are needed to convince anglers that muskellunge can survive after release.

In 1982, a MDC brochure was prepared and distributed at sport shows, public meetings and MI functions; it includes a section promoting the release of legal muskellunge, as well as proper release methods. MDC is planning a program that will recognize anglers releasing muskellunge 40 in and larger, with the objective of stimulating increased release of muskellunge that are not personal trophies.

Economic impacts of muskellunge programs have been estimated by three states within the northern range of the muskellunge. In New York, the Chautauqua Lake fishery is valued at $1.8 million annually (Mooradian and Shepherd 1973); the Minnesota fishery at $13 to $26 million annually (Porter 1977); and in Wisconsin there was an estimated expenditure of $19 million annually by anglers in 1960 (Oehmcke 1969). One published economic study within the southern range of the muskellunge was completed by Belusz and Witter (1986); they estimated the value of the Pomme de Terre Lake muskellunge fishery to Missouri to be $479,000 in 1981. High-release rates of legal muskellunge can increase these overall values by providing anglers with opportunities to catch fish more than once.

An example of the cost effectiveness of a successful release program is available from data from Lac Court Orielles (Johnson 1975) and presented in a Muskies, Inc. brochure (Latvaitis, no date). According to those data each muskellunge released in the 30 to 36-in size range equals thirteen fingerling muskellunge that would not need to be stocked. The average cost per 10 to 12-in fingerling for six states and one province equals $5.56 (Ragan et al. 1986). Therefore, the total amount of money saved for each 30 to 36-in muskellunge released equals $72.28 in fingerling costs. For Pomme de Terre Lake, for the years 1976-83, 580 legal muskellunge were released; so the total savings for these fish, using the cost figures of Ragan et al. (1986), would be $41,922. With higher release and good survival rates more muskellunge could be available for stocking new waters.

With increased fishing pressure for muskellunge, rising hatchery costs and budgetary constraints, fishery agencies should not overlook management options that can lower harvest rates of legal muskellunge and recycle these fish into trophy-size fish. Surprisingly, the majority (14 states and one province) of the 19 states and three provinces responding to a questionnare by Ragan et al. (1986), view catch and release fishing as a viable management option, yet few states include a voluntary release program in their publicity or management programs. A large number of catch and release fisheries could be produced and serve more anglers than high-harvest fisheries which do not support release.

The catch and release program at Pomme de Terre Lake has decreased the harvest of legal muskellunge and maintained a quality fishery in the face of increased fishing pressure and a limited stocking program. The benefits of an agency-sponsored and highly publicized catch and release program can be a useful and cost-effective tool that can supplement fisheries management programs. The promotion and participation in catch and release by members of fishing clubs provides valuable needed support in these efforts.

ACKNOWLEDGEMENTS

I appreciate the assistance of Lee C. Redmond, Ken R. Perry and Kevin R. Richards for their critical review of the manuscript and John E. McPherson, Kevin R. Richards, Pam Haverland and Clyde de Jarnette who assisted with data collection and analysis, and Melanie A. Benitz for her patience and understanding in typing the manuscript. I wish to thank all the members of the Pomme de Terre Chapter of Muskies, Inc. for cooperating with the program and to Roger and Kathy Metz of State Park Marina who provided the catch records.

REFERENCES

Anonymous. 1982. Wisconsin Muskellunge Waters. Wisconsin Department of Natural Resources, Publication 1-3600, Madison, Wisconsin.

Axon, J.R. 1981. Development of a muskellunge fishery at Cave Run Lake, Kentucky, 1974-1979. North American Journal of Fisheries Management 1:134-143.

Beggs, G.L. 1977. Some physiological consequences of angling stress in muskellunge, *Esox masquinongy*. Master's thesis, University of Toronto, Toronto, Canada.

Belusz, L.C. 1978. An evaluation of the muskellunge fishery of Lake Pomme de Terre and efforts to improve stocking success. American Fisheries Society, Special Publication 11:292-297.

Belusz, L.C. and D.J. Witter. 1986. A survey of muskellunge angler characteristics, expenditures and benefits. Pages 39-45 *in* G. Hall and M. Van Den Avyle, editors. Reservoir Fisheries Management: Strategies for the 80's. Proceedings Reservoir Fisheries Management Symposium, Lexington, Kentucky, June 1983.

Clark, R.D. 1983. Potential effects of voluntary catch and release of fish on recreational fisheries. North American Journal of Fisheries Management 3:306-314.

Crossman, E.J. 1978. Taxonomy and distribution of North American esocids. American Fisheries Society, Special

Publication 11:13-26.

Davis, B. 1983. Muskies, Inc. members only fishing contest results 1970-1892. Muskie 17. St. Paul, Minnesota.

Dent, R.J. 1986. Methods and parameters used in evaluating bass length limits on Pomme de Terre Lake, Missouri. Pages 65-72 *in* G. Hall. Reservoir Fisheries Management Symposium, Lexington, Kentucky. June 1983.

Hanson, W.D. 1977. The tailwater fisheries of Lake of the Ozarks and Pomme de Terre Lake, Missouri. Missouri Department of Conservation. D-J Project F-1-R-25. Final Report.

Johnson, L.D. 1975. How many muskies aren't there anymore? Wisconsin Conservationist Bulletin 40:20-21.

Latvaitis, B. (No date). Why release muskies? Muskies, Inc., St. Paul, Minnesota.

Miller, M.L. 1983. The status of muskellunge management in North America (results of a mail survey conducted during fall 1981). Division of Wildlife, Ohio Department of Natural Resources, Columbus, Ohio.

Mooradian, S.R. and W.F. Shepherd. 1973. Management of muskellunge in Chautauqua Lake. New York Fish and Game 20:152-157.

Oehmcke, A.A. 1969. Muskellunge management in Wisconsin. Wisconsin Department of Natural Resources Report No. 19.

Porter, L.R. 1977. Review of selected literature on muskellunge life history, ecology and management. Minnesota Department of Natural Resources, Special Publication 119. 81 pp.

Ragan, J., T. Steinwand, and G. Van Eckhout. 1986. A synopsis of results from a questionnaire on muskellunge - resource agency survey. American Fisheries Society Special Publication 15:370-372.

Strand, R.F. 1986. Identification of principal spawning areas and seasonal distribution and movements of muskellunge in Leech Lake, MN. American Fisheries Society Special Publication 15:62-73.

PANEL 2 - PRIVATE SECTOR PARTICIPATION
The Role of Anglers and Private Organizations in Muskellunge Management

ARTHUR A. OEHMCKE, CHAIRMAN

ABSTRACT

A conciliatory state is set on which anglers, managers and professional fishery workers can unite. Four basic suggestions for achieving harmony between these groups and the responsibility of the angler and the professional were underscored: share common goals, listen and be open-minded, exercise patience and have mutual respect for professional and laypersons opinions. The presence of increased numbers of elite anglers and their sophisiticated equipment poses a management dilemma which can only be solved by supportive fishery professionals. Musky clubs and organizations were non-existent prior to 1952 when the first club was formed in Wisconsin. Phenomenal growth since then is revealed in national survey data of Muskies, Inc. on 50 currently responding musky clubs.

Organizations and anglers, especially in Wisconsin, have made substantial contributions to the protection and improvement of musky populations and fishing. Although New York has required musky anglers to report their catches annually, the rule is ignored by 80% of these special licensees. Volunteer angler assistance used during the past 10 years indicates that state agencies need to assign biologists to educate and coordinate anglers in the purposes and benefits from cooperating in required study methods. A similar threat among all panelists was the requisite that musky clubs and anglers can best serve their sport by demanding and supporting scientifically founded management focused on impinging social, economic and harvest objectives.

The Angler's Role

ARTHUR A. OEHMCKE

Musky Clubs Alliance of Wisconsin, Inc.
Spooner, Wisconsin 54801

The basic guide for the role of the angler was laid down in a book written by Izaak Walton, who died in London at the age of 90 on December 16, 1683. Entitled *THE COMPLEAT ANGLER or The Contemplative Man's Recreation,* it was first published in 1653, and since then has been reprinted in 162 editions. The angler's role, from documentation in the 17th century until early in the last decade, has been largely exploitive. In addition to the pleasureable aspects of this contemplative sport, a full creel or bag limit has, until recently, always been the goal. Now, 300 years after Walton's pronouncement, the musky angler's role is being revised from purely harvest to recycling; from filling the creel to catch-and-release; from fishing for food to fishing for fun. The role is now an effort to maximize the angler's gratification by a switch from quantity to quality.

But what Izaak Walton said, about the time the USA was colonized, is still applicable to the musky angler: "He that hopes to be a good angler must not onely(sic) bring an inquiring, searching, observing wit, but he must bring a large measure of hope and patience and a love and propensity to the art itself."

To paraphrase Ernie Swift, former Director of the previous Wisconsin Conservation Department, fish management is "a subject wrapped up in a complexity of human emotions, and, at times, misguided energy. It is analagous to Josh Billings' observation when he said 'It ain't that people are ignorant, it's just that they know so many things that ain't so."

Feeling the need to promote a better angler understanding of a complex fishery management program, this panel drew upon the knowledge and expertise of a journalist, a sportsman, a government agency administrator, an aquatic bioloist and an angler's group representative for clues and direction. It was hoped this would stimulate sound thought and action on all fish management issues, not just those affecting the management of muskellunge.

Responsibilities of the Angler and the Professional

DOUGLAS STANGE

Managing Editor, THE IN-FISHERMAN magazine
Brainerd, Minnesota 56401

Clearly, the future of musky fishing faces problems. While no problems are beyond resolution, they need the attention of concerned anglers, private organizations like Muskies, Inc. (MI) and, of course, professional fisheries personnel. But these groups must work together. There must be cooperation and communication at all points. With this in mind, let's consider these suggestions.

Suggestion #1: Anglers and fisheries personnel, please constantly remind yourselves that you're in this together and that you share common goals.

Anglers -- realize that fisheries professionals have dedicated their lives to proper management of our fisheries. Professionals have a serious love for, and desire to maintain, fishing and fish populations. Their reasons for choosing management techniques are usually based on this desire. This is not to say that fisheries professionals are always correct. But whatever is done, it's intended for the good of the fisheries and fishing.

Fisheries professionals -- realize that anglers, too, usually have the same goal as you. Even though an "education gap" exists between your groups, which may cause communication problems, please don't remain aloof. Try not to assume a know-it-all position which results in trying to solve problems all by yourselves. The public can be a problem, but realize that most fisheries goals are best accomplished after anglers understand goals and are enlisted in an attempt to achieve them.

Suggestion #2: Take time to listen to each other.

Anglers -- you violate this piece of advice too often. Don't assume that your on-the-water experience gives you all the answers concerning the technical problems involved in managing a fishery. Management techniques are too often suspect for no other reason than that they are already common. Everything can't always appear logical. Listen and be open-minded. Question in a sincere manner. If you must disagree, try to disagree agreeably. If biologists are on the defensive all the time, nothing will get done. Biologists mean only the best for our waters. Our goals are the same; we're in this together.

Fisheries professionals -- angler conflict will always be a part of your work. Try to keep it from hardening your hearts and ruining your overall relationship with them. Like the school teacher or preacher, your job seems hopeless at times, but you have to keep plugging away. However, fisheries science is a fledgling by comparison to others. The simple fact is, you don't know it all, yet. Acknowledge mistakes along with your accomplishments. You do have most of the knowledge concerning management, and it will ultimately be your job to make the decisions, but you can learn from anglers, too... Take time to listen.

Suggestion #3: Be patient.

Anglers -- Most of you know only too well that your fishing knowledge was not attained instantly. The same will also be true of conversation-related knowledge. Listen, read and learn. If you are patient and sincere in your desire to learn, it will happen. If you assume to know all the answers, it can't. If fisheries biology is a fledgling science, then give it time. Help it develop by providing your support.

Fisheries professionals -- patience is needed on many fronts, but above all be patient and unfailing in attempting to bring more advanced fisheries education to anglers. The history of most worthwhile endeavors suggests a substantial time period before something new catches on. You'll often be faced with a long-term effort before changes occur. Don't give up.

Suggestion #4: Keep a perspective on who you are and what you know.

Anglers -- The fact of the matter is that you do not have years of specialized training with direct emphasis on learning about fish population dynamics, genetics, etc. However, because you use enacted regulations and observe their results, certainly you can and should convey whether or not you think they are working. Many of you have extensive fishing time on a particular body of water (or perhaps on many different waters), and you can offer a viewpoint that can't easily be acquired by a biologist.

Fisheries professionals -- as mentioned before, guard against raising your defenses because of angler criticism, and guard against allowing yourself to become the know-it-all, aloof professional, and above all, your display of patience is part of the fisheries professional you are. Keep considering that groups of anglers can provide a viewpoint unique from your own. Of course, you know that 50 different anglers can, and will, often suggest 50 different ways to handle a problem. Still, you must realize that you are generally better off for having heard those different ideas. Now, because you are the one with most of the technical knowledge, and because it's ultimately your job to make the decisions, you must blend those ideas with accepted scientific facts and come up with workable solutions to problems.

Anglers -- it's your job to initiate biological inquiry by mentioning problems you observe. Like a law officer, biologists are always looking for problems that need solving. But just as law officers cannot identify every problem without the help of concerned citizens, so too must anglers help fisheries professionals. However, their observations are only a small part of a larger picture that must be filled in by science.

We are discussing the role of anglers and private organizations and their relationships to fisheries professionals. The role can't be played properly without communication and cooperation. This symposium was testimony to the communication and cooperation efforts of anglers, MI, and fishery organizations. However, things are not perfect. Let's dwell for a moment on "the reliability of angler information."

The fact is that angler information is likely to be biased. The reasons are no revelation. Anglers usually prefer that the region where a musky is caught not be known; if the region's known, they'd rather the lake wasn't; if the lake's known, they'd rather the weedbar wasn't; if the weedbar's known, they'd rather the lure wasn't; if the lure's known, they'd rather the exact retrieve wasn't, and on and on.

Frustration and intense pressure to release muskies may also influence the reliability of catch-and-release, and catch-and-keep, information. "All fishermen are liars except you and me, and I'm not so sure about you?". That's not true, but the frustration of fishing for 40 hours before hooking a musky, and then losing the musky several feet from the boat, may often result in a fish counted as a caught-and-released fish. Perhaps it should be counted!

Likewise, intense peer pressure to release fish may result in releasing terribly injured fish that are going to die. Even though released, these fish are not true releases. On every front, from lure types and techniques used, to length and weight of fish caught, to whether fish are released or not released, we can expect this bias in musky fishing information from anglers.

But reliable data is absolutely necessary in order for management decisions to be made. Ways must be found to work with, or around, bias. Mandatory-reporting regulations may work in some instances. Another approach may be to use smaller, but more reliable samples. In other words, working with a small number of anglers on a personal level, and with the promise of confidentiality, may result in a smaller, but more accurate sample, and thus more accurate information.

And what of the future of musky fishing? Is there danger that our fishing has become, or will become, too advanced? Will stricter regulations be necessary to protect musky populations?

There's no question that angler ability to harvest fish has grown tremendously. Boats, motors and tackle have all been refined and are used by an ever-increasing army of anglers.

While most musky catches are correctly attributed to a small group of ardent musky anglers, today's technology alone allows even mediocre anglers to catch an occasional musky. These catches are usually killed. The sheer number of anglers killing an occasional catch may have an impact on musky populations in many waters.

But what of the small group of elite musky anglers who indeed catch the most muskies? Actually, this group is no longer small. A growing number of elite anglers target almost all their fishing effort at catching the limited number of muskies available in any environment.

Today's elite angler has also benefited from boat, motor and tackle technology. Further, today's advanced angler understands intricate details of musky location and is well versed in fishing methods that make muskies strike. While these anglers release most of their fish, there is always some mortality associated with fishing. Some fish are still killed either for food, photos or as trophies, and there's always some post-release mortality. And another consideration is that much of the kill pressure by advanced anglers is directed at the big fish segment of a musky population. This part of the population can least withstand harvest.

Serious musky fishermen are faced with a dilemma. Musky fishing is so wonderful that everyone should enjoy it. But the more people who enjoy it, the fewer muskies there are to go around. Should MI strive to increase membership and the number of ardent, informed musky anglers at a time when there really aren't enough muskies to go around? It's a tough but altogether appropriate question.

The goal of catch-and-release efforts is to increase musky size and numbers. This is indeed a worthwhile goal. However, while things would be much worse without a serious catch-and-release effort, the effort seems to be overwhelmed by sheer odds. Incidental catch/kills by anglers seeking other fish, plus occasional kills by serious musky anglers who usually release fish, seem to overwhelm the catch-and-release effort. It isn't that catch-and-release efforts aren't doing any good; they definitely are. Still, it seems like our goal isn't achievable only through voluntary catch-and-release efforts. As a result, in many waters -- probably most -- we are faced with reduced average fish size, fewer fish and a slim chance of ever taking a truly huge fish.

There isn't likely to be a retreat from the trend toward more advanced fishing and more anglers. New regulations are necessary. I believe innovative management approaches would be well received by ardent musky anglers, although perhaps not by the general fishing public.

Will new regulations need to be on tackle and technique? At this point history seems to indicate that anglers always refine the techniques available to them, and are always able to affect musky populations. Therefore, no matter what new regulations are tried, the finest among many answers remains ever-increasing angler knowledge of what it takes to conserve muskies.

We're all in this together; we have the same goals. We must be patient and we must realize who we are, and what we can do to insure a musky fishing future.

Anglers it's our responsibility to provide reliable information to fisheries personnel; proper management depends on it. We should also provide plenty of support for management personnel and the management process. Finally, let's consider new regulations that may help to conserve muskies. But no matter what the regulations, the basis for attaining our goals rests with both anglers and fishery professionals committed to, and working toward, a musky fishing future.

Musky Club Contributions To Muskellunge Management and the Sport Fishery

KEITH OGDEN

Muskies, Inc.
Cavalier, North Dakota 58220

The muskellunge is a truly amazing fish, sought by a remarkable breed of fishermen. These anglers are so intent with their love affair that they begin to wonder if they live alone with this mania. In search of musky affection, two maniacs become ten. Ten fishermen become a group having fun, sharing musky skills and musky waters where each has been successful. Ten becomes twenty, with two convincing four that there is more to the sport than just catching the freshwater king. The group expands to one hundred and now has the manpower to press on for the betterment of the sport fishery. Money can be raised, fish stocked, legislation enacted to preserve the musky. But one of the best of these efforts is to achieve public awareness that the release of nontrophy muskies must be practiced to insure the king's future.

Money is raised by this musky group and donated to state projects for stocking, telemetry and electrophoretic studies and many other worthwhile musky improvements. Recognition is given to members for the successful release of nontrophy muskies. These fish would now live to fight again and perhaps provide another hatch, a valuable asset to the fisherman's favorite lake.

Musky fishermen in other regions hear of this group's success and begin to form organizations throughout the musky range. Hundreds of club members bloom to thousands and an awareness of the musky and its followers is becoming known. Legislation pushed by these groups is passed by several states increasing the minimum legal length of musky, decreasing the bag limit from three fish a day to two; then one, and in some jurisdictions to only two muskies per year. Yes it happened! Power in numbers once again prevailed.

The growth of musky clubs and organizations has been phenomenal! A Muskies, Inc. (MI) survey has determined that approximately 11,700 musky fishermen have chosen to join forces to accomplish what one lone fisherman cannot. They have gone one step further than just concern about fishing; they have chosen to share their "time on the water" knowledge with others, to work toward benefiting the musky fishery and to educate the public on the value of the musky and its proper management.

MI has surveyed 50 responding musky clubs, which represent 84% of all musky clubs in North America. These organizations represent 9,825 concerned musky fishermen. Their accomplishments are numerous and they need to be recognized! These 50 groups range in size from 12 to 2,600 (average 196), with dues from $1.00 per year to $20.00 (average $15) per year. Most clubs have a monthly publication. Whether large or small, each has its own specialty. Together they have a remarkable story.

For most clubs there is an umbrella organization that assists with governing and directing them toward the betterment of the sport fishery. Muskies Canada, Inc. (MCI) and MI are the two most predominant national umbrella organizations. MI, the largest group, boasts nearly 4,000 members in 37 chapters covering 13 states; it was founded in 1966 by the late Gil Hamm. These two organizations each offer a monthly publication, national awards, public-relations direction, national release education programs for non-members, merchandise at discount non-profit prices, group clout for state, provincial and local issues, slide and film libraries for program use, national release tournaments and biannual board meetings that provide discussion and national business, sharing of ideas and camaraderie. MI offers a color cover, forty-page monthly magazine with feature articles, appropriate advertising, and a chapter report section that allows all chapters to relay their events, success and downfalls.

An alliance of musky clubs is taking place in many states. Twelve musky clubs make up the Musky Clubs Alliance of Wisconsin, Inc.; they offer solid evidence that numbers can achieve more than individual fishermen. Eight Illinois musky organizations have teamed to form the Illini Muskies Alliance.

Musky clubs have, in their own way, made many impacts on the sport fishery. Their various efforts are too numerous to mention. However, to prevent slighting any club inadvertently, mention is made here of only a few accomplishments that are too large to be overlooked.

Fifty active musky clubs are currently joining forces to accomplish state-wide and province-wide goals involving the future of the musky. They are using their numbers and voices to encourage legislation to increase the legal length of muskies and decrease bag limits. In Minnesota, Muskies, Inc. clubs have been directly involved in passage of laws increasing the minimum-legal length of muskies from 30 to 36 in and delay of season-opening dates to allow completion of musky spawning. In Wisconsin, the Musky Clubs Alliance has succeeded in urging and supporting its Bureau of Fish Management to establish a Memorial Day weekend musky season opening, a 32-in size limit, a musky waters classification booklet, and adoption of policies prohibiting motor trolling on small, prime musky lakes, where damage to populations and habitat can be confirmed.

Musky clubs are blowing the whistle on pollution matters affecting local lakes and rivers. In West Virginia, The Elk River Musky Club has been directly involved in dump-site cleanup along one of their prime musky rivers. The Musky Clubs Alliance of Wisconsin is credited with assisting the DNR in halting a potential discharge to the Wisconsin River. A Swedish company had applied for a permit to release silicic acid and sodium ions into this stream. The Alliance opposed this permit and requested intensive studies of the chemical's potential impact on aquatic life in the river. Since thousands of musky fingerling had been stocked in the river by the Alliance, the group asked the DNR to halt issuance of the permit. The company withdrew its request.

Involvement in rearing ponds has been an undertaking of seven musky clubs in the United States. A Minnesota MI chapter can boast the rearing and release of over 100,000 musky fingerling. Clubs in Wisconsin, North Dakota and Missouri have reached a musky stocking level of five-digits, for a total of 182,777 musky fingerling that have been stocked by clubs! The average length of fingerling muskies stocked is approximately four inches. The only private hatchery in the United States specializing in raising muskies for sale asks $0.60 per inch for fingerlings. Converting this figure, musky clubs can easily boast they have spent $438,664 toward stocking of this famed freshwater king!

In 1969, MI chapters from five states joined forces to purchase a fish hatchery in Minnesota. It was to be used for only musky hatching and rearing purposes. Before being resold the clubs raised approximately 40,000 fingerlings. The First Wisconsin Chapter of MI cooperated with its DNR is a unique rearing project in 1977 utilizing a sewage treatment pond. This club has raised over 50% of the privately stocked musky fingerling in Wisconsin from 1978 to present; it was the first organization in that state to tackle such a project. Another unique project was undertaken in Wisconsin by Muskies Today Ltd. The Consolidated Paper Company donated $12,500 to the Musky Clubs Alliance of Wisconsin, which transferred the funds to Muskies Today, Ltd. to raise hybrid muskies with a dry-pellet food process. It became the second club on record in Wisconsin to manage a rearing pond and the first to feed dry food to hybrids and achieve a high survival rate. A Minnesota chapter, Twin Cities MI, has assisted the Minnesota DNR with 7,000 man hours, mainly in stocking projects. Three Wisconsin chapters donated 1,000 to 5,000 manhours toward stocking efforts. Eighteen other North American musky chapters have donated 25,600 man-hours to assist our government fish agencies!! Using a $5.00 per hour wage, this could be converted to a sizeable $128,000 donation.

One of the surest ways of preserving the musky fishery is through release of nontrophy muskies. All 50 responding musky chapters are practicing the release of nontrophy muskies. Two of the newest chapters are currently at 100% release, the average is 85%. From known records, a total of 22,030 legal muskies have been released by club members.

We cannot overlook the value of the released fish itself. It is nearly impossible to put an appropriate price tag on a healthy 45-inch female musky. However, the average musky released was 36 in, and using the same $0.60 per inch purchase figure, we could easily convert these released fish to a donation of over $475,000 to our favorite lakes. Some state agencies have paid up to and exceeding $100 for prime brood stock!! Try that figure!

The following were uncovered as other sizeable dollar contributions from responding clubs. Muskies Today, Ltd. has invested $20,000 each year toward Wisconsin's musky fishery by purchasing automatic feeders, aerating equipment and feed for rearing purposes. A North Dakota musky chapter, Fargo-Moorhead, has donated over $50,000 to its DNR stocking and telemetry projects! Other clubs have donated funds toward dump site cleanup, stocking, telemetry studies, forage fish purchases to feed reared muskies, brood stock procurement, blood test studies, rearing pond funding and endless similar musky projects. Seventeen musky organizations have donated a total of $164,800 toward worthwhile state and provincial projects!! Many of the other responding musky clubs are actively assisting their natural resource agencies in ways that can be translated into dollars. An average of 512 man-hours per organization have been donated to state agencies in stocking operations, musky-tagging, creel-census studies and waste clean-up; just to mention a few.

Tagged muskies will give state management agencies valuable statistics such as age and growth rates, recapture frequency, home-range data and musky-population trends. Sixteen musky clubs have members involved in tagging legal muskies. To date, the participating clubs have tagged 1,988 muskies.

MI and MCI are believed to be the first organizations to promote public-release programs for muskellunge. Both organizations ask fishermen who have successfully released a legal musky to fill out a brief form which provides these organizations with valuable informations on the released fish and gives recognition to the fishermen. MCI gives the angler a release certificate and patch, while MI awards the angler a certificate, patch, musky information packed and, with the cooperation of the Berkley Tackle Company, a spool of high-quality line. These relatively new programs have attracted over 1,000 entries!

MI Chapters have also been involved with Minnesota's Shoe Pact strain investigation, which led to a discontinuance of stocking offspring from this strain where Mississippi-strain muskies are present. The Shoe Pac-strain muskies were found to be slower-growing, and field tests are now underway to determine if stocking efforts of private and state groups can be redirected to other worthwhile activities.

In 1965, it was revealed in the New York Fish and Game Journal that the Chautauqua Lake muskellunge fishery was worth $1,800,000 to the surrounding area. At today's present rate of inflation this value would now be over $4,000,000. Musky clubs are very much a part of this donation to their communities.

Anglers in today's musky-fishing society are becoming an aware group. The musky club allows for an atmosphere whereby its members can contribute to the fishery in many ways.

The Effectiveness Of Anglers' and Organizations' Contribution In Improving Muskellunge Populations

JAMES T. ADDIS
Wisconsin Department of Natural Resources
Madison, WI 53707

Are angling organizations and anglers contributing to the protection and improvement of muskellunge populations? Muskellunge fishing? What is their role? How do professional fisheries managers view the role of the angler; the angling organization?

I suspect that these questions are repeated time and again, by both anglers and managers, faced with decisions on where to best allocate the limited time, energy, and money available for improving musky fishing. In part, agencies have acknowledged the worth of the anglers' advice, by increasingly inviting their participation in developing long-range fishery plans. Fishery managers recognize that they must understand what anglers want and need from their fishing experience, before they can design effective programs to deliver "good recreational fisheries."

For their part, anglers continue to have a keen interest in affecting the ways professionals manage the fishery. They are increasingly being organized for both collaboration with agencies and for opposition to those agency policies they dislike. In my view, both collaboration and constructive resistance can be positive; but the key to this is how both the professionals and the activist angler feel about each other after the debate is finished and decisions are rendered.

Remember 40-60 years ago, when angling and the environment were under massive assault? Resource agencies were largely staffed by a patronage system that had little respect for professional resource managers. Entire fisheries were being destroyed by pollution, large-scale overharvest, and other abuses. Activist anglers as early as the 1930's, fought for the elimination of political, patronage-oriented resource programs. They demanded management based on resource principles founded in ecology and fisheries biology. Activist anglers of the '50's lobbied to establish the Federal Fish Restoration Act, better known as the Dingell-Johnson Act, to help pay for new fishery programs.

Armed with research findings, and a renewed awareness that the very future of recreational angling depends on an abundant supply of unpolluted water, anglers formed the core of what was to become the environmental movement of the 60's. Again, angler activists, wearing their environmental hats along with many new-found friends, pushed a series of legislation through U.S. Congress, setting as a national goal, the restoration of the nation's waters to a fishable and swimmable status by 1983.

Today we enjoy the chance to fish in many thousands of acres of lakes and thousands of miles of streams that have been restored to support fish and aquatic life. A feat beyond the imagination of most, a few decades ago, but demanded by those few who viewed the significance of water pollution to fisheries. Today's fishing opportunities, whether on the seacoasts, on the Great Lakes, or in inland waters far exceed the fishing available during the 1950's. For instance, in Wisconsin alone, the Upper Wolf River, the Wisconsin River, and the Fox River, once polluted and devoid of fish, now support thriving fisheries. Today's muskellunge angler not only enjoys fishing for musky in a vastly expanded range, but also has the benefit of highly-sophisticated tackle, boats, and scientific fishing techniques.

Musky anglers relish the chance to exchange stories, debate issues and be partners in shaping the future of musky management. Without doubt these folks are serious about their sport. They want management agencies to continue making scientifically sound decisions. They recognize that the increased angling demand for muskies may adversely affect the fishery, and, because they do, organized musky anglers increasingly advocate tighter restrictions on harvest and impose highly ethical standards of behavior on themselves. Certainly, this is a critical need, for as Aldo Leopold so appropriately stated, "mass use tends to dilute the quality of organic trophies like game and fish".

Few fishing pursuits depend more on the contemplative and aesthetic aspects of fishing than does muskellunge fishing. Muskellunge populations will never support mass harvest, but will provide challenge to many anglers if protected and managed. No greater tragedy could unfold than for the increased interest and demand generated by anglers becoming organized to lead to a de-emphasis of the aesthetic, contemplative and challenging aspects of musky fishing and an emergence of emphasis on mass use. To lose the challenge of musky fishing would amount to losing musky fishing. I have concluded from experience with Wisconsin's musky anglers that most recognize that these values often referred to as quality are essential to perpetuate this sport. I feel sure that musky club members will continue to foster high ethical standards and wise use of musky resources. Present emphasis on catch and release shows the concern over how club activities affect the future and the willingness of musky fishermen to sacrifice harvest today to ensure good fishing tomorrow.

In Wisconsin, we have been blessed with a strong collaborative working relationship between fishery managers and muskellunge anglers. Wisconsin musky clubs donate thousands of dollars and thousands of hours of personal effort to our fish management program. The following are some projects they supported and their significance to Wisconsin's fishery program.

Initially, the newly organized musky clubs oriented their donations toward enhancing and expanding our stocking program. Fish plants sponsored by musky clubs are made in collaboration with fish managers. Projects ranged from cooperative rearing ponds to the outright purchase of fish from private hatcheries. Wisconsin DNR usually provided eggs or fry to the private facility to ensure that strains were consistent with its own stocks, and clubs paid for rearing and distribution costs. Beyond this, grants were made to Wisconsin DNR to buy experimental equipment needed in our musky hatcheries and for improving forage operations. All these club-sponsored plantings were made in waters that could not be stocked because of state production limitations.

Clubs donated over $10,000 to research studies measuring the impact of angling pressure on Wisconsin musky populations. Wisconsin musky clubs are active supporters of other research activities relating to genetic strains and musky behavior. Although these studies have limited direct impact for anglers, their value to fishery managers in determining critical spawning habitat, preferred cover types and other facets of life history enhances managers' ability to protect and improve muskellunge fishing.

Musky clubs are cooperating with fishery managers to better define and address the needs and desires of the musky fishing public. Fishery science provides a basis for determining technically feasible management alternatives, but only limited insight into which objectives best meet the needs and desires of anglers. Simplistic application of fisheries science could fail to address ethical concerns, and could lead to ignoring of important issues, about which users might be most knowledgeable.

Wisconsin DNR often involves anglers in its long-range planning and rule-making processes. Frequently, managers' views of the fishery, along with future management objectives, are discussed with user-groups, prior to their being incorporated into plans. Wisconsin has an organized statewide Conservation Congress that advises the Natural Resources Board and the Department of Natural Resources on fishing rules. During recent years, the Musky Clubs Alliance of Wisconsin, collaborated with DNR fish managers and the Conservation Congress to increase the muskellunge size limit from 30 to 32 in and to move the season in the northern musky zone back nearly a month. The latter rule was easily adopted because of the direct effort clubs made to inform their constituents why pursuing this course of action was desirable.

Experience shows that anglers are capable and willing to work with government to protect and restore the environment. However, even with this past record of success, I have some concerns that anglers — like other citizens — are becoming less willing to consider the judgement of professionals. This trend, which emerged in the late 1960's and accelerated in the post-"Watergate" era, is fairly well documented in the popular literature. It is a situation that may shake the foundation of our previously effective working relationship. It results from a distrust of big government, and of the fragmentation of the fishing community into single-issue interest groups. Angler demand is increasingly framed in relation to specific, single-species concerns. For example, groups like BASS, Trout Unlimited, Muskies, Inc., the Great Lakes Sports Fishermen or Walleye Unlimited, generally emphasize single-species management concerns. In the past, broader-based views of groups like the Izaak Walton League or the National Wildlife Federation were more prevalent. Thus, men and women, who previously stood together to defend recreational angling in general, now compete for the limited dollars and human resources that fishery agencies have for managing fisheries. The resulting competitive atmosphere often puts fisheries professionals in the middle, where they become the focus for the frustration and disappointment of groups caught up in single-issue conflicts — conflicts which often demand mutually-exclusive management practices.

The increasing fragmentation of fisheries interests and a possible decline in the rapport between professionals and anglers could be a serious future problem. This trend is further exacerbated by the extraordinary growth in lay anglers' technical skills and scientific knowledge relative to their specialty fishery. In many cases, it seems that the angler is more knowledgeable about the movements and feeding patterns of fish in specific lakes, or streams, than the local fish manager. I'm sure that professional managers have felt threatened when debating an enlightened, excited and extremely confident constituent at a local club meeting. This can result in communication failures and tensions that heighten the chance for misunderstanding and conflict. These pitfalls should and can be avoided. Even the most informed angler seldom has the underlying knowledge, experience, and, hopefully, resulting wisdom that trained managers bring to their job. Fishery managers, though, must continue to acknowledge the credibility of the ideas of specialized anglers and use them in deliberations leading to agency decisions. Above all else, agency heads and fishery chiefs must continue to demand that agency professionals communicate openly and candidly with the public, so that citizens know they are playing a part in shaping decisions.

What does the future hold? Although I am optimistic that the future of recreational fishing will be bright, the road to that future will be rough. I am convinced that we, both professional fishery managers and anglers, must recognize that the very fabric of our society is changing and that we must change with it. I see, as John Naisbit (author of *Megatrends*) and others have pointed out, a re-emergence of the United States' commitment to the rights and desires of the individual, that will require greater decentralization and localization of government decision-making. I see a massive information overload that will overwhelm us all during the next decades. I see citizens increasingly organizing and splintering into special-interest groups that fight single-mindedly for their own issue despite how that may affect the broader interest of the community or of society.

If so much change and such potential stridency will exist, why am I still optimistic? Simply because we have a lot of things going for us; our successful past, a continued deep commitment to the environment by us all and a willingness to sacrifice the present harvest to ensure future use. To continue our commitment to a quality fishery, I recommend several things. First, I urge musky anglers to keep sight of their role as part of the larger community of anglers and environmentalists. Fishing depends entirely on a clean and healthy aquatic environment. Organized anglers need to continue their efforts to promote quality environmental values to the general public, for not everyone believes that what's good for the environment is good for society.

Secondly, I would emphasize teamwork with professionals and avoiding the common trap of using conflict and crisis—real or fabricated—to rally your cause, generate a constituency and build club membership. This is used too often and it tends to isolate anglers from the professional managers. Rather than being crisis-oriented, build relationships with fish managers and the nonangling public by seeking team efforts and collaborative solutions.

Of course, you can still expect a lot from your fish managers. For example, expect management programs to provide public service reflect excellence, show commitment to scientific management and result in good fishing. However, try to understand the tough job managers face, and don't expect them to solve all conflicts between a variety of users. Organizations may be better suited to working through some of these tough problems together than an agency. If the angling community can unite, we will all continue to succeed as we have in the past.

The angler's role is a complex one—as user, protector of the resource and judge of management's performance. However, I believe individual anglers and angler organizations have been enormously effective at improving sport fishing. Past efforts have resulted in an increase in professional management and a deep and forceful public commitment to a clean and healthy environment. Fishery managers will continue to view anglers as both equal partners and tremendous resources of knowledge and support for fisheries programs.

Cooperative vs. Mandatory Assistance From New York State Muskellunge Anglers

STEPHEN R. MOORADIAN
*New York State Department of Environmental Conservation
Olean, NY 14760*

New York State has required mandatory reporting from the musky anglers of Chautauqua and Cattaraugus counties since 1941. Chautuaqua Lake is the principal water within this area and the primary reason for instituting the reporting system. The department requires that the special musky license, with the appropriate catch data, be returned to the management office no later than ten days following the close of the season (October 31). Although the special license states that failure to return same by the 15th day following closure is a misdemeanor, many anglers totally ignore this threat. During the initial year of reporting (1941), 85% of the anglers submitted a report; by 1945, only 21% of these reports were returned to the agency. Through the 1950's and 1960's, license reports were returned at a rate ranging between 22 and 28%.

In 1974, with an all-time low of only 12.5% returned licenses, New York decided to change the special-license format. Beginning in 1975, the special license was printed on a post card with a postage-paid return address. The objective was to increase the return rate up to 50% of the total licenses sold. In 1975, the number of returns doubled from the previous year to 25.3%, but, to the dismay of the agency, by 1977 the return rate had dropped to 21.4%. During the past six years, the return of licenses has averaged only 20.4%.

Many wonder why a significant number of anglers fail to return their report as mandated in the New York Conservation Law (71-0921(8)). That section states failure to return the license is subject to a $200 fine, or 3 months imprisonment, or both fine and imprisonment. Needless to say, not a single case of enforcement of this law is recorded. There are three reasons for this non-action:

1) Approximately one-third of the special license holders are non-residents. These anglers can be apprehended only while in New York State, not at their residence.

2) With 6,000 residents holding a special license, apprehending all the violators is unrealistic. When thoughts of enforcing this law for a small sample of anglers was entertained, the department attorney disdained such action as "selective enforcement."
3) The cost of apprehending all the violators, both residents and non-residents, would far outweigh the benefits of the effort.

Probably a more effective action to encourage return of reports would be to withhold the new license until the individual has returned the previous license. Not only would this task be an administrative headache, but the volunteered angling data would be suspect when reported by hostile anglers.

Catch data on this mandatory report includes date of catch, length, tag number (if any) and water; also, the number of days fished, sub-legal and legal musky released. An enormous amount of data has been collected and cataloged since 1941, but its relative value in managing the Chautauqua musky fishery is questsionable. Examination of the reports and comparing angler catch/effort with the catch/effort in department survey nets for a given number of years suggests an inverse relationship. Also, there appears to be some variability in the angler's catch/effort and the percentage of anglers submitting reports. The angler's length data seem to be consistent with changes in legal-size limits, but these data do not segregate length by sex. The most significant contribution from the special license report is a reliable index of the angling catch, even though we cannot verify any relationship between the reported catch and a musky abundance. Because of the historic data base and the nominal administrative cost, New York intends to continue the special musky license program.

Within the past ten years, cooperative angler assistance has also been used for Chautauqua Lake. Utilization of the members of a newly-formed Muskies, Inc. chapter, who volunteered to maintain a diary of angling activities, resulted in a dismal experience. A training session recruited 18 diarists, but only 2 diaries were recovered following three written requests for their return. The major problem in this cause was the lack of true commitment by the chapter, which was a result of poor objectives established by agency biologists. A positive experience of cooperative diarists for Chautauqua was realized during the New York State Largemouth and Smallmouth Bass Study (1977-80). Anglers were asked to record their muskellunge catch data in addition to data for bass. Diary returns varied each year, but the result was a 70% to 75% response after four written requests and at least one telephone contact. This excellent return was the result of sound study objectives and a commitment by the B.A.S.S. Clubs. In general, cooperators comprised only 5% of the angling effort in a study water, but accounted for 50% of the tag returns from that water. Catch data was superior to that of untrained, mandatory responses.

The need to train anglers at a meeting, in lieu of written instructions only, is a necessary requirement. Written instructions are not read through by most cooperators, much less understood without questions. This became apparent in earlier cooperative programs and at training sessions by the number of meaningful questions asked. Topics covered at training sessions included:

1) Issuance of diaries; assignment of code numbers; securing correct names, addresses and phone numbers of cooperators.
2) Reading through the written instructions.
3) Covering diary entries; how to measure and record data; and the importance of unbiased records.
4) Probably the most critical subject — a listing and discussion of the objectives of the study and how it will work.

Even with this intensive individual training, about 11% of the diaries had mistakes through the four-year study period.

In comparing the two sources of angling activities, and their relative significance in managing musky populations, it is important to identify (1) needs, (2) resources, (3) commitment and (4) limitation.

(1) Prior to recruiting any angler assistance, the agency must have a plan outlining objectives as to what information regarding the fishery is needed to answer management questions. In most cases, a cooperative venture will provide useful data, but will not be on a par with data collected by agency biologists. If a tag recovery program is part of a study, then there would be a critical need for cooperative anglers to record their catch data. For Chautauqua Lake, it would have required 311 man-hours per year of electrofishing to capture the number of largemouth and smallmouth bass reported by our cooperators, which equates to 934 man-hours of agency electrofishing time over the study period.
(2) If agency resources are adequate, and biol-

ogists have the necessary time and funding, the use of angler assistance to collect data is not necessary. When some assistance is needed, then trained cooperators will collect superior data as compared to mandatory reporting.

(3) If the needs and resources mandate the assistance of anglers, the success of the program will depend on a commitment from both anglers and agency biologists. Anglers must be prepared to attend one or more training sessions and maintain accurate records throughout the study. The agency must assign a biologist as program coordinator to meet with angler cooperators as often as necessary to attain the study objectives. Without this total commitment, both the agency and anglers may be collecting useless data. When the Bass Study was first discussed, New York B.A.S.S. agreed to be the lead agent to recruit the cooperators. While some chapters fulfilled this commitment, many chapters failed to do any recruiting and the agency met the additional obligation.

(4) Whenever the needs and resources deem assistance by anglers as necessary, and if there is a total commitment, then a meeting is necessary for both groups to understand the limitations which exist. While some musky anglers are very capable of collecting and recording both simple and sophisticated data, others will never meet the requirements. Prior to any study, a trial year should be planned. Angler records must be checked throughout the year with corrections made as needed. Only those anglers that appear able to maintain satisfactory records should be used for the actual study. During the Chautauqua Lake Bass Study, 10% of the cooperators caught 60% of the basses; the top 30% of the cooperators caught 86% of the basses.

In summary, angler assistance can be helpful in completing a data need, but only if the agency cannot do it with its own resources. If anglers are used, then the agency and anglers must be totally committed and maintain a constant communication throughout the study period. The key to good angler assistance, whether mandatory or cooperative, is continuous communication from the agency, including written reports on study progress and data summaries.

Panel Discussion Summary

ARTHUR A. OEHMCKE

In outlining the present role of musky anglers and their organizations. I believe this panel has revealed at least three important developments relative to this fishery: 1) the unprecedented growth in the numbers of musky anglers and the establishment of hundreds of clubs dedicated specifically to muskellunge fishing and management; 2) the acceptance and practice of a new concept in angling involving the voluntary release of many large muskellunge; 3) support of governmental programs in muskellunge management through agreement on more restrictive regulations, a willingness to underwrite part of the growing management and research costs and increased cooperative assistance in such management activities as rearing, stocking, creel census, tagging and environmental watchdogging.

However, although it's possible to recycle muskies, it's doubtful if we can recycle the environment! So the future of muskellunge "lebensraum" will, of necessity, be focused on the protection of what habitat we have left, the maintenance of that which we are using and the rehabilitation of what we have lost—in that sequence. A huge, costly order but a high priority for all those sincerely interested in the future of the muskellunge; one that will require new behavioral patterns for musky anglers.

To help perpetuate the species, habitat and sport, anglers and their organizations can work with management in two ways, The first route would provide the basic medium for muskies to thrive. In the best US muskellunge populations there are fewer than 0.5 musky per acre over 30 inches in length. In many waters, there is virtually no natural reproduction and numbers of muskies are limited or rare. If reproduction can be re-established, it will be a significant breakthrough, but, in the interim, we will have to be satisfied with more expensive culture and stocking programs. The present outlook is very encouraging for bridging this developmental gap! Muskellunge anglers and their groups are demonstrating a willingness to aid in the improvement of their sport by forfeiting some of the privileges they previously had with liberal seasons, bag and size limits.

The second route, as the panel has underscored, must be one of solid support for scientifically-founded management through increased cooperation with managers and professionals. A great start has been made in that direction with this symposium; continued progress on this path can only be realized through conscientious commitment by the angler and a respect for the angler's contributions on the part of the manager or the professional. If the ardent musky 'hunters', who currently report a 100% catch-and-release record, are earnestly interested in promoting that philosophy and technique, they should readily accept new fishing methods and restrictions designed to reduce the overall muskellunge harvest and to favor increase in its numbers. True advocates of the procedures and goals for attaining prime, trophy-muskellunge fishing should not hesitate to file off, or flaten, all of the barbs on their lures; nor should they object to reducing all their double- and triple-gang hooks to singles, or throwing away all gaff hooks, if research evidence indicates rewarding returns to muskellunge populations from such procedures. Also, would these catch-and-release purists be willing to move on to much greater challenges for enhancing the life of *Esox masquinongy* by accepting a 45-inch, minimum-size limit, a lower season bag limit and the creation of "no-take" lakes? Sincere involvement with management must include an honest effort by the musky angler to assent to revolutionary changes that retulatory agencies deem essential for muskellunge fishery improvement and maintenance.

If the two routes of habitat preservation and angler-management team effort are followed, how much encouragement should be given to increased participation in fishing for this thinly-populated species? If limitations must be placed on future numbers of anglers to prevent over-exploitation of the muskellunge fishery, could their numbers be curbed more effectively by mandate or voluntarily? Some direction on this question is offered by a parallel situation in the management of the sturgeon fishery in Wisconsin. Slow maturation and low populations of sturgeon dictate tight regulation. The desired biological limits of the annual sturgeon harvest are controlled by the number of license tags issued and information gleaned from mandatory reporting at the end of a short (one month) season. This sturgeon fishery currently guarantees a sustained yield through a 45-inch, minimum-size limit and a single fish, annual bag limit. Under this conservative management policy, stocking has not been necessary to maintain adequate sturgeon numbers for an excellent seasonal sport. Unprecedented beneficial changes could also become a reality in muskellunge management, if the present attitude of acceptance of further harvest restrictions by muskellunge anglers continues.

PANEL 3 - FUTURE RESEARCH NEEDS

Improving Musky Management: A Review of Management and Research Needs

DAVID A. HANSON[1]

Department of Natural Resources
Box 309
Spooner, Wisconsin 54801

JAMES R. AXON

Department of Fish and Wildlife Resources
#1 Game Farm Road
Farnkfort, Kentucky 40601

J.M. CASSELMAN

Ministry of Natural Resources
Box 50
Maple, Ontario L0J 1E0

ROBERT C. HAAS

Department of Natural Resources
33135 South River Road
Mt. Clemens, Michigan 48045

ALBERT SCHIAVONE

Department of Environmental Conservation
317 Washington Street
Watertown, New York 13601

MARTIN R. SMITH

Lakeland Chapter, Muskies, Inc.
Box 45
Minocqua, Wisconsin 54548

ABSTRACT

The muskellunge is currently regarded as a trophy fish for angling in many regions. Reproductive success had declined in many waters, presumably due to deteriorating habitat, excessive angler harvest and increased competition from other species. Consequently, many muskellunge populations are maintained solely through hatchery propagation. The objective of this manuscript is to discuss areas of management and research which limit agencies from managing the species effectively.

Management agencies need to clearly define their management philosophy. This includes both why they manage muskellunge (to preserve a self-sustaining fishery, to increase angling opportunity, etc.) and setting quantitative management goals (i.e. defining desired population densities, age and size structure, etc.). Simply stating the agency philosophy as trophy management does not provide sufficient guidelines for field managers. Major research needs for improved management include identification of genetic diversity of muskellunge populations, expanded data bases on populations obtained from more-advanced sampling techniques, better knowledge of muskellunge-northern pike interactions, identification of spawning and nursery habitat requirements and evaluation of propagation and stocking techniques.

The long range well-being of this species is dependent on improved management practices. Intensive muskellunge management often places excessive demands on the resources of agencies. Private organizations can provide a significant role in obtaining solutions to current problems with their input of money, labor and data.

---

[1] Panel summarizer

Public sentiment about muskellunge has undergone a remarkable change in the past 100+ years. Its status was originally that of a commercially harvested "meat fish" (Crossman 1986). Subsequently, the muskellunge was regarded as a sport fish which was often ritually shot in the head prior to being brought into the boat (Graff 1986). Today, it is considered a trophy fish (Wingate 1986), which anglers frequently release alive to aid in the preservation of the sport (Dent 1986; Gasbrino 1986; and Richards and Ramsell 1986). Yet, despite its current favorable esteem, the muskellunge is one of the least understood coolwater species.

The interchange of knowledge and ideas at the symposium will certainly increase our ability to understand, appreciate, and manage muskellunge; however, as many participants have noted, many management problems remain to be solved. The panel's objective was to discuss areas of management and research which limit our ability to manage muskellunge effectively. Emphasis was placed on problems and issues which have a long-range impact over broad geographic areas, rather than on short-term or site-specific problems.

MANAGEMENT PHILOSOPHY

Wingate (1986) discussed past and present philosophies for muskellunge management. Management goals generally fall into one of the following catagories:

1) preservation of a self-sustaining natural population,
2) restoration of a formerly self-sustaining natural population,
3) managing a fishery for maximum recreational opportunity, or
4) increasing predator populations to control the abundance of a prey species.

Under categories 1 and 2, a healthy, natural population of muskellunge is the primary concern, and angler use is allowed only to the extent that it does not harm the resource. Management activities under category 3 frequently will manipulate the fishery to maximize opportunity, and stocking is considered an acceptable cost for doing business. Under category 4, the muskellunge (or hybrid muskellunge) is merely a tool in the management of some other specie(s). The first three goals are justifiable uses under the right circumstances. Management under category 4 has been frequently attempted, but rarely has a successful application been documented.

Management activities in most states and provinces are limited by significant gaps in information on muskellunge populations and their sport fisheries. Consequently, there is a tendency to incorporate research results and management practices from other geographic areas without assessing whether the results are applicable to that situation. This becomes especially evident when there are differences in what would seem to be the relevant management philosophy (i.e. preservation or restoration of a resource versus management of a fishery for recreational opportunity). Management agencies should use a systematic approach to clearly define their management goals for muskellunge; otherwise, increasing angling pressure on a limited resource will force agencies to inadvertently shift their decisions to reflect management under category 3.

The muskellunge has a limited geographic distribution (Crossman 1978) and is maintained by supplemental stocking throughout much of its range (Ragan et al. 1986 and Porter 1977). It is therefore imperative to preserve what is left of the "wild" muskellunge resource and not allow it to be available only as a domesticated fishery.

Most states/provinces manage muskellunge for "trophy" size fish. The definition of trophy size varies among anglers, depending on past experiences. While anglers have the luxury of making personal definitions, managers need a clear biological definition of "trophy management." Three possible definitions for consideration are:

1) management for the maximum density of 30-in fish;
2) management for a large percentage of individual fish reaching their potential ultimate sizes;
3) management to produce a world record size fish.

Management decisions under these definitions will differ greatly. For instance, under definition 3, genetic manipulation through selective breeding or bioengineering may be desirable, whereas, under definition 2, management will emphasize maximum production within existing genetic constraints. Management for a high density may lead to reduced growth rates and would conflict with definition 2 and 3. Limited angling pressure in the past has allowed adequate management in the absence of a clear definition of the program goal; however, the future will not be so kind. Managers will need specific program objectives concerning the management philosophy as well as management goals regarding the desired density, size and age structure of populations.

RESEARCH NEEDS

Genetics

The subject of identification of genetic stocks of muskellunge has been of historical interest and is discussed by Crossman (1986). Despite this interest, little work has been done on stock identification and evaluation of stock characteristics until recently (Hanson et al. 1983, and Koppelman and Philipp 1986), and both need a great deal more study. Once identified, genetic diversity of self-sustaining populations needs to be preserved. The impact of stocking practices on the genetic diversity in natural populations is generally unknown. This problem needs attention, particularily since the inability of hatcheries to handle large numbers of adults from many populations makes the muskellunge a candidate for inbreeding and inadvertent selective breeding.

Management agencies need to develop a policy for preservation of native stocks and their genetic suitability. Issues which should be addressed include documentation of present diversity, evaluation of the effect of management practices on diversity and identification of conditions under which selective breeding and genetic engineering should be considered.

Population Assessment

Management options are often limited by the inability to obtain sufficient samples of muskellunge at various life stages. Presently, the best sources of data are from samples of fall fingerlings by electroshocking, spawning adults captured in trap nets and the angler's creel. These data are frequently inadequate for detailed statistical analyses. The ability to sample populations needs to be improved so that management can be based on quantitative analysis. Current management is reflexive, responding after changes in adult stocks have occurred. Development of improved methods of sampling juveniles (ages I through pre-spawning adults) might allow better predictions of recruitment and its effect on adult abundance.

Muskellunge anglers frequently handle the species more than biologists; therefore, they should be asked to assist with the collection of certain types of data, especially some aspects of the life history. This type of cooperation already exists in some regions (Casselman and Crossman 1986); however, biologists must incorporate experimental designs which will minimize the potential bias from such sampling schemes.

Muskellunge-Northern Pike Interactions

There has been considerable concern regarding the decline of muskellunge populations following the establishment of northern pike (Dombeck et al. 1986; Inskip 1986; Inskip et al. 1986; Caplin 1982; Johnson 1981; Threinen and Oehmcke 1952). The expansion of northern pike into new waters can pose a real threat to many native muskellunge populations; yet, in many waters the two species seem to co-exist. The negative interaction seems most prevalent in waters where muskellunge were native and northern pike were introduced. There is evidence that in waters where both species are endemic, the two species may co-exist by spatial separation (Harrison and Hadley 1978). This separation may be facilitated by muskellunge spawning at greater depths (Haas 1978, Strand 1986) than in the shallow muddy bays where it has more often been described (Dombeck 1979 and 1986; Hess and Heart well 1979; and Scott and Crossman 1973). In waters where both species are endemic, stocks of northern pike and muskellunge may be adapted genetically for co-existence. More information is needed about the interactions between these two species where they co-exist naturally, as well as where they do not.

Self-Sustaining Populations

There has been a sharp decline in the number of waters supporting self-sustaining populations of muskellunge during the past century. Maintenance of the remaining ones should be a priority. Basic knowledge of muskellunge biology is lacking for most of these waters. Obtaining this information, along with habitat identification and protection are the major components needed for their management.

Biological information needed includes population assessment and factors affecting recruitment. Although not known, it seems probable that year-class strength for muskellunge is most dependent on spawning and nursery conditions. For many populations the location of the spawning grounds is not known, and these must be identified and protected. Dombeck et al. (1984) characterized environmental conditions of selected spawning grounds. There has been recent evidence of homing to a spawning ground (E.J. Crossman, pers. comm.). Craig and Black (1986) identified nursery habitat in southern Georgian Bay. These studies are a beginning, but we need considerably more effort on a broad geographic basis. Since relatively few waters in the U.S. contain self-sustaining populations of muskellunge, each population deserves special attention. For further discussion of habitat needs, see Casselman (1984) and Dombeck (1986).

Water quality and habitat in streams and rivers is often subject to greater changes than lakes. Knowledge of critical habitat for early life stages and adults is needed for both large and small streams.

Culture and Stocking

Methods of muskellunge propagation have been reviewed by Klingbiel (1986) and Graff (1978). Hatchery personnel anticipate that an artificial diet for propagation would greatly reduce production costs. But, successful rearing on artificial diets will require more information on the feeding behavior of young muskellunge as well as nutritional requirements (Orme 1978). Also, the survival and behavioral characteristics of stocked fingerlings reared on artificial diet should be compared with minnow-reared fingerlings prior to large scale use of artificial rations.

Hatchery personnel usually prefer to propagate fingerlings by tank or trough culture rather than by pond culture. In one stocking evaluation in Chautauqua Lake, pond-reared muskellunge fingerlings had a higher survival than tank-reared fingerlings (John Forney, pers. comm.). Spawning characteristics, mortality and spatial distribution of hatchery-reared lake trout has been observed to differ from native lake trout in Lake Superior despite similar genetic origin (Krueger et al. in press). The effects of various propagation techniques on the survival and behavioral characteristics of stocked muskellunge in the wild needs further evaluation (Carline et al. 1986).

Studies documenting post-stocking mortality of muskellunge and tiger muskellunge are numerous (Andrews 1983; Beluzs 1978; Beyerle 1984a, 1984b, and 1981; Carline et al. 1986; Hanson et al. 1986; Johnson 1978; Serns and Andrews 1986; Stein et al. 1981). Handling stress, water temperature at stocking, predation on stocked fingerlings and prey availability have all been implicated as factors influencing post-stocking survival. Future studies must determine the mechanisms causing mortality of stocked fish under a variety of stocking and environmental conditions. Development of quantitative relationship between hatchery-controlled constraints (i.e. size, time and location of stocked fish, stocking rates, etc.) and post-stocking survival would enable improvements in the cost-effectiveness of propagation programs (Hanson et al. 1986).

Muskellunge management programs which rely heavily on stocking have frequently allowed hatchery logistic and economic contraints to dictate the size and time of stocking, broodstock source and rearing methods. This strongly resembles the cliche "the tail has been wagging the dog". For management programs, whose goals are to restore formerly self-sustaining populations, this approach may be a significant factor preventing successful rehabilitation.

Hybrid Muskellunge

Although natural hybrid muskellunge do occur (Buss et al. 1978), discussion of these fish here is limited to waters where they are maintained by stocking. Hatchery managers have had greater success rearing hybrids than muskellunge on artificial diets (Bender and Graff 1986; Jorgensen 1986; Orme 1978; and Pecor 1978). Survival of stocked fingerlings reared on an artificial diet has been less than minnow-fed fingerlings (Beyerle 1984b; Andrews 1983; and Johnson 1978). The relationships between both cultural methods and stocking practices versus post-stocking survival need much more study.

Hybrids generally lack the spawning urge; therefore, population assessment of their adults is more difficult than for muskellunge (Brege 1986 and Hesser 1978). The best samples of this fish are generally of fall fingerlings from electro-shocking and adults from angler creels. At present, these samples do not provide sufficient data for management. Effective methods of sampling for hybrids need to be developed.

Hybrid muskellunge are generally stocked in waters where muskellunge are not endemic. The stocking of either hybrids or muskellunge into new waters frequently raises fears among anglers that these introductions will result in decimated populations of other fishes. This fear stems more from the general public's image of these animals as an "eating machine" (Graff 1986) than past management experience. Much of this apprehension could be alleviated by a few good case history studies following a new introduction, and additional investigations on predator-prey relationships (Engstrom-Heg et al. 1986).

THE ROLE OF PRIVATE ORGANIZATIONS IN RESEARCH

Private organizations have become increasingly more involved in the management of fish populations. Many musky clubs have assisted various research studies with donations of money, labor and data (Oehmcke et al. 1986). In addition to providing guidelines for management agencies, this discussion of management and research needs is intended to further assist such organizations in future decisions of where to put their available resources.

The panel did not attempt to assign a priority to the suggested studies. However, those which assist in the preservation of self-sustaining populations and identification of present genetic diversity are of extreme importance. Funding organizations are encouraged to assess their goals in relation to the categories of research outlined here or consider top-priority site-specific studies. The relative importance of site-specific studies will have to be assessed locally.

We remind organizations that there are other ways to help in research without directly funding any particular study. For instance, the money spent by Muskies, Inc. Co-sponsors of this symposium could have been spent on a single research study. Instead, it was spent on the disemination of information from previously conducted studies. This communication is considered much more valuable than the funding of a single research study, and it is strongly recommended that endeavors to facilitate technology transfer of research findings and management experiences continue.

Organizations wishing to fund research should seek out technical-assistance advisors to help screen potential projects, if club members lack research and statistical expertise. Furthermore, results of research should be subject to scientific peer review, prior to widespread dispersion and application of the results. For a further discussion on the role of anglers and private organizations in muskellunge management see Oehmcke et al. (1986).

CONCLUSIONS

The number of self-sustaining muskellunge populations has declined, presumably due to the musky's limited biological adaptations for reproduction (Dombeck et al. 1984) in conjunction with deteriorating habitat and excessive harvest. The potential for continued habitat deterioration and increased angler use will provide significant challenges to fishery managers to maintain self-sustaining populations and, hopefully, to restore some of the former ones. Both the skill and desire of management agencies must improve in order to prevent these fisheries from becoming a domesticated resource maintained solely for recreational purposes. Development of some of the needed skills will provide a hefty challenge to researchers; however, some improvements in management can be readily made given a will to do so (i.e. a clear definition of management philosophy would be a good start).

ACKNOWLEDGEMENTS

Prior to preparing this manuscript, we contacted many individuals from the states and provinces which manage muskellunge to obtain an understanding of what they preceived as major problems. In addition, a questionaire was circulated to attendants of the symposium for further input. We wish to thank the many individuals who took the time to respond to our inquiries. Particular thanks go to E. Crossman, R. Strand, H. Westers, and P. Wingate for their useful suggestions. Additional thanks go to G. Busacker, M. Dombeck, A. Forbes, L. Jacobson, J. Lyons and H. Snow for their comments on an earlier draft.

REFERENCES

Andrews, S.J. 1983. Survival and growth of tiger muskellunge reared on live versus artificial diets. Mimeograph report. Fisheries Section, Indiana Department of Natural Resources, Indianapolis, Indiana, USA.

Belusz, L.C. 1978. An evaluation of the muskellunge fishery of Lake Pomme de Terre and efforts to improve stocking success. American Fisheries Society Special Publication 11:292-297.

Bender, T.R. Jr., and D.R. Graff. 1986. Pennsylvania practices for intensive culture of hybrid muskellunge. American Fisheries Society Special Publication 15:270-284.

Beyerle, G.B. 1984a. Survival and growth of early- and normal-plant tiger muskellunge stocked in a small lake with forage fish and largemouth bass. Michigan Department of Natural Resources Fisheries Research Report 1923. Ann Arbor, Michigan, USA.

Beyerle, G.B. 1984b. Comparative survival of pellet-reared muskellunge stocked as fingerlings in bluegill ponds with and without largemouth bass. Michigan Department of Natural Resources Fisheries Research Report No. 1920. Ann Arbor, Michigan, USA.

Beyerle, G.B. 1981. Comparative survival and growth of 8.9- and 17.8-cm (3.5- and 7.0-inch) tiger muskellunge planted in a small lake with forage fishes. Michigan Department of Natural Resources Fisheries Research Report No. 1894. Ann Arbor, Michigan, USA.

Brege, D.A. 1986. A comparison of muskellunge and hybrid muskellunge in a southern Wisconsin lake. American Fisheries Society Special Publication 15:203-207.

Buss, K., J. Meade III, and D.R. Graff. 1978. Reviewing the esocid hybrids. American Fisheries Society Special Publication 11:210-216.

Caplin, D.L. 1982. An experimental study of interactions between young of the year pike *(Esox lucius)* and muskellunge *(Esox masquinongy)*. M.S. Thesis. University of Wisconsin, Madison, Wisconsin, USA.

Carline, R.F., R.A. Stein, and L.M. Riley. 1986. Effects of size at stocking, season, largemouth bass predation and forage abundance on survival of tiger muskellunge. American Fisheries Society Special Publication 15:151-167.

Casselman, J.M. 1984. Muskellunge research needs for the 1980's...a Canadian prospective. Muskie 18:10-13.

Casselman, J.M., and E.J. Crossman. 1986. Size, age and growth of trophy muskellunge- the cleithrum project. American Fisheries Society Special Publication 15:93-110.

Craig, R.E., and R.M. Black. 1986. Nursery habitat for muskellunge in south Georgian Bay, Lake Huron, Canada. American Fisheries Society Special Publication 15:79-86.

Crossman, E.J. 1986. The nobel muskellunge: a review. American Fisheries Society Special Publication 15:1-13.

Crossman, E.J. 1978. Taxonomy and distribution of North American esocids. American Fisheries Special Publication 11:13-26.

Dent, R.J. 1986. A case history of voluntary catch and release of muskellunge with management implications for fishery programs. American Fisheries Society Special Publication 15:316-322.

Dombeck, M.P. 1986. Muskellunge habitat with guidelines for habitat management. American Fisheries Society Special Publication 15:208-215.

Dombeck, M.P. 1979. Movement and behavior of muskellunge determined by radio telemetry. Wisconsin Department of Natural Resources, Technical Bulletin 113. Madison, Wisconsin, USA.

Dombeck, M.P., B.W. Menzel, and P.N. Hinz. 1984. Muskellunge spawning habitat and reproductive success. Transactions of the American Fisheries Soeicty 113:205-216.

Dombeck, M.P., B.W. Menzel, and P.N. Hinz. 1986. Natural muskellunge reproduction in midwestern lakes. American Fisheries Society Special Publication 15:122-134.

Engstrom-Heg, R., R.T. Colesante, and G.A. Stillings. 1986. Prey selection by three esocid species and a hybrid esocid. American Fisheries Society Special Publication 15:189-194.

Gasbarino, P. 1986. Catch and release of muskellunge: philosophy and methods. American Fisheries Society Special Publication 15:300-308.

Graff, D.R. 1986. Muskie management- a changing perspective from past to present. American Fisheries Society Special Publication 15:195-198.

Graff, D.R. 1978. Intensive culture of esocids: the current state of the art. American Fisheries Society Special Publication 11:195-201.

Haas, R.C. 1978. The muskellunge in Lake St. Clair. American Fisheries Society Special Publication 11:334-339.

Hanson, D.A., M.D. Staggs, S.L. Serns, L.D. Johnson, and L.M. Andrews. 1986. Survival of stocked muskellunge eggs, fry, and fingerlings in Wisconsin lakes. American Fisheries Society Special Publication 15:216-228.

Hanson, D., B. Strand, D.D. Post, W.H. Le Grande, and S. Fillback. 1983. Muskellunge electrophoresis study. Muskie 17:9-13.

Harrison, E.J. and W.F. Hadley. 1978. Ecological separation of sympatric muskellunge and northern pike. American Fisheries Society Special Publication 11:129-134.

Hess, L. and C. Heartwell. 1979. Literature review of large esocids (muskellunge, northern pike, and hybrid muskellunge). Pages 139-175 *in* J. Dubé and Y. Gravel, editors. Proceedings of the 10th warm water workshop. Northeastern Division, American Fisheries Society. Ministère du Loisir, de la Chasse et de la Pêche du Québec, Direction de la recherche faunique. Montréal, Québec, Canada.

Hesser, R.B. 1978. Management implications of hybrid esocids in Pennsylvania. American Fisheries Society Special Publication 11:302-307.

Inskip, P.D. 1986. Negative associations between abundances of muskellunge and northern pike: evidence and possible explanations. American Fisheries Society Special Publication 15:135-150.

Inskip, P.D., and J.J Magnuson. 1986. Fluctuations in growth rate and condition of muskellunge and northern pike in Escanaba Lake, Wisconsin. American Fisheries Society Special Publication 15:176-188.

Johnson, L.D. 1981. Comparison of muskellunge *(Esox masquinongy)* populations in a stocked lake and unstocked lake in Wisconsin, with notes on the occurrence of northern pike (Esox lucius). Wisconsin Department of Natural Resources Research Report 110. Madison, Wisconsin, USA.

Johnson, L.D. 1978. Evaluation of esocid stocking in Wisconsin. American Fisheries Society Special Publication 11:298-301.

Jorgenson, W.D. 1986. Iowa culture of muskellunge on artificial diet. American Fisheries Society Special Publication 15:285-287.

Klingbiel, J.H. 1986. Culture of purebred muskellunge. American Fisheries Society Special Publication 15:273-278.

Koppelman, J.B., and D.P. Philipp. 1986. Genetic applications in muskellunge management. American Fisheries Society Special Publication 15:111-121.

Krueger, C.C., B.L. Swanson, and J.H. Selgeby. In press. Evaluation of hatchery-reared lake trout for re-establishment of populations in the Apostle Islands Region of Lake Superior, 1960-84. *In* R.H. Strand, editor. Role of Fish Culture in Fisheries Management. American Fisheries Society Fish Culture and Fish Management Sections. Bethesda, Maryland, USA.

Oehmcke, A.A., J.T. Addis, S.R. Moordian, K. Ogden, and D. Stange. 1986. The role of anglers and private organizations in muskellunge management. American Fisheries Society Special Publication 15:323-334.

Orme, L.E. 1978. The status of coolwater fish diets. American Fisheries Society Special Publication 11:167-171.

Pecor, C.H. 1978. Intensive culture of tiger muskellunge in Michigan during 1976 and 1977. American Fisheries Society Special Publication 11:202-209.

Porter, L.R. 1977. Review of selected literature of muskellunge life history, ecology and management. Minnesota Department of Natural Resources Special Publication 119. St. Paul, Minnesota, USA.

Ragan, J.T. Steinwand, and G. Van Eeckhout. 1986. A synopsis of results from a questionnaire on muskellunge-resource agency survey. American Fisheries Society Special Publication 15:370-372.

Richards, K., and R. Ramsell. 1986. Quantifying the success of muskellunge catch and release programs: a summary of cooperating angler-tagging programs. American Fisheries Society Special Publication 15:309-315.

Scott, W. and E. Crossman. 1973. Freshwater fishes of Canada, Bulletin 184, Ottawa. Fisheries Research Board of Canada, 966pp.

Serns, S.L., and L.M. Andrews. 1986. Comparative survival and growth of three sizes of muskellunge fingerlings stocked in four northern Wisconsin lakes. American Fisheries Society Special Publication 15:229-237.

Stein, R.A., R.F. Carline, and R.S. Hayward. 1981. Largemouth Bass predation on stocked tiger muskellunge. Transactions of the American Fisheries Society 110:604-612.

Strand, R. 1986. Identification of principle spawning areas and seasonal distribution and movements of muskellunge in Leech Lake, Minnesota. American Fisheries Society Special Publication 15:62-73.

Threinen, C.W. and A. Oehmcke. 1952. The northern invades the musky's domain. Wisconsin Conservation Bulletin 15. Madison, Wisconsin, USA.

Wingate, P.J. 1986. Philosophy of muskellunge management. American Fisheries Society Special Publication 15:199-202.

POSTER CONTRIBUTIONS

ABSTRACTS ONLY

Nine presentations were made with posters at the Symposium. Three of these were later submitted as papers for publication. After peer review they were accepted and appear in their entirety further on in this section. Abstracts of the other six poster presentations are as follows:

The Muskellunge as a Multiple Spawner: An Adaptive Strategy for Production of a Large Number of Eggs

B. LEBEAU and G. PAGEAU

Department of Biological Sciences
University of Montreal

E.J. CROSSMAN

Royal Ontario Museum
University of Toronto
Toronto, Ontario, Canada M5S 2D6

Knowledge of reproduction in muskellunge is limited, and even recent interpretations of spawning behaviour are probably in error. The pattern of egg development and egg laying is not the same in the muskellunge and the northern pike. A study of the development of ovaries and oocytes and of fecundity and spawning of the muskellunge was carried out in two lakes in Quebec - Lac Barriere (Papineau County) and Lac Saint-Louis (near Montreal).

Some results of the study were as follows: In females 5-15 years of age oocytes maturing in the current year exist in two characteristic modes of size and pass progressively and in synchrony from stage IV to stage V (modified Kesteven stages). In esocids, unlike some other fishes, there is a maximum ovarian volume. Therefore, the larger number of eggs produced by the muskellunge, in contrast to that of the northern pike, cannot all reach maximum size simultaneously. By some adaptive mechanism in muskellunge virtually all egg development is halted at the end of stage V until a few days before spawning. At stage VI (start of spawning) the larger oocytes of the more advanced mode become free from the follicle and are at stage VII when deposited. Following this, the eggs in the less-advanced mode complete vitellogenesis and are deposited several days later.

This bimodal development pattern of oocytes is considered to be an adaptation, which allows an individual to carry and deposit a larger number of eggs than would otherwise be possible with the morphological limitations of the ovary. This confers on the muskellunge the status of a multiple spawner. In older females the process is not synchronized, and a large number of eggs are delayed in their development. For the period from mid-September to November 30 (adjusted to statistical mid-point, October 8) females 5 years old have approximately 80 percent of their oocytes in stage IV, compared with 18-year-old females, which have less than 50 percent of their eggs in stage IV.

Several types of information suggest that senility in muskellunge occurs between 20 and 23 years of age. These are: (1) proven decreasing vitellogenesis; (2) gradual change of niche (drift niche); (3) the fact that the maximum age of fish in the spawning run was 18, yet larger (= older?) fish were encountered in the lake; and (4) proven decrease in relative fecundity with age. Recorded fecundity ranged from 50,000 (±) eggs at 5 years to 400,000 at 20 years. The maximum number (est. by volume) was established at 454,717 for a fish 16 years of age. This large egg production is considered to be an adaptive and compensatory mechanism which allows the species to adjust egg number to maintain ecosystem balance.

As age increases relative fecundity and vitellogenesis decrease. These changes can be explained by physiological changes in individuals resulting from increased energy demands in the gonads and decreased efficiency in metabolism.

Joint Participation of Government Agencies, an Angling Organization and Riparians in Management of a Lake Fishery

ROBERT R. JACKSON

Wisconsin Department of Natural Resources
Box 440
Woodruff, Wisconsin 54568

Implementation of intensive investigations to assess fish stocks and species interactions to facilitate both the development of management strategies and their evaluation requires a substantial commitment of logistical resources. Fiscal restraints, declining revenues and inflation erode the capability of management to meet this need. The problem is most acute in those regions where there is an abundance of lakes and streams. By working together, management agencies, other government units, angling organizations, and property owners can achieve objectives that unilaterally would be difficult or unattainable.

An example of such an effort can be found in the Kentuck Lake (Vilas and Forest Co., Wis.) project, a lake with an excellent reputation for walleye and musky fishing. In response to angler allegations that the walleye population had severely declined, a comprehensive fishery invesigation of the lake was initiated in the spring of 1983 to assess fish stocks, species interactions and angler use. This study is a cooperative effort between the Wisconsin Department of Natural Resources, U.S. Forest Service, Headwaters Chapter of Muskies, Inc., and Kentuck Lake property owners. Project activities and data collection include muskellunge and walleye tagging, population estimates, length frequencies, age, growth and length-weight relationships; a voluntary creel census and muskellunge stomach analysis.

A concerned, informed and involved citizenry is vital to the development and implementation of management strategies. The Kentuck Lake experience revealed that there are groups, organizations and people willing to help Wisconsin Department of Natural Resources manage the state's fisheries. Effective mobilization of these interest groups rests with the local fish manager/biologist.

Tiger Muskellunge as a Trophy Species in Iowa Lakes

BRUCE C. ADAIR

Iowa Conservation Commission
Cold Springs State Park
Lewis, Iowa 51544

Approximately 206,000 tiger muskellunge (F_1 hybrid female muskellunge x male northern pike) fingerlings have been stocked into 34 Iowa lakes since 1978 to provide a unique trophy species in lakes dominated by centrarchids and ictalurids. These are mainly impoundments less than 500 surface acres, concentrated in the five southern tiers of Iowa counties. Hatchery production costs for 6-7 in fingerling tiger muskellunge are currently $129 per 1,000 fish delivered to the lakes. Survival of the stocked hybrids is good in most instances, although sampling difficulty occurs in some lakes. Recruitment to the minimum-legal length (30 in) typically occurs in age IV fish. The popularity of the tiger muskellunge with the Iowa angler ranges from enthusiasm, to indifference, to dislike.

Muskellunge Transportation Units

ZENAS BEAN

Pennsylvania Fish Commission
Pleasant Mount, Fish Culture Station
Pleasant Mount, Pennsylvania 18453

Several fish transportation units were examined and numerically rated for their ability to transport muskellunge and muskellunge-northern pike hybrids. Various life support systems including air blowers, bottom-draw aerators, surface agitators, and oxygen injection were installed on tanks of varying capacity, construction, and locomotion. The factors used in establishing the ranking were loading density, initial cost, and operation experience. Of the factors considered, loading density has the greatest significance in establishing the numerical rating. Insulated fiberglass construction tanks using bottom-draw aerators with oxygen injection received the highest rating. Recommended safe loading levels were established for two units currently in use by the Pennsylvania Fish Commission.

The Esocid Cleithrum--Its Removel, Preparation, and Interpretation for the Acquisition of Age-Growth and other Biological Data

J.M. CASSELMAN

Ontario Ministry of Natural Resources
Research Section, Fisheries Branch
Box 50, Maple, Ontario L0J 1E0

The esocid cleithrum is a superficially located, calcified structure that is part of the pectoral girdle; it can be easily removed by hand for detailed examination. Field techniques for removing and providing an undamanged, relatively clean bone for a preliminary age assessment in less than 1 1/2 min were illustrated.

Both simplified and detailed cleithral terminologies were presented and described. The methods of examining (e.g., type and direction of illumination) and interpreting age from the optical zonation in this flat bone were included. Precise definitions of annuli and pseudoannuli were provided and illustrated in terms of degree and type of translucent and opaque zonation. Validation of the cleithral method of age assessment has been achieved with cleithra from pike of known age and from pike tagged and injected with a fluorochrome label (tetracycline), which forms a temporal and spatial mark recognizable in ultraviolet light.

Cleithral mensuration was illustrated and described in terms of body length. Procedures for precisely locating the origin were illustrated. The relation between linear measurements and actual curvilinear cleithral growth was demonstrated. The anterior cleithral radius in fish of this family is approximately 1/10 of the fork length of the body, which makes the cleithrum a quick and easy tool for grossly estimating body length; more precise estimates can be obtained by using growth rate, which can also be extracted from the cleithrum. The size of the cleithrum relative to the size of the body is correlated with body condition and can be used to estimate available food and nutritional level of the population. From detailed interpretation of the optical zonation, it is possible to determine sex of the fish, at least for northern pike cleithra. Hence, the cleithrum is a calcified indicator structure, which represents a unit of harvest that alone can provide the necessary basic biological data for general monitoring of exploitation.

Under certain types of live handling, the medial cresta in the cleithrum is cracked and broken, and after the fish is released the break heals to form a bone callus. The frequency of occurrence of this callus can be used as an indication of angling pressure through the frequency of capture, injury, and release.

The Muskellunge -- What's in a Name

E.J. CROSSMAN, STEPHEN CAMPBELL, and LESLIE E.M. MUNRO

Royal Ontario Museum
University of Toronto
Toronto, Ontario, Canada M582DC

The scientific and common names of the muskellunge have an interesting history. The accepted scientific name, *Esox masquinongy,* is attributed to S.L. Mitchill, a New York physician. The description of this new species and the introduction of this scientific name are supposed to have been published in 1824. No copy of the publication has been seen, at least since 1842, and there is doubt that the 1824 description and use of the name ever existed. The scientific name has passed through stages during which people defined the muskellunge, largely on the basis of color pattern, as one species, three species, and three subspecies of one species. The concensus now would appear to be that the three semi-distinct groups of populations represent three races of one variable species. It is generally assumed now that these three races, like the three ecotypes of *Salmo gairdneri,* do not warrant separate scientific names.

Much of the literature in the 1800's must be read with great care since the muskellunge and the northern pike were regularly confused and the scientific and common names interchanged. Over the years, there have been at least 94 common names applied to this species. The two most frequent names are muskellunge and maskinonge. The first is now the approved one, but the second is still in use in Canada. Many attempts have been made to establish the derivation of these names from Indian words via French pronunciations. It would appear now very likely that the name is Algonquin in origin and represented a spotted kind of pike.

The situation is clouded further by the general use today of the name "tiger muskellunge" for the hybrid between the muskellunge and northern pike. That name was first used for the western race of the muskellunge on the basis of the very bold vertical to oblique "stripes" characteristic of muskellunge in that area.

POSTER PAPERS

Angler Catch, Growth, and Hooking Mortality of Tiger Muskellunge in Small Centrarchid-Dominated Impoundments

DENNIS L. NEWMAN

Illinois Natural History Survey
R.R. 1, Box 233
Charleston, IL 61920

TED W. STORCK

Illinois Natural History Survey
R.R. 1, Box 157
Sullivan, IL 61951

ABSTRACT

Angling vulnerability, growth, and hooking mortality of age I and II tiger muskellunge were examined at Ridge Lake, Illinois. Data were obtained from a complete creel census, supplemented by spring and autumn electrofishing samples. Angler catch rates of yearling tiger muskellunge were highest in the spring. Angler catch of the 1981 year class in 1982 and 1983 (N=202) exceeded the number of stocked as young-of-the-year fish (N=151), indicating multiple catches of some fish. Growth of tiger muskellunge was most rapid in spring and autumn and slowest during mid-summer. The majority of tiger muskellunge stocked in Ridge Lake as young-of-the-year fish in 1981 are not expected to reach harvestable size (762 mm) until 1985. Ten percent of sublegal tiger muskellunge caught by anglers and 13% of the number stocked in 1981 died as a result of the hooking experience. Eighty-five percent of this mortality occurred in June, July, and August, even though only 33% of the catch occurred during that period.

Esocids are stocked in many impoundments in an effort to add diversity to the creel in the form of a trophy species and to control forage fish populations (Weithman and Anderson 1977a; Hesser 1978). The tiger muskellunge (F_1 hybrid of female muskellunge and male northern pike) may be more valuable in these roles than either parent species because it is (1) better adapted to artificial propogation (Hesser 1978); (2) intermediate in angling vulnerability (Beyerle 1973; Weithman and Anderson 1977b); (3) more tolerant of high temperatures (Scott 1964); and (4) has a higher growth and survival rate (Beyerle 1973; Weithman and Anderson 1977a; Johnson 1978). This study examines angler catch, growth, and hooking mortality of tiger muskellunge and its contribution to the sport fishery of small, centrarchid-dominated impoundments in central Illinois. Our purpose was to evaluate the adaptability of the tiger muskellunge to a combination of warm summer temperatures typical of central Illinois and to a forage base that was composed primarily of centrarchid species.

METHODS

Ridge Lake, a 6.1-hectare impoundment located in central Illinois, was the principal study site. Mean and maximum depths were 2.4 m and 7.6 m, respectively. The lake was drained in October 1980 and restocked during 1981 (Table 1). Two other central Illinois impoundments containing established centrarchid populations, Allerton Lake (6.1 hectares) and Lake of the Woods (10.1 hectares), were also sampled to provide ancillary data on the growth of tiger muskellunge.

Pellet-reared tiger muskellunge were obtained from hatcheries in Iowa in July 1981 and June 1982. These fish were stocked in nursery ponds for the remainder of each summer at mean lengths (TL) of 160 mm and 108 mm in 1981 and 1982, respectively, and fed a diet of minnows and bluegills. Those fish recovered from nursery ponds in September and October of each year were stocked in the study impoundments (Table 1). Fish from the 1981 year-class were given fin clips, to distinguish them from the 1982 stocking.

All fishing at Ridge Lake was from boats and by permit only. During 1982 and 1983 the lake was open to fishing from late April through mid-October and fishing was permitted daily (except Mondays and Tuesdays) from 6-10 am and 3-8 pm. No restrictions were placed on the type of bait used by fishermen. For each fishing party a complete record of catch, harvest, and hours spent fishing was kept by the creel clerk. All fish caught were held in live

Table 1. Stocking summary for Ridge Lake, Allerton Lake, and Lake of the Woods in 1981 and 1982.

Lake and species	Year	Number stocked	Number/ hectare	Mean total length (mm)	Length range (mm)
RIDGE LAKE					
Largemouth bass	1981				
Yearlings		471	77.6	170	--
Adults		416	68.5	278	--
Bluegill	1981	887	169.0	--	--
Channel catfish	1981				
Juvenile		435	71.7	204	--
Adults		179	29.5	517	--
Tiger muskellunge	1981	151	25.0	238	160-327
Tiger muskellunge	1982	75	12.5	221	188-268
ALLERTON LAKE					
Tiger muskellunge	1982	150	25.0	225	190-292
LAKE OF THE WOODS					
Tiger muskellunge	1982	250	25.0	193	153-226

wells (except tiger muskellunge) and returned to the check station at the end of each fishing session. Each fish was measured and weighed and then released if sublegal. Minimum-length limits of 355 mm for largemouth bass and channel catfish and 762 mm for tiger muskellunge were enforced.

Fishermen were given flags with which to signal the catch of a tiger muskellunge. A creel clerk retrieved those fish from fishermen as quickly as possible and transported them in a live well to the laboratory, where they were measured and weighed. Immediately after measurement, tiger muskellunge were placed in a holding cage to determine hooking and handling mortality. The holding cages (0.7 m x 1.2 m x 1.0 m) were constructed with ACE-style netting (4.7-mm mesh) attached to a frame of PVC pipe which floated at the surface. All fish were released the next morning following their capture. Tiger muskellunge caught by project personnel (N=12) were also included in the hooking mortality study; therefore, the total sample used to determine hooking mortality exceeded the number recorded in the creel census.

Water temperature at the 1-m depth was monitored daily at Ridge Lake with a recording thermograph in 1982 and with a maximum-minimum thermometer in 1983.

For growth estimates, tiger muskellunge were collected with a 230-volt (AC) boat-mounted electrofishing unit in autumn 1982 and spring 1983 at Ridge Lake, and in autumn 1983 at Ridge Lake, Allerton Lake, and Lake of the Woods. These fish were weighed, measured, and returned to the water. Growth increments for each age-group were estimated as the difference between mean lengths of fish caught in successive collections.

RESULTS AND DISCUSSION

Angler Catch

Anglers caught 87 and 19 yearling tiger muskellunge in Ridge Lake in 1982 and 1983, respectively. In both years, catch rates were highest in the late summer and early autumn (Fig. 1). The annual catch rate of yearlings was approximately six times greater in 1982 (0.022/hour) than in 1983 (0.003/hour), despite only a two-fold difference in stocking rates as young-of-the-year fish (Table 1). This difference in catch may have been due to predation losses of young-of-the-year fish, since tiger muskellunge stocked in 1982 were subjected to higher densities of large predators than those stocked in 1981 (Newman and Storck 1984), and individuals in the 1982 year-class were on average slightly smaller (Table 2). Beyerle (1971) and Axon (1978) also reported that the first stockings of esocids were more successful than subsequent stockings.

In 1983, anglers caught 115 age-II tiger muskellunge at Ridge Lake. In contrast to the pattern observed for yearling fish, catch rates were highest during the first few weeks of the fishing season, declined to lower levels for most of the summer, and

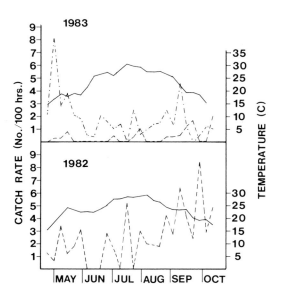

Figure 1. Catch rates (no./100 angler hours of age-I (dashed line) and age-II (dash-dot line) tiger muskellunge and mean weekly water temperatures (solid line) at Ridge Lake during the 1982 and 1983 fishing seasons.

increased moderately in late August as water temperature began to decline (Fig. 1). Weithman and Anderson (1977b) also reported that tiger muskellunge were less vulnerable to angling during midsummer.

Angler catch of the 1981 year-class in 1982 and 1983 ($\underline{N}$=202) exceeded the number stocked ($\underline{N}$=151). Clearly, some individuals were captured at least twice during the 2-year study period. Furthermore, assuming a total annual mortality rate of 30% (Anderson and Weithman 1978) for the 1981 year-class, more age-II fish were caught by anglers in 1983 ($\underline{N}$=115) than were present at the beginning of that fishing season. In contrast, the relatively small catch of individuals of the 1982 year-class in 1983 probably reflected poor survival rather than low vulnerability. Although tiger muskellunge are substantially less vulnerable to angling than northern pike (Beyerle 1973; Weithman and Anderson 1977b), the catch statistics for age-I and -II fish from Ridge Lake indicate that protective regulations are needed to produce trophy-size individuals and to maintain predation pressure on panfish populations.

Although the catch of tiger muskellunge of the 1981 year-class in Ridge Lake exceeded the number stocked, their contribution to the total angler catch in 1982 and 1983 was small relative to the contribution of other sportfish. For example, the catch of largemouth bass (kg/ha) was 12 and 5 times greater in 1982 and 1983, respectively, than that of tiger muskellunge; the catch of bluegill was approximately two times greater in both years (Table 2). In addition, anglers were aware that tiger muskellunge had not reached legal size. Nevertheless, 13% of the anglers questioned at Ridge Lake in 1983 (N=1,014) indicated that they were fishing for that fish. Thus, angler interest in tiger muskellunge was high relative to potential rewards and is expected to increase as they approach harvestable size.

Growth

Growth of tiger muskellunge in the three study impoundments compared favorably with their growth in centrarchid-dominated impoundments in Indiana (Andrews 1983) and Michigan (Beyerle 1981; Table 3). Annual growth increments of yearlings exceeded 200 mm at Ridge Lake in 1982 and 1983 and at Lake of the Woods in 1983. The smallest annual growth increment for age-I fish (190 mm) occurred at Allerton Lake. Approximately 90% of that population was heavily infected with black-spot

Table 2. Angler catch and harvest of the four principal species in Ridge Lake during 1982 and 1983. Angler effort (man-hours/hectare) in parentheses.

	1982	(648)			1983	(964)		
	Catch		Harvest		Catch		Harvest	
Species	Number/ hectare	kg/ hectare	Number/ hectare	kg/ hectare	Number/ hectare	kg/ hectare	Number/ hectare	kg/ hectare
Largemouth	221.1	65.9	2.5	2.1	249.8	91.4	21.1	15.0
Bluegill	110.5	10.3	28.2	3.5	443.6	36.5	162.4	17.2
Tiger muskellunge	14.3	5.2	--	--	22.1	17.9	--	--
Channel catfish	14.8	16.9	7.6	13.8	21.9	31.0	20.1	30.0

Table 3. Mean lengths and annual growth increments of age-I and -II tiger muskellunge. Angler catch and electrofishing samples from early October were pooled to estimate mean lengths.

Lake	Surface area (hectares)	Year	Age I			Age II		
			N	Mean total length (range)	Annual growth increment (mm)	N	Mean total length (range)	Annual growth increment (mm)
Ridge Lake	6.1	1982	27	456 (394-515)	218			
	6.1	1983	14	460 (403-506)	239	12	586 (491-670)	130
Allerton Lake	6.1	1983	9	415 (392-465)	190			
Lake of the Woods	10.1	1983	6	412 (366-477)	219			
Blue Lake, Indiana (Andrews 1983)	97	1982	11	403	201	2	683	
Round Lake, Indiana (Andrews 1983)	53	1982	5	371	168	5	638	
Daggett Lake, Michigan (Beyerle 1981)	6.1	1980	33	450		53	577	

disease, which has been shown to suppress growth of northern pike (Harrison and Hadley 1982).

Tiger muskellunge stocked in Ridge Lake in autumn of 1981 increased in length during the winter period (October-April) of both years (Fig. 2), but selective mortality of smaller individuals may have inflated the estimate of growth over the first winter interval. Growth during winter has also been reported for tiger muskellunge in Missouri impoundments (Goddard and Redmond 1978) and in Indiana (Andrews 1983). Both age-groups in Ridge Lake continued to grow throughout the spring, summer, and autumn (Fig. 2), but growth was slowest during mid-summer. A period of slow growth in mid-summer has previously been reported for tiger muskellunge in Missouri (Weithman and Anderson 1977a) and Indiana (Andrews 1983).

The largest of the age-II tiger muskellunge captured ($N = 12$) at Ridge Lake in October of 1983 was 670 mm long, and at the current rate of growth, few individuals of this age group will reach harvestable size (762 mm) as age-III fish. In contrast, some tiger muskellunge reached harvestable size in the spring of their third year (age II) at Stockton Lake, Missouri (10,072 hectares), where gizzard shad was the principal forage species (Goddard and Redmond 1978). Although growth of tiger muskellunge in our small centrarchid-dominated impoundments was substantially below the potential demonstrated by this hybrid in impoundments containing shad and

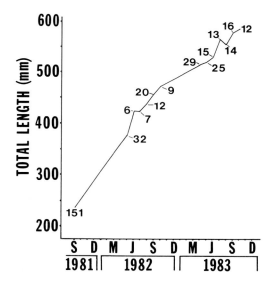

Figure 2. Growth history of tiger muskellunge from the 1981 year-class caught in Ridge Lake by angling and electrofishing. Numbers on the growth curve represent sample size.

other soft-rayed forage, the mean weight of age-II tiger muskellunge in October (1,055 g) at Ridge Lake was greater than the mean weight of largemouth bass harvested by anglers during the same interval (740 g). Thus, tiger muskellunge provided a comparatively large sportfish for anglers well in advance of its reaching harvestable size.

Hooking Mortality

Hooking mortality accounted for the loss of 20 tiger muskellunge from the 1981 year-class and one from the 1982 year-class (Table 4). Most of the mortality of both age-groups (85%) occurred in June, July, and August, even though only 33% of the total catch occurred during that interval. By comparison, sublegal largemouth bass (≤355 mm) exhibited a similar seasonal pattern of hooking mortality but a substantially smaller fraction of the bass died as a result of the hooking experience (Table 4).

Weithman and Anderson (1977b) hypothesized that as esocids grow larger they are more difficult to handle, and therefore suffer a higher hooking mortality, but a substantially smaller fraction of the bass kellunge at Ridge Lake was not significantly greater for age-II than for age-I individuals ($X^2=0.86$, $P>0.25$). It appears that temperature at the time of capture may have been a more important factor influencing hooking mortality than was size of the fish, since mortality was much greater in summer than in spring or autumn.

Thirteen percent of the tiger muskellunge stocked in 1981 died from hooking in 1982 and 1983; that mortality may be significant, especially because it is likely that additional losses will occur before these fish reach legal size.

Hooking losses of the 1981 year-class were high at Ridge Lake, partly because each tiger muskellunge was exposed to the hooking experience an average of 1.3 times. Heavy fishing pressure at Ridge Lake (Table 2) and the inherent vulnerability of the hybrid to angling were responsible for this high catch.

These estimates of hooking mortality should be interpreted with some caution because fish were subjected to the additional stress of being weighed and measured before being placed in the holding cage. However, most of the fish that died in the holding cage had lost equlilbrium by the time they were retrieved from fishermen; therefore, stresses associated with the struggle to escape capture, hook damage, and angler handling were probably the most important factors contributing to mortality. Many anglers at Ridge Lake were inexperienced with handling tiger muskellunge and feared personal injury; consequently, they removed hooks with extreme caution, increasing the time that the fish were out of water.

CONCLUSIONS

The results of this study indicate that tiger muskellunge can make a useful contribution to the sport fishery of small, centrarchid-dominated, warmwater impoundments. However, seasonal changes in catch rates, growth rates, and hooking mortality of tiger muskellunge at Ridge Lake suggest that warm summer temperatures can limit that contribution. In spite of the temperature effect, and the absence in Ridge Lake of soft-rayed forage preferred by tiger muskellunge (Moody et al. 1983), growth and survival were adequate to allow a substantial number of large, although sublegal, fish to enter the catch during their third growing season. Growth, survival, and harvest of tiger muskellunge over the next two years will be measured to evaluate more fully the management potential of this hybrid in small warmwater lakes.

ACKNOWLEDGMENTS

This study was supported in part by Federal Aid in Fish Restoration under Illinois Dingell-Johnson Project F-40-R. R.D. Davis, W.F. Dimond, S.J. Miller, J.A. Tranquilli, and J. Waite reviewed the manuscript and offered many helpful suggestions.

REFERENCES

Anderson, R.O. and A.S. Weithman. 1978. The concept of balance for coolwater fish populations. American Fisheries Society Special Publication 11:371-381.

Table 4. Hooking mortality of tiger muskellunge and sublegal largemouth bass in Ridge Lake during 1982 and 1983.

	Number caught	Number dead	Percent mortality	Mean total length (mm)
Tiger muskellunge				
1982				
Age I	71	6	8.5	424
1983				
Age I	22	1	4.5	436
Age II	124	14	11.3	540
Largemouth bass (<355 mm)				
1982	163	6	3.8	292
1983	188	2	1.1	296

Andrews, S.J. 1983. Survival and growth of tiger muskellunge reared on live versus artificial diets. Indiana Department of Natural Resources, Indianapolis, Indiana, USA.

Axon, J.R. 1978. An evaluation of the muskellunge fishery in Cave Run Lake, Kentucky. American Fisheries Society Special Publication 11:328-333.

Beyerle, G.B. 1971. A study of two northern pike-bluegill populations. Transactions of the American Fisheries Society 100:69-73.

Beyerle, G.B. 1973. Comparative growth, survival and vulnerability to angling of northern pike, muskellunge and their hybrid tiger muskellunge stocked in a small lake. Michigan Department of Natural Resources Fisheries Research Report.

Beyerle, G.B. 1981. Comparative survival and growth of 8.9- and 17.8-cm (3.5- and 7.0-inch) tiger muskellunge planted in a small lake with forage fishes. Michigan Department of Natural Resources Fisheries Research Report 1894.

Goddard, J.A. and L.C. Redmond. 1978. Northern pike, tiger muskellunge, and walleye populations in Stockton Lake, Missouri: a management evaluation. American Fisheries Society Special Publication 11:313-319.

Harrison, E.J. and W.F. Hadley. 1982. Possible effects of black-spot disease on northern pike. Transactions of the American Fisheries Society 111:106-109.

Hesser, R.B. 1978. Management implications of hybrid esocids in Pennsylvania. American Fisheries Society Special Publication 11:302-307.

Johnson, L.D. 1978. Evaluation of esocid stocking program in Wisconsin. American Fisheries Society Special Publication 11:298-301.

Moody, R.C., J.M. Helland and R.A. Stein. 1983. Escape tactics used by bluegills and fathead minnows to avoid predation by tiger muskellunge. Environmental Biology of Fishes 8:61-65.

Newman, D.L. and T.W. Storck. 1984. Evaluation of the introduction of tiger muskellunge into impoundments dominated by the bass-bluegill combination, evaluation of a partial creel and size-specific survival of stocked channel catfish. Illinois Natural History Survey. Federal Aid in Sport Fish Restoration Act. F-40-R, Annual Report, submitted to the Illinois Department of Conservation, Springfield, Illinois, USA.

Scott, D.P. 1964. Thermal resistance of pike (*Esox lucius* L.), muskellunge (*E. masquinongy* Mitchell), and their F_1 hybrid. Journal of the Fisheries Research Board of Canada 21:1043-1049.

Weithman, A.S. and R.O. Anderson. 1977a. Survival, growth and prey of Esocidae in experimental systems. Transactions of the American Fisheries Society 106:424-430.

Weithman, A.S. and R.O. Anderson. 1977b. Angling vulnerability of Esocidae. Proceedings of the 30th Annual Conference of Southeast Association Game and Fish Commissioners 30:99-102.

Managing Water Heating Costs for Intensive Culture of Tiger Muskellunge

WILLIAM F. KRISE and JOSEPH T. FUSS

U.S. Fish and Wildlife Service
National Fishery Research and Development Laboratory
R.D. #4, Box 63
Wellsboro, Pennsylvania 16901

ABSTRACT

Energy costs for rearing tiger muskellunge were compared using water heated with oil, gas, or electric heat pump. An oil-fired boiler proved to be the most expensive, costing approximately 50% more than natural gas. A water-to-water heat pump was most efficient, costing 40% less than gas. Electric rates, however, are most susceptible to local fluctuations, particularly "electric demand" charges. In the worst case, operation of a heat pump would equal gas heating costs.

The cost of culturing tiger muskellunge was compared at temperatures of: 15.6 C, 20 C, and 22 C. Heating ambient water (7.8 C0 to 20 C and 22 C cost 57% and 85% more, respectively, than heating water to 15.6 C. By establishing culture water temperature regimes, hatchery managers should be able to exercise better control over biological factors such as fish growth, feed conversion, incidence of disease, and cannibalism that may interfere with optimum production.

Heating water supplies for intensive culture of tiger muskellunge is a necessity at northern latitudes. Because tiger musky growth is generally poor at temperatures less than 16 C (Pecor 1978), rearing water must be heated during their early life stages, and production cycles generally begin indoors. When the fish reach lengths of 6-10 cm, outdoor rearing, water-supply temperatures have risen to 14-20 C (Harry Westers and Zenas Bean pers. comm.), and the fish can be stocked in outdoor raceways. Although growth and survival are improved if indoor water temperatures are increased above water-supply temperatures, heating of that water can be expensive. Actual heating costs depend on several factors, including number of fish, flow rate, and type of fuel used, but it is possible to predict these costs and budget accordingly.

The National Fishery Research and Development Laboratory (NFRDL) has used three types of water heating systems and has comparable cost figures for each system. Boilers fired with oil and with natural gas have been used to heat the building as well as the fish culture water supplies, and a water-to-water heat pump (Fuss 1983) has been used to extract heat from discarded culture water and transfer it back to fresh culture water. The purpose of this paper is to describe the NFRDL water heating system and a method of calculating water-heating costs. Costs of rearing tiger muskellunge and biological factors associated with their growth and survival in early culture stages are included.

HEATING SYSTEM DESCRIPTION

The only significant difference between the culture water heating system at NFRDL (Fig. 1) and a typical water heating system is the incorporation of a liquid-to-liquid heat pump. Fresh water enters the building at about 8 C and is then divided between cold and hot water distribution systems. The cold water enters a degassing tank where low pressure air is used for agitation to drive off excess dissolved gases. Water then passes to a blending valve where the desired fish-rearing temperature is automatically maintained.

Water to be warmed passes through a heat exchanger supplied by the heat pump. The warmed water then can pass directly to a degassing tank, similar to that in the cold water system, or it can first be routed to the building's boiler room. A second heat exchanger in the boiler room can supply any additional heat that may be needed, or it is capable of supplying the entire heating load. Within the boiler room are four boilers, two of which are fired by No. 2 fuel oil and two by natural gas. Once water is heated, it is returned to the wet laboratory, where it is degassed. From the degassing tank it passes through a blending valve, where it is mixed with cold water to maintain a constant supply temperature to culture tanks. The final temperature at individual culture tanks is determined by manually setting hot and cold water valves. Wastewater from the culture units is collected and routed through the heat pump

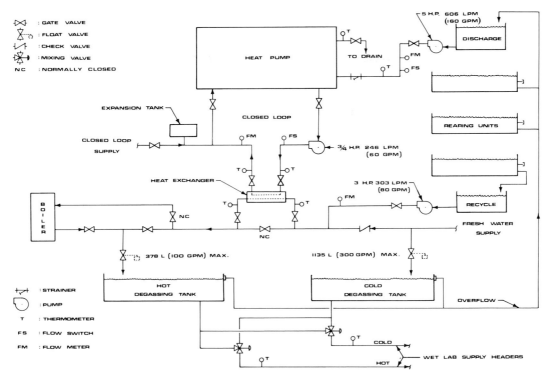

Figure 1. Schematic drawing of NFRDL water heating system.

where the heat is extracted and transferred to incoming water. Heat pump wastewater is then dumped to drain. Flowmeters and thermometers are strategically located in all the piping systems to determine energy consumption and efficiency.

METHOD FOR DETERMINING WATER HEATING COSTS

Heating costs were determined for three energy sources: 1) a natural gas-fired boiler, 2) an oil-fired boiler, and 3) an electrically powered heat pump. The first step in calculating costs for any energy source is to determine the heating load, or demand, which is done by applying the following equation:

$$Q = (K)(T)(F), \text{ where:}$$

Q = heating load in Joules/hour; K = a constant, 248,815; T = temperature rise in °C; F = flow rate in liters/minute.

As an example, to calculate the heating load for a flow rate of 3.8 liters/minute from an initial temperature of 7.8 C to a final temperature of 15.6 C, the equation becomes:

$Q = (248,815)(15.6-7.8)(3.8)$
$= 7.4 \times 10^6$ joules/hour
$= 1.77 \times 10^5$ kilojoules/day

(The incoming water supply at NRFDL is a nearly constant 7.8 C; for other facilities, the appropriate temperature should be substituted.) A single cubic meter of gas contains 3.90×10^4 kilojoules. A typical boiler operating efficiency would be 80%, resulting in the heated water receiving $0.80 \times 3.90 \times 10^4$ kilojoules = 3.12×10^4 kilojoules from each cubic meter of gas consumed. Thus, the 24-hour load of 1.77×10^5 kilojoules would require $1.77 \times 10^5 \div 3.12 \times 10^4$ or $5.67 m^3$ of gas to heat the water to 15.6 C. At an average cost for natural gas of $0.194/m^3$, the cost to heat water flowing at 3.8 L/minute with a 7.8 C rise in temperature would be $1.10 per day.

The calculations for an oil-fired boiler system are similar. The heating load (Q) remains the same, and the typical operating efficiency of the boiler is again 80%. The No. 2 fuel oil that is generally used for heating averages 1.45×10^5 kilojoules per 3.8 liters. After efficiency losses are accounted for, the water will receive 1.16×10^5 kilojoules. With the assumption that oil costs $1.10 per 3.8 liters, the cost of heating the 3.8 L/minute, 7.8 C is $(1.77 \times 10^5) \div (1.16 \times 10^5) \times (1.10) = \$1.68$ per day. However, oil-fired boilers require considerably more maintenance than gas-fired systems since the heat exchanger surfaces within an oil-fired unit can quickly lose their heat-transferring ability due to soot deposits. If the fuel-to-air mixture is incorrect, efficiency often drops to less than 60%. Boiler operating efficiency needs to be checked and adjusted at least once every two months during periods of heavy use.

Calculating the heating costs when using an electrically powered heat pump is more difficult. To determine an average or typical cost of electricity, several assumptions must be made. Three factors are involved: 1) the base electrical rate for the geographic area, 2) the rate structure of the individual company, and 3) the "demand" charges, if any. (Demand charges are an extra fee paid by large consumers such as industries, and can often increase a bill by as much as 50%.) Since rates vary considerably, two rates (one of which includes demand charges) were used for calculating electrical costs. One represents a high rate area ($0.09 per kilowatt hour) and the second a more typical charge ($0.055 per kilowatt hour). For the application in which only 7400 kilojoules per hour are needed, the NFRDL heat pump consumes approximately 0.5 kilowatt of electricity per hour, or 12 kilowatts per day. Thus, the daily cost for the high rate is 12 x 0.09 or $1.08, while the lower electrical rate is 12 x 0.055 or $0.66.

Using the same equation, efficiencies and equipment, it is possible to calculate the different heating costs per 1° C rise in water temperature. Heating 3.8 L/minute of water for 1 hour at NFRDL costs $0.0090/°C for oil; $0.0059/°C for gas; $0.0058/°C for electricity (high rate); and $0.0036/°C for electricity (typical rate).

These variable heating costs have been used in Table 1. When greater flows than 3.8 L/minute are required, the costs are simply multiplied by the new flow. However, when different temperature rises are encountered, the heating load must be recalculated since these costs are only valid for a rise of 7.8°C. Additional costs, such as the initial purchase of heating units and secondary pieces of equipment (pumps, valves), and maintenance have not been considered here.

Table 1. Daily water heating costs for 3.8 liters of water per minute from 8 C ambient temperature to selected fish rearing temperatures of 15.6 - 23.9 C.

Culture temperature (°C)	Fuel		Heat pump Electricity Costs	
	Gas	Oil	High	Typical
15.6°	$1.10	$1.68	$1.08	$0.66
20°	1.73	2.64	1.70	1.04
22.2°	2.04	3.12	2.01	1.23
22.8°	2.12	3.24	2.08	1.27
23.9°	2.28	3.48	2.24	1.37

DISCUSSION

By using biological data for cultured fishes, hatchery managers are able to establish rearing characteristics necessary to achieve fish production goals. One of the major factors that directly affects production cost is heating of rearing water. It is generally advantageous to heat water supplies, but, due to the high associated costs, selecting the best operating temperature is often a compromise between budgetary constraints and biological factors such as survival, cannibalism, growth, feed conversion, and disease.

Studies by Meade et al. (1983) showed that survival of fish 3-4 cm long is nearly the same when they are cultured at different temperatures in the range of 14-22 C. Tiger muskellunge can be grown at these temperatures without jeopardizing survival. Mortalities at these small sizes can generally be attributed to cannibalism (Pecor 1978), especially if feed is inadequate. According to Meade et al. (1983) mortality due to cannibalism increased sharply at water temperatures above 22 C, unless feed amounts were increased. Cannibalism occurring at these higher temperatures resulted in a negative net production, even though individual fish were larger than those cultured in cooler water. Other culture conditions affecting cannibalism such as fish density, feeding rate, feeding technique, and grading can be dealt with independently of water temperature.

It is well documented that fish growth is directly related to water temperature, the same being true of feeding efficiency as expressed by feed conversion. Optimum rearing temperature for growth of tiger musky was determined by Meade et al. (1983) to be 20-22 C for fish 3-4 cm long, and 23 C for fish 12-13 cm long, with their optimum feed conversion range from 20-22 C. Additionally, Pecor (1978) noted that growth and feed conversion rate improved markedly when water temperatures were raised from 15.4 to 17.9 C; Westers (1979) recommended culturing tiger muskellunge at 18-20 C. Rearing fish at temperatures outside the range of 16-24 C is less cost effective (Meade et al. 1983). Efficient feed conversion increases fish production at a reduced feed cost. The cleaner rearing conditions that result from less feed waste also reduce the possibility of pathogen contact by eliminating the presence of fungal or bacterial substrates.

Diseases can be controlled in part by culturing tiger muskellunge at water temperatures below 22 C. Incidence of disease increases as water temperatures are warmed in excess of 20 C (Z. Bean and H. Westers pers. comm.). To reduce the possibility of disease, many hatchery managers choose rearing temperatures between 17 and 20 C, avoiding conditions that optimize fungal or bacterial growth. When all these factors are considered, the selection of rearing temperatures from 18-22 C appears to be the best compromise.

After considering these and possibly other biological factors, the hatchery manager can select a rearing water temperature and, by choosing a constant loading level (weight of fish/water-flow rate, or kilograms/liter/minute), can predict fish growth and heating costs. With the exception of first feeding fish, our calculations are based on a loading level of 0.5 kg/L/min. Table 2 reflects the costs of culturing 10,000 tiger muskellunge with 0% mortality at 20 C using daily growth data obtained by Meade et al. (1983). Costs of culturing fish at any temperature can be calculated by estimating fish length and weight plus water flows needed throughout the rearing cycle. Table 3, 4, and 5 were calculated as examples of growth and costs of rearing 10,000 tiger muskellunge at 18 C, 20 C, and 22 C respectively.

The cost of heating water is a linear function of temperature rise. For example, costs for heating the NRFDL water supply to 18 C would be 84% of the magnitude of the figures given in Table 2 (7.8 °C to 18°C versus 7.8°C to 20°C). Heating 3.8 L/min with gas would cost about $1.42 compared with $1.70 at 20 C. The cost of heating water at 3.8 L/min with gas to 22 C would be 116% of the cost of heating to 20 C, or about $1.98 per day.

Using growth data by Meade et al. (1983) and an initial fish length of 1.5 cm, 90-day growth predictions would be: 13.4 cm when cultured at 18 C; 15.2 cm at 20 C; and 16.4 cm at 22 C. With these same growth rates, flow rates, and loading levels for gas

Table 2. Daily cost of heating culture water from 8 to 20 C for 10,000 tiger muskellunge over a 3-month time period, calculated at 30-day intervals.

Day	Flow (lpm)	Weight of fish (kg)	Fuel Gas	Fuel Oil	Heat pump electricity costs High	Heat pump electricity costs Typical
1	3.8		$ 1.70	$ 2.64	$ 1.70	$ 1.04
30	13.8	6.9	6.21	9.52	6.21	3.73
60[a]	105.2	54.6	47.34	72.60	47.34	28.40
90	289.8	144.9	130.41	199.96	130.41	78.24

[a]Growth rate from 61-90 days is calculated from Table 2 of Meade et al. (1983) on the basis of a daily growth of 0.14 cm. For smaller fish, growth is calculated for a length increase of 0.18 cm per day.

Table 3. Cost of heating culture water with gas from 8 to 18 C for 10,000 tiger muskellunge over a 30 month period, calculated at 15-day intervals.

Day	Flow Liters Per Minute	Weight of fish (kg)	Length of fish (cm)	Daily heating cost	Total cost at end of 15-day interval
1	3.8	1.0	1.5	$ 1.42	21.30
15	3.8	1.0	2.6	1.42	21.30
30	10.0	5.0	5.0	3.70	55.50
45	33.0	16.5	7.4	12.21	183.15
60[a]	77.2	38.6	9.8	28.56	428.40
75	128.0	64.0	11.6	47.36	710.40
90	198.0	99.0	13.4	73.26	-----
				TOTAL COST	$1420.05

[a]Growth rate from 61-90 days is calculated from Table 2 of Meade et al. (1983) on the basis of a daily growth of 0.12 cm. For smaller fish, growth is calculated for a length increase of 0.16 cm per day.

Table 4. Cost of heating culture water with gas from 8 to 20 C for 10,000 tiger muskellunge over a 3-month period, calculated at 15-day intervals.

Day	Flow Liters Per Minute	Weight of fish (kg)	Length of fish (cm)	Daily heating cost	Total cost at end of 15-day interval
1	3.8	1.0	1.5	$ 1.70	25.50
15	3.8	1.0	2.8	1.70	25.50
30	13.8	6.9	5.6	6.21	93.15
45	46.6	23.3	8.2	21.00	315.00
60[a]	105.2	54.6	11.0	47.34	710.10
75	183.2	92.6	13.0	82.44	1236.60
90	289.8	144.9	15.2	130.41	-----
				TOTAL COST	$2405.85

[a]Growth rate from 61-90 days is calculated from Table 2 of Meade et al. (1983) on the basis of a daily growth of 0.14 cm. For smaller fish, growth is calculated for a length increase of 0.18 cm per day.

Table 5. Cost of heating culture water with gas from 8 to 22 C from 10,000 tiger muskellunge over a 3-month period, calculated at 15-day intervals.

Day	Flow Liters Per Minute	Weight of fish (kg)	Length of fish (cm)	Daily heating cost	Total cost at end of 15-day interval
1	3.8	1.0	1.5	$ 1.98	29.70
15	3.8	1.3	3.2	1.98	29.70
30	19.4	9.7	6.2	10.09	151.30
45	63.6	31.8	9.2	33.07	496.05
60[a]	149.2	74.6	12.2	77.58	1163.70
75	241.0	120.5	14.2	125.32	1879.80
90	363.6	181.8	16.4	188.76	-----
				TOTAL COST	$3750.25

[a]Growth rate from 61-90 days is calculated from Table 2 of Meade et al. (1983) on the basis of a daily growth of 0.14 cm. For smaller fish, growth is calculated for a length increase of 0.20 cm per day.

fuel, water heating costs over a 90-day season would be: $1420.05 at 18 C; $2405.85 at 20 C; $3750.25 at 22 C (Tables 3-5). These projections reflect costs of increased water use a higher temperatures necessary to maintain loading levels at 0.5 kilogram/liter/minute flow.

Managing water heating costs can include two basic choices mentioned here. First, selection of a heating system and calculation of heating cost estimates at different water temperatures can provide a basis for budgeting. These calculations can be made for a system already in operation or to estimate costs for a proposed system. Second, comparison of water heating costs with biological requirements of species cultured at different water temperatures can assist in decision making. Evaluation of these parameters can lead to production of suitable-sized fish at the most efficient cost.

REFERENCES

Fuss, J.T. 1983. Evaluation of a heat pump for an aquacultural application. Progressive Fish-Culturist 45(2): 121-123.

Meade, J.W., W.F. Krise, and T. Ort. 1983. Effect of temperature on production of tiger muskellunge in intensive culture. Aquaculture 32:157-164.

Pecor, C.H. 1978. Intensive culture of tiger muskellunge in Michigan during 1976 and 1977. Pages 202-209 in R.L. Kendall, editor. Selected coolwater fishes in North America. American Fisheries Society, Washington, D.C., Special Publication No. 11.

Westers, H. 1979. Controlled fry and fingerling production in hatcheries. Pages 32-52 in E.A. Huisman and H. Hogendoorn, editors. European Inland Fishery Advisory Commission workshop on mass rearing of fry and fingerlings of fresh water fishes. The Hague, The Netherlands.

Mortality of Muskellunge Fingerlings Attributed To A Multiple Tapeworm Infection

JOHN H. SCHACHTE, JR.

New York State Department of Environmental Conservation
Fish Disease Control Unit
Rome, New York 13440

GLENN L. HOFFMAN

U.S. Fish and Wildlife Service
Fish Farming Experimental Station
Stuttgart, Arkansas 72160

ABSTRACT

Muskellunge fingerlings under intensive culture at the Chautauqua State Fish Hatchery, Mayville, New York, experienced unusually high mortality following feeding of live fathead minnow forage during June 1982. Three hatchery troughs of muskellunge fingerlings being fed an experimental dry diet were unaffected. Examination of the intestine of muskellunge revealed the presence of adult tapeworms belonging to the genera *Triaenophorus* and *Proteocephalus* in affected fish only. No other pathogens were detected. Fathead minnow forage fish were also found to be infected with adult and immature forms of both parasites. The observations described illustrate the hazards associated with the use of live, wild forage as food for cultured muskellunge; this practice should be avoided where possible.

In late June 1982, the Fish Disease Control Unit of the New York Department of Environmental Conservation (NYDEC) was contacted by the Chautauqua State Fish Hatchery manager regarding unusually high mortality in muskellunge fingerlings. Temperature and dissolved oxygen levels were within acceptable ranges (temperatures ranged from 16 to 18 C; dissolved oxygen ranged from 7.8 to 8.3 mg/L). Gross examination of fingerlings by the manager revealed tapeworms which were being expressed from the anus. This report describes the ensuing investigation and detection of a multiple tapeworm infection in muskellunge fingerlings. It also indicates the dangers associated with feeding infected forage fish during the early rearing phase of muskellunge culture.

METHODS AND MATERIALS

During the hatchery examination, affected fish in the raceways appeared lethargic. Many individuals crowded the surface of the tank and were not evenly distributed throughout the water column as is normal. Affected fish failed to respond normally to hatchery personnel approaching rearing troughs. All affected fish were being fed fathead minnow fry reared in outdoor earthen hatchery ponds. Three raceways were unaffected and contained muskellunge being fed an experimental dry diet.

Muskellunge fingerlings were examined both macroscropically and microscopically at the hatchery for the presence of external parasites and gill disease bacteria. Internal examination included a culture attempt of body cavity contents on tryptic soy agar (TSA) and the removal and dissection of the gut with the aid of the dissecting microscope. A duplicate examination was conducted on fathead minnow fry taken from a holding tank in the hatchery. A subsequent examination was also conducted on fathead minnow adults from earthen ponds at the NYDEC Randolph State Fish Hatchery, one source of forage fish. All other forage was obtained from wild sources. In all cases samples of affected or non-affected fish of both species from both hatcheries were collected and preserved in 10% formalin. Live specimens were transported to the NYDEC Fish Disease Control Unit Laboratory at Rome. The gut was dissected and live worms removed and inactivated in hot (80 C) water. This procedure was used to relax the parasites and facilitate staining for identification to species. Additional complete gut samples, which had been fixed in formalin, were sectioned at $6\mu m$ and processed routinely with hematoxylin and eosin stains for histological examination.

Finally, vials containing fathead minnow fry and muskellunge from Chautauqua State Fish Hatchery and fathead minnow adults from Randolph State Fish Hatchery preserved in 10% formalin, as well as the tapeworm specimens inactivated in hot water, were forwarded to the U.S. Fish and Wildlife Service Fish Farming Experimental Station at Stuttgart, Arkansas.

RESULTS

Parasites found in the two species of fish from laboratory examinations are given in Table 1. Muskellunge fed on fathead minnow fry were found to contain one to three adult parasites per fish in the gut. Fathead minnow fry contained four to eight immature parasites per fish, mainly in the body cavity. Fathead minnow adults from the Randolph State Fish Hatchery were infested with varying numbers of metacercariae of the trematode *Posthodiplostomum minimum minimum*. In muskellunge only the gut was infected with the parasites and it was the opinion of personnel at both the Fish Disease Control Unit and the Fish Farming Experimental Station that the magnitude of infection was sufficient to plug the gut (Fig. 1) and cause abnormal behavior and eventual death of the host fish. Histological sections of the muskellunge gut (Fig. 2) substantiated the extent of impaction due to the parasitic load in an individual fish. It could not be determined if one of the two tapeworms was more responsible for the problem.

All microbiological assay procedures were negative for the presence of any other parasitic or bacterial pathogens. Initially, prophylactic 1-3% salt bath treatments were recommended until the extent of the problem was determined. It was recommend that feeding of infected fathead minnow fry cease; however, because of a lack of availability of alternate forage at that time, this action was not entirely possible, and feeding of infected minnows continued.

DISCUSSION

The genus *Esox*, which includes muskellunge in North American waters, is described as the definitive host worldwide of *T. nodulosus* (Kuperman, 1981). Hoffman (1967) indicated that the muskellunge is included among those species of fish parasitized by cestodes belonging to the genera *Proteocephalus* and *Triaenophorus*. The presence of

Table 1. Species of fish, parasites present and their location in the fish.

HOST	PARASITE (stage of life cycle)	LOCATION
Muskellunge	*Proteocephalus* sp. (adult)	Intestine
	Triaenophorus nodulosus (adult)	Intestine
Fathead Minnow	*Triaenophorus nodulosus* (adult)	Body Cavity
	Proteocephalus sp. (plerocercoids)	Body Cavity

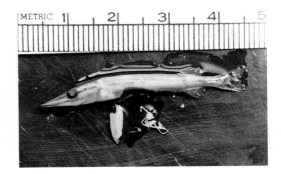

Figure 1. Muskellunge fingerlings with excised intestine and adult tapeworms (arrow).

Triaenophorus and *Proteocephalus* in the fathead minnow fry and in the muskellunge provides evidence that fathead minnow was the source of the infection. This is further substantiated by the fact that the unaffected muskellunge at Chautauqua State Fish Hatchery were in three hatchery troughs on the same water supply receiving an artificial diet. There had been some concern that the Randolph State Fish Hatchery broodstock of fathead minnows, which were the source of fish in Chautauqua State Fish Hatchery earthen ponds, were a possible source of infection. The destruction of these fish was considered: however, failure to find the parasites in these fathead minnows modified that decision.

It is difficult to pinpoint the source of fathead minnow infection except to say that it occurred at Chautauqua State Fish Hatchery. If wild harvested fish were included in the forage, then these could have been a source of the parasites. Meyer (1970)

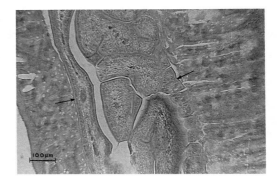

Figure 2. Section of muskellunge intestine impacted with adult tapeworms (arrow denote margin of intenstinal lumen).

and Hoffman (1970) have reported the risk of parasitic infection when introducing wild fish with unknown disease history into culture systems with uninfected fish.

At the present time, there is no effective treatment for these cestode infections. Draining and drying of the ponds used to rear fathead minnows during the off-season may break the cycle of infection by killing the first intermediate copepod host.

REFERENCES

Hoffman, G. L. 1967. Parasites of North American Freshwater Fishes. University of California Press, Berkley, CA

Hoffman, G. L. 1970. Control and treatment of parasitic diseases of freshwater fishes. U.S. Fish and Wildlife Service, Fish Disease Leaflet No 28, Washington, D.C.

Kuperman, B.I. 1981. Tapeworms of the genus *Triaenophorus*, parasites of fishes. New Dehli: Amerind Publishing Co. Pvt, Ltd. Translated from Russian by B.R. Sharma.

Meyer, F. P. 1970. Seasonal fluctuations in the incidence of disease on fish farms. American Fisheries Society Special Publication 5: 21-29.

SUMMARY

Review of Technical Contributions

ROBERT F. STRAND

Minnesota DNR

Compiled from the contributions of each identified Session Chairman

SUMMARY

The muskellunge is a rare species, rare in the sense that it usually exists at low population densities, and that individuals are solitary and quite secretive most of the time. The challenge that these characteristics present to anglers needs no elaboration; they are an important part of the fish's mystique and popular appeal. But to the point of this summary, these characteristics also greatly complicate efforts to obtain knowledge of the species' biology and to manage populations in nature. Because of this, there is probably less detailed biological and ecological knowledge of the muskellunge than for any other major freshwater game fish in North America. Although this symposium was not organized in a way intended to draw together all that is known about this fish, it has been instrumental in helping us to understand what type of research is presently being done, and in pointing to needed knowledge.

Biology and Life History - Bruce Menzel, Iowa State University, Chairman

Throughout the symposium, there was a recurring call for more information on aspects of reproduction and early life history. It is comforting to learn that a number of researchers are dealing with this gap in existing knowledge. The contribution on sperm production and spawning success (Gammon) suggested that although male muskellunge may devote relatively less metabolic energy to sperm production than male northern pike, this does not necessarily detract from egg-fertilization success. However, he found that the viability of spawned eggs depends on local factors, such as presence of egg predators and the microenvironment. The latter point was emphasized (Dombeck *et al.*) in the Community Interactions Session as well. Also complementing Gammon's report was the poster paper (Lebeau *et al.*), which described the female reproductive cycle and made the thought-provoking suggestion that fecundity, within local populations, may be an element of life history adaptive strategy.

Biological knowledge on muskellunge has recently been expanded through applications of biotelemetry, as evidenced in four studies reported here. The technique was used (Strand) to identify spawning sites in Leech Lake, Minnesota, which he described in some detail. Based on his findings, it seems that an important factor contributing to the co-habitation of muskellunge and northern pike in that lake is spatial isolation of their spawning and nursery areas. Telemetry was also used (Schiavone, not published) to identify spawning areas in the Thousand Islands region of the St. Lawrence River. Descriptions of these areas, plus earlier information gathered by Gammon, Crossman, Dombeck and others is helping us to understand the spawning habitat puzzle, but research in other environments is necessary before we can be satisfied with our knowledge.

Moreover, there is evidence that suitable muskellunge nursery habitat differs from spawning habitat in some features. St. Lawrence River research has provided some information on this and promises more. The detailed study of Georgian Bay nursery habitat (Craig and Black) suggests that water depth and type and quantity of aquatic vegetation are important factors in defining habitat quality. This information was translated into useful guidelines for identifying, managing and protecting the habitat.

Several telemetry studies contributed to knowledge of behavior of wild muskellunge. Strand discovered distinct warm and cold-season home ranges among Leech Lake fish. He also identified physical habitat features that seem to be associated with home range size and location. Similarly, Schiavone reported annual shifts in habitat use among St. Lawrence River fish, and Miller and Menzel reported warm-season shifts in habitat use and movement behavior among adults in an Iowa Lake. The latter authors noted that these behavioral patterns were associated with seasonal changes in water temperature, transparency, and vegetation growth. They placed these observations into the context of an optimum foraging strategy, suggesting that the fish change from active, searching predators in the early season to sedentary, ambush predators later. In total, this new knowledge about seasonally variable behavior has important implications for muskellunge angler and manager alike.

Finally, a useful management application of telemetry was demonstrated (Gilbertson). By observing the winter movements of telemetered juveniles in winterkill-prone nursery lakes, responses to decreasing dissolved oxygen concentrations were determined. This information was then used to develop a more efficient winter-rescue operation.

Knowledge of the age and growth of fishes is a basic managerial tool, but it was pointed out (Casselman and Crossman) that standard aging techniques employing scale measurements produce erroneous results when applied to tropy-sized muskellunge. They described an aging technique based on cleithrum measurements, which has proven accurate for all ages and sizes of the species. With it, they showed that trophy fish from Canadian waters tend to live to greater ages and sizes than those from U.S. waters. They also showed that although the hybrid muskellunge may exhibit good growth up to about 8 years in age, its maximum age and size is probably less than that of muskellunge.

In the report on lymphosarcoma in esocids (Sonstegard and Chen) we were reminded that otherwise successful management programs can be devastated by infectious diseases. At this time, there is little remedy for controlling lymphosarcoma once it is established in a population. It is encouraging, however, that there is accumulating knowledge on the viral transmission of the disease and on environmental factors that promote it. Based on this, there are prophylactic management approaches that can be recommended for areas where disease outbreaks occur. They cautioned against establishing new populations by transfer of adults, since this could introduce lymphosarcoma also. In concluding, they also noted that knowledge of genetic factors associated with cancer in fishes may have important implications for human cancer research and that advances in genetic technology offer promise for growth promotion in cultured animals, perhaps including muskellunge.

One of the most important recent conceptual advances in freshwater fisheries management has been the recognition that genetically distinct stocks often exist within species, which represent evolutionary adaptations to local environments. The existence of separate genetic stocks of muskellunge is implied by the nomenclatural history of the species, reviewed here (Crossman), but direct genetic evidence to this point has been lacking until this symposium. Tissue protein electrophoresis is a convenient method of stock delineation because it permits direct quantification of genetic characters. The technique was applied (Koppleman and Phillip) to samples of natural muskellunge populations and hatchery animals from several midwestern and northeastern states. Their results indicated that considerable genetic variation exists within the species, possibly indicative of several genetic stocks. Interestingly, there was greater genetic similarity between samples from the St. Lawrence River and Wisconsin lakes than between the St. Lawrence samples and those from hatcheries in proximity to that river.

A similar electrophoretic survey (Fillbach et al. not published) was performed on populations from Wisconsin and Minnesota lakes and a North Dakota hatchery. They particularly sought evidence on whether observed growth differences between these populations might be genetically determined. Their results suggested the existence of three genetically distinct stocks, one of which is exclusively composed of slow-growing populations. DNA analyses, while implying a somewhat different pattern of genetic distinction between populations, also tended to support this conclusion. These findings alone do not extablish that the slow-growth characteristic has a genetic base, but they suggest that the slow-growth populations might be managed as a separate genetic stock. Together, these studies emphasize the utility of biochemical genetic approaches to muskellunge management and establish a foundation for future studies.

Community Interaction - E.J. Crossman, Royal Ontario Museum, Chairman

The need to address this topic has long been obvious in the concern over the effect on the muskellunge of natural changes to, and the introduction of other species into, a habitat in which the muskellunge was previously, or presently, well established and thriving. Of major concern in this long-recognized problem are the aspects of interaction between muskellunge and northern pike, and between muskellunge and those species comprising the food resource. The latter interaction operates by changing the availability of food to the muskellunge and/or changing the predatory control of the number of individuals of species in that resource. Of somewhat newer concern is the interactive effect of the addition of the walleye as a second or third piscivorous predator in territory previously occupied by the muskellunge alone, or shared by the muskellunge and the northern pike. The newest concern is the need to understand the potential consequences of artificial "sympatry" created when tiger muskellunge are introduced into native or introduced populations of muskellunge.

While the session title suggests a holistic approach, the whole symposium centered on the muskellunge. It is, therefore, not unusual that the contributions used a rather restricted definition of community and that the species of interest in the community was the muskellunge, nor that any recommended adjustments to the interactions were intended to favor the management of muskellunge, and to human advantage.

The information presented heavily emphasized the interaction between the muskellunge and one other species, either the northern pike or the walleye. It did, however, treat experimental results on esocid species with different combinations of prey species, the effects of the muskellunge and the tiger muskellunge in competition for a common food resource, prey availability and muskellunge size as factors in survival of stocked muskellunge, and the outcome of introducing muskellunge into an established bass-panfish community.

In general, all of the papers made it obvious that our understanding of the results of short-term and long-term interactions of various types is far from complete. There is presently better evidence for effects on individuals than there is for effects on population trends through time. The factors are complex, probably not simply additive, and geographically variable. We cannot depend on information derived from populations in one area having universal application over the range of the muskellunge. Management of specific stocks (single populations or regional groups of populations) requires precise information on interactions in those stocks. It was indicated (Dombeck et al.) that within one region, in three lakes, there was a reversal of interactions between muskellunge and northern pike.

We have now been provided with a diagnostic tool (Dombeck et al.) which makes possible a prognosis for the success of self-reproducing populations in various bodies of water. That tool is derived from the analysis of morphoedaphic and community factors which are readily available and easily assimilated. That tool provides an opportunity for value judgements during the planning process when it comes to expending research time and money. Use of that tool suggests that northern pike are the worst single factor threatening muskellunge populations. Large bodies of water providing a heterogeniety of habitat, riverine situations, drainage lakes, and regimes of rising water levels in the spring apparently assure the best chances of survival of muskellunge forced to interact with northern pike. However, we must be cautious that in all-out management for muskellunge we do not seriously damage another present or future resource. It is possible that many resource users might prefer the higher rate of catch provided by a thriving population of northern pike.

It is always useful for us to carefully examine our attitudes for interfering biases. Such an overview (Inskip) indicated that we have too long looked at interactions as having only black and white tones. The existence of a grey area in between is suggested by the fact that evidence of negative interaction was said to be persuasive but all too often circumstantial. It was suggested that negative results, from the perspective of muskellunge abundance, might arise without direct interaction. We were cautioned to remember that only slight differences in tolerances or in optima for environmental variables could lead to the consequences we attribute to interaction. The need to constantly keep in mind the influence of human activities, such as changing land-use patterns, eutrophication, shoreline loading, removal of aquatic vegetation, increasing pressure and changing angling philosophy, was emphasized.

It is possible that the introductions of new piscivorous competitors into habitats containing established populations of muskellunge (Mooradian et al.) do not always have the serious adverse effects we may previously have attributed to those introductions. Such effects do occur, but are variable and not universal. Changes in muskellunge demography, over the period of an expanding walleye population, may have been quite independent of that population of competitors. In addition, no small muskellunge were apparent in the ration of the walleye, and no clear impact on food available to the muskellunge was demonstrated. It would appear that, in that example, both species were responding to intraspecies interactions rather than one or both responding to interspecies interaction. What we derive from this particular example of an interaction must be tempered by the knowledge that the muskellunge population in that lake is now almost completely attributable to fish cultural support (fingerlings) and may constitute a special situation.

Prey selection by muskellunge has important implications for survival and growth of muskellunge, and for the way in which the food resource may be shared by various species in the community. Experimental work (Engstrom-Heg et al.) comparing feeding behavior and food selection indicated several differences between the muskellunge, the northern pike and the tiger muskellunge. The results suggested that, at least for the choice of prey provided, there was a gradient in degree of selection with tiger muskellunge highest, northern pike intermediate, and muskellunge lowest. The hybrid also exhibited the greatest tendency to select soft-finned prey over spiny-finned prey, and a greater tendency to select white suckers over other soft-finned species. For all three esocids, there was a suggested correlation between the prey chosen and the position in the water column generally occupied by the esocid. Tiger muskellunge, at least under these experimental conditions, actively selected muskellunge as food in spite of the availability of adequate numbers of other suitably-sized prey. The results permitted the suggestion that large populations of large and small, soft-finned fishes would favor northern pike over muskellunge, would provide a buffer between northern pike and spiny-rayed sportfishes in the same habitat, but may not provide a buffer in communities including muskellunge and/or the hybrid. One definite implication was that it would be unwise to combine muskellunge and tiger muskellunge in any body of water.

Tests with fingerling tiger muskellunge in ponds and small impoundments (Carline et al.) indicated that where largemouth bass were potential predators success was better assured by the presence of an alternate prey which could be selected by the largemouth bass, by stocking larger fingerlings as late in the season as possible and by holding as low as possible the level of stress associated with the stocking process. Time of stocking seemed critical, since a manager apparently has to trade off the benefits to be derived from reduced predation and opportunity to build up adequate energy stores for overwinter survival. These results also suggested that hatchery diets of pelleted food may contribute to an apparent inability of the hybrid to prey on small bluegills in natural situations.

There is a possibility that there is a weak predatory capability for predation on bluegills by all esocid species. In a Wisconsin situation (Snow, not published) natural variations in success of bluegill year-classes had far greater effect on numbers of bluegills than did the introduction of muskellunge into the established bass-bluegill community. Even the introduction of a combination of muskellunge and walleyes did not achieve the desired control of bluegill numbers. It was suggested that the bluegill population has gotten too large before the new predators were introduced. This implies that appropriate timing might achieve the control. Pumpkinseeds are the dominant food of muskellunge in Nogies Creek, Ontario, but a long-established and unexploited combination of muskellunge and big largemouth bass has never reduced the population of that prey either.

Soft-finned, native forage species seem the only suitable candidates for costly and labor-intensive introduced populations of muskellunge. The results of the work presented in this session also amplified earlier observations that the investment of rearing muskellunge, or tiger muskellunge, to as large a size as possible before introduction will show benefits in survival over the first winter and later.

The contributors to this session identified a number of areas that could be addressed by both research and management. It seems highly appropriate that we consider the need to add to this analysis the desires of the resource user, or at least what we as biologists predict resource users can expect to get from managed communities which include muskellunge, tiger muskellunge, or muskellunge and northern pike.

Management Considerations - Charles Johnson, Wisconsin DNR, Chairman

Muskellunge management in the immediate future will be more critical to the species than it has been in any other time period. Yes, the species is threatened. Declining natural reproduction has been documented in the past, at present, and is predictible in the immediate future. Therefore, for at least a portion of the range, the muskellunge is not a renewable resource, and it will require pin-point management emphasis to maintain native populations at self-sustaining levels within the native range.

Biologists have available a number of management strategies to maintain, improve and develop specific fisheries, including such options as stocking, regulations, habitat modifications, surveys and population manipulation. Most commonly employed strategies are stocking and specific fishing regulations. Evaluations have documented that stocked muskellunge are contributing significantly to the creel. Regulations have been changed to reflect new biological data to maximize age and growth potential and to enhance egg deposition for one to two years before removal by anglers is legal. Harvest data clearly show that the majority of creeled muskellunge are young fish which are, therefore, prevented from reaching their size and reproductive potential. Other concerns facing managers are loss of habitat, declining water quality and public acceptance of muskellunge management programs. As more data are collected, answers to these concerns will provide new direction for attaining our goals.

The message is emphatic: habitat protection, preservation and maintenance are essential to perpetuate the species. Loss or alteration of habitat, especially for spawning, is considered to be a major cause for declining native muskellunge populations. Habitat management guidelines were provided (Dombeck) and discussed in the context of reproductive habitat, management of watersheds to enhance water quality, design of dams to facilitate water level management, riparian zone vegetation including timber and control of northern pike. Axon and Kornman indicated that degradation of the north fork of the Kentucky River, resulting from poor watershed management, impacted 63 miles of stream. Loss of spawning areas due to impoundment, sedimentation or acidic water conditions were identified as problems associated with stream degradation. Their study documented physical stream conditions that were attractors to fish and concluded that zoning laws to regulate industry are critical to protect stream fisheries in Kentucky.

Stocking evaluations (Serns and Andrews) have shown the importance of rearing large fingerlings. Stockings of small muskellunge fingerlings (4 in) were effective in lakes with a low density of predators and a good supply of suitable forage. They recommended that only large fingerlings (12 in or larger) be stocked where there is a moderate to large northern pike population. Their recommendation was supported by other findings (Hoff and Serns) in Escanaba Lake, Wisconsin; stocked yearlings were nine times more effective than fingerlings at adding fish to the creel. It was also reported (Hanson et al.) in another Wisconsin investigation that the greatest success resulted from stocking the largest fingerlings; however, they also suggested that similar results might have been achieved by stocking greater numbers of small fingerlings. They also found in four trials in Wisconsin Lakes that fry-stocking was unsuccessful in creating a year-class of muskellunge, but that two of three trials of planting fertilized eggs on firm substrate was successful in that respect. Their success with egg planting, compared with fry stocking, suggests further research is needed on the time of fry release. Results to date indicate egg planting is more practical than fry release for maintaining or introducing muskellunge in a few waters. Limitations in collecting large numbers of fertilized eggs prohibits broad application of this technique on many waters.

A long-term management survey of Green Lake, Wisconsin (Brege) indicated that hybrid muskellunge offer significant advantages over true muskellunge in producing legal-sized fish. The hybrid had lower production costs, faster growth rates and a better return to the creel. He also reported that the hybrid is more susceptible to bait fishing. Muskellunge harvested from Green Lake, which receives heavy fishing pressure, were larger than hybrids from that lake.

The muskellunge population in Escanaba Lake, Wisconsin was unregulated by angling size, season or bag limits from 1946 through 1981 (Hoff and Serns). Evaluation of changes in that muskellunge population under those conditions indicated that a 30-in minimum-length limit theoretically would have reduced the harvest by 77% from 1946 through 1981. A daily bag limit of one and a closed season from December 1 through the first Friday in May would have theoretically reduced harvest an additional 18% for that time period.

We are not always able to accurately predict the outcome of management actions. An introduction of muskellunge in Iron Lake, Michigan produced unusual results (Siler and Beyerle). They reported that natural reproduction, in addition to regular stocking, resulted in an abnormally high population density of 30-in and larger muskellunge. Associated changes in that community were a severe decline in previously substantial populations of black crappie and white sucker, and emigration of some muskellunge into two popular brook trout streams. Public displeasure with the resulting Iron Lake fishery required corrective action by the Michigan DNR that included removal of adult male muskellunge. The corrective measures resulted in a significant decline in net catches of muskellunge, the mean annual growth of adults increased and survival of naturally produced young was higher. However, the population structure of other species did not revert to former levels. The Iron Lake experience stresses the importance of carefully examining the consequences of any actions being considered and of intensely monitoring a fish population being subjected to novel management techniques.

For an intensive management program fall estimates of the number of fingerling muskellunge can now be made employing a simple equation (Johnson, not published). A data base was developed from 56 trials on 24 Wisconsin lakes to derive the equation. The catch per mile of shoreline with a boom shocker, or a single electrofishing circuit of the entire shoreline during fall, can provide the catch per unit effort for determining fingerling production per acre.

A philosophy of muskellunge management (Wingate) provides an excellent opportunity for a public forum on participative management. Public input is essential to develop a concensus of demand for fishing tournaments, muskellunge stamps, special regulations, and gear restrictions, as well as angler expectations. Public attitude in some areas has to be dealt with realistically to eliminate the negative image of muskellunge as a "voracious predator" that eliminates stocks of wanted species. Biological data needs to be effectively communicated to develop an accurate awareness of the interaction of the muskellunge with other species.

Culture - Delano Graff, Pennsylvania Fish Commission, Chairman

The session emphasized that, in certain areas, artifically propagated muskellunge are and will, at least in the short term, remain a very important part of what muskellunge management is all about. The Wisconsin experience (Klingbiel) provided a comprehensive summary of extensive and intensive culture of muskellunge utilizing natural foods. The art of culture has reached a point where, even though we are looking primarily at intensive culture and a breakthrough with artificial diet, reliable results can still be achieved and a management program supported by extensive culture. Klingbiel effectively summarized the pros and cons of both extensive and intensive culture. He pointed out that with control, and considering the causes of variation in production, extensive culture of muskellunge has been and will continue to be successful. It is reliable, consistent and will support a fine sport-fishing program. Until some breakthrough occurs with artificial diet, musky culture will primarily remain with this method of production.

Successful rearing of muskellunge with artificial diet was demonstrated at Spirit Lake, Iowa (Jorgenson). Sufficient numbers were produced at low cost and with good survival to the size required by Iowa for management needs. Several years work with artificial diet were successful and provided consistent results. Even though that state (Iowa) was producing a slightly smaller fish than most other states, this was an indication that we may be on the threshold of being able to produce muskellunge on artificial diet.

Even though extensive culture seems to be the accepted practice in rearing muskellunge, that is not the case with hybrids. A complete synthesis of current techniques for rearing hybrid muskellunge in Pennsylvania (Bender and Graff) in intensive culture units indicated we are at the point where no major breakthrough is needed. What we need to be doing now, as indicated by them, is refining of techniques.

The need for refinement through improvement of techniques was addressed by Meade and Lemm. They discussed research in small rearing units involving controlled experiments on hybrid muskellunge and provided guidelines for optimum production under conditions encountered in an intensive-culture situation using artificial diet.

An integral part of fish culture is the quality of the hatchery product, i.e., the health of the animal. This topic was examined (Harvey) from a case history of diagnostic pathology that was carried out under standardized and rigorous procedures encountered with hatchery-produced fish under intensive culture, particularly on artificial diet. Systemic bacterial infections cause problems in hatcheries, and several kinds of bacterial gill diseases were identified. We can now identify these problems and be looking for them, but prevention is better than treatment. Hatchery management which avoids disease is extremely important. Large numbers of fish produced by the hatchery are meaningless if the fish are not healthy. If muskellunge are already stressed when stocked, chance of survival is lessened.

Frequently, propagators are criticized for overemphasizing the numbers they produce and the cost effectiveness of that production. The mistaken impression is that once the truck rolls through the gate hatchery people consider they have completed their task. The presenters of this session seem to have effectively dispelled that myth by recognition of the role of culture in musky management.

Panel Discussions - Robert Strand, Minnesota DNR, Chairman

Catch and Release

The quest for a new world record is a common thread among muskellunge anglers, and it generates numerous discussions concerning the reasonableness of that expectation. Organized muskellunge fishing clubs express concern about the future of the muskellunge and feel an obligation to contribute to its well-being. Of common concern to these groups is that trophy fish remain available. To that end, they feel that voluntary live release is the most direct and unselfish act an angler can practice to ensure future spawners and trophies.

The importance of proper handling of fish was described as the most important aspect of live release by Gasbarino. He appropriately pointed out that anglers need to understand what to do from the time the muskellunge is hooked until after it is released, and he subsequently described proper handling methods for the technique or equipment employed in each. Gasbarino, speaking for Muskies Canada, expressed the hope that fishery agencies in Canada and the U.S. would undertake further studies to document the best release-handling methods.

The credibility of release methods expounded by Gasbarino was reinforced by the contributions of other panel members. A summary was provided of tagging studies (Richards and Ramsell) since 1976, in which muskellunge fishing-club members from seven states participated in tagging and releasing over 1,600 fish. Approximately 17% of these angler-tagged muskellunge were recaptured by anglers. Stratification of the results indicated higher recapture rates for fish caught, tagged and released by experienced anglers, with return rates ranging from 22% to 44%, and averaging 28%. They emphasized that these rates approximate recapture rates by anglers in many studies, where muskellunge were tagged by state fishery crews during netting or electrofishing operations, thereby indicating normal survival rates for angler-caught muskellunge that are properly handled and released. Their summary discouraged indiscriminant tagging of muskellunge and recommended that future programs, based on needs for specific information, be closely coordinated with fishery agencies and be designed to statistically validate correlations and new or old observations.

Similar findings were evident in the contribution (Dent) from a case history of voluntary catch and release of muskellunge in a Missouri reservoir. Dent concluded that the catch and release program at Pomme de Terre Lake has decreased the harvest of legal muskellunge and has maintained a quality fishery, in spite of increased fishing pressure and a limited stocking program.

There is a definite trend toward increased participation in catch and release by muskie anglers. Muskie Inc. members reported a 94% release rate for legal muskellunge caught in 1984 and other clubs seem to be following suit. The final chapter on the value of catch and release remains to be written; however, available information indicates that survival of angler-caught and released muskellunge is high, and catch and release programs do contribute to the future quality of the fishery. As pointed out by Dent, fisheries agencies should not overlook management options which can lower harvest rates of legal muskellunge and allow some individual fish the opportunity to become trophy-size. An important consideration is that many fish are released which are larger than any length limit agencies would want to place on the general public, or one which would not be politically feasible to enact.

Role of Anglers

The special panel on The Role of Anglers and Private Organizations brought a timely message to angler and professional alike. We are in this together, share common goals, and can best achieve those goals through cooperation. In our constructive efforts to understand, educate and communicate, it seems appropriate to suggest that we not overlook the associated tourism industries.

We were reminded of the importance of angler information (Stange) but cautioned that this information is likely to be biased. His examples chosen to illustrate forms of bias were realistic, and since reliable data are necessary for proper management decisions, he suggested that ways must be found to work with, or around, bias. Mandatory reporting regulations may work in some instances, but results are inconsistent at this time. We were informed (Mooradian) that New York State has required mandatory reporting from the musky anglers of two counties since 1941. He reported that, essentially, that system has not worked. In contrast, the highly regulated lake sturgeon fishery in Wisconsin, which includes mandatory reporting (Oehmcke), has been successful. Those who are considering enlisting/requiring angler reporting can profit from the specific suggestions and guidelines (Mooradian), based on the New York experience.

Some thought-provoking ideas on the future of musky fishing (Stange) were injected. He emphasized that most musky catches are attributed to a small group of anglers, but that today's technology allows even mediocre anglers to catch an occasional musky and these fish are usually killed. This can, and probably does, have an impact on populations in many waters. He also pointed out that the once small group of elite musky anglers is no longer a small group. It has experienced phenomenal growth, and their members target most of their fishing effort at catching the limited number of muskies available in any environment. He also touched on a dilemma faced by serious musky anglers. The scenario described (Ogden) on club formation accurately portrayed the dilemma inferred by Stange. Ten fishermen become a group having fun, ten become twenty, the group expands to one hundred etc., sharing musky skills and musky waters where each has been successful. Should organized musky clubs practice unlimited recruitment of members into a sport based on a very limited resource? This could prove to be a double-edge sword. Incidental catch/kills by the sheer number of anglers out there, plus occasional kills by serious musky anglers who usually release fish, can potentially overwhelm the worthwhile goal of catch-and-release.

Musky club formation, and the contribution of the resulting organizations, is a success story of which their memberships can be proud. Their accomplishments, quite emphatically, go beyond the direct and obvious benefit of rearing/stocking fish. Recognition of their accomplishments must also include the categories of sizeable contributions of money and manpower in support of research and management projects, legislative influence, public relations, education and involvement in environmental issues. They actively promote and practice catch and release, and publicly recognize angler achievements. From known records (Ogden) a total of over 22,000 legal muskies have been released by club members. That figure becomes even more meaningful in light of the information presented on survival of released fish.

The question was posed (Addis) on whether angling organizations and anglers contribute to the protection and improvement of muskellunge populations. He answered by providing a brief, but informative, summary of the kinds of things these people have accomplished in general terms, and elaborated in greater detail on examples from the Wisconsin scene. Interest in participation by these groups has not waned. They are increasingly being organized both for collaboration with agencies and for opposition to those agency policies they dislike. Addis felt that both collaboration and constructive resistance can be positive, but the final answer will depend on how both the professionals and the activist angler feel about each other after the debate is finished and decisions are rendered. He also expressed the concern that anglers - like other citizens - are becoming less willing to consider the judgment of professionals. This trend results from a distrust of big government and of the fragmentation of the fishing community into single-issue interest groups, and it is a situation that may alter previously effective working relationships. He urged musky anglers to keep sight of their role as part of the larger community, and to avoid the common trap of using conflict and crisis to further a cause.

Three important developments were listed (Oehmcke) relative to this fishery which were identified by the panel. He suggested that to help perpetuate the species, habitat and sport, anglers and their organizations work with management in two ways. The first was habitat preservation to provide the basic medium for muskies to thrive; the second was solid support for scientifically founded management through increased cooperation with managers and professionals. His challenge is well taken; true advocates of the procedures and goals for attaining prime trophy muskellunge fishing should not hesitate to adopt changes, if research evidence indicates likely rewarding returns to muskellunge populations from such procedures. Fishery professionals are all too familiar with situations where there is general agreement, an idea is perceived as good, only to encounter the attitude "but don't do it on my lake" when the time comes for implementation.

Future Research Needs

Several authors elaborated on the remarkable change in public sentiment about muskellunge in some areas. The Future Research Needs Panel pointed out that, despite its current high esteem, the muskellunge is one of the least understood coolwater species. That panel touched on many problems which still remain, and discussed them in terms of limitations of the ability to effectively manage the species. The panel emphasized issues which have long-range impact over broad geographic areas rather than on short-term or site-specific problems. Panel members felt that management activities by most agencies were limited by deficiencies in information; they cautioned that there is a tendency to incorporate results and practices from other geographic areas without proper assessment of the local situation

The panel pointed out that the muskellunge has a limited geographical distribution and is maintained by supplemental stocking throughout much of its range. They urged that we preserve what is left of a wild resource and not allow the muskellunge to become a domesticated animal.

Most agencies manage the muskellunge as a "trophy" fish. The definition of trophy size is variable among anglers, and anglers have the luxury of setting personal goals and standards. The panel emphasized that managers need a clear biological definition of "trophy management" and offered three possible definitions for consideration. We were reminded that limited angling pressure in the past has allowed adequate management in the absence of a clear definition of the program goal, but that the future will not be so kind.

While the subject of genetic stocks of muskellunge has been of historical interest, little work has been done on stock identification and evaluation of stock characteristics until recently. Clearly, this subject is a research need and, once identified, genetic diversity of self-sustaining populations must be preserved. The impact of stocking practices on genetic diversity in natural populations is also a need. Hatcheries are generally unable to handle large numbers of adults from many populations, which makes the muskellunge a candidate for inbreeding and inadvertent selective breeding. The panel recommended that agencies develop a policy for management of stocks and genetic diversity, and outlined issues that should be addressed.

The inability to obtain sufficent samples of muskellunge at various life stages from most waters prevents detailed statistical analysis. This sampling problem must be overcome to improve management through quantitative analysis. Anglers can, and should, be asked to assist with certain types of data, and biologists should incorporate experimental designs which minimize the potential bias from such sampling schemes. With the current and growing popularity of catch-and-release, a concerted effort should be made to obtain as much information as possible from each fish killed.

There is considerable concern regarding the decline of muskellunge populations following the establishment of northern pike. A cause and effect relationship is often implied, but the information is not conclusive. It was pointed out that in many waters the two species seem to co-exist. Definitive information is needed about the interactions between these two species where they successfully co-exist, as well as where they do not.

There has been a sharp decline in the number of waters supporting self-sustaining populations of muskellunge during the past century. The panel identified as a priority the maintenence of the remaining self-sustaining populations. Suggestions for accomplishing this include obtaining detailed information about muskellunge biology from these waters, habitat description, and habitat protection. Specific examples of types of work were given, but much more effort is needed on a broad geographic basis. Relatively few waters remain which are in this category and each deserves special attention.

The panel identified a need to evaluate the effects of various propagation techniques on the survival and behavioral characteristics of stocked muskellunge in the wild. Ideally, this would include comparison of fingerlings reared on artificial diet with minnow-reared fingerlings prior to large scale use of artificial rations. Future work relative to culture, and to stocked populations, must determine the mechanisms causing mortality of stocked fish under a variety of stocking and environmental conditions.

The discussion of tiger muskellunge was limited to those situations where they are maintained by stocking. Success of rearing, and advantages of using, hybrids were extensively discussed by several authors. The relationships between cultural methods and stocking practices versus post-stocking survival, as with muskellunge, need to be investigated. Hybrids pose an additional problem relative to sampling. Since they may lack the spawning urge, population assessment of adults is more difficult than for muskellunge. Present sampling methods do not provide sufficient data and others need to be developed. Hybrids are generally stocked in waters where muskellunge are not endemic. This frequently raises fears that these introductions will result in decimated populations of other valued fishes. This on-going conflict, which is true for muskellunge also, could be alleviated by good case history studies following a new introduction, coupled with additional information on predator-prey relationships.

Private organizations have played a role in research, and will become increasingly more involved in the management of fish populations. It is hoped that the discussion of management and research needs will assist these organizations in deciding where to put their efforts. The research needs panel did not assign priorities to the studies suggested, but did state that the categories of preservation of self-sustaining populations and those which identify present genetic diversity are of extreme importance. They encouraged funding organizations to assess their goals in relation to research needs outlined, as well as to consider top priority site-specific studies when deciding where to place their resources. They also recommended that endeavors to facilitate technology transfer of research findings and management experience continue. Organizations wishing to fund research are encouraged to seek technical assistance from advisors. Furthermore, all were reminded that results of research should be subject to scientific peer review, prior to widespread dissemination and application of the results.

Poster Session Summary - Kevin Richards, Missouri Conservation Department, Chairman

The Poster Session brought together nine diverse and informative papers which were presented in an appealing visual format. Collectively the nine posters addressed the nomenculture, biology, culture and management of muskellunge. The Poster Session provided a colorful spectrum of muskellunge information and served as an extension of each of the other sessions.

At least 94 common names have been applied to the muskellunge (Crossman et al.). The accepted scientific name, *Esox masquinongy*, and a description of the species were supposed to have been published in 1824. The scientific name has passed through stages during which people defined the muskellunge as one species, three species, and three subspecies of one species, largely on the basis of color pattern. The present concensus appears to be that the three semi-distinct groups of populations represent three races of one variable species.

Although the muskellunge has been recognized as a separate species for approximately 160 years, we are just starting to understand some aspects of muskellunge biology and life history. A study of the development of ovaries and oocytes, and of fecundity and spawning (Lebeau et al.) indicated the muskellunge is a multiple spawner. This appears to be an adaptation which allows individual females to carry and deposit a larger number of eggs than would otherwise be possible.

While we continue to learn more about the natural reproduction of muskellunge we strive to increase our efficiency of hatchery-rearing of muskellunge and hybrid (tiger) muskellunge. Three posters addressed hatchery operations. Observations at the Chautauqua State Fish Hatchery in New York (Schachte and Hoffman) pointed out a potential hazard associated with the use of live, wild forage fish as feed for muskellunge. Tapeworm infections, apparently transmitted from fathead minnows, caused unusually high mortality of muskellunge fingerlings. A heating-cost comparison (Krise and Fuss) conducted at the National Fishery Research and Development Laboratory in Pennsylvania indicated that a water-to-water heat pump may reduce water-heating costs for tiger muskellunge culture. The Pennsylvania Fish Commission (Bean) developed a ranking system for fish transportation units used for hauling muskellunge. Insulated fiberglass tanks using bottom draw aerators with oxygen injection received the highest rating.

Hatchery production has allowed managers to utilize the tiger muskellunge as an introduced trophy species in many waters. Production costs for hybrids are typically lower than for muskellunge. In Iowa (Adair) it costs about $0.13 for each 6 to 7-in hybrid stocked. At Ridge Lake, Illinois angler interest in stocked hybrids is high. The ongoing study at Ridge Lake (Newman and Storck) is providing needed data concerning the growth, hooking mortality and angling vulnerability of tiger muskellunge.

The role of the angler and the importance of interagency cooperation in muskellunge management (Jackson) was highlighted in a study of Kentuck Lake, Wisconsin. The state Department of Natural Resources worked with the U.S. Forest Service, a group of local property owners and a local muskellunge club chapter to collect needed data for muskellunge and other species. Another example of the importance of angler participation in muskellunge management is the cleithrum study (Casselman) being conducted by the Ontario Ministry of Natural Resources and Royal Ontario Museum. Anglers and taxidermists have been instructed on the removal of the cleithrum bone. The cleithrum can be sent to the local fisheries management office to be used for age assessment of trophy muskellunge. The cleithrum provides more accurate age assessment of large muskellunge than scale analysis.

APPENDIX

A Synopsis Of Results From A Questionnaire On Muskellunge - Resource Agency Survey

JAMES RAGAN, TERRY STEINWAND and GENE VAN EECKHOUT

North Dakota Game and Fish Department
100 North Bismarck Expressway
Bismarck, North Dakota 58501-5095

Questionnaires were sent to 55 states and provinces with a resultant 96% return rate. Of the respondents, 30 indicated they were not currently active in any type of muskellunge management. The following data was derived from 22 positive responses.

The Great Lakes region has the largest number of lakes and streams supporting muskellunge populations. This area coincides with the natural range for the species. Moving south and west the number of water bodies supporting muskellunge fisheries progressively decreases or is nonexistent. The data indicate substantial expansion beyond the natural range, although the number of lakes/streams are limited (1-20) in those situations. At least nine agencies outside the natural range have a muskellunge program. Twenty-three percent of the agencies with a muskellunge fishery have waters specifically designated as "musky waters", presumably for greater protection of the species, and all of these agencies (5) are in areas where the species occurred naturally. None of the agencies managing only introduced populations have designated waters.

All states/provinces (22) used hatchery stocks in managing their muskellunge fisheries with 12 agencies (54.5%) relying exclusively on hatchery stocks. None rely solely on naturally reproducing stock and only four emphasize natural reproduction. Ten respondents (45%) indicated they use natural stocks from lakes in their culture, two use brood fish (presumably captive stock), and the remainder use combinations of the above and also rely on other agencies for their product. Both fry and fingerlings are utilized in muskellunge stocking programs, and the majority of agencies (86%) indicated the use of fish greater than six inches in their programs. Only two agencies rely on fry stocking only, and both indicated limited programs and no plans for expansion or change. Time of stocking is directly related to size of fish. The earliest fry stocking occurs in May and the latest fingerling stocking in October.

Stocking densities for fry ranged from 100 to 1,000 per surface acre, and the rates for fingerlings ranged from 0.05 to 15 per surface acre. At least half of the agencies indicated they would stock more muskellunge if they had access to increased hatchery stock and/or fish culture facilities.

Twice the number of agencies (36%) emphasize stocking of only muskellunge than those which emphasize stocking hybrids only (18%), and 46% stocked both. Only one agency (Pennsylvania) stocks both hybrids and muskellunge in the same waters.

An extremely wide variety of other major species of sport fish was indicated where muskellunge occur or have been introduced. The forage bases indicated for those waters also show a wide variety. Introduced populations of muskellunge are utilizing many different forage species. The responses indicated that the main forage base included soft-rayed species of large size, either coregonids or catostomids, or both, in those waters having naturally occurring muskellunge populations.

Cost of muskellunge culture is highly variable and directly relates to the size of fish at time of stocking. Cost comparisons of rearing pure and hybrid muskellunge by the same agencies (5), in addition to average cost for all respondents, clearly indicate that hybrids are less expensive to raise. Average cost of hybrid production was $0.45/fish, with a range of $0.04 to $1.25. Muskellunge culture costs ranged from $0.55 to $17.84 per fish and averaged $5.56. That average cost for muskellunge was strongly influenced by the costs reported by two agencies with figures considerably higher than the norm.

With few exceptions, 30 inches appears to be the standard minimum-length limit for both hybrid and muskellunge. One state and one province indicated they have no size limit. There was a general trend for state- or province-wide application of size limits, but lake-specific regulations were indicated by three agencies to protect brood stock. Muskellunge attain legal size in 3-6 years (mean 4.5) and hybrids in 2.5-5 years (mean 3.3), approximately one year earlier.

Daily creel and possession limits are fairly consistent, being either one or two. Thirteen agencies reported a daily creel limit of one, and seven allow a daily limit of two. Possession and daily limit are the same for 11 agencies and seven have a possession limit double that of the daily bag limit. One state (Illinois) has no possession limit and one (North Carolina) has a possession limit 3 times the daily bag limit (2 and 6). Most agencies apply regulations equally to muskellunge and hybrids.

Five agencies indicated bag-limit regulation changes in recent years. Reasons cited were to increase protection on stocks and to distribute the catch. Of the respondents that have changed regulations in recent years, all but one (Kentucky) have increased minimum lengths for protection of brood stock.

Various techniques were used by agencies to evaluate effectiveness of regulations. Only one used an indirect method (abundance of white sucker) of sampling forage. Nearly all indicated that objectives of regulations were to provide trophy potential and/or protect brood stock.

Most agencies view catch and release (15 yes, 6 no) as a viable management tool, although survival estimates of fish released under this scheme have been attempted by only two agencies.

Spearfishing as a method of harvest is allowed in only four states. Angling harvest estimates were provided by 11 agencies. Of those, nine reported total harvest ranging from 50 to 61,000 fish, and two had partial data. Eleven (50%) respondents indicated these data were not available. Fish per angler data were provided by 7 of 22 agencies and ranged from 0.01 to 1.2 fish per angler annually. Primary sources of harvest data were creel survey (5), mail-in survey (5) and musky club data (3). Other sources of data were mandatory reporting (1), angler reports of trophy fish (1) and license sales (1). Muskies, Inc. appeared to be a dominant force in obtaining voluntary reports. Mandatory reporting of harvest by muskellunge anglers was required by only one agency (New York).

Twelve of 21 agencies believed incorrect identification of muskellunge as northern pike was a problem, and nine agencies indicated it was not a problem.

A large majority of agencies (19 of 22) manage the species as a trophy fishery; the remaining 14% manage for maximum sustained yield.

The economic data provided were quite limited. Six agencies (27%) had data on angler expenditures. Recreation estimates were similar with 7 agencies (31%) providing data, one of which was an estimate for one specific wildlife reserve only.

One state (New York) and one province (Ontario) issue a special "musky stamp." The revenue generated by this stamp in New York is not earmarked for musky enhancement programs. Ontario requires nonresidents to purchase a musky stamp, since only nonresidents are required to be licensed.

Five agencies (24%) utilize some sort of fish tagging in their muskellunge management program. Of these, four obtain fish for tagging by netting or electrofishing. Only one (Missouri) utilizes angler-caught fish. All rely on voluntary returns from anglers, and two of five also rely on net or electrofishing recaptures.

Fish management emphasis by those agencies with a muskellunge program reflects a wide range of species and combinations of those species. Fourteen agencies indicated they emphasize centrarchids, 13 percids, 13 salmonids, 10 esocids and 7 place emphasis on ictalurids.

The following were considered to be significant problems in muskellunge management: 1) limited habitat which includes water pollution and inhibitors of natural reproduction; 2) education-public perception of muskellunge being voracious predators capable of decimating existing fish populations, and anglers misidentifying muskellunge as northern pike; 3) a lack of basic biological data, including population dynamics, rates of exploitation and survival; 4) data on angler use and economics; 5) propagation costs and inadequate hatchery/rearing facilities. The numerical sequence above does not represent a descending or ascending order of importance.

Ten respondents (45%) indicated possible changes in management strategies. Anticipated changes included but are not limited to: 1) development of brood stock and expansion of propagation capabilities; 2) regulation changes (size-bag limits); 3) increased number of waters stocked and size of fingerlings stocked; 4) habitat improvement; 5) collection and development of basic information; 6) identification of genetic stocks and evaluation of performance; 7) reduced stocking because of cost.

Numerous categories were listed as topics having the most potential for future research. These included, but are not limited to: 1) exploitation, stocking and harvest; 2) muskellunge culture and survival of stocked fish; 3) effects of muskellunge introductions on other species; 4) natural reproduction, spawning habitat protection and enhancement; 5) cost/benefit analysis of programs; 6) population dynamics and life history data; 7) use of muskellunge as a management tool to control other species.

Sixty-eight percent of the states and provinces reported active musky fishing clubs. Most indicated awareness of club contributions in areas of finances, manpower, legislative influence, public relations, and angler education. Five states and two provinces reported no active clubs.

Fifty percent of the respondents provided data on muskellunge-program budgets. The annual budgets ranged from $5,000 to $200,000, and averaged $76,800. Based on the survey, 27 states and 3 provinces do not have muskellunge programs.

The varied responses to problems in managing a muskellunge fishery indicate much work is yet to be accomplished. The musky provided an important recreational fishery in certain areas while ranking low in others. This importance is undoubtedly a driving force for the level of management employed.

Some questions received limited response or were of a subjective nature. Anyone desiring a complete copy of the questionnaire answers should write to:

North Dakota Game and Fish Department
100 North Bismarck Expressway
Bismarck, ND 58501-5095